The Frontal Cortex

Organization, Networks, and Function

Strüngmann Forum Reports

Julia R. Lupp, series editor

The Ernst Strüngmann Forum is made possible through the generous support of the Ernst Strüngmann Foundation, inaugurated by Dr. Andreas and Dr. Thomas Strüngmann.

This Forum was supported by the Deutsche Forschungsgemeinschaft

The Frontal Cortex

Organization, Networks, and Function

Edited by

Marie T. Banich, Suzanne N. Haber, and Trevor W. Robbins

Program Advisory Committee

Amy Arnsten, Marie T. Banich, Mark D'Esposito, Suzanne N. Haber, Julia R. Lupp, John O'Doherty, and Trevor W. Robbins

The MIT Press

Cambridge, Massachusetts
London, England

Series Editor: J. R. Lupp
Editorial Assistance: A. Gessner, C. Stephen
Lektorat: BerlinScienceWorks

The book was set in TimesNewRoman and Arial.
Printed and bound in the United States of America.

Library of Congress Cataloging-in-Publication Data is available.

ISBN: 978-0-262-54953-0

10 9 8 7 6 5 4 3 2 1

Contents

List of Contributors

Averbeck, Bruno B. Laboratory of Neuropsychology, NIH/NIMH, Bethesda, MD 20892, U.S.A.

Badre, David Dept. of Cognitive, Linguistic, and Psychological Sciences, and Carney Institute for Brain Science, Brown University, Providence, RI 02912, U.S.A.

Balleine, Bernard Dept. of Psychology, University of New South Wales, Kensington, N.S.W. 2052, Australia

Banich, Marie T. Dept. of Psychology and Neuroscience, Institute of Cognitive Science, University of Colorado Boulder, Boulder, CO 80309, U.S.A.

Beste, Christian Cognitive Neurophysiology, University Neuropsychology Center and Dept. of Child and Adolescent Psychiatry, Faculty of Medicine, Technische Universität Dresden, 01309 Dresden, Germany

Buschman, Timothy J. Princeton Neuroscience Institute and Dept. of Psychology, Princeton University, Princeton, NJ 08033, U.S.A.

Constantinidis, Christos Dept. of Biomedical Engineering and Program in Neuroscience, Vanderbilt University, and Dept. of Ophthalmology, Vanderbilt University Medical Center, Nashville, TN 37235, U.S.A.

Cools, Roshan Centre for Cognitive Neuroimaging, Donders Institute for Brain, Cognition and Behaviour, Radboud University Medical Center, 6500 HB Nijmegen, The Netherlands

Curtis, Clayton E. Dept. of Psychology and Neural Science, New York University, New York, NY 10003, U.S.A.

Datta, Dibyadeep Dept. of Psychiatry, Yale School of Medicine, New Haven, CT 06510, U.S.A.

D'Esposito, Mark Helen Wills Neuroscience Institute and Dept. of Psychology, University of California, Berkeley, Berkeley, CA 94720-3190, U.S.A.

Duncan, John MRC Cognition and Brain Sciences Unit, University of Cambridge, Cambridge CB2 7EF, U.K.

Fellows, Lesley K. Dept. of Neurology and Neurosurgery, Montreal Neurological Institute, McGill University, Montreal, QC H3A 2B4, Canada

Fiebach, Christian J. Dept. of Psychology and Brain Imaging Center, Goethe University Frankfurt, 60323 Frankfurt am Main, Germany

Friedman, Naomi P. Institute for Behavioral Genetics and Dept. of Psychology and Neuroscience, University of Colorado Boulder, Boulder, CO 80309, U.S.A.

Gratton, Caterina Dept. of Psychology, Florida State University, Tallahassee, FL 32306, and Dept. of Psychology, Northwestern University, Evanston, IL 60208, U.S.A.

Haber, Suzanne N. Dept. of Pharmacology and Physiology, University of Rochester, Rochester, NY 14642, and Dept. of Psychiatry, Harvard Medical School, McLean Hospital, Belmont, MA 02478, U.S.A.

Halassa, Michael M. Dept. of Neuroscience, Tufts University School of Medicine, Boston, MA 02111, U.S.A.

Iordan, Alexandru D. Dept. of Psychiatry, University of Michigan, Ann Arbor, MI 48109, and Dept. of Psychiatry, Research Program on Cognition and Neuromodulation Based Interventions, University of Michigan, Ann Arbor, MI 48105, U.S.A.

Izquierdo, Alicia Behavioral Neuroscience Area, Dept. of Psychology, University of California Los Angeles, Los Angeles, CA 90095, U.S.A.

Jaeggi, Susanne M. Dept. of Psychology, Center for Cognitive and Brain Health, Northeastern University, Boston, MA 02120, U.S.A.

Koechlin, Etienne Dept. of Cognitive Studies, Ecole Normale Supérieure/INSERM, 75005 Paris, France

Ladwig, Zach Dept. of Psychology, Northwestern University, Evanston, IL 60208, U.S.A.

Liston, Conor Dept. of Psychiatry, Brain and Mind Research Institute, Weill Cornell Medicine, New York, NY 10021, U.S.A.

Luna, Beatriz Dept. of Psychiatry, University of Pittsburgh, Pittsburgh, PA 15213, U.S.A.

Mitchell, Anna S. School of Psychology, Speech and Hearing, University of Canterbury, Christchurch, 8041, New Zealand

Murray, Elisabeth A. Laboratory of Neuropsychology, National Institute of Mental Health, Bethesda, MD 20892, U.S.A.

Murray, John D. Dept. of Psychological and Brain Sciences, Dartmouth College, Hanover, NH 03755, U.S.A.

O'Doherty, John Division of the Humanities and Social Sciences, California Institute of Technology, Pasadena CA 91125, U.S.A.

Palomero-Gallagher, Nicola Institute of Neuroscience and Medicine (INM-1), Research Center Jülich, 52428 Jülich; and C. and O. Vogt Institute for Brain Research, Heinrich Heine University Düsseldorf, 40225 Düsseldorf, Germany

Perez, Diana C. Dept. of Psychology, Northwestern University, Evanston, IL 60208, U.S.A.

Rasmussen, Steven A. Dept. of Psychiatry and Human Behavior, Brown University, Butler Hospital, Providence, RI 02906, U.S.A.

Rich, Erin L. Dept. of Neuroscience, Icahn School of Medicine at Mount Sinai, New York, NY 10029, U.S.A.

Robbins, Trevor W. Dept. of Psychology, University of Cambridge, Cambridge CB2 3EB, U.K.

Roberts, Angela C. Dept. of Physiology, Development, and Neuroscience, University of Cambridge, Cambridge CB2 3DY, U.K.

Rowe, James B. Dept. of Clinical Neurosciences and Cambridge University Hospital NHS Trust, Cambridge University, Cambridge CB2 0SZ, U.K.

Rudebeck, Peter H. Nash family Dept. of Neuroscience, Icahn School of Medicine at Mount Sinai, New York, NY 10029, U.S.A.

Salmi, Juha Dept. of Neuroscience and Biomedical Engineering, Aalto University, 00076 Espoo, Finland

Schuck, Nicolas Institute of Psychology, Universität Hamburg, 20148 Hamburg, Germany

Shenhav, Amitai Dept. of Psychology, University of California, Berkeley, Berkeley, CA 94720, U.S.A.

Sinha, Rajita Yale Stress Center, Depts. of Psychiatry, Neuroscience, and Child Study, Yale University School of Medicine, New Haven, CT 06519, U.S.A.

Smaers, Jeroen B. Dept. of Anthropology, Stony Brook University, Stony Brook, NY 11794-4364, U.S.A.

Vertes Robert P. Center for Complex Systems and Brain Sciences, Florida Atlantic University, Boca Raton, FL 33431, U.S.A.

Wang, Xiao-Jing Center for Neural Science, New York University, New York, NY 10003, U.S.A.

Weiner, Kevin Dept. of Psychology, University of California, Berkeley, Berkeley, CA 94704, U.S.A.

Preface

Science is a highly specialized enterprise—one that enables areas of enquiry to be minutely pursued, establishes working paradigms and normative standards, and supports rigor in experimental research. All too often, however, "problems" encountered in research fall outside the scope of any one area of study and to progress, new perspectives are needed to expand conceptualization, increase understanding, and define pathways for research to pursue.

The Ernst Strüngmann Forum was established in 2006 to address these types of topics. Founded on the tenets of scientific independence and the inquisitive nature of the human mind, we provide a platform for experts to scrutinize topics that require input from multiple areas of expertise. Our gatherings, or Forums, take the form of intellectual retreats: existing perspectives are questioned, gaps in knowledge exposed, and strategies are collectively sought to fill these gaps. To ensure access to the emerging insights, the results of the entire process are disseminated through the Strüngmann Forum Report series.

This volume reports on the organization and function of frontal lobe networks. Proposed by Marie Banich, Suzanne Haber, and Trevor W. Robbins, they were eager to create a cross-disciplinary discourse aimed at integrating information across limbic, cognitive, social, and motor control subregions of the prefrontal cortex. After the proposal's acceptance by our Scientific Advisory Board, Amy Arnsten, Mark D'Esposito, and John O'Doherty joined us on the Program Advisory Committee to transform the proposal into a framework that would support an extended, multidisciplinary discussion. The committee worked together to delineate discussion topics, identify potential participants, and finetune the overarching goal: To examine the circuitry, neuronal mechanisms, and computations by which different regions and associated networks in the prefrontal cortex mediate key component mental operations (e.g., limbic-affective, cognitive, social, and motoric) that enable higher-level thought and behavior in health and neuropsychiatric disorders. Further, the committee defined focal themes and guiding questions for the working groups:

- Group 1: Evolutionary perspectives: Homologies and analogies
- Group 2: Functional fractionation and integration: Physiology, networks, and behaviors
- Group 3: Integrative psychological, computational, and mechanistic approaches to frontal lobe function
- Group 4: How can understanding of the PFC be translated to the bedside and society at large?

Given the wide-ranging expertise involved in the Forum (e.g., behavioral neuroscience, cognitive neuroscience, computational neuroscience, evolutionary biology, neuroanatomy, neurobiology, neurophysiology, psychopharmacology, systems neuroscience), invited "background papers" presented information in advance, and from July 9–14, 2023, researchers gathered in Frankfurt for a most lively discussion. This volume synthesizes the ideas and perspectives that emerged from the entire process.

An endeavor of this kind, especially one developed amidst COVID lockdowns, creates unique group dynamics and puts demands on everyone. I wish to thank each person who participated in this Forum for their time, efforts, and positive attitudes. A special word of gratitude goes to the Program Advisory Committee (Amy Arnsten, Marie Banich, Mark D'Esposito, Suzanne Haber, John O'Doherty, and Trevor W. Robbins) as well as to the authors and reviewers of the background papers. In addition, the moderators of the discussion groups (Trevor W. Robbins, Mark D'Esposito, John O'Doherty, and Suzanne Haber) and rapporteurs (Kevin Weiner, Elizabeth A. Murray, Amitai Shenhav, and James B. Rowe) deserve special recognition, for to enable lively debate and transform this into a coherent, multiauthor report is never a simple matter. Finally, I extend my appreciation to Marie Banich, Suzanne Haber, and Trevor Robbins, whose expertise and leadership were essential to the entire project.

Through the generous backing of the Ernst Strüngmann Foundation, established by Dr. Andreas and Dr. Thomas Strüngmann in honor of their father, the Ernst Strüngmann Forum is able to conduct its work in the service of science and society. The efforts of our Scientific Advisory Board are also gratefully acknowledged, as is the partnership with the Ernst Strüngmann Institute, which shared its vibrant intellectual setting with us during the Forum.

It is never easy to extend the boundaries to knowledge, and long-held views are often difficult to put aside. Yet once such limitations are recognized, the act of formulating strategies to move past this point becomes a most invigorating activity. On behalf of everyone involved in this 35th Ernst Strüngmann Forum, we hope this volume will inform future analysis of this critically important brain region and spur further investigation, notably into the translation of findings from animal to human models, the role of connectivity in frontal function, and the unique aspects of human cognition supported by the frontal lobes.

Julia R. Lupp, Director, Ernst Strüngmann Forum
Frankfurt Institute for Advanced Studies
Ruth-Moufang-Str. 1, 60438 Frankfurt am Main, Germany
https://esforum.de/

1

Introduction

Marie T. Banich, Suzanne N. Haber, and Trevor W. Robbins

It is generally acknowledged that understanding the human brain would represent a pinnacle of scientific achievement. A major part of that goal is to identify the causal relationships between neural mechanisms and behavior and cognition, the ultimate functions of the brain. Key to this crucial issue in mammalian cognition are the functions of the frontal lobes. These lobes, which are most expanded in humans as compared to other species, include the prefrontal cortex (PFC) and associated structures, such as anterior cingulate cortex (ACC), and are known to play a key role in higher-order thinking, executive function, and cognitive control, processes by which an organism can effortfully guide adaptive behavior.

Like most of the cerebral cortex, there is evidence of quite extensive specialization within frontal brain regions. The PFC is typically divided into subregions that mediate these functions (Friedman and Robbins 2022; Haber and Behrens 2014; Monosov and Rushworth 2022; Rudebeck and Izquierdo 2022). The orbitofrontal cortex (OFC) is involved in value encoding (revaluation and devaluation) of reward-related sensory input and outcomes. The caudal OFC and ventrolateral (vl) PFC have been implicated in a number of functions, including credit assignment (determining the previous events that resulted in a specific outcome) and behavioral flexibility. The dorsolateral (dl) PFC is in a central position to mediate cognitive control, working memory and higher-order thinking. Finally, frontopolar regions are linked to dlPFC and vlPFC regions and allow for meta-cognitive abilities. As such, there is a matrix of subregions and their interactions that characterize PFC organization.

Clearly these regions, which mediate goals, values, choices, and actions, interface extensively to develop appropriate adaptive behavior. Yet how the interactions between these complex frontal regions result in behavioral choice is less well understood, as research has become "siloed" with researchers focusing on a particular portion of frontal cortex. For example, several functions have been attributed to the ACC with theories ranging from

- conflict detection (Botvinick et al. 2004),
- interfacing between motivation, action, and effort (Rushworth 2008),
- detecting whether an action led to the expected outcome (Alexander and Brown 2011), and

- late-stage selection, often motor response-related (Banich 2009).

However, since ACC is a rather large frontal subregion with complex connections, it is unlikely to have a single function (Tang et al. 2019). Likewise, computational models of PFC function often focus on a specific region or network and typically do not take into account interactions between and across multiple networks. Moreover, behavioral paradigms often fractionate underlying mental operations, each of which is often associated with specific brain regions or circuits.

We now know that a central feature of cognitive function is that behavior results from an intricate interplay between localized processing within specific brain regions and patterns of connectivity across regions (Geschwind 1965; Haber et al. 2022; Mesulam 1998). Networks are characterized by specialized subsystems as well as "hubs," which represent regions for integrating and distributing information from multiple cortical and subcortical regions. The ability for subsystems, hubs, and connectivity to be flexibly reorganized enables the brain to meet a wide variety of cognitive and computational demands.

Yet it remains unclear exactly how such higher-order cognitive processing is achieved by these organizational principles across heterogeneous PFC subregions and their connectivity between each other, as well as to more distant brain regions. Currently, a variety of methods across different levels of analysis, used singly and in conjunction, is starting to provide a framework for understanding this organization. These methods span anatomical approaches, including circuitry studies, to neurophysiological, neuroimaging and neurochemical studies of brain morphology and functional activation. To help better address how the complexity of cortical regions interact, computational methods such as Bayesian learning models and graph theory are being utilized to develop network interactive models.

However, the different approaches needed to address these interactions are, to some extent, species-specific (i.e., optogenetics in mice, resting-state MRI in humans, invasive electrical recordings in nonhuman primates). This leads to major issues when findings from animal models are extrapolated to inform drug discovery programs or identify better treatment targets for human mental disorders. Another major contemporary question is that of development: How does the PFC develop during human childhood and adolescence in relation to the rest of the brain?

This Forum was convened to examine the circuitry, neuronal mechanisms, computations, and potential treatment targets of different PFC areas that mediate key component operations (e.g., social, affective, cognitive, and motor control) and when and how such regions act locally or collectively. To address these goals, the Forum brought together scientists with expertise in different disciplines (ranging from psychology to computational modeling to all the main branches of neuroscience, including neuroimaging, neurophysiology and neuropharmacology) who work with different approaches (from animal models to humans and the clinic).

We felt that it was especially beneficial to convene this Forum for several reasons. Research on the frontal lobes is an active area of research across various fields, yet interdisciplinary interactions have been minimal as there has been no major meeting on the frontal lobes since the five-day Baycrest-Rotman/Berkeley conference held in 2010 (Stuss and Knight 2012), which followed on from an earlier meeting in 2002 (Stuss and Knight 2002). Despite the enormous conceptual and technical advances that have been made over the last decade, no further meetings were planned in that series. Smaller symposia or meetings have focused on restricted issues (e.g., the OFC, dlPFC, cortico-striatal systems, rodent prefrontal function) or some of the processes implemented by frontal regions, such as cognitive control (e.g., the Control Processes meeting held at Brown University in 2019). None, however, has addressed specifically how the organization of the frontal lobe enables such processing nor attempted an integrative approach, such as proposed in this volume. Since the Royal Society meeting in the mid-1990s (Roberts et al. 1998), the community has lacked a published discussion among the main leaders of the field. Hence, to realize the important implications of this knowledge for health and disease, effective discussion and collaboration across fields was required and could readily be provided by the Ernst Strüngmann Forum.

Framework for Discussion

The unique aspect of an Ernst Strüngmann Forum lies in its think-tank approach. For a week, experts from around the world interact in a residential setting to scrutinize issues and questions posed in advance by the organizing committee. To initiate this discourse, the committee commissioned papers to introduce key topics. Then, during the week, intense interactions emerged between the invited, multidisciplinary experts, centered around the following themes:

Group 1: Evolutionary Perspectives: Homologies and Analogies

Discussion in the first working group examined key principles that determine homologies and cross-species functional similarities of PFC. Aided by Chapters 2 (A. Izquierdo) and 3 (R. P. Vertes et al.), the group addressed the following questions:

- What are the major regions and circuits observed across species within PFC?
- What are the structural and functional homologies of the PFC across species?

- How did the PFC evolve and how has this evolution led to produce higher-order cognition, including social and moral reasoning elements in humans?
- What are the mechanisms by which these major circuits exert control? By patterns of anatomical connectivity, neural synchronization and oscillations, chemical neuromodulation, excitatory/ inhibitory balance, plasticity and long-term potentiation, or other mechanisms specific to the prefrontal cortex?
- How can the functions of micro-circuit approaches defined by optogenetics and multiple unit electrophysiology be linked to macro-circuits as revealed by human imaging modalities?

The resulting discussion was informed by experts from behavioral neuroscience, cognitive neuroscience, cross-species comparisons, evolutionary neuroscience, neuroanatomy, neurophysiology, neuropsychology, neuropsychopharmacology, and systems neuroscience: Bernard Balleine, Michael Halassa, Alicia Izquierdo, Nicola Palomero-Gallagher, Trevor W. Robbins (moderator), Peter Rudebeck, Jeroen Smaers, and Kevin S. Weiner (rapporteur). Together, they worked toward generating a consensus statement regarding the ambitious goal of determining the homologies of PFC, as well as functional similarities across species, providing evolutionary, cognitive, and translational insights (see Chapter 4 by K. S. Weiner et al.).

Group 2: Functional Fractionation and Integration: Physiology, Networks, and Behaviors

The second working group focused on the functions of subregions of the PFC, their associated circuitry, and interactions. Three papers were commissioned—Chapter 5 (E. Rich and B. Averbeck), Chapter 6 (J. D. Murray and C. Constantinidis), and Chapter 7 (D. Badre)—to initiate the debate, which addressed the following issues:

- To what degree is the PFC composed of discrete functional regions, and if so to what degree do they overlap?
- How can interactions between PFC regions best be understood (e.g., are there specific hubs that coordinate PFC function)?
- Is the organization of these functional regions hierarchical or can it be conceptualized in some other organizational mode?
- To what degree do subcircuits within the PFC converge map onto non-frontal regions?
- How are PFC circuits modified by genetic expression and environmental input?

Input from behavioral, cognitive and systems neuroscience, including neuroimaging, neurophysiology and neuropsychopharmacology, informed the debate between Bruno B. Averbeck, David Badre, Christos Constantinidis, Roshan Cools, Clayton E. Curtis, Mark D'Esposito (Moderator), Lesley K. Fellows, Anna S. Mitchell, Elisabeth A. Murray (Rapporteur), John D. Murray, and Erin L. Rich. Together, they considered evidence for functional fractionation of the frontal lobes, discussed whether the organization of the frontal lobes should be conceptualized in terms of functional and anatomical gradients, instead of discrete areas with well-delineated boundaries. Their report also highlights critical gaps in knowledge for future research (see Chapter 8 by E. A. Murray et al.).

Group 3: Integrative Psychological, Computational, and Mechanistic Approaches to Frontal Lobe Function

In this working group, experts considered whether the frontal lobes have a unitary function that works across multiple demands or has a set of more diverse functions, and the computational bases of such PFC function(s). Aided by three papers—Chapter 9 (J. Duncan and N. P. Friedman), Chapter 10 (X.-J. Wang and E. Koechlin), and Chapter 11 (C. Gratton et al.)—group discussions centered around the following questions:

- To what extent is there unitary versus diversity of fronto-executive functions?
- Are concepts of cognitive or executive control outmoded? Should there be a new taxonomy?
- How can computational approaches enhance our understanding of the concept of cognitive control or executive function?
- How can such computational approaches be linked to PFC function?
- How best can information about PFC anatomy, function, connectivity, and computational modeling of behavior be combined to produce new insights into how the PFC enables higher-level cognition?

Informing the debate were experts from cognitive and computational neuroscience, as well as functional neuroimaging and neurophysiology: Marie T. Banich, Christian Beste, Timothy J. Buschman, Naomi P. Friedman, Caterina Gratton, Etienne Koechlin, John O'Doherty (Moderator), Nicolas Schuck, Amitai Shenhav (Rapporteur) and Xiao-Jing Wang. In their report (Chapter 12), Shenhav et al. propose a new neurocomputational modeling framework for conceptualizing PFC function and discuss critical directions needed to validate or falsify this account. They also considered whether neurocomputations are processed at a lower (cellular) level or emerge from network organization.

Group 4: How Can Understanding of the PFC Be Translated to the Bedside and Society?

In the fourth working group, participants considered important clinical, translational, and societal issues that arise from our understanding of the PFC and its networks. Informed by three background papers—Chapter 13 (A. Roberts and C. Liston), 14 (S. M. Jaeggi et al.) and 15 (S. A. Rasmussen)—the group explored the following questions:

- How can an understanding of the functional anatomy of the PFC inform our understanding of psychiatric and neurological disorders, including comorbidities?
- How can animal models involving PFC function be enhanced to address salient clinical issues?
- What are the surgical, pharmaceutical, or cognitive-behavioral interventions that can produce improvements in cognitive control and executive function?
- What are the mechanisms by which such interventions are likely to act?
- What general principles can we learn about the functions of the PFC that have broader societal implications (e.g., philosophy of volition).

Experts in the group included behavioral, cognitive and systems neuroscientists, as well as clinician-scientists (neurologists, psychiatrists and psychologists): Dibyadeep Datta, Christian J. Fiebach, Suzanne N. Haber (Moderator), Susanne M. Jaeggi, Conor Liston, Beatriz Luna, Steven A. Rasmussen, Angela C. Roberts, James B. Rowe (Rapporteur), and Rajita Sinha. Given the lack of one-to-one mappings between clinical syndromes, their underlying pathophysiology, and root neurobiological causes, Rowe et al. (Chapter 16) propose a multilevel framework in which syndromes can be linked to symptom profiles, symptoms to cognitive processes, and cognitive processes to neurochemical, neurophysiological, and computational processes embedded in PFC and its associated networks. They also consider prefrontal disorders in the context of global opportunities for education, health, and social policy.

Moving Forward

The issues presented in this volume have enormous societal implications. Understanding PFC function is highly relevant to many neurological and psychiatric diseases and disorders, including anxiety, obsessive-compulsive disorder, posttraumatic stress disorder, depression, frontal lobe dementia, schizophrenia, addiction, and autism. Virtually all neuropsychiatric disorders involve malfunctioning of PFC circuits as core impairments, yet currently we do not know whether general fronto-executive impairments contribute to these clinical phenotypes or whether each have a distinct "signature." This distinction has implications for treatments based on circuits or neuromodulation. How nature has solved the problems posed by executive control may also inspire new approaches to artificial intelligence and business organizations.

The final step of any Ernst Strüngmann Forum involves turning over the ideas that emerged from the debate to others for further consideration. To this end, the invited papers and group reports have been reviewed, finalized, and carefully edited to provide an account of where the debate stands. In addition, in Chapter 17, we have attempted to synthesize and draw together the main themes.

We cannot assert that answers to each and every one of our questions were provided definitively with consensus in these discussions. Most certainly, some controversies persist! It is our hope, however, that this volume captures some of the excitement of the interactions we experienced at the Forum, facilitated by the excellent infrastructure provided by Julia Lupp and her team.

2

A Cross-Species Analysis of Prefrontal Cortex Homology Based on Anatomical Connectivity, Behavior, and Cell Types

Alicia Izquierdo

Abstract

The steepest rise in publications on prefrontal cortex (PFC) function over the past decade has been in mouse studies. If we adhere to cell layer organization criteria for what constitutes PFC, rodent researchers may be studying a different PFC to primate PFC. Indeed, this chapter reviews several unique aspects of primate brain: primate cortical evolution favored a clustering of cell types more than rodent; primate PFC is more specialized in the expression of interneurons compared to rodent; and where comparative transcriptomic studies of different cell types in PFC have been conducted, they reveal unique similarities only within primate species. In contrast to these differences between species, strong similarities are also reviewed: connectivity patterns across rodent and primate PFC, specifically agranular orbitofrontal cortex and anterior cingulate cortex, as well as common features of foraging with some innovations that may have contributed to PFC specializations in primate. The study of cell types should be better integrated in the study of PFC across species, and this integration should, in principle, be closely related to a characterization of the cells along a spatial and behavioral gradient that reflects phylogenetic refinement. Currently, few studies combine neural activity with molecularly defined cell types within a species, and even fewer take a comparative approach. Combining transcriptomically defined cell-type information with other characteristics, such as task-related signaling in PFC and their connectivity patterns across rodent and primate species, represents a major challenge to the field, but would be an impactful way forward.

Introduction

The zeitgeist of present-day neuroscience involves a fascination with the "central executive" which oversees and coordinates all behavior. Executive function is an umbrella term that includes the many different functions of the prefrontal cortex (PFC), including planning, self-ordered memory and monitoring, attentional set shifting (Fuster 1989; Luria 1966b; Robbins 1996), and cognitive control (Friedman and Robbins 2022), to name a few. The recent emphasis on mimicking executive functions is not surprising, given the rise in interest in using artificial intelligence and neural network architecture to support these functions (Tsuda et al. 2020). There is presently no shortage of research on PFC, probed with increasingly powerful tools and analyses in service of understanding the functions of this complex and heterogeneous region.

In a recent analysis of the prevalence and common misconceptions of what constitutes rodent frontal cortex, Laubach et al. (2018) found that while human PFC still held the lion's share of publications, followed by rats, then mice and monkeys (Figure 2.1a), the steepest rise in publication prevalence on PFC during the past couple of decades was actually in mouse (Figure 2.1b). Given this growing bias in model systems for PFC function, it begs the question: When one studies frontal cortex in rodents, how does this knowledge, *if at all,* translate to our understanding of PFC in primates? This is an established topic of consistent, heavy debate (Carlen 2017; Laubach et al. 2018). This debate arose many years ago as the Rose and Woolsey definition of PFC centered around anatomical connectivity to mediodorsal thalamus, with major input from MD thalamus into a clearly visible granule cell or "granular" Layer IV (L4) in PFC (Preuss 1995; Rose and Woolsey 1948). On the basis of this laminar or cell-layering (i.e., cytoarchitecture) criterion alone, rodents undisputedly lack a PFC. If we consider other criteria, such as connectivity, gene expression, electrophysiological properties, and behavior, we may make better comparisons across species.

I begin here by highlighting key limitations in the rodent model, especially those related to how findings from rodents may translate to human PFC, that have to do with gross anatomical differences in brain structure and shape. Unsurprisingly, primate brains have greater neuron numbers simply as a result of their folded-ness (i.e., deep sulci). This enhanced neuron number and

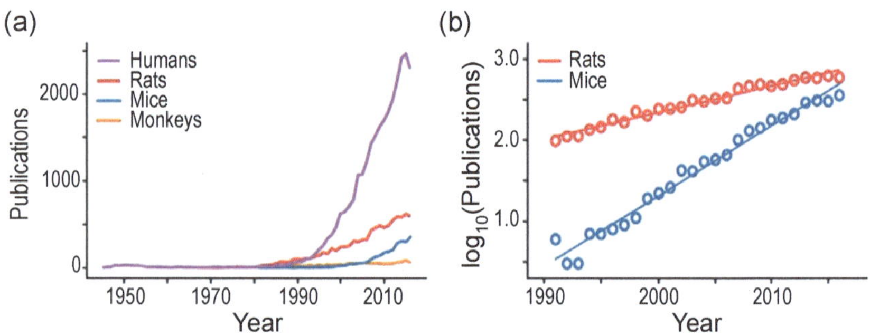

Figure 2.1 Publication trends (a) across species in (pre)frontal cortex and (b) number of published works per year. Reprinted with permission based on Laubach et al. (2018).

cortical expansion is thought to be largely due to the proliferation of progenitor cells in the outer subventricular zone, OSVZ (Kriegstein et al. 2006; Lui et al. 2011). There are a few proposed mechanisms for this expansion, including radial glial cells acting as migratory guides for columnar distribution of neurons and intermediate progenitor cells contributing to increases in cell layers within each layer (i.e., lateral expansion of cortex, in primates). Indeed, humans have a "scaled up" primate brain (Herculano-Houzel 2009), meaning that the ratio of brain weight and neuron number is in accord with other primate brains of similar mass. Critically, however, rodent neocortex is nonfolded (i.e., lissencephalic, not gyrencephalic) so the ability to model human neocortical PFC developmental evolution is quite limited, perhaps especially using analyses of OSVZ expansion. A more complete integration of the cellular dynamics that subserves cortical development and evolution to include the molecular basis of the neural stem and progenitor cell diversity is generally lacking in the literature and should be explored across different primates, with full acknowledgment that it may have limited application to rodents. Another gross anatomical difference between rodent and primate PFC was highlighted by Vogt and Paxinos (2014) and Laubach et al. (2018), detailing that, comparatively, rodent brains lack curvature. They suggest that a primate specialization that contributed to this curvature is the expansion of the midcingulate cortex (MCC) wrapping around the genu of the corpus callosum, leading to the displacement of anterior cingulate cortex (ACC) more rostrally and ventrally. This primate expansion may have given rise to compression of tissue in the form of gyri in primates, but not rodents. Hence, this circles back to the difference between folded and nonfolded brains in primates and rodents, respectively.

Such substantial anatomical differences call into question if we are comparing similar brains across rodent and primate, especially since their nearest common ancestor occurred approximately 70 million years ago. In this vein, I recently reviewed evidence with Peter Rudebeck (Rudebeck and Izquierdo 2022) and concluded that along with comparative anatomy, there needs to be more thoughtful consideration of what different brains must do to obtain food in their natural environments, since different species evolved in unique foraging niches (Murray et al. 2011). In that review, we took an approach inspired by Cisek (2019): instead of highlighting functions that we think could have been subserved by different fronto-cortical systems (e.g., decision making, set shifting, flexible learning, working memory), we tried to follow the footsteps of evolution and assumed there was "phylogenetic refinement" of *clusters of functions* based on foraging niches. Below, I begin by reviewing commonalities across species having to do with neuroanatomical connectivity patterns and foraging behaviors that may have given rise to PFC specializations, and then consider more divergent results across species; information that may be most needed for a deeper understanding of comparative function across species *cell types* in PFC.

Comparative Connectivity

The most anterior parts of macaque frontal cortex are either dysgranular or completely granular cortex, having a discernible granule cell L4. This includes orbital areas 13, 11, and 12 (ventrally), 10 and 9 medially, and areas 46 and 6 more laterally. More caudal areas of macaque orbital cortex (area 13) and areas 25, 32, and 24 more medially are agranular. Thus, similar to rats and mice, macaque caudal orbitofrontal cortex (OFC) and medial ACC are completely agranular. The medial wall of frontal cortex in rodent has historically been referred to as prelimbic (PL), infralimbic (IL), anterior cingulate (Cg1 and Cg2), but now more often referred to as areas 32, 25, and 24 with clear reference to their anterior-posterior (A-P) positioning. There is also the most ventrolateral portion of rodent frontal cortex that includes agranular insular, which I have suggested previously is not as well studied or understood as other subregions (Izquierdo 2017). Importantly, all these agranular subregions are shared by mammalian brains (Murray et al. 2011).

There are established "rules" about what could be considered PFC. First, as mentioned above, rodent frontal cortex is completely agranular, and the existence of a granule cell L4 has been for decades the primary definition of PFC in primate, with rats lacking any kind of homologue to prefrontal areas of primates (Preuss 1995). More importantly, moving away from this strict cytoarchitectonic criterion, we may better rely on other dimensions for species comparisons. Uylings et al. (2003) outlined five criteria for cross-species comparisons of PFC:

1. Cytoarchitectonic similarities
2. Connectivity patterns considering the density of those connections
3. Neurochemical distribution and receptor expression
4. Embryological development
5. Functional properties, including electrophysiological and behavioral similarities

Here, I emphasize criterion 2 (connectivity patterns) and criterion 5 (function), the latter with a focus on behavior. Electrophysiological comparisons are critically important as well, yet others have written on this topic (e.g., Seamans et al. 2008); for an updated comprehensive review, see Rich and Averbeck (this volume).

Laubach et al. (2018) provided a summary of the diversity of expert opinions in answer to the question: "What, if anything, is the rodent prefrontal cortex?" Their meta-analysis showed that there has been an overemphasis of the functions of the medial wall of frontal cortex compared to more lateral areas in rodents. Further, they reported that much of the diversity of expert opinions as to what constitutes rodent PFC may be partly due to the use of multiple and often inconsistent sets of anatomical nomenclature and acronyms to refer to the same subregions. Parcellation of subregions of rodent PFC may be conducted by using a similar framework as primate ACC: centered on gray matter location around the *genu* of the corpus collosum (i.e., along the A-P gradient). This proposal is substantiated by cross-species connectivity data, which I review next.

Several groups have suggested more attention should be given to the A-P axis as well as lateral over medial frontal cortex comparisons across species (Barreiros et al. 2021a, b; Izquierdo 2017; Rudebeck and Rich 2018; Wallis 2011). The most posterior and medial segments of nonhuman primate PFC are agranular and more similar in terms of connectivity to rodent (Heilbronner et al. 2016; Wise 2008). It is unclear, however, whether these map onto the most widely used anatomical atlases in rodents. For example, a recent review by van Heukelum et al. (2020) directly compared the structural and functional distinctiveness of cingulate cortex from both human (based mostly on diffusion tensor imaging) and rodent neuroanatomical tracing studies (Figure 2.2a, b). For this they used two different parcellations of cingulate cortex: a definition based on ACC and MCC along the A-P plane (Figure 2.2b) or the more widely used rat atlas Cg1/Cg2 designations that vary instead along the dorsal-ventral (D-V) plane (not shown). They found that the former, but not the latter, segmentation better reconciled functional results across species, referring to the A-P defined ACC as "homologous" and D-V Cg1/Cg2 segmentation as characteristically "nonhomologous." Connectivity of primate and rodent "homologous" ACC is strong with autonomic brainstem nuclei, amygdala, OFC, hippocampus, hypothalamus, and thalamus (van Heukelum et al. 2020), largely consistent with comparative studies (Floyd et al. 2001; Freedman et al. 2000). It should be noted, however, that there are ACC connections to autonomic regions in rat (e.g., nucleus of the solitary tract, magnocellular neurosecretory cell groups in the hypothalamus) that have not been observed or reported in macaques (Freedman et al. 2000).

In seminal work by Heilbronner et al. (2016), investigators used anatomical cases with anterograde tracers in rat and macaque ACC and OFC to study the extent to which cortico-striatal terminal inputs overlapped. They found that terminals into striatum overlapped extensively along with the medial wall and that area 25 in rats was most similar to area 25 in monkeys, similar to what van Heukelum et al. (2020) later reported in their metanalysis (Figure 2.2). Importantly, Heilbronner et al. (2016) also found similar patterns of connectivity across medial versus lateral OFC, with what they refer to as "homologous" segmentation along the striatum (Figure 2.3). Thus, using this striatal-based connectivity approach to study networks across species, Heilbronner et al. revealed largely conserved fronto-cortical inputs in rats and macaques. Since the sample in this study included various nonhuman primate species (*Macaca fascicularis, M. mulatta, M. nemestrina*) and rat strains (*Rattus Norvegicus*: Sprague-Dawley, Wistar, hooded strains), these findings are likely robust and generalizable in their conclusions of cross-species topography. Taken together, these studies (Heilbronner et al. 2016; van Heukelum et al. 2020) reveal the value of studying connections with striatum,

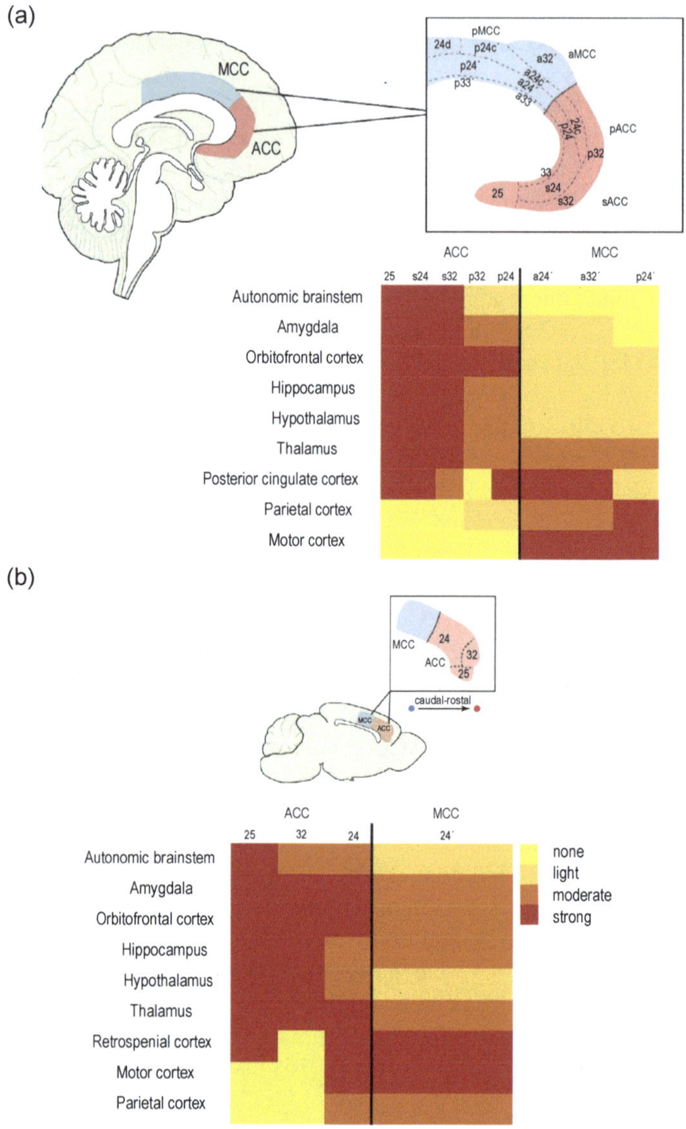

Figure 2.2 Similarity in primate and rodent anterior cingulate cortex (ACC) connectivity along the A-P axis. (a) ACC and mid-cingulate cortex (MCC) in humans. Midsagittal view of individual Brodmann's areas (top) and connectivity patterns (density) with other areas based mostly from diffusion tensor imaging (bottom). (b) The connectivity profile for the ACC/MCC nomenclature closely resembles the connectivity found in humans. See Brodmann's areas 25, 32 and 24 which are most anterior, different from a dorsal-ventral segmentation in Cg1 and Cg2 (not shown). Similarly strong connectivity can be found with amygdala, orbitofrontal cortex, hippocampus, hypothalamus, and thalamus. Modest-to-no appreciable connectivity with parietal and secondary motor cortex. Adapted with permission based on van Heukelum et al. (2020).

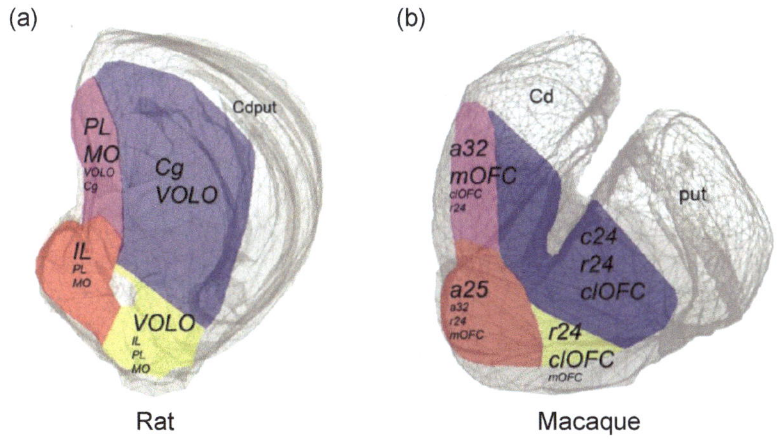

Figure 2.3 Similarity in parcellation of anterior cingulate cortex and orbitofrontal cortex inputs to striatum in rat and rhesus macaque. Striatal "segments" have unique combinations of fronto-cortical inputs and they are largely similar across rat (a) and nonhuman primate (b). Abbreviations: prelimbic (PL), infralimbic (IL), ventrolateral orbital (VOLO) cortex, medial orbital (MO) cortex, caudate (Cd), putamen (put), medial orbitofrontal (mOFC), cingulate (Cg), caudate-putamen (Cdput), caudolateral orbitofrontal (clOFC). Other divisions are also shown in Macaque: area 32 (a32), central area 24 (c24), rostral area 24 (r24), and area 25 (a25). Adapted with permission based on Heilbronner et al. 2016.

amygdala, other fronto-cortical regions, hippocampus, hypothalamus, and thalamus to assess rostral-caudal patterns of connectivity in PFC circuits, even if cytoarchitecture may not reveal as much similarity across species. Though recent anatomical tracing studies have been rigorously conducted within a single species (Barreiros et al. 2021a, b; Izquierdo 2017; Rudebeck and Rich 2018; Wallis 2011), it is a more powerful approach to study these patterns across species. Related to connectivity analyses, the Allen Brain Mouse Atlas (2011) provides a high-resolution, freely available anatomical reference along with a deep catalog of projection mapping experiments detailing axonal projections labeled by viral tracers. Similarly, with inclusion of the NIH Blueprint NHP atlas, there is a growing repository of selected gene analyses across the adult macaque brain that includes cellular marker genes with cortical area specificity as well as families of genes important to specific neural functions. Unfortunately, the Allen Brain Atlas lacks rat connectomics in their open-source atlas, which is by my estimation a missed opportunity in understanding the *functional* consequences of such comparative connectivity between rodents and nonhuman primates (NHP) because many classic, theory-driven behavioral experiments have been directed at understanding the substrates of these pathways in rats, not mice.

Optogenetic and Chemogenetic Manipulations In Rodents and Primates

In addition to the many papers on rodent frontal cortex (Figure 2.1), in parallel there has been steady growth in NHP electrophysiological studies with high-channel count recordings (Berger et al. 2020; Mitz et al. 2017) and correspondingly sophisticated computational analyses to describe the neural correlates of high-order behavior and cognition. For an important discussion of the results of electrophysiological studies in PFC, see Rich and Averbeck (this volume). Importantly, viral-mediated technology to target the brain with cell-type specificity is increasingly more commonplace in NHP, including optogenetic manipulations. These perturbations work by precise light-gated excitation or inhibition of neural activity made possible by introduction of a viral vector in neurons to express light-sensitive channels, which are then responsive to different wavelengths of light (Deisseroth 2015). Optogenetic studies probing cortical circuits have demonstrated feasibility in NHPs (Diester et al. 2011) and have been especially useful to date in interrogating sensorimotor functions (El-Shamayleh and Horwitz 2019), though optogenetic efficacy in studying higher-order cognitive function and its potential application to psychiatry remains to be fully determined (Bliss-Moreau et al. 2022).

By comparison, chemogenetic techniques working through viral expression of mutant G-protein coupled receptors or designer receptors exclusively activated by designer drugs (DREADDs) (Armbruster et al. 2007) have been more widely applied to diverse behaviors in NHP than optogenetics. There is now evidence that success in using this technique depends on transduction level of the receptor and the ligand (i.e., actuator) used to activate these receptors in rodents, but perhaps most especially in NHPs (Eldridge et al. 2016; Grayson et al. 2016; Nagai et al. 2020; Roseboom et al. 2021; Upright and Baxter 2020; Upright et al. 2018). Conditional, pathway-specific DREADDs, often used in rodents, are also now being used in NHPs (Oguchi et al. 2021b; Oyama et al. 2022; Vancraeyenest et al. 2020; Wood et al. 2023). Yet despite these advances in tools in NHP experiments, due in part to the ease of working with the technology in rodent species (along with other critical factors such as lower cost of research, shorter lifespan of rodents, as well more limited access to training in working with NHPs), there have been steeper increases in the use of rodent models to study PFC function. Thus, at present there is not enough of a critical mass of papers for a thorough comparison of rodent and NHP studies using these techniques, but there is expected to be in the near future. In the interest of cross-species comparison with rodent, OFC and ACC connectivity with striatum, amygdala, and midbrain dopamine in macaque would be beneficial, as a great deal of pathway dissection has been conducted in rodent. On the other hand, in the interest of translation to the human primate, it may be best for macaque work to emphasize uniquely granular PFC region connectivity (e.g., to/from ventrolateral PFC, dorsolateral PFC).

Foraging Innovations, Prediction, and Primate Specialization of PFC

Cue- and action-based learning in naturalistic environments requires a diverse set of neural processes. PFC functions that support flexible learning and decision making in such environments evolved in freely moving animals, yet these systems are frequently assessed in head-fixed animals. Head fixation enables precise cue presentation and the collection of data from hundreds and thousands of trials; thus, often better than tasks involving freely moving behavior at testing of computational models of neural responses and behavior. Conversely, learning paradigms in freely moving animals simulates more naturalistic foraging behavior with some amount of control, while animals have more options to engage in the required behaviors (or not), like in the real world. Along with recording and imaging thousands of neurons across long periods of time and multiple brain regions, the ecological validity of the behavior should be considered (Izquierdo 2021). Freely moving and head-fixed experiments may reveal the same underlying patterns of results, but there are also differences. For example, macaques are risk-seeking in head-fixed settings when tested in computerized gambling tasks, but risk averse while freely moving and foraging (Eisenreich et al. 2019). Given that pose estimation in freely moving rodents (Lauer et al. 2022; Mathis et al. 2020; Segalin et al. 2021) and NHPs (Bala et al. 2020) is an increasingly common and accessible method, this advance is predicted to better enable the incorporation of behavior as a correlate data stream to neural data than years before.

According to optimal foraging theory (Charnov 1976; Pyke 1984; Pyke et al. 1977), several factors contribute to an organism's enhanced fitness and profitable reward procurement. These include, but are not limited to, knowledge of a high-yielding food source (or "patch"), the nature of the food available in the current patch in comparison to others, when it is best to leave a patch, and the degree to which mobility is possible or an account of the travel time to different patches (Pyke 1984). Additional factors to the original theory include whether the animal has a central home or nest, the impact of uncertainty about the profitability of the reward environment (McNamara et al. 2013), and species "risk proneness" (Pyke 1984). Using the marginal value theorem, one can predict foraging behavior on the basis of energy-maximizing strategies across species (Charnov 1976) as well as time-minimizing strategies if there is greater risk of predation (Kie 1999). However, both rodent and primate species exhibit similar biases, leading to seemingly paradoxical or "suboptimal" behaviors in laboratory settings. For example, both species demonstrate myopic behavior when foraging, harvesting locally beyond what is predicted by optimal foraging theory, and exhibit a preference for immediate versus delayed rewards (Kane et al. 2019). Both rodents and primates also exhibit a paradoxical preference for information about the likelihood of obtaining reward, even if the information cannot change the outcome and when it comes at a cost (Bromberg-Martin and Hikosaka 2009; González et al. 2023; Jezzini et al. 2021; White et al. 2019). Mice, rats, and humans also have similar sensitivity to "sunk costs" (i.e., time dedicated to pursuing reward), resulting in a resistance to giving up on a reward once they have committed to pursuing it (Sweis et al. 2018). It is still unclear what these deviations from optimal behavior mean. The observed foraging "errors" or departures from optimal strategies observed across species could reflect cognitive or computational constraints or more covert learning processes that have yet to be fully understood (Harhen and Bornstein 2023).

A critical function of PFC may be to incorporate new behaviors into the species' repertoire. As recently reviewed (Rudebeck and Izquierdo 2022), there may be several compelling explanations for PFC expansion and specialization, including foraging (Dunbar and Shultz 2017), and specifically "foraging innovations" that result in the organism's enhanced ability to procure reward such as sampling of new foods or adopting new strategies to obtain reward. Some of the most relevant factors in this regard include species' "time horizon" needed for successful foraging (i.e., not in the order of seconds or minutes, but rather days and weeks required to keep track of seasonal changes in resources), a rapidly accelerating metabolism that comes with a larger brain size (Pontzer et al. 2016), as well as species perceptual and physical capabilities (i.e., smaller body size and odor-guided navigation of rodents compared to larger body size and visually guided navigation of primates). Foraging innovations can be considered behaviors that enhance *prediction*, *evaluation*, and *action* (Figure 2.4). These behaviors involve assessment of stimuli, outcome, and possible actions that map nicely onto learning theory (Balleine and O'Doherty 2010; Holland 2008).

Murray et al. (2011) summarized the literature on the subregions of frontal cortex as performing either the "top-down biasing of behavioral control systems" or the "flexible alterations of foraging strategies." For example, ACC biases competition among multiple stimuli and actions, the medial wall (PL and IL, or areas 32 and 25) bias behavior toward goal-directed choices versus habits, respectively, and OFC biases behavior in favor of higher-valued rewards. What is absent from rodent frontal cortex that is uniquely present in primate (granular) PFC is what authors referred to as the ability to generate a representation of "valueless" reward. Specifically in human and NHPs, rewards can guide goal-directed behavior independent of their biological value, perhaps as a rich, visual representation of the food item made possible by the greatly expanded visual capacity in primate brains compared to rodents. In primates, these result from robust connectivity with vision association areas such as inferior temporal cortex and perirhinal cortex that relay the unique properties of reward and objects to granular PFC. As an empirical example of this "valueless" reward in primate brain, Murray and Rudebeck provide compelling evidence that macaque ventrolateral PFC mediates knowledge about the availability of reward, apart from its desirability (Murray and Rudebeck 2018; Rudebeck et al. 2017b).

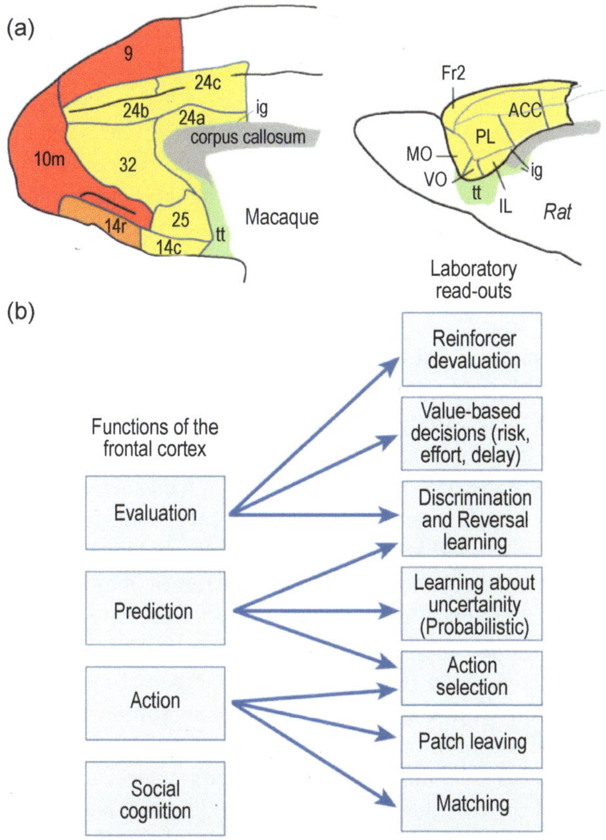

Figure 2.4 Foraging factors and PFC specialization in primates and rodents. If relying solely on cytoarchitectonic comparisons, rat frontal cortex is most similar to posterior and medial macaque PFC in that they are both agranular (yellow), as opposed to granular (red) or dysgranular (orange). There is allocortex in both species (green). These map onto ACC and OFC across species, though only a medial, not ventral, view is shown here (A). Foraging innovations were described in Rudebeck and Izquierdo (2022) as behaviors that enhanced Evaluation, Prediction, Action, and Social cognition (B). These functional categories (and the laboratory readouts of these categories) could be considered part of a "phylogenetic refinement" of brain and behavior. Social cognition is an important function of the frontal cortex, especially in primates, but outside the scope of the present review. Adapted with permission based on Rudebeck and Izquierdo (2022).

We extended the explanation in Rudebeck and Izquierdo (2022), highlighting that primates use highly developed visual capabilities to forage over larger ranges and longer time horizons, enhancing *prediction*: maintaining representations of both value and "valueless" reward would be essential for planning ahead for the right time to harvest different foods. Conversely, rats rely on olfactory capabilities to forage locally; thus foraging innovations in rodents may have arisen from adaptively enhancing the assessment of cues in their immediate environments, with short time horizons (i.e., favoring the *evaluation* function). We surmised that the *action* function may be most similar and overlapping across rodents and primates since it involves a convergence of pathways with a single outcome among many possible alternatives. Note that the laboratory readouts (Figure 2.4b) do not neatly map onto different subregions of either rodent frontal cortex or primate PFC, but the evidence suggests there is more shared support and less specialization of *evaluation* (i.e., reinforcer devaluation) than *prediction* (i.e., stimulus-based reversal and probabilistic learning). As an example, we recently found evidence in favor of less specialization in rat PFC during stimulus-based reversal learning. We conducted an experiment to compare local field potentials (LFPs) directly, specifically theta oscillations in OFC and ACC in rat frontal cortex during reversal learning (Ye et al. 2023b). We found strong support for OFC theta signaling of accuracy in reversal learning, unperturbed by chemogenetic inhibition of ACC (which expectedly did disrupt the ACC theta signal). Thus, we observed parallel, redundant signals of accuracy in both subregions of rat frontal cortex. Importantly, inhibition of ACC resulted in an impairment of early stimulus-based reversal learning, similar to those that follow OFC lesion or inhibition. This stands in contrast to a more specialized division of labor for stimuli (OFC) and actions (ACC) in primate PFC (Camille et al. 2011b; Rudebeck et al. 2008b). As another example, much like the finding of impaired confidence report in perceptual decisions following OFC muscimol inhibition (Lak et al. 2014), we find DREADDs inhibition of ACC also impairs confidence report in rats (Stolyarova et al. 2019). Collectively, these results suggest a lack of specialization of rat frontal cortex subregions for reversal learning and decision making under uncertainty, in behaviors requiring *prediction* (see also Jahn et al. 2014).

One of the more controversial perspectives has been that homology of PFC across species can be derived only if one finds the right level of PFC to make the comparison (Carlen 2017), with the recent strong suggestion that cellular-structural distinctions (cell types and morphology) are the most relevant dimensions of comparison of PFC across species (Le Merre et al. 2021). Though it will appear that I take a similar perspective here, I note that behavior and connectivity need to be incorporated as constraining factors (i.e., moderators) to these cell-type and morphological accounts.

Cell Types in PFC

Projection neurons make up 80% of all cortical neurons. Pyramidal neurons are the major class of these neurons, characterized by their triangular, pyramid-like, shape with both apical and basilar dendrites combining input from different cortical layers and sending this information to other brain regions. By comparison, intrinsic neurons, or interneurons, form synapses only within a particular brain region. Thus, the definitions of pyramidal and interneurons have mostly to do with their projections and not the source of their inputs (Masland 2004). Recent anatomical comparison of structural features of mouse and rhesus macaque Layer III (L3) pyramidal neurons in primary visual (V1) and frontal association areas shows that L3 neurons are broadly generalizable across these two areas in mouse, but not in monkey. In macaques, L3 lateral PFC neurons are much larger in size than V1 neurons and differ in their dendritic topology, but these neurons do not differ along these dimensions in mice (Luebke 2017). Thus, pyramidal neurons may not be the generalizable building blocks of cortical networks across species, at least if the classification is solely based on structural features.

A critical question is how transcriptome-defined cell types in PFC relate to their targets and functions across species. Yet single-cell transcriptomics (the collection of all the genetic readouts or expressed mRNA molecules in a single cell) and systems (behavioral) neuroscience have progressed largely as separate fields, rarely converging until recently (Lui et al. 2021). Single-cell RNA sequencing (scRNA-Seq) enables assessment of cortical or any type of cells clustered on the basis of morphological and physiological criteria (Yuste et al. 2020). Perhaps the first question to consider is what is meant by "cell type"? Yuste et al. (2020) suggest this definition should be based on data obtained from different methodological approaches, developmental stages, and species. According to these and other authors (Kepecs and Fishell 2014), cortical cell-type definition criteria could be based on (a) cell morphology, (b) connectivity motifs of interneurons with pyramidal cells, (c) molecular marker subtypes (i.e., parvalbumin, PV; somatostatin, SST; vasoactive intestinal peptide, VIP), and (d) intrinsic physiological properties of the neurons (e.g., fast-spiking, regular spiking non-pyramidal) (Figure 2.5).

Cortical inhibitory neurons can also be further classified as subtypes or subclasses via scRNA-Seq: *Pvalb, Sst, Lamp5, Vip, Sncg* (Bugeon et al. 2022). For example, Bugeon et al. (2022) report that modulation of responses to visual stimuli differ by subclass and activity can even be predicted by their transcriptional clustering. Ostensibly, this method could go beyond modulation of V1 activity by stimuli (Bugeon et al. 2022; Knoblich et al. 2019) and be extended or applied to behaviors more closely linked to PFC function, like reinforcer devaluation, set shifting, and reversal learning. Yuste et al. (2020:1464) also advise that transcriptomically similar cell types should in principle be related to the "proper levels of the anatomical structure"; in other words, a definition of the cells along a spatial gradient that corresponds with evolutionary distance between species. This overlaps considerably with a phylogenetic refinement mechanism proposed by Cisek (2019). Yet to align these datasets quantitatively in this way would require, as these authors describe, a "serious community effort," but would prove very worthwhile.

Cellular diversity afforded by interneurons may be a crucial evolutionary strategy to provide both stability and complexity (Kepecs and Fishell 2014)

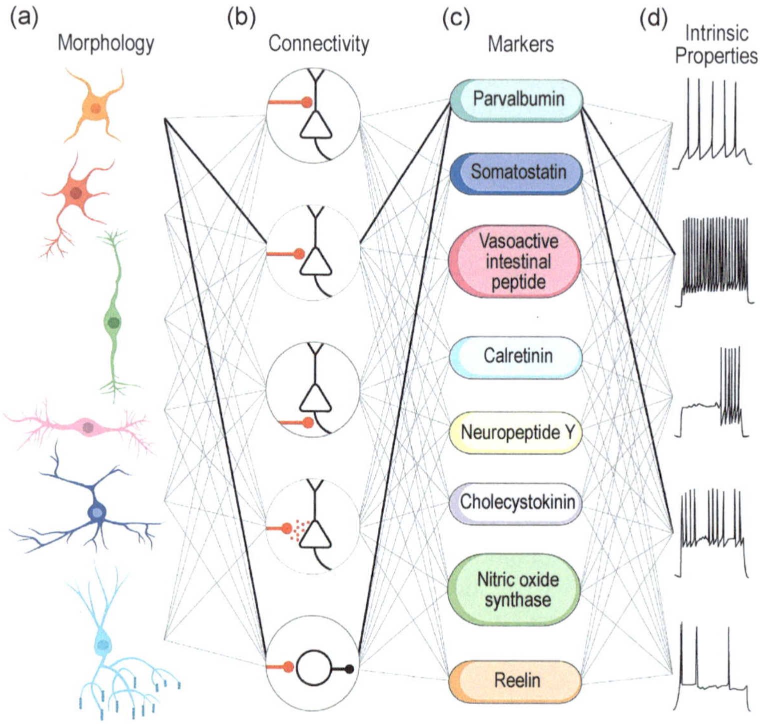

Figure 2.5 Classification of cell types that may be useful in guiding future cross-species transcriptomic studies. Cell types can be defined based on (a) morphology, (b) connectivity motifs, (c) molecular markers, and (d) intrinsic electrophysiological properties. Figure credit to Julia Kuhl, reprinted with permission based on Kepecs and Fishell (2014).

of neuronal firing patterns, especially relevant to PFC function that requires both. The diversity of interneurons in primate PFC may enable higher-dimensional neural representations important for behavior (Rigotti et al. 2013) and the dynamics of learning (Najafi et al. 2020). Programmed cell death of interneurons has been demonstrated to be a critical mechanism for adjusting the excitatory-inhibitory ratio, necessary for the assembly of neocortical circuits in mice (Wong et al. 2018). Though more cross-species studies are needed, several groups have conducted the important work of comparing cell types and gene expression patterns across rodents and primates (Hodge et al. 2019; Krienen et al. 2020). Different from the conclusion based on dendritic topology across pyramidal cells by Luebke (2017), Hodge et al. (2019) used scRNA-Seq in mice and a comparable single nucleus RNA-Seq (snRNA-Seq) method in humans and found largely conserved cortical cellular architecture across species and found similar functional gene families that discriminate inhibitory neuron types in both humans and mice, and homologous clusters of excitatory neuron projection targets. Where there was clear divergence across species was in the expression of genes associated with connectivity and signaling in homologous cell types. For example, gene families with the most divergent expression patterns included neurotransmitter receptors (especially serotonin), ion channels, and cell adhesion molecules. Hodge et al. suggest these differences likely impact *microcircuit* function, and even offer the possibility that this divergence could be one of the causes for the failure of preclinical studies in mice to translate to effective therapeutics in humans. Notably, an important limitation of this study as it relates to PFC is that the mouse tissue samples were obtained from V1 and a premotor area, anterior lateral motor cortex, not PFC per se. Nevertheless, these results highlight the need for more human and NHP studies to understand human brain disease as well as more investigation of local or microcircuit function.

Although the origins of interneurons may be conserved across species, the extent of homology of interneuron or interneuron subtypes within and across rodent and primates was poorly understood until recently. To study this, Krienen et al. (2020) conducted scRNA-Seq to profile expression of interneurons across brain regions, including neocortex and specifically PFC, across three primates (human, macaque, and marmoset) and rodent (mouse). They found that the same genes (*Sst, Pvalb, Vip, Lamp5*) were expressed in nonoverlapping neocortical interneurons across species and that their origins are similar: interneurons arise from medial and caudal ganglionic eminences (MGE and CGE) with MGE giving rise to *Sst* and *Pvalb* interneuron types, and CGE giving rise to *Vip* and *Lamp5* types. Interestingly, mouse frontal cortex contained these interneurons in proportions similar to those found in V1, but primates have significantly higher proportions of interneurons in PFC relative to V1. Additionally, there is evidence of "homologous interneuron types" readily identified by their RNA-expression patterns across species, with only a small fraction of "marker" genes being shared in another species. These marker genes vary less among primates and also show spatial expression gradients in primates more than rodents. Altogether, this suggests there is more specialization in the expression, but not the origin, of primate PFC interneurons compared to rodents.

Another cross-species transcriptomic analysis of the two cortical subtypes, glutamatergic (Glu) projection neurons and GABAergic interneurons, yields similar conclusions. This analysis included human, chimpanzee, and rhesus macaque (Kozlenkov et al. 2020) and revealed a pattern of cell-type evolution of gene regulatory elements (GREs), such as promoters and enhancers that drive and stabilize mRNA transcription. Using a combination of methods to isolate Glu and GABA nuclei in rhesus macaques, chimps, and humans, Kozlenkov et al. found several GREs in support of similar "concordant" evolutionary gene expression changes. Importantly, they found that GREs undergo subtype-specific changes more than GREs that are shared by different cell types. Similar results have been obtained by Khrameeva et al. (2020), showing that astrocyte and oligodendrocyte progenitor cells exhibit more differences than neurons across macaques, bonobos, chimpanzees, and humans and that the unique expression differences found in the human brain fall along neocortical and subcortical networks, similar to those revealed by neuroimaging studies. Though large-scale cell transcriptomic analyses have been conducted in different tissues in macaques (Han et al. 2022), there is no such transcriptomic atlas for macaque central nervous system or specific subregions of PFC. There is, however, a transcriptomic atlas of marmoset central nervous system (Lin et al. 2022) which could provide a useful resource to compare the evolution of PFC in New and Old World monkeys. Taken together, these findings suggest that primate cortical evolution favored a clustering of cell types.

Few studies have directly compared rodent and NHP cell-type function, either behaviorally or electrophysiologically. In a rare example, Povysheva et al. (2008) compared anatomical and physiological characteristics of PV-positive basket interneurons (multipolar GABAergic interneurons) in PFC of macaques and rats. Whereas there were several similarities (such as soma size, dendritic length, axonal horizontal, and vertical arbor span), macaque PV basket cells were found to be generally more excitable yet the frequency of the miniature excitatory postsynaptic potentials was higher in rats than macaques. Povysheva et al. deduced that these structural differences translate to differences in electrophysiological properties of the cortical networks, and ultimately may contribute to species differences in PFC function. This is reminiscent of the idea suggested earlier that there is species divergence in local, or microcircuit, function in PFC networks.

Laminar and Functional Patterns Among Cell Types

While rodent frontal cortex Layer I (L1) contains pyramidal neurons and GABAergic interneurons, Layer II and III (L2/3) contain cortical-projecting cortical cells or intratelencephalic neurons. L5 is the major output layer that contains both cortical-projecting and pyramidal tract cells targeting subcortical regions, and finally Layer VI (L6) mainly constitutes cortico-thalamic relay cells (Anastasiades and Carter 2021). Optogenetic inhibition of L2/3 pyramidal neurons in mouse medial frontal cortex results in intact behavioral flexibility as measured by probabilistic reversal learning. Conversely, selective silencing of deep layer pyramidal cortico-striatal and cortico-thalamic neurons (L5/6) does impair performance on this task (Nakayama et al. 2018). Interestingly, inhibition of interneuron-mediated "local" pyramidal neurons in mouse medial frontal cortex (in VGAT-ChR2 mice) produces enhanced premature responding and choice bias but intact reversal learning, suggesting dissociable roles of cell types on behavior that depend on laminar location. In NHPs, projections from agranular cortices (e.g., caudal orbitofrontal cortex) terminate mostly in upper layers of granular cortices (e.g., lateral PFC), and projections from granular cortices terminate mostly in the deep layers of agranular cortices (Rempel-Clower and Barbas 2000). As described above in rodent, laminar organization in NHP PFC—whether the cells influence local or long-range projections—may similarly be tightly associated with their putative roles in behavior, but this has yet to be fully elucidated.

Gao et al. (2022, 2023) have conducted tour de force studies on the spatial gradients of cell types in mouse frontal cortex. To determine whether single neurons project to specific targets, they reconstructed the projection patterns of genetically identified cell types, generating a "single-neuron projectome" in mouse. They found that the same transcriptome subtype corresponds to multiple projectome subtypes in different fronto-cortical regions (Gao et al. 2022) and identified morphological scaling of soma-dendrite combinations across lamina and subregions of frontal cortex. Combinations of dendrite-axon organization corresponded to cytoarchitecture and revealed a columnar organization of projection neuron subtypes in mouse frontal cortex (Gao et al. 2023). These are important studies; however, it will be important to integrate a comparative approach in the future since it is unclear if rat, macaque, and human frontal cortex follow similar principles of organization as mouse.

Few studies have combined electrophysiological recordings or calcium imaging data—either single cell, population, or LFPs—with molecularly defined cell-class information. Combining transcriptomically defined cell-type information with other characteristics, such as task-related signals in PFC as well as their connectivity patterns, represents a major challenge in the field. Using miniscope Ca^{2+} imaging in mice, Pinto and Dan (2015) found that pyramidal neurons exhibited much more functional heterogeneity in terms of task-related signaling on a go/no-go task than interneurons, and pyramidal neuronal responses varied across lamina. Interestingly, even though interneurons of the same subtype (SST+, PV+, VIP+) were more similar to each other, each subtype signaled different task-related events.

Returning to one of the classifications or criteria of cell types, the *connectivity motifs*, this classification may be especially informative as it relates most directly to integrated systems and microcircuit function. Using scRNA-Seq, Lui et al. (2021) studied the laminar distribution of cells expressing cluster-specific marker genes across both ventromedial and dorsomedial frontal cortex in mouse and found largely similar ratios for those marker genes. Of all the cell types they studied, the most specific marker genes for L5 were *Npr3* and *Tshz2*. Liu et al. discovered a great deal of redundancy in the projection targets of those neurons from multiple cell types. Not surprisingly, there was a complex collateralization pattern of various cell types in mouse frontal cortex to several target regions important in reward and cognition, such as amygdala and nucleus accumbens, which they referred to as "a many-to-one and one-to-many" mapping of cell type and projection targets. Specifically, they found that different cell classes signaled diverse aspects of task encoding as measured by calcium imaging, indicating that each transcriptomic type makes different contributions to behavior. In fact, connectivity patterns can be highly heterogeneous even within narrowly genetically defined cell clusters. It will be very useful to approach such an investigation in a comparative way in the future, to apply this technique to rat and macaque circuit dissection. Another powerful technique is multiplexed analysis of projections by sequencing, or MAPSeq (Kebschull et al. 2016). This high throughput method maps the projections of (thousands to millions of) single neurons by labeling sets of neurons with random "barcode" RNA that can then be extracted and sequenced from the putative projection zone or area. To my knowledge, only one group has used this approach in macaques (Zeisler et al. 2023), thus suggesting that this is a nascent approach.

Aside from these transcriptomic methods, other methods to study pathways are still commonly used, such as fMRI (Schaeffer et al. 2020) and mesoscopic mapping of pathways using tissue clearing methods (Xu et al. 2021). More traditional tract-tracing approaches fill neurons with proteins, often virally, so that their connectivity can be revealed using microscopy after the experiment. These methods often include retrograde Cre and Cre-dependent DREADDs. However, many limitations exist with these techniques, including lack of uniformity of expression, collateralization, and unpredictable transsynaptic viral expression. Transcriptomic methods to identify pathways identified by cell type across species offer a powerful way forward.

Stability and "Combinatorial Complexity" in PFC

Together, pyramidal and interneuron activity in PFC provide stability and "combinatorial complexity" (Kepecs and Fishell 2014), both critical for adaptive behavior in rodent and primate species. A purely excitatory network consisting of only pyramidal neurons would be unstable. Interneurons not only provide balance; they normalize local excitatory circuits and can provide feedforward inhibition, as a sort of "gain control," allowing for more temporal precision in neural activity. Superficially, this overlaps with the idea that mixed selectivity in PFC is important in generating high-dimensional representations for adaptive behavior (Fusi et al. 2016) that can be refined by learning and experience, shaped by excitatory and inhibitory (sub)networks (Najafi et al. 2020). As an example of this, our group performed bidirectional chemogenetic activation (hM4Di and hM3Dq-mediated) studies of pyramidal neurons in ACC on behavior, targeted with DREADDs on a calcium/calmodulin-dependent protein kinase IIα (CaMKII) promoter (Hart et al. 2020). Surprisingly, we found that *either* increases or decreases in ACC population activity produced impairments on effort-based choice in rats. In fact, a heterogeneous population would be more susceptible to perturbation by bulk inhibition or excitation, as demonstrated by the results of our DREADD manipulations. More interestingly, 1P calcium imaging (with GCaMP also driven by a CaMKII promoter) in freely behaving rats revealed that population activity was most predictive of choice, not individual cells. It may be an excitatory/inhibitory ratio in frontal cortex that computes (in our case here) relative cost-benefit, sending appropriate outputs to downstream targets that are more or less influential based on their laminar location. It could also be that not targeting specific cell types may serve to introduce noise and decrease signal-to-noise ratio in value-based choices. A caveat is that recent studies have determined that CaMKII and synapsin promoters exhibit more similar cell-type preferences than previously thought (Radhiyanti et al. 2021; Veres et al. 2023; Watakabe et al. 2015), transducing both excitatory and inhibitory neurons. In the future, it will thus be important to target interneuron function selectively in these cognitive processes, for direct comparison with pyramidal neuron involvement.

Conclusions

In this chapter, I have reviewed connectivity patterns across rodent and primate PFC and highlighted ways in which foraging behaviors may have given rise to PFC specializations. I also provided evidence in support of increasing efforts to study PFC cell types across species, with an appreciation for laminar and behavioral gradients that have undergone "phylogenetic refinement." Few studies combine neural activity with molecularly defined cell types within a species, and even fewer take a comparative approach. Across rodent and primate species, connectivity motifs likely provide the stability and complexity needed for the myriad executive functions of PFC. The field needs more studies that combine transcriptomically defined cell-type information with connectivity patterns and behavior-related signals in PFC across species. Collectively, this requires an integrative approach that incorporates the study of genes, neurophysiology, and behavior in both rodents and primates. These studies could be aimed at studying the evaluation function of PFC (i.e., value, value-based decision making), as there is substantial cross-species concordance of findings in this domain.

Acknowledgments

This work was supported by UCLA's Division of Life Sciences Retention fund, National Institutes of Health grants R01 DA047870 and R21 MH122800. I thank Drs. Laura DeNardo and Andrew Wikenheiser for their helpful comments and fruitful discussions on cell types and foraging, respectively.

3

The Two Prefrontal Streams

Evidence for Homology Across Species

Robert P. Vertes, Nicola Palomero-Gallagher, and Michael M. Halassa

Abstract

The prefrontal cortex (PFC) plays a critical role in human cognition, but the precise mechanisms by which its circuitry accomplishes its proposed functions are unclear. Nonhuman animals are indispensable in revealing such mechanisms, as the ability to monitor and manipulate their circuitry provides necessary insights. A major impediment to linking the growing progress in animal research to insights for human cognition and applications to human health is the lack of consensus on how the PFC is homologous across species. In this perspective, we follow the classification of human PFC into medial and lateral streams, with the medial being primarily evaluative and the lateral being executive. Based on anatomy, physiology and function, we advance the proposal that the rodent medial prefrontal cortex contains elements of both streams, with functional parallels between primate ventromedial and dorsolateral PFC with rodent infralimbic and prelimbic areas, respectively. To support this argument, we highlight the granular nature of the prelimbic cortex in *Tupaia belangeri*, a basal primate whose PFC macrostructure is rodent-like. Our perspective may help provide additional input to the debate on PFC homology and lead to new testable hypotheses.

Introduction

The prefrontal cortex (PFC) is a complex and highly interconnected region that engages in a wide variety of cognitive functions, including attention, working memory, decision making, and social behavior (Miller and Cohen 2001; Soltani and Koechlin 2022). In the human brain, the PFC has shown great expansion compared to even the closest primate relatives (Preuss and Wise 2022), a process thought to be key to the unparalleled cognitive expansion seen in our species. However, both the principles by which PFC circuits contribute to cognition as well as their origin/emergence are poorly understood.

Nonhuman animal research is poised to help fill this knowledge gap because, in addition to its basic scientific value, it offers important insights into human health given the involvement of PFC dysfunction in several neurological and psychiatric illnesses (Liston et al. 2011; Smucny et al. 2022). Given the mechanistic accessibility afforded by newer monitoring (Tian and Looger 2008; Wu et al. 2022; Xu et al. 2017) and causal tools (Kim et al. 2017; Rabut et al. 2020; Roth 2016), there has been an explosion in PFC animal research over the last decade focused on rodent PFC. Yet despite this progress, it is considerably challenging to relate these advances into insights applicable to understanding the human (and nonhuman primate) PFC given the considerable differences in macro- and microarchitecture. Specifically, while the human PFC has a large number of well-differentiated areas (Haber and Robbins 2022)—von Economo and Koskinas (1925), for example, identified 39 cytoarchitectonically distinct areas on the cortex covering the lateral, medial, and orbital portions of the frontal lobe—the rodent PFC is far less differentiated, thus making homology assignments very challenging.

Here, we follow the general two-stream human PFC classification (Domenech and Koechlin 2015) as a starting point. Specifically, this functional classification suggests that the lateral stream, which is largely composed of the lateral PFC (lPFC) is involved in executive control and rule-based behavior (Friedman and Robbins 2022). In contrast, the medial stream, which is composed of the ventromedial PFC (vmPFC) and dorsomedial PFC (dmPFC), is involved in adjusting behavioral strategies based on reinforcement and self-monitoring (Domenech and Koechlin 2015). According to the definition of Domenech and Koechlin (2015), the lPFC encompasses Brodmann's (1909) areas 44 and 45, as well as the lateral portion of areas 8, 9, and 10 (although those authors do not mention areas 46 or 47 which are commonly included in the lateral stream). Their vmPFC covers Brodmann's areas 11, 12, 14, 25, the medial part of 10, rostral part of 24, and ventral portion of 32, whereas the dmPFC encompasses the caudal and dorsal parts of 24 and 32, respectively, as well as the medial portion of areas 6, 8, and 9.

We present evidence that the rodent medial prefrontal cortex (mPFC) exhibits homology to both streams. Specifically, our thesis indicates that the rodent infralimbic cortex (i.e., area IL) is most closely related to the primate vmPFC based on both connectivity and function. On the other hand, the rodent prelimbic cortex (i.e., area PL) exhibits gradients of connectivity that makes it a likely precursor of several regions found in the primate PFC. Specifically, the evidence reviewed here supports that PL is a precursor of areas belonging to the primate medial and lateral stream regions such as dmPFC area 32, and dorsolateral PFC (dlPFC) areas 10, 9, and 8. The notion of a single rodent-like precursor of several primate cortical areas is not new and has been utilized to explain evolutionary expansion and differentiation in the sensorimotor system (Kaas 2004). Here, we extend the notion of an evolutionary precursor to prefrontal circuitry, providing a clearer context for relating rodent functional data to primate cognition. Consistent with our proposal, we point to *T. belangeri*, an evolutionary intermediate whose prelimbic cortex contains an area that is granular, a microcircuit feature that establishes its correspondence to primate dlPFC.

The Prelimbic Cortex As a Precursor of Dorsomedial and Dorsolateral Prefrontal Cortex

The cerebral cortex has undergone significant changes and differentiations throughout evolution, providing space for the development of distinct cortical areas with specialized functions. The evolution of somatomotor control, for example, from simple reflexive movements to highly coordinated and precise voluntary actions, is associated with a significant cortical expansion and segregation as well as neuronal specialization. Indeed, the *Bauplan* of the brain of opossums resembles that of small-brained placental mammals in all but one aspect: it contains a "somatosensory-motor amalgam," with a complete overlap of somatosensory representation and motor control maps (Dooley et al. 2014; Karlen and Krubitzer 2007; Wong and Kaas 2009a). Since marsupials diverged from placental mammals around 130 million years ago, Kaas (2004) proposed that this somatosensory-motor amalgam could be considered a "precursor area" of the architectonically distinct sensory and motor areas found in the brains of the latter infraclass. Small placental mammals, including tenrecs (Krubitzer et al. 1997), hedgehogs (Catania et al. 2000), or rats (Haghir et al. 2023), present a distinct primary motor cortex (M1), and in most cases their somatosensory region encompasses four areas: a primary (S1) and a secondary (S2) somatosensory area as well as rostral and caudal somatosensory belt areas. A secondary motor area has also been described in the rat brain, and some of these species present a further somatosensory area located ventrocaudally to S2 (for a comprehensive review, see Kaas 2004). In addition to these two motor and five somatosensory areas, the brain of tree shrews (the closest relatives of primates) presents a rudimentary somatosensory posterior parietal area (Wong and Kaas 2009a). A further differentiation occurs in the brains of small primates such as galagos (Wu and Kaas 2003) and slow lorises (Carlson and Fitzpatrick 1982), which display additional somatosensory areas located in the lateral fissure. In macaque monkeys, but not in marmosets, the caudal somatosensory belt area developed further into areas 1 and 2 (Kaas 2004), and three subfields can be identified within M1 (Rapan et al. 2023). This cortical segregation reaches its apex in humans, where both the motor and somatosensory cortex have expanded significantly in terms of size and complexity to enable finer control of movements, including intricate finger and hand movements, as well as the production of speech, and enhance the individual's capacity for motor planning and decision making. The gradual changes in cytoarchitecture associated with the phylogenetically related emergence of multiple areas from the marsupial somatosensory-motor amalgam are in line with the "gradation theory" postulated by Sanides (1962) to explain cortical differentiation in the human PFC. Specifically, his systematic analysis revealed that segregation in the human PFC is associated with discontinuous step-wise changes of cyto-architectonic features which not only follow phylogenetically related cortical expansion (i.e., when moving medio-laterally from allocortical through meso-cortical to neocortical areas), but also when moving in a poleward direction throughout the prefrontal neocortex (Sanides 1962). Below, we present both structural and functional evidence in support of the framework that rodent area PL could be considered a precursor of primate dmPFC area 32 *and* of areas belonging to the primate dlPFC.

Structural Studies

The prelimbic cortex occupies a very large area of the prefrontal cortex in rodents. In rats, the PL extends rostro-caudally for about 3 mm, from the anterior pole of the PFC, sitting above the medial orbital cortex, to caudally situated dorsal to IL (Swanson 2004). While PL has generally been regarded as a single entity, recent evidence leads us to propose that PL may anatomically and functionally consist of two major divisions: rostrodorsal and caudoventral divisions. Specifically, there are notable anatomical differences between these two parts of PL with respect to both their inputs and outputs. For instance, in an early examination of PFC projections to the striatum, Berendse et al. (1992) reported that the dorsal part of PL projected to mid-regions of the dorsal striatum, whereas ventrally PL selectively distributed to the nucleus accumbens (ACB), and we could confirm this distinction (Vertes, pers. comm.; see also Figure 3.1).

As is well established, the mediodorsal nucleus (MD) of the thalamus is strongly connected reciprocally with the mPFC. However, the caudoventral PL distributes specifically to the medial segment of MD, whereas the rostro-dorsal PL projects selectively to the lateral segment of MD (Groenewegen 1988; Vertes 2004). Taken together, this pattern indicates that the rostrodorsal PL communicates primarily with action/premotor-associated structures and may therefore serve a role in executive control, similar to areas of the primate dlPFC. On the other hand, caudoventral PL is strongly interconnected with limbic structures and may accordingly be involved primarily in affective behaviors, comparable to those of area 32 of primates.

With respect to limbic connections, the caudoventral PL receives pronounced projections from the hippocampus, mainly originating from CA1 and the subiculum of the ventral hippocampus. Thalamic afferents to this division of PL arise predominantly from medial/central regions of the thalamus including MD (as mentioned above), rostral intralaminar nuclei, and the midline nuclei: the paraventricular, paratenial, rhomboid, and reuniens (RE) nuclei (Hoover and Vertes 2007; Vertes 2004, 2006). Finally, the caudoventral PL

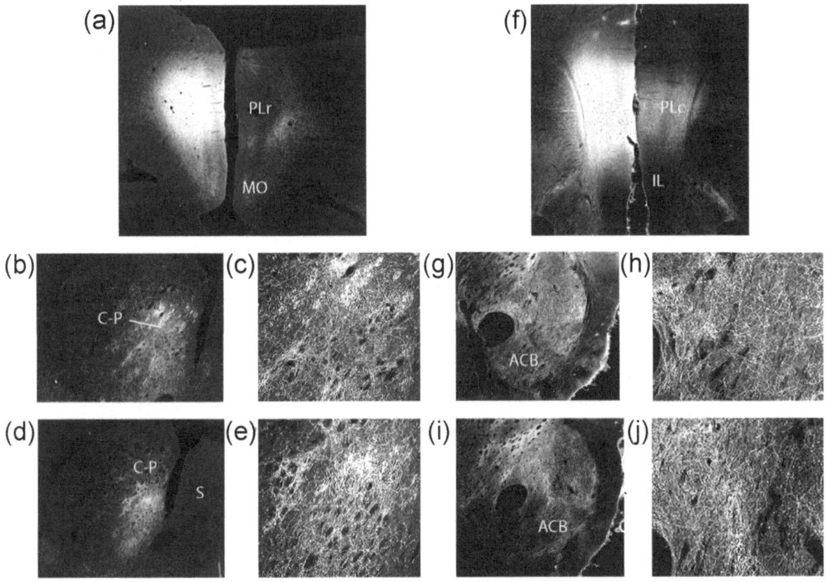

Figure 3.1 Pattern of distribution of labeled fibers at rostral (b, c) and caudal (d, e) levels of the dorsal striatum (C-P) at low (b, d) and high (c, e) magnification produced by a PHA-L injection in the rostral part of the prelimbic cortex (a). Pattern of distribution of labeled fibers at rostral (g, h) and caudal (i, j) levels of the nucleus accumbens (ACB) at low (g, i) and high (h, j) magnification produced by a PHA-L injection in the caudal part of the prelimbic cortex (f). Note that projections from the rostral prelimbic area (PLr) distribute selectively to medial aspects of C-P, whereas those from the caudal prelimbic area (PLc) project selectively to the ACB. IL: infralimbic cortex; MO: medial orbital cortex; S: septum.

receives significant projections from the basal nuclei of the amygdala as well as from monoaminergic nuclei (e.g., dopaminergic, noradrenergic and serotonergic) of the brainstem. It is well recognized that the monoaminergic nuclei exert pronounced modulatory effects on PL in affective and cognitive functions (Friedman and Robbins 2022).

With some exceptions, the output of caudoventral PL parallels its input (Hoover and Vertes 2007; Vertes 2004). Cortically, this caudoventral PL strongly targets other prefrontal cortical regions, including the medial orbital cortex, the dorsal and ventral agranular insular cortex, the anterior piriform cortex, and the entorhinal cortex. Subcortically, caudoventral PL distributes heavily to (a) the ACB, olfactory tubercle, and claustrum of the basal forebrain; (b) the central and basal nuclei of the amygdala; (c) the MD, intermediodorsal, paraventricular, paratenial, reuniens, and centromedial thalamic nuclei; and (d) the substantia nigra, pars compacta, ventral tegmental area, and dorsal and median raphe nuclei of the midbrain. In summary, the inputs and outputs of the caudoventral PL largely mirror those of area 32 of primates.

Functional Studies

While the debate on the rodent homologue of the dlPFC of primates may never be resolved to everyone's satisfaction, primates (especially humans) possess abilities that undeniably exceed those of rodents, and this undoubtedly is tied to cortical evolution including that of the dlPFC. Still, it must be acknowledged that rodents exhibit executive functions that are classically attributed to primate dlPFC. In addition to the anatomical evidence discussed above, behavioral evidence suggests that rostrodorsal PL is a "functional homologue" of primate dlPFC.

Granon and Poucet (2000) were among the first to make this proposal. Specifically, they reviewed evidence showing that alterations of PL in rodents (but not other mPFC regions) produced severe impairments on various spatial and nonspatial delay tasks. This indicated a profound working memory deficit—a hallmark of damage to the dlPFC. The working memory deficits were part of a constellation of cognitive impairments produced by alterations of PL that included attentional deficits. In addition, Granon and Poucet pointed out that rostrodorsal PL is reciprocally connected to the lateral subdivision of the MD, paralleling primate dlPFC projections to the lateral MD (Granon and Poucet 2000). Several other studies described similar reciprocal connections between PL and lateral MD in rodents (Bolkan et al. 2017; Mukherjee et al. 2020; Schmitt et al. 2017; Wolff et al. 2008). Granon and Poucet (2000:235) concluded that "in both species [rodents and primates], the prefrontal cortex, seems to share some common function in those aspects of cognitive processing that, in humans, are usually referred to as executive functions. Within the rat prefrontal cortex, the prelimbic area appears to play a central role in such processes."

Several subsequent reports have confirmed the role of PL of rodents in working memory and in several additional cognitive functions including attentional processes, set shifting behavior, reversal learning, and decision making (for reviews, see Chudasama 2011; Friedman and Robbins 2022). Specifically, these are all functions that in primates are associated with activation of the dlPFC.

Physiological evidence also supports the idea that the rostrodorsal PL and dlPFC are homologous. Classical work by Fuster, Goldman-Rakic, and others (Funahashi et al. 1993b; Fuster and Alexander 1971) have shown that neurons in the dlPFC exhibit persistent increase in spike rates in the context of working memory, which has been considered to be a cellular correlate for this cognitive process (Fuster and Alexander 1971). Newer studies have corroborated these observations, albeit they emphasize a persistent network activity pattern (rather than individual neurons) and perhaps temporally sparser patterns of working memory correlates at the level of single neurons (Lundqvist et al. 2016). Consistent with these latter observations, and with the PL homology, multiple studies have found evidence for persistent network activity patterns in the context of working memory tasks. For example, Bolkan et al. (2017) found evidence for a sequential PL activity pattern in the context of a spatial working memory task. Interestingly, this activity pattern was not spatially specific, potentially reflective of the PL's function in the generation of abstract rules, which are a known attribute of dlPFC. This was corroborated by data from Schmitt et al. (2017), who trained mice on a cross-modal attentional control task where mice selected between visual and auditory target stimuli based on a cue that varied on a trial-by-trial basis. Out of several cortical areas inactivated in the PFC, including orbitofrontal cortex, anterior cingulate cortex, and premotor cortex, only the PL showed a delay period specific effect (Wimmer et al. 2015). Recordings from the PL showed a persistent network activity pattern over the delay, where single neurons exhibited a temporally precise increase in firing rate tiling the delay period (sequential activity pattern). These network patterns where "rule specific" (Rikhye et al. 2018; Schmitt et al. 2017), consistent with the finding from primate dlPFC which showed the highest proportion of neurons encoding abstract rules in working memory tasks (Wallis et al. 2001). Perhaps the most compelling link to the specificity of these observations to the rostrodorsal PL is the work by Nakajima et al. (2019), which showed that neurons in this particular region project to the dorsal striatum (Figure 3.2a) and exhibit activity patterns consistent with attentional modulation (Figure 3.2b, c).

Lastly, in studying the architectonic subdivisions of the neocortex of the tree shrew, *T. belangeri*, a close relative of primates, Wong and Kaas (2009a) found that the PL of that species (and which they designated as area MF) contained a well-developed layer 4, which was densely populated with granule cells. This suggests that area PL of rodents, which occupies the same relative position as area MF of tree shrews, dorsally on the medial wall of the PFC, could be the antecedent of the granule cell layer of primates. Consistent with this notion, we show comparative sections of this region across rats, Tupaia, and macaques (Figure 3.3).

Homology between Infralimbic Cortex and vmPFC

Whereas the rodent homologue to the dlPFC of primates remains controversial, there appears to be a general consensus that ventral parts of the mPFC of rodents are anatomically and functionally equivalent to the agranular ventral medial PFC (vmPFC) of primates. More specifically, area IL of rodents appears anatomically homologous to area 25 (A25) of primates.

For instance, the IL of rodents and A25 of primates serve well-recognized roles in autonomic, visceral, and affective functions. IL has been described as a visceromotor cortex. The projections of IL reflect its involvement in visceral/affective functions. Specifically, Vertes (2004) examined IL projections in rats and showed that IL distributes to several sites of the forebrain and brainstem

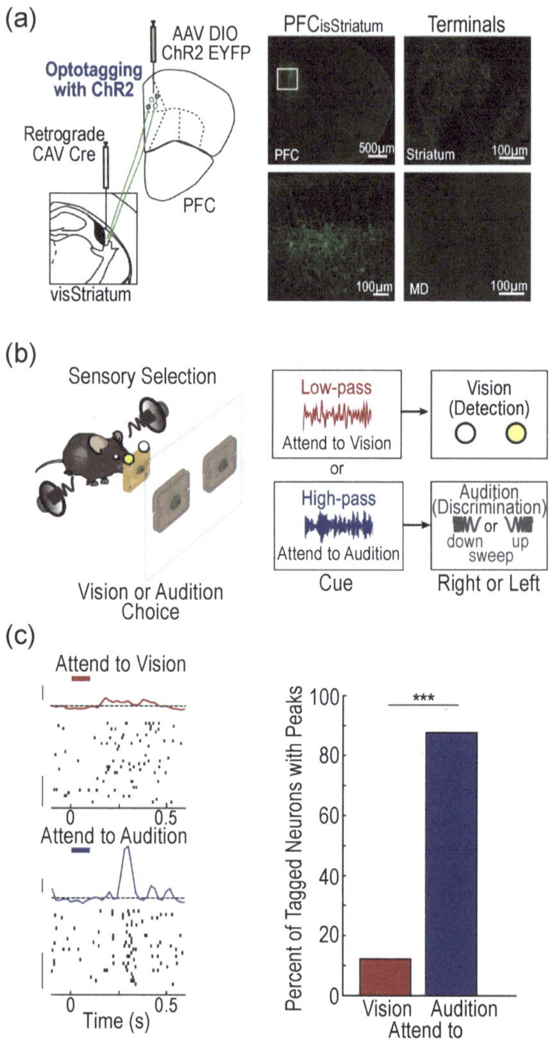

Figure 3.2 Rostrodorsal prelimbic neurons project to the dorsal striatum and show attentional modulation. (a) Schematic of the strategy of intersectional canine associated virus 2 (CAV2)-Cre based retrograde labeling of PFC neurons projecting to visual striatum. Expression of channel rhodopsin 2 (ChR2) in these neurons allows for optogenetic tagging. (b) Cartoon of the 2AFC cross-modal attention task (Wimmer et al. 2015). (c) Left: Example raster and peri-stimulus time histograms (PSTHs) of the response of an optogenetically tagged PFC neuron projecting to the visual striatum recorded in the cross-modal two alternative forced choice (2AFC) task. Zero time indicates cue presentation (100 msec duration, LP–Red bar, HP–Blue bar, PSTH y-axis scale bar: 1 Zscore, Raster y-axis scale bar: 10 trials). Right: The majority of tagged neurons showed peaks only in attend to audition (blue) but not during attend to vision trials (red) (N = 2 mice per condition, n = 112 neurons; *** $p < 0.001$ pairwise binomial test). Figure adapted from Nakajima et al. (2019).

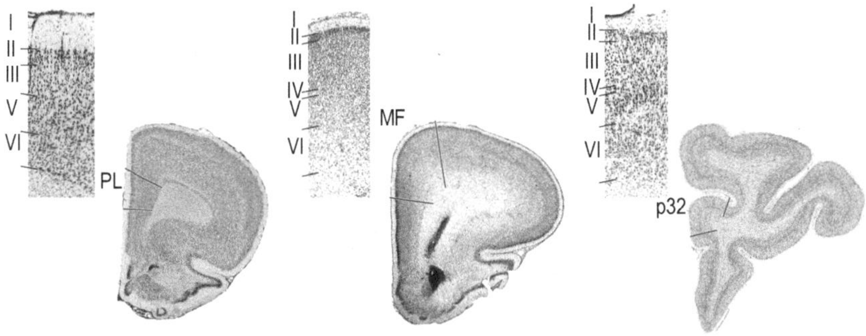

Figure 3.3 Coronal sections through the rat (left), Tupaia (middle) and a macaque (right) prelimbic region and processed for the visualization of cell bodies. Insets provide a detailed view of the cytoarchitecture of prelimbic area in each species. Note the lack of an inner granular layer (layer IV) in the rat prelimbic area (PL) and the presence of a few scattered granule cells indicative of an incipient layer IV in prelimbic area MF of Tupaia. Prelimbic area p32 of the macaque brain presents a dysgranular layer IV. Roman numerals indicate cortical layers.

linked to autonomic and affective behavior. These included orbitofrontal cortices, shell of nucleus accumbens (sACB), lateral septum, bed nucleus of stria terminalis (BST), medial and lateral preoptic nuclei, central nucleus of the amygdala, lateral and posterior nuclei of the hypothalamus, and the periaqueductal gray, parabrachial nucleus and solitary nucleus of the brainstem. Each of the structures has been shown to modulate autonomic/visceral activity, and thus emotional behavior, and importantly as a group, these nuclei receive input almost exclusively from IL and little from PL.

Although fewer reports have examined vmPFC (or A25) projections in primates, A25 projections in the monkey appear to directly parallel those of IL in rodents. Specifically, an early report by Chiba et al. (2001) compared the efferent projections of A25 (IL) and A32 (PL) in the Japanese monkey and showed that the output of A25, like that of IL in rodents, strongly targeted sites involved in autonomic/visceral control, primarily including the sACB, the preoptic area, BST, central nucleus of the amygdala (CeM) and the periaqueductal gray and parabrachial nucleus of the brainstem. They thus concluded that their findings "support the hypothesis that IL is a major cortical autonomic motor area." Several subsequent examinations of A25 projections in monkeys and have similarly demonstrated that A25 prominently distributes to several "visceral-related" subcortical structures of the basal forebrain, amygdala, hypothalamus and brainstem (Barbas et al. 2003; Ghashghaei et al. 2007; Heilbronner et al. 2016; Joyce and Barbas 2018; Rios-Florez et al. 2021; Roberts et al. 2007). Major targets included the ACB, BST, central nucleus of the amygdala, posterior and lateral nuclei of the hypothalamus, periaqueductal gray and parabrachial nucleus.

Barbas et al. (2003) described projections from mPFC in primates, including A25, to discrete nuclei of the amygdala and hypothalamus that directly distribute to (autonomic) brainstem and spinal cord nuclei which innervate peripheral autonomic sites. This system of connections linked mPFC/A25 with autonomic effector sites in the modulation of visceral functions and emotional behavior. However, in subsequent studies Barbas and colleagues have suggested that the connections of posterior OFC with the intercalated cell masses of the amygdala more resemble rodent IL, than primate A25 (Zikopoulos et al. 2017).

In contrast, Heilbronner et al. (2016) compared the projections to the striatum from A25 in macaques and IL in rats. Specifically, they first identified a region of the sACB (termed the "striatal emotion processing network" or EPN) and conserved across these species. The EPN is a convergence zone of projections from the amygdala and hippocampus to the sACB. Importantly, they showed that both IL and A25 distributed heavily to the striatal EPN, whereas other prefrontal cortical areas (of both species) projected at best weakly to EPN. They concluded that "consistent with prior literature, the infralimbic cortex and area 25 are likely homologous" (Heilbronner et al. 2016:509). Future studies should perform whole brain connectivity fingerprints across species for a more comprehensive comparison. However, it should be noted that even if rodent IL and primate A25 show overall similar connectivity patterns, the evolutionary expansion of the PFC may endow primate A25 with unique interregional connectivity patterns and divergent functions.

Recently, Roberts and colleagues (Alexander et al. 2023) comprehensively reviewed the structural and functional properties of the vmPFC across species (rat, monkey, human) and cited evidence showing that (a) the IL of rats and A25 of primates show some functional homology/analogy in the regulation of behavior in the reward domain but not in the punishment domain. Specifically, they showed that A25 overactivation in marmosets blunted Pavlovian approach and motivated responding, comparable to that reported following similar manipulations in rodents. In marked contrast, the same manipulation heightened behavioral and cardiovascular responsivity to both proximal and distal threat, opposite to that reported in rodent IL. This suggests that IL and A25 may act similarly within reward networks but their roles may have diverged within threat networks illustrating the complexity of cross-species functional comparisons. Roberts and colleagues also showed (b) that IL/A25 and PL/A32 predominantly serve distinct and separable functions, with A25 mainly involved in cardiovascular and affective functions and A32 in cognitive functions. A cytoarchitectonically informed meta-analysis of functional imaging studies in humans provides further evidence for this functional segregation of A25 and A32 (Palomero-Gallagher et al. 2015). For instance, with respect to differences between A25 and A32, Wallis et al. (2017) demonstrated that inactivation of A25 produced pronounced cardiovascular changes, whereas inactivation of A32 had no cardiovascular effects, and further that A25 and A32 mediated opposite effects on a Pavlovian fear conditioning and extinction paradigm: A25 inactivation decreased fear-elicited behavior responses promoting extinction, whereas A32 inactivation enhanced these responses thereby suppressing extinction.

Lastly, Diehl and Redish (2023) have performed comprehensive recordings across the rat mPFC in the context of a foraging task termed "restaurant row." This task combines multiple cognitive elements including associative learning, working memory, switching, and value-based judgments. Although they found that all prefrontal areas encode the various relevant task variables, there was clear specialization, with the IL clearly encoding more value-related cognitive variables than executive or sensorimotor ones. This is consistent with an earlier report, in which Hardung et al. (2017) examined the neural substrates for response inhibition across areas of the rodent frontal cortex using both optogenetic inactivation and electrophysiological recordings. Strikingly, inactivation of the PL and IL had opposite effects on the behavior, where PL inactivation increased and IL inactivation decreased premature responses. Electrophysiological recordings were also consistent with opposing roles for these two subregions, again, consistent with the idea that PL shares functional homology with the primate lateral stream whereas the IL is medial (and evaluative).

Conclusions

Building on the two-stream notion of human (or generally primate) PFC, the collective evidence reviewed in this chapter argues for homology with the two major divisions of rodent PFC: the PL and IL. The argument implicitly makes a prediction about how the rostrodorsal PL may have disconnected from the IL throughout evolution, and subsequently pushed laterally to form what is currently recognized as dlPFC of primates. The fact that *T. belangeri* MF is granular is consistent with this idea. Overall, we hope this synthesis will stimulate further discussion and motivate the design of new experiments to test this hypothesis directly.

4

Evolutionary Perspectives

Homologies and Analogies

Kevin S. Weiner, Bernard Balleine, Michael M. Halassa,
Alicia Izquierdo, Nicola Palomero-Gallagher,
Peter H. Rudebeck, Jeroen B. Smaers,
and Trevor W. Robbins

Abstract

Determining homologies and analogies of brain structure and function across species is of major interest in systems neuroscience, comparative biology, and brain mapping. Prefrontal cortex (PFC) is a continued target of such analyses because it has expanded considerably throughout evolution. It is heavily differentiated and expanded in primates compared to mouse, rat, tree shrew, and marmoset brains, and it performs computational functions that are more complex than other association cortex.

This chapter reviews the major regions and circuits observed across species within PFC. It looks at the evolution of PFC and how this could produce higher-order cognition, including social behavior, as well as language elements in humans. It provides a synopsis of some main organizational principles of PFC as well as potential mechanisms by which major circuits in PFC exert control. It then reviews how unique contributions of optogenetics, chemogenetics, large-scale electrophysiology, and calcium imaging contribute to understanding PFC function. It also addresses the utility of animal models for understanding the structure and function of PFC.

The discussions that contributed to this chapter provide a modern foundation for the ongoing goal of generating a consensus statement regarding the ambition of determining the homologies and analogies of PFC, as well as the cognitive, developmental, and translational insights gleaned from the promise of such an eventual consensus statement.

Group photos (top left to bottom right) Bernard Balleine, Trevor Robbins, Kevin Weiner, Alicia Izquierdo, Michael Halassa, Jeroen Smaers, Peter Rudebeck, Trevor Robbins, Kevin Weiner, Trevor Robbins, Bernard Balleine, Jeroen Smaers, Alicia Izquierdo, Kevin Weiner, Nicola Palomero-Gallagher, Bernard Balleine, Alicia Izquierdo, Michael Halassa, Peter Rudebeck, Trevor Robbins, and Nicola Palomero-Gallagher

Introduction

The definition of what comprises prefrontal cortex (PFC) has depended on several criteria, including simple location (i.e., regions of anterior cortex), cytoarchitectonic characteristics (notably granularity associated with lamina 4 innervation), and connectivity (e.g., mediodorsal thalamic input). None of these criteria is decisive, especially when comparing across species, specifically when considering human, nonhuman primate (NHP), and rodent. Although PFC has been classically defined as the granular cortex in the frontal lobe, how can we say that granularity is of particular importance, if we do not fully understand its biological significance?

What is understood is that PFC is a nexus for higher cognitive function and dysfunction in humans and may be the cause of numerous psychiatric disorders. Consequently, understanding PFC function is a critical aim for basic research. While some would opine that PFC can only be studied in primates or tree shrews (Preuss and Wise 2022), there are limits to the research that can be ethically and/or practically accomplished if we take this position. Thus, to make faster headway, it is reasonable to ask how best to compare and model human PFC subregions across species beyond primates. This involves issues of homology (i.e., shared ancestry between a pair of structures or genes in different taxa). One of the main aims in our discussions was to prioritize the various criteria for homology, based on micro-architectonics (including cytoarchitecture and the architecture of neurotransmitter receptors), connectivity with other brain regions, and development. Another criterion, which cannot be considered as homology in the formal sense, is based on analogy. In this chapter, we consider analogy as resemblances in function across species between organs (e.g., different regions of PFC) that may have different evolutionary origins. These may reveal essential building blocks in rodents of more complex executive processes in primates. Work in each species is in itself a significant scientific problem of great utility, with impact in areas such as artificial intelligence and human health. Specifically, insights from nonhuman animal species may ultimately inform the understanding of clinical conditions. Here, we attempted to take all these considerations into account when discussing the evolution of the PFC and its possible drivers, for example, increasing complexity of information processing required for foraging and social behavior as well as ultimately the capacity for language and moral reasoning.

We consider whether there is anything "special" about the PFC and its organization, including regional localization of function, whether there is hierarchical organization across species and dorsal-ventral or medial-lateral gradients. Allied to this analysis, we also consider whether there are unique aspects of neuronal activity of the PFC that confer its higher-order functioning (e.g., neuronal synchrony and oscillation), its plasticity and possible capacity for fast learning, as well as its top-down controllability of neurochemical modulation by the ascending monoamine and cholinergic systems. We also address whether the network organization of prefrontal-related circuits, as defined in human studies, is represented in other animals and how this relates to concepts of goal-directed control.

Finally, we discuss the unique opportunities for delineating functional neural circuitry involving PFC in nonhuman animals using modern neurobiological techniques, such as optogenetics and chemogenetics. These methodologies can be used to establish causal relationships at nodes within circuits, including PFC, as well as the interactions and sequencing of recruitment among prefrontal regions themselves to guide behavior. Furthermore, they can potentially be used to simulate states and mechanisms of treatments associated with clinical disorders, with implications for animal models of human clinical disorders. Of course, these are bold goals to achieve in one chapter, and while we appreciate that we will fall short from achieving these goals, we are hopeful that this discussion will motivate future experiments, models, and quantifications that come closer to understanding the evolution of neural circuits underlying the complexity (Rigotti et al. 2013) of prefrontal cortical structure and function linked to higher-level aspects of cognition that have critical insights for better understanding the neural underpinnings of neuropsychiatric disorders.

What Are the Major Regions and Circuits Observed across Species within PFC?

To answer this question, we found it necessary to define the relevant species and areas of focus. We chose to focus on widely used animal models for humans across subdisciplines in the broad fields of neuroscience and medicine: rodents and NHPs. Ultimately, one way to organize the quest for homology would be to take human PFC as the starting point and work "backward" through the evolutionary tree. Taking these issues into consideration, we consider a parcellation of PFC based on connectivity patterns and roles in cognitive and emotional processes focusing on a tripartite division involving orbitofrontal cortex (OFC), dorsolateral PFC (dlPFC), and ventrolateral PFC (vlPFC). Within these anatomical locations, the main cytoarchitectonic areas that we focus on (using Brodmann's/Walker's nomenclature) in this chapter are 10, 11, 13, and 14 (Brodmann 1909; Walker 1940). While the cingulate cortex is classically not considered to be part of PFC, it is closely interconnected (structurally) and interacts (functionally) with prefrontal areas. Thus, during our discussions, we adopted/tolerated the view that anterior cingulate cortex (ACC, areas 25, 32 and 24) are part of the PFC, and are specifically located within the medial PFC. In addition to these decisions, we also considered classic questions such as: Where, if anywhere, is PFC in rodents? To what extent are the organizational principles of the NHP PFC, specifically in macaque monkeys, comparable to those of the human PFC?

As reviewed by Izquierdo (this volume) and elsewhere (Le Merre et al. 2021; Uylings et al. 2003; Vogt and Paxinos 2014), there are criteria for defining PFC in the rodent brain. The presence of the internal granular cell layer, layer IV (LIV), has been considered the primary definition of primate PFC. LIV is a microcircuit feature of isocortical areas considered to be especially critical in cortical regions that have expanded the most throughout evolution in association cortex such as PFC. As granularity of dorsal frontal cortex in rodents is a matter of debate and their OFC areas are agranular, and thus lack this LIV, classic research widely purported that rodents lacked any homologues to areas in primate PFC (Laubach et al. 2018; Preuss 1995). More recent criteria have been proposed beyond cytoarchitectonic features, such as functional properties (similarities in behavior) and electrophysiological neural signatures, neurochemical distribution and receptor expression, and/or architecture, embryological development (which we briefly discuss in this section), and connectivity (both the patterns as well as the density of connections) (Seamans et al. 2008; Uylings et al. 2003; Rich and Averbeck, this volume). Again, it is worth noting that the term homology refers to shared ancestry. Thus, it may be better to characterize these additional proposals of PFC features as indicators of an area being analogous to human PFC.

Adding to this complexity of the PFC homology/analogy debate is what the pioneering neuroscientist Charlie Gross once referred to as the "alphabet soup" of cortical areas (Gross 1994). That is, the inconsistency of anatomical nomenclature and the use of multiple terms/acronyms for the same subregion of the brain not only across species but also within species. This is not a new issue. It stems all the way back to the late 1800s, when Burt Green Wilder (1881, 1896) and Wilhelm His (1895) led different teams to address it, and still persists today, not only for cross-species comparisons but also within species (Weiner 2019; Weiner and Zilles 2016). For brevity, we refer the reader to Izquierdo (this volume) for a review of this issue between rodent and primate; for discussion of the different criteria recently proposed, see Barreiros et al. (2021a, b), Heilbronner et al. (2016), Izquierdo et al. (2017), Rudebeck and Rich (2018), Wallis (2011), and Wise (2008).

In addition to the tripartite parcellation of PFC noted above, we also included the inferior frontal cortex and frontopolar areas in our discussions as they are likely not homologous between rodent, marmoset, macaque, and human: areas 44 and 45 (inferior frontal cortex or "Broca's region"), the vlPFC encompassing area 12/47, and areas FP1 and FP2 in the frontal pole within Brodmann's area 10 (Bludau et al. 2014). This aspect of our discussion led logically to the next question: What are the most important criteria for similarity between species? This is especially critical considering the massive differences in brain size and the complexity of cortical convolutions across species. For example, the mouse brain is about 4,000 times smaller than the human brain and contains about 71 million neurons, whereas the macaque brain is about 15 times smaller than the human brain and contains about

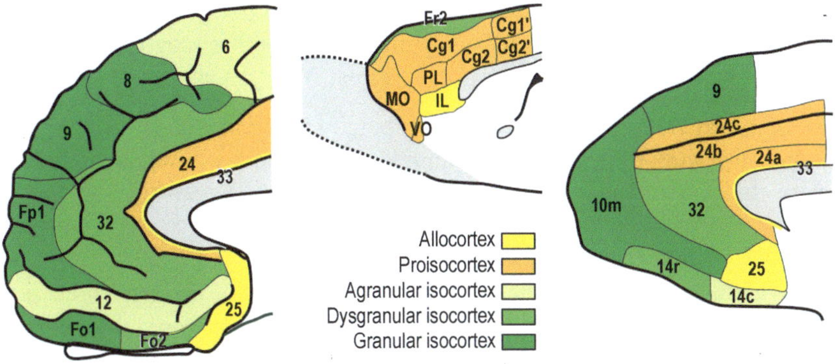

Figure 4.1 Comparative neuroanatomy of regions covered in this chapter. Medial views of the human (left; modified after Brodmann 1909), rat (middle; modified after Haghir et al. 2023), and macaque monkey (right; modified after Morecraft et al. 2012) frontal cortex. The cytoarchitecture of each area is indicated by different color shading: granular (dark green), dysgranular (light green), agranular (yellow/green), proisocortex (orange), and allocortex (yellow). Please refer to the main text for our discussion about disagreements regarding the exact parcellation of each area in this cortical expanse in each species, as well as the variously proposed combination of numbers and letters used to refer to each area since the 1800s. Note, in the schematic representation of the macaque brain we highlight the position of cortical borders in relation to the *fundus* of the cingulate sulcus (i.e., area 24c is located on the dorsal bank of the cingulate sulcus), whereas in the human brain we do not show cortex buried in the sulci.

6,376 million neurons (Azevedo et al. 2009; Herculano-Houzel 2009). For an overview of the comparative neuroanatomy of regions that are a focus of this chapter, see Figure 4.1.

What Are Important Criteria of Homology?

In terms of semantics, in neuroimaging and cognitive neuroscience, the term homology has a different definition than in comparative and evolutionary biology—a difference that can be traced back to Owen's definitions in 1843 (Gross 1993). As such, it is important for us to define homology in the context of this chapter. Here, homology refers to a shared structure among species. Through extensive discussion, it was concluded that for the purpose of determining potential homologies (or not) among these main PFC regions between rodent and primate, cytoarchitecture and connectivity were the two most critical criteria. Our analysis of the PFC is necessarily constrained by the evidence of homology in a number of areas in the medial and orbitofrontal cortex in rodents with similar structures in primates. These regions include:

- The rodent ACC areas Cg1 and Cg2, mainly considered to be homologous with primate area 24
- Prelimbic area, discussed as homologous to area 32 or more controversially, the dlPFC in primates

- Infralimbic area, mainly thought of as homologous to primate area 25
- Lateral and medial OFC—agranular regions that may correspond to posterior lateral and medial OFC (area 14) in primates, respectively

Rodent frontal areas Fr1 and Fr2 are thought to contain areas that are functional analogues of primate premotor and supplementary motor areas as well as of the frontal eye field (Donoghue and Wise 1982; Neafsey et al. 1986). Below, we integrate and highlight prominent features of each to consider homology (or not) among species.

Cytoarchitectonic mapping is based on the fact that the cerebral cortex presents a laminar organization that consists of six horizontal layers that run parallel to the cortical surface and vertical columns. The most important criteria followed in classic cytoarchitectonic studies include:

- Absolute cortical thickness
- Thickness of a given layer relative to that of the remaining layers and of the cortical ribbon (a roughly 3 mm strip of gray matter on the outer surface of the cerebral cortex[1])
- Size and packing density of neuronal cell bodies
- Presence of vertical columns and/or of sharp borders between layers
- The distribution pattern of cell bodies throughout the layers (homogeneous or clustered)
- The presence of special cell types such as the giant cells of Betz

With the advent of immunohistochemistry and receptor autoradiography, modern neuroanatomists have been able to make use of the heterogeneous distribution of cytoskeletal elements or enzymes, as well as of neurotransmitters and their receptors (Palomero-Gallagher and Zilles 2018) to quantify differences directly in micro-architecture between adjacent pieces of cortical tissue. The presence of LIV, together with its thickness, has been the cytoarchitectonic definition used to segregate PFC from the rest of cortex (Table 4.1). Thus, PFC encompasses areas that are granular or lightly granular. Within this region, some areas have a broader, and others a narrower, LIV. In some cases, LIV is particularly thin and invaded by layer III and layer V pyramids so that it appears as a discontinuous layer within the cortical ribbon. Areas with such an inconspicuous LIV are classified as being dysgranular in nature. As stated above, during our discussion, we found it necessary to consider a tripartite parcellation of PFC, which also considered agranular areas in OFC and portions of ACC (which could be considered controversial) but resulted in fruitful conversations regarding homologous and analogous areas across species.

Brain connectivity also provides another means by which to assess the structural similarities and differences of PFC between species. In both rodents and NHPs, the PFC is reciprocally connected with the mediodorsal thalamus

1 For a perspective of scale, 3mm is about how much your fingernail grows in one month.

Table 4.1 Most prominent cytoarchitectonic features of Walker's (1940) areas (modified from Rapan et al. 2023).

Area	Cytoarchitecture
8A	Pale layer III Granular; broad and densely packed layer IV
8B	Densely packed layer II Small-sized pyramids in layer III, particularly its upper portion Dysgranular
9	Gradient in cell size within layer III Granular Layer V divided into sublayers Va and Vb
10	Prominent layer II Small-sized layer III pyramids Granular; broad and densely packed layer IV Small-sized layer IV pyramids
11	Granular Layer V divided into sublayers Va and Vb
12	Most rostral and caudal portions are dysgranular Centrolateral portion is granular Sublamination of layer V in the centrolateral but not the rostral and caudal portions
13	Caudal portion is dysgranular. Rostromedial portion is granular Layer V divided into sublayers Va and Vb
14	Pale but clearly identifiable layer II Caudal portion is agranular Rostral portion is dysgranular Columnar pattern in layers V and VI
46	Prominent layer II Scattered middle-sized pyramids in lower layer III Granular Layer V divided into sublayers Va and Vb
45	Middle-sized layer III pyramids Granular. Thin, relatively inconspicuous layer IV
44	Dysgranular Single larger pyramids scattered throughout layer V

(Ray and Price 1992, 1993). Ventral and medial PFC in both species also receive extensive connections from the amygdala, hippocampus, and sensory areas in the temporal lobe, indicating that rodents and NHPs broadly share similar connectivity (Öngür and Price 2000). While these broad similarities exist, there are key differences in the patterns of connections, which we will highlight in subsequent sections as we cover each part of the PFC.

The emergence of novel high throughput connectomic approaches may enable future studies to better reveal just how different or similar these patterns of connections are between rodents and NHPs (Kebschull et al. 2016; Zeisler et al. 2023). In addition to differences in the patterns of connections from one brain area to different parts of the PFC, there are major differences in the routes that projections take to their targets in the PFC. For instance, white matter pathways that carry connections to and from the PFC are organized into large bundles, such as the cingulum bundle. The presence and physical organization of these bundles in macaques are highly similar to those in humans (Lehman et al. 2011), but the correspondence between rodents and humans is much less clear. This relationship has been essential for modeling the impact of deep brain stimulation delivered to white matter to treat psychiatric disorders (e.g., Mayberg et al. 2005).

Cytoarchitecture and Connectivity

Infralimbic in Rodent and Area 25 in Primate

The term infralimbic (IL) is used in rodent but is much less common in primate research. It is widely accepted that area IL is generally homologous to primate area 25 (e.g., Preuss 1995; Room et al. 1985; Saper and Stornetta 2015; Vogt and Paxinos 2014). IL is agranular and part of the allocortex. Primate area 25 is also allocortical, and there are clear similarities in the connections of the primate area 25 and rodent IL, especially those to striatum (Heilbronner et al. 2016).

Prelimbic in Rodent and Area 32 (or dlPFC) in Primate

The term prelimbic (PL) is used in rodent but much less so in primate research, and the issue of which area in the primate brain is homologous to PL remains the subject of intense debate. Some consider PL to be homologous to cingulate area 32 (e.g., Preuss 1995; Room et al. 1985; Saper and Stornetta 2015; Vogt and Paxinos 2014), whereas others consider it to be equivalent to primate dlPFC (e.g., Kesner and Ragozzino 2003), with still others to cingulate area 24 (Milad and Quirk 2012). Rodent area PL is agranular and part of the proisocortex (transition from allocortex to isocortex), as is primate area 24. However, primate area 32 and areas of the dlPFC are all isocortical. LIV is inconspicuous in area 32 (dysgranular cortex) but clearly visible in the dlPFC (granular cortex). Because area 32 has a thin LIV, while areas in dlPFC have a prominent LIV, and PL is proisocortical, this is stronger evidence for the theory that PL in rodent is homologous to area 32 in primate. On this basis, rodent PL cannot be homologous to dlPFC in primates as they do not share a common ancestry. However, evidence from connectivity is not as clear, and results from more recent functional studies in rodents indicate that PL could be considered analogous or similar to primate dlPFC (see Vertes et al., this volume). A possible explanation for this apparent discrepancy could be that PL is a precursor of both primate area 32 and dlPFC (Vertes et al., this volume). Thus, depending on the aspects analyzed, researchers have uncovered the characteristics of PL that are more similar to those of 32 or of dlPFC.

Expanding beyond the potential similarities of these cortical areas across species, there is also evidence of two prefrontal "streams" across species (Vertes et al., this volume), although we note that some members in our group preferred tripartite organization for the frontal cortex. Vertes et al. (this volume) also incorporate findings from a close relative of primates, tree shrews. When considering homologies discussed in their chapter, an interesting piece of the evolutionary puzzle is that tree shrews contain a well-developed LIV in an area located within a topographical position comparable to that occupied by PL (Wong and Kaas 2009b). This further suggests that rodent PL could be a precursor to the granular dlPFC of primates.

We also highlight that PL is not likely one area, as indicated by connectivity data, and may have rostral/caudal and dorsal/ventral components. As further discussed by Vertes et al. (this volume), in an experiment that demonstrated the differences in retrograde labeling following tracer injections into the ventral versus dorsal-ventral striatum (VS), labeled cells following injections in the ventral VS were found in both the IL and PL. However, labeled cells following injections in the more dorsal VS were found primarily in PL. Closer inspection of the PL-labeled cell distributions showed a possible rostrocaudal and dorsoventral distinction. There appeared to be fewer labeled cells in the caudal PL. Moreover, the density of labeled cells from the ventral VS were found in the ventral part of the PL compared to the density of cells following injections in the dorsal VS.

The dorsoventral distinction may be critical for linking homologous PL regions with the monkey cingulate cortex. Comparing the projections from areas PL (in rodents) and 32 (in primates) to the striatum demonstrated that the PL terminates along the medial border of the striatum, similar to the projection zone of area 32 in the monkey. However, importantly, PL extends more laterally into the striatum, compared to the monkey, into the regions occupied by pregenual, area 24 in the monkey (Heilbronner et al. 2016). This may indicate that part of PL may be homologous to rostral area 24 in the monkey, as proposed by Milad and Quirk (2012) based on functional similarities with respect to threat expression. In contrast, CG (expanded on further below) projections in the rodent terminated dorsal and lateral to the PL-striatal projections. The striatal space in primates is not the main recipient of cingulate projections but is the main recipient from dlPFC and premotor projections (Heilbronner et al. 2016).

Anterior Cingulate Areas: ACAd, ACAv in Rodent and Parts of Area 24 in Primate

Cortex dorsal and caudal to PL contains proisocortical areas dorsally (ACAd) and ventrally (ACAv), which are characterized by the absence of a LIV and by a broad layer V with relatively large neurons (Swanson 2018). Area ACAd encompasses areas Cg1 rostrally and Cg1′ caudally, and ACAv areas Cg2 and Cg2′ (Haghir et al. 2023; Vogt and Paxinos 2014). Hereby Cg1′/Cg2′ constitute the midcingulate cortex, which is not considered relevant to this survey. Areas Cg1 and Cg2 are thought to be homologous to primate areas 24b and 24a, respectively (Vogt and Paxinos 2014). Thus, primate area 24c, located within the cingulate sulcus, would not have a homologue in the rodent brain.

Nonhuman Primate Areas 11, 13, and 14

The initial parcellation of macaque ventral frontal cortex was completed by Walker (1940). Macaque OFC area 11 is granular and can be divided into medial and lateral components based on differences in layer V (Carmichael and Price 1994; Rapan et al. 2023). Areas 13 and 14 can each be subdivided based on rostrocaudal differences in the appearance of their LIV, which in both areas becomes less prominent when moving caudally (Rapan et al. 2023). Thus, area 14r is dysgranular whereas caudal to it, area 14c is agranular. Rostral area 13b is granular, whereas caudal area 13a is dysgranular. The reason for this apparent discrepancy is that, topologically, area 13a of Rapan et al. (2023) corresponds to area 13b of Carmichael and Price (1994). Further, other research groups have subdivided area 13 into medial and lateral segments based primarily on differences in SMI-32 and parvalbumin staining (Carmichael and Price 1994). Area 12 is also granular and can be subdivided into four subregions—12r, 12l, 12m, 12o—based on differences in myelin, ACHe, calbindin, and parvalbumin stains. A similar parcellation of marmoset ventral frontal cortex has also been produced (Burman and Rosa 2009). These areas also differ in their receptor architecture (for a summary of receptor densities, see Table 4.2).

In their analysis of human ventral frontal cortex, Öngür and Price revealed homologous areas to those identified in the macaque (Öngür et al. 2003; see also Wise 2008). Humans have a clear anterior to posterior gradient: posterior areas 13b, 13l, and 13m are dysgranular and more anterior areas including areas 11m and 11l are granular. All parts of area 12, like those in macaques, are also granular and split into a number of different subdivisions. The most posterior areas on the ventral surface of the frontal lobe, like those in macaques, are agranular (Öngür et al. 2003). Thus, there are clear homologues of human ventral frontal areas in macaques.

In rodents, OFC is agranular. Thus, there are no clear homologues of primate granular or dysgranular areas 11, 13, or 14 in rodent OFC (Preuss 1995; Preuss and Wise 2022; Wise 2008). Based on position and cytoarchitecture, it is reasonable to consider the rodent OFC to be similar to the agranular parts of the human and macaque ventral frontal cortex (Wise 2008). If we take the approach advocated by Wise, then rodents likely share areas 13a and 14c as well as the agranular insula areas with primates. There are other reasons to think that the OFC in rodents is similar to the OFC in primates. Like macaque

Table 4.2 Mean (±s.d.) densities in fmol/mg protein of receptors for the classical neurotransmitters glutamate (AMPA, kainate, and NMDA receptors), GABA ($GABA_A$ and $GABA_B$ receptors, $GABA_A$ associated benzodiazepine binding sites ($GABA_A$/BZ), acetylcholine (muscarinic M_1, M_2, and M_3 receptors), noradrenaline (adrenergic α_1 and α_2 receptors), serotonin (5-HT_{1A} and 5-HT_2 receptors) and dopamine (D_1 receptors) in cytoarchitectonically identified subdivisions of macaque areas 11, 12, 13 and 14 (from Rapan et al. 2023).

	11m	11l	12r	12m	12l	12o	13b	13m	13l	14r
AMPA	604	623	659	598	630	670	489	753	713	470
	(100)	(111)	(122)	(136)	(112)	(165)	(44)	(67)	(95)	(81)
Kainate	771	807	854	799	840	817	820	856	756	818
	(65)	(123)	(120)	(55)	(73)	(97)	(103)	(111)	(60)	(107)
NMDA	1585	1562	1406	1533	1400	1527	1548	1499	1498	1442
	(139)	(113)	(121)	(175)	(126)	(158)	(223)	(122)	(187)	(255)
$GABA_A$	1762	1876	1843	1792	1494	1579	1615	1622	1683	1427
	(142)	(235)	(283)	(246)	(221)	(267)	(120)	(126)	(180)	(162)
$GABA_B$	2476	2644	2412	2222	2010	2142	2311	1908	2057	2482
	(466)	(478)	(312)	(353)	(483)	(414)	(452)	(429)	(240)	(424)
$GABA_A$/BZ	1975	2066	1991	1873	1789	2102	1901	1864	2052	1715
	(218	(247)	(307)	(421)	(417)	(436)	(431)	(269)	(303)	(542)
M_1	1094	1050	1026	1152	824	888	1039	1059	1054	921
	(200	(228)	(301)	(262)	(347)	(174)	(263)	(121)	(148)	(385)
M_2	159	159	180	202	182	209	166	206	223	134
	(64)	(54)	(72)	(74)	(75)	(64)	(57)	(94)	(78)	(35)
M_3	965	944	922	918	780	832	897	918	826	833
	(132)	(101)	(96)	(108)	(132)	(149)	(104)	(130)	(108)	(118)
α_1	473	462	439	481	491	484	480	485	461	497
	(50)	(46)	(38)	(48)	(82)	(32)	(73)	(21)	(15)	(109)
α_2	342	351	306	379	320	401	350	417	404	297
	(40)	(45)	(52)	(71)	(43)	(66)	(75)	(21)	(26)	(95)
5-HT_{1A}	549	529	540	504	531	541	562	527	460	583
	(167)	(116)	(88)	(103)	(163)	(87)	(206)	(138)	(107)	(119)
5-HT_2	357	357	350	354	351	384	355	357	351	323
	(60)	(51)	(51)	(45)	(48)	(61)	(57)	(50)	(43)	(44)
D_1	92	96	86	86	71	89	93	78	70	86
	(27)	(29)	(9)	(22)	(6)	(20)	(22)	(11)	(4)	(15)

OFC, parts of the rodent OFC receive inputs from all of the sensory modalities as well as mediodorsal thalamus, amygdala, and hippocampus (Öngür and Price 2000; Rudebeck and Izquierdo 2022). Indeed, similar to macaques, there are similar medial to lateral gradients in the patterns of amygdala and hippocampal connections, where connections from the basolateral amygdala (BLA) complex primarily target more lateral parts of OFC, whereas hippocampal connections are relatively stronger in the more medial areas. Further, Barreiros et al. (2021b) have identified anterior to posterior gradients of connections in rat OFC, which indicate that, like macaques and humans, there may also be anterior-posterior distinctions. Taken together, these patterns of connections and cytoarchitecture indicate that rodent OFC bears many of the features of primate OFC.

It is also important to note that while macaque and human ventral frontal cortex is highly similar, there are also differences. For instance, on the basis of connectional fingerprinting, Neubert et al. (2015) found that no area in the macaque frontal cortex has a similar connectivity profile to anterior lateral OFC in humans. The area that they identified likely corresponds to area 111 in humans; this highlights that there are human anatomical specializations in ventral frontal cortex.

Subdivisions of Primate vlPFC, Area 47/12

The cortex on the ventral and lateral convexity of the PFC in humans was identified by Brodmann as area 47 (Brodmann 1909), and this cortical area contains both granular and dysgranular cortical areas (Rapan et al. 2023). A similar vlPFC area in macaques was also identified by Walker (1940) as the cytoarchitecture of the area made it distinct from the medially adjacent OFC and the more dorsally situated dlPFC. In their comparative analysis of macaques and humans, Petrides and Pandya (2002) designated this part of the PFC as area 47/12. Careful cytoarchitectonic analysis of this area by different investigators (Carmichael and Price 1994; Rapan et al. 2023) further subdivided the vlPFC into four main subdivisions: 12l, 12r, 12o, and 12m. Areas 12l and 12m are granular, whereas 12o and 12r are dysgranular. Analysis of the marmoset vlPFC found the same subdivisions of area 12 with the exception of 12r, which did not appear to be present.

Subdivisions of Primate Area 10

The frontopolar cortex is occupied by Brodmann's area 10, characterized by a broad and densely packed LIV (Brodmann 1909). In humans, quantitative cytoarchitectonic analysis revealed the existence of lateral and medial parts of BA10—areas Fp1 and Fp2, respectively (Bludau et al. 2014): Fp1 has a broader LIV as well as more densely packed layer II and IIIc than does Fp2. Differences in the densities of multiple receptor types confirm this mediolateral segregation (Palomero-Gallagher and Zilles 2018). In the macaque monkey, four cyto- and receptor-architectonically distinct subdivisions of area 10 have been identified (Rapan et al. 2023):

- 10d (on the dorsolateral surface of the frontal pole)
- 10o (on the most ventral aspect of the frontal pole)
- 10mv (medial surface, ventrally)
- 10md (medial surface, dorsally)

As in humans, all subdivisions of area 10 have a prominent LIV, though it is slightly broader in 10d and 10o than in 10md or 10mv. The marmoset, too, has a clearly defined area 10, although unlike macaque and humans, it is not really subdivided (Burman and Rosa 2009). The rat (and mouse) does not have an architectonic correlate of area 10, although we discuss functional homologues of area 10 in rodents below.

Broca's Region

In humans, Broca's region is considered to be the cytoarchitectonic correlate of Brodmann's areas 44 and 45 (Brodmann 1909; Amunts et al. 1999). However, receptor architectonic analyses have demonstrated a more complex picture, with dorsal and ventral subdivisions of 44 (44d and 44v) as well as anterior and posterior parts of 45 (45a and 45p) (Amunts et al. 2010). Areas 44d and 44v are both dysgranular: 44d has a higher acetylcholine M2, but lower glutamate AMPA receptor density, than 44v (Amunts et al. 2010). Areas 45a and 45p are granular: 45a has a higher acetylcholine M1, but lower glutamate kainate receptor density, than does 45p. Given the dominance of the left hemisphere in language production, it is not surprising that Broca's region has been subject of numerous studies aiming to link this functional asymmetry with an anatomical one (Sprung-Much et al. 2022). In this framework, extraordinary competence in language performance was found to be associated with cytoarchitectonic alterations in areas 44 and 45 and differences in interhemispheric asymmetries (Amunts et al. 2004).

The lateral PFC of macaques contains areas 44, 45a, and 45b, which are thought to be the homologues of Broca's region in humans (Petrides and Pandya 2002). Area 44 is located mainly on the ventral wall of the inferior arcuate sulcus, close to the fundus, and encroaches onto its dorsal wall, where it is followed by area 45b (Petrides and Pandya 2002; Rapan et al. 2023). Area 45a occupies the prearcuate convexity and its border with 45B was consistently found at the tip of the inferior arcuate sulcus (Rapan et al. 2023). As in humans, macaque area 44 is dysgranular and 45 is granular (Petrides and Pandya 2002; Rapan et al. 2023). In 45b, LIV is narrower than in 45a, and LIII pyramids tend to build clusters. As in humans, macaque areas 45a and 45b differed in their M1 and kainate receptor densities. Interestingly, area 44 presents one of the highest, and 45a the second lowest, 5-HT1A receptor densities within macaque PFC (Rapan et al. 2023). In contrast, the marmoset has a single area 45 with no A and B subdivision and no identified area 44 (Paxinos et al. 2012).

Finally, functional connectivity analysis of macaque areas 44, 45a, and 45b revealed a strong intercorrelation of 45a and 45b as well as their association with the auditory core region within the temporal cortex. Whereas 45a is correlated with areas of the OFC, 45b presents a widespread connectivity throughout the medial and inferior parietal cortex. The connectivity pattern of area 44 resembles that of 45b, although it does not include the primary auditory cortex: it does, however, show a strong correlation with the somatosensory cortex and area 4p of the primary motor cortex (Rapan et al. 2023). In accordance with these findings, electrical intracortical microstimulation of area 44 was found to elicit somatomotor responses in the orofacial musculature of macaque monkeys (Petrides et al. 2005).

Are There Functional Similarities of the PFC Across Species?

Whether putative homologous regions across species exhibit comparable functionality would appear to be an important consideration for understanding the evolution of PFC, but it does raise several potential problems. For example, suppose a region is defined to be homologous between rodent and primate, but then appears to have different functions. This problem arises when considering the IL and PL cortex in rodents, hypothesized to be homologous to area 25 and 32 in primates, respectively, based on their cytoarchitecture and connectivity patterns (Vogt and Paxinos 2014). However, comparison of their functional contributions to threat regulation in the rat and marmoset is inconsistent with this. Using a similar Pavlovian-conditioned threat paradigm to that used in rats, inactivation of marmoset area 25 increased the rate of extinction of a behavioral and cardiovascular conditioned threat response, whereas inactivation of area 32 produced the opposite effect, at least with respect to the behavioral response, thus decreasing the rate of conditioned threat extinction (Wallis et al. 2017). Consistent with this, area 25 overactivation induced generalization of the conditioned threat response and heightened anxiety-like behavior to uncertain threat (Alexander et al. 2020). This is diametrically opposite to that demonstrated in rats in which inactivation of IL decreases extinction of the conditioned freezing response while inactivation of PL accelerates extinction (Sierra-Mercado et al. 2011). Thus, at the level of the regulation of conditioned threat responses, these regions across primates and rats do not appear functionally analogous. In contrast, when considering the regulation of appetitive responses, there is greater correspondence between rat IL and marmoset area 25. Both regions, when activated, reduce aspects of reward processing (Alexander et al. 2019; John et al. 2012) via their effects on the nucleus accumbens (Wood et al. 2023). Thus, there is no simple functional correspondence between these regions across marmosets and rats.

At the level of cognitive function, distinct from emotional function, three of the main domains of human executive function have been defined as working memory, inhibition, and cognitive flexibility (Miyake et al. 2000). Simulations of each of these have been tested in rodents and NHPs, allowing possible behavioral similarities in PFC function to be explored across species. In such comparisons, there is always the issue of whether superficially similar performance of humans and other animals is determined by similar psychological processes. If it can be shown that homologous areas contribute to such performance across species, this provides evidence that they are likely to be employing at least the building blocks of more complex human executive functions.

Cognitive Flexibility

An early study by Dias et al. (1996a) showed that excitotoxic lesions of the OFC and vlPFC selectively impaired reversal learning and extra-dimensional set shifting in the marmoset, a double dissociation of function that has also been shown in the rat (Birrell and Brown 2000) and mouse (Bissonette et al. 2008), using an odor/tactile set-shifting task. The role of the medial PFC in rodents in extra-dimensional shifting is also consistent with work on so-called strategy shifting in rats, for example from visual to spatial cues or vice versa (Floresco et al. 2006). A study in humans has shown that resting-state functional connectivity between PFC regions including, lateral (12/47) PFC and caudate nucleus, correlated with deficits in extra-dimensional set shifting in patients with obsessive-compulsive disorder (Vaghi et al. 2017). Hence, there appears to be a degree of cross-species similarity in this capacity.

Reversal Learning

OFC has been heavily implicated in cognitive flexibility due to the effects that lesions have on this part of the frontal lobe in reversal learning paradigms. Reversal learning impairments have been consistently reported in rodents, new world primates, old world primates, and humans. There are, however, species differences in the nature of the tasks that may affect recruitment of OFC. For example, reversal learning tasks in rats and mice use spatial/action in their response (Barlow et al. 2015; Boulougouris et al. 2007; Dalton et al. 2016; Groman et al. 2019) more than stimulus/cue (Clarke et al. 2004; Izquierdo et al. 2013; Schoenbaum et al. 2003), whereas macaques and marmosets are most often tested using instrumental visual tasks.

As reviewed by Izquierdo et al. (2017), several subprocesses captured in most reversal learning tasks include rule implementation and reinforcement learning. For the sake of brevity, we highlight cross-species concordance of findings on reinforcement learning and the related function of "credit assignment." Credit assignment (i.e., the ability to assign an outcome to its contingent stimulus, cue, or action so that the most reliable prediction of future reward) relies on OFC in rodents and primates (Akaishi et al. 2016; Hervig et al. 2019; Izquierdo et al. 2013; Noonan et al. 2010; Schoenbaum et al. 1999; Walton et al. 2010). In addition, OFC and distinct circuits involving OFC across species (Aguirre et al. 2023; Dalton et al. 2016; Groman et al. 2019; Hervig et al. 2019; Lee and D'Esposito 2012; Wallis 2007) support multiple facets of reinforcement learning, including the maintenance of value across delay and/or trials, which is often probed in reversal learning tasks. Reversal learning tasks with probabilistic outcomes, in particular, permit estimation of choice behavior based on trial history using reinforcement learning algorithms (Sutton and Barto 2018), which provide estimates for how different features (e.g., learning rate, exploration) drive behavior. Importantly, reversal learning tasks differ in their engagement of reinforcement learning processes, which is likely a feature that determines OFC involvement and should be systematically compared across species in the future.

In recent years, there has been a point of controversy about the role of OFC in reversal learning in NHPs. In macaques, an old world NHP, aspiration lesions of the OFC were consistently found to produce a profound effect on reversal learning performance (Butter 1969; Iversen and Mishkin 1970; Izquierdo et al. 2004). This mirrors the effects seen in humans after damage to the OFC (Fellows and Farah 2003; Rahman et al. 1999) as well as marmoset with excitotoxic lesions of OFC (centered on BA 11) (Clarke et al. 2008; Dias et al. 1996b). In the marmoset and rat, there is also evidence that selective serotonin depletion from the OFC impairs reversal learning (Alsiö et al. 2020; Barlow et al. 2015; Clarke et al. 2004). Moreover, similar excitotoxic OFC lesions in the marmoset impaired the reversal of a Pavlovian-conditioned appetitive task in terms of both autonomic and behavioral responding (Reekie et al. 2008).

Recent work, however, found that excitotoxic lesions of the OFC in macaques (including Walker's areas 11, 13, and 14) do not cause deficits on instrumental deterministic reversal learning tasks (Rudebeck et al. 2013b). Follow-up studies using more complex three-choice probabilistic reversal learning tasks also failed to find any effect of excitotoxic OFC lesions on performance of the reversal or credit assignment (Rudebeck et al. 2017b). Instead, the deficits caused by aspiration lesions to OFC in macaques appear in part to be caused by damage to white matter pathways (Rudebeck et al. 2013b). Further, data from multiple modalities, including excitotoxic lesions (Rudebeck et al. 2017b), focused ultrasound (Folloni et al. 2021), and fMRI in macaques (Chau et al. 2015), indicated that the vlPFC (Walker's area 12)—and not OFC—is essential for credit assignment during reversal learning paradigms regardless of whether they include reversals or not.

One way to think about this apparent discrepancy between macaques and rodents, as well as macaques and marmosets, is to appreciate the point that we made earlier—namely, that reversal learning tasks probe two related, but distinct, functions: reinforcement learning and rule implementation. Viewed as a task that probes reinforcement learning, it appears that this function in macaques has become the purview of the vlPFC. Data from positron emission tomography (PET) studies of humans performing stimulus-reward learning tasks also appear to support this role for vlPFC. As people learn new stimulus-reward mappings, there is greater activity in Brodmann's area 47/12 in vlPFC (Rogers et al. 2000; Zald et al. 2005) instead of in OFC areas 11 and 13. In the latter study, participants showed more robust activation when humans were learning the rule versus after they learned the rule; this provided human evidence that supported the findings from macaques on the role of area 47/12 in reinforcement learning. Increased dlPFC activity emerged during delayed spatial alternation but not delayed object alternation, whereas orbitofrontal activations emerged in both alternation tasks. Moreover, the use of PET to image human OFC avoided the susceptibility artifacts when imaging OFC with fMRI. Thus, in macaques and humans, it appears that functions that were solely the purview of OFC in rodents (and potentially marmosets) are now subserved by area 47/12.

This leaves open the role of central OFC in macaques (Walker's areas 11 and 13) and humans (Brodmann areas 11 and 13) and how this compares to rodents. Here, there may be a clear functional similarity; namely, the updating of specific stimulus-reward associations. This function is classically assessed using reinforcer devaluation tasks (Holland and Straub 1979; Málková et al. 1997). Across a range of approaches and species, OFC appears to be essential for learning and updating specific stimulus-reward associations (Gottfried et al. 2003; Izquierdo et al. 2004; Malvaez et al. 2019; Ostlund and Balleine 2007). Thus, this computation appears to be a possible core function of OFC across species. It might be useful to determine whether the change in value is accompanied by a reduction of autonomic response to the appetitive conditioned or unconditioned stimuli in macaques, or to an uncoupling of such visceral responses with the behavioral response, as occurs in the marmoset following excitotoxic lesions of the OFC (Reekie et al. 2008). In rodents and marmoset, however, it is clear that OFC plays an important role in several forms of reversal learning; perhaps this is related to more caudal agranular regions in the primate ventral frontal cortex.

Inhibition

Behavioral inhibition can be measured in several different ways, which may indicate that this construct can be fractionated into precise behavioral processes and neural substrates (see Dalley and Robbins 2017). One prominent test paradigm is the stop signal reaction time (SSRT) procedure, which measures the ability to stop an initiated response. This can be effected in humans (Logan et al. 2014), monkeys (Schall et al. 2017), and in rodents (Eagle et al. 2008b) using either oculomotor or limb responses, respectively, in the SSRT task. There is evidence that SSRT performance in humans is dependent on a network that includes the right inferior PFC (areas 44, 45) (Aron et al. 2014;

Cai et al. 2014), probably in conjunction with the adjacent insular cortex. The latter may mediate the salience component of the SSRT task, whereas the "motor braking" inhibitory element is thought to depend on a network that includes not only ACC and PFC regions such as 44/45, but also the hyper-direct pathway to the subthalamic nucleus (Aron et al. 2014).

The involvement of the right inferior frontal gyrus has been substantiated by fMRI studies that also include a pharmacological intervention; atomoxetine (a noradrenergic reuptake inhibitor) enhanced SSRT performance in healthy volunteers and was associated with a larger BOLD activation in the right inferior frontal gyrus (Chamberlain et al. 2009). Of relevance to the issue of comparable behavioral findings, Eagle et al. (2008b) showed that large excitotoxic lesions of the lateral OFC in rats severely impaired performance by selectively prolonging SSRT whereas medial PFC lesions, perhaps surprisingly, had no effect. Bari et al. (2011) extended these results by demonstrating that temporarily inactivating the rat ACC/dorsal PL region lengthened the SSRT. However, atomoxetine infused into the rat lateral OFC improved performance, as it had done so following systematic administration in humans, whereas intra-dorsal PL infusion had a smaller effect. The functional significance lies in considering whether areas 44 and 45 would exhibit homology in the rat brain. From many considerations, it would appear that such lateral PFC structures are not, in fact, represented in rats (Preuss and Wise 2022). However, this apparently common behavioral inhibitory function does appear to be mediated by structures in the medial PFC of the rat (i.e., ACC/dorsal prelimbic) as well as by the rodent lateral OFC (and perhaps the adjacent insula), possibly simulating the inferolateral frontal cortex involvement in humans. What is clear is that further anatomical and behavioral studies are required to understand whether and how rodent OFC can be used as a model for the role of human vlPFC in behavioral inhibition.

Closely related to response inhibition is the ability to wait or tolerate delays. There is significant evidence that subregions of PFC across species, including OFC, contribute to explicit timing (Bakhurin et al. 2017), in making decisions in delay discounting tasks (Hosokawa et al. 2013; Roesch et al. 2006; Sellitto et al. 2010; Winstanley et al. 2004), and in temporal wagering tasks as models of decision confidence (Lak et al. 2014; Sosa et al. 2021; Stolyarova et al. 2019).

Working Memory and Attentional Control

When considering the analogy between certain functional properties of rodent PL cortex and primate dlPFC, it is important to point to the engagement of both networks in working memory and attentional control. With respect to working memory, there is correspondence between delayed alternation tasks across primates and rodents with respect to the selective engagement of dlPFC and PL, respectively. In macaques, lesions of the dlPFC impair several types of delayed alternation tasks (Goldman and Rosvold 1970; Stamm and Weber-Levine

1971), and neural recordings in this region show delay period activity patterns reflective of the working memory correlates (Kubota and Niki 1971). This is mirrored in delayed saccade tasks (Funahashi et al. 1989), which have inspired several neural models of working memory (Compte et al. 2003). Spatial alternation tasks have been extensively implemented in rodents, consistently implicating the engagement of PL. For example, work by Brito et al. (1982) showed the impact of PL neurotoxic lesions on delayed alternation in the rat, and more recent optogenetic inactivation of area PL in the mouse shows similar effects (Bolkan et al. 2017). Interestingly, both rodents and NHPs show delay period activity in these cortical areas as well as in their connected mediodorsal thalamic regions (Bolkan et al. 2017; Funahashi et al. 1989).

Working memory is closely linked to the endogenous control of attention. Classical work by several investigators has implicated the dlPFC in attentional control (e.g., Lebedev et al. 2004), which provides a complementary interpretation to its role in short-term memory maintenance (Fuster and Alexander 1971). Building on the primate task design of a cross-modal attentional task by McAlonan et al. (2006), Wimmer et al. (2015) developed an attentional control task in rats and mice. Here, a freely behaving animal chooses between two target stimuli (either a visual or an auditory target) on single trials in a cued manner; at the beginning of each trial, it receives one of two learned cues that varies on a trial-by-trial basis. Multiple performance metrics and manipulations have corroborated that mice use a rule-based strategy across most trials (Rikhye et al. 2018; Schmitt et al. 2017; Wimmer et al. 2015). This sets the stage for interpreting temporally precise optogenetic manipulations: out of several cortical areas inactivated in the PFC, including OFC, ACC, and premotor cortex, only area PL showed a delay period-specific effect (Wimmer et al. 2015). Recordings from PL showed a persistent network activity pattern over the delay in which single neurons exhibited temporally precise increase in firing rate during the delay period (sequential activity pattern). These network patterns were “rule specific” (Rikhye et al. 2018; Wimmer et al. 2015), consistent with the finding from primate dlPFC that shows the highest proportion of neurons encoding abstract rules in working memory tasks (Wallis 2011). In addition, Bolkan et al. (2017) found evidence for a sequential PL activity pattern in the context of a spatial working memory task. Interestingly, this activity pattern was not spatially specific, potentially also reflective of PL’s function in the generation of abstract rules.

Beyond our main regions of interest here, a common area targeted to study working memory in the macaque is the frontal eye field (FEF). Some neurons in dlPFC tend to maintain an elevated rate of spiking, relative to pretrial baseline firing rates, during working memory retention intervals (Fuster and Alexander 1971; Kubota and Niki 1971). Funahashi et al. (1989) demonstrated that activity persists in the principal sulcus of the PFC during memory-guided saccade delays, and experimental lesions that presumably abolish this persistent activity impact memory for the spatial location of targets in the contralesioned hemifield (Funahashi et al. 1993a). Given the potential impact of these findings on theories of working memory, researchers launched attempts to translate these findings to humans. However, and contrary to expectations, the first neuroimaging (PET) study of spatial working memory (Jonides et al. 1993) found delayed activity in superior precentral sulcus, not dlPFC. Then, the failure of several studies to find spatial working memory-related delay period activity in a homologous part of human dlPFC became the norm rather than an exception (Courtney et al. 1998; Rowe et al. 2000; Smith et al. 1996; Zarahn et al. 1999). On the other hand, fMRI measurements during memory-guided saccade delays consistently provided evidence of persistent activity in the human superior precentral sulcus (Curtis and D'Esposito 2006; Curtis et al. 2004; Duffau 2011; Hallenbeck et al. 2021; Jerde et al. 2012; Rahmati et al. 2020; Saber et al. 2015; Schluppeck et al. 2006; Sprague et al. 2014; Srimal and Curtis 2008; Tark and Curtis 2009). Moreover, dlPFC lesions that spare the precentral sulcus in humans do not impact working memory, whereas lesions that do encroach on the precentral sulcus cause memory-guided saccade errors (Mackey et al. 2016). In a follow-up study, Mackey and Curtis (2017) found that transcranial magnetic stimulation to the precentral, but not a more anterior, part of the putative homologue of monkey principal sulcus perturbs the accuracy of memory-guided saccades (Mackey and Curtis 2017). There are different ways to think about these findings with respect to interspecies PFC homologies. Anatomically, they represent a clear difference: in the monkey, but not human, dlPFC neural activity persists and is necessary for working memory. Functionally however, the findings align, albeit in a slightly different part of the dlPFC. In addition, the human superior precentral sulcus is thought to be the human homologue of the monkey FEF (Paus 1996). Lesions to the monkey FEF impairs working memory performance (Dias and Segraves 1999; Sommer and Tehovnik 1997), and neurons in monkey FEF show persistent activity during working memory delays (Bruce and Goldberg 1985; Sommer and Wurtz 2001).

Goal-Directed Action

The PFC has long been implicated in executive control generally and in goal-directed action in particular (Stuss and Benson 1984). Consistent with this, early experiments investigating PL in rats found that lesions that occur before training abolished the acquisition of a goal-directed action, such as lever pressing for a food reward, the performance of which depends on (a) encoding the relationship between specific actions and their consequences—that is, action-outcome (AO) associations—and (b) the value of those consequences (Balleine and O'Doherty 2009). This conclusion was based on the failure of lesioned animals to pass specific tests: a contingency degradation test, which assesses sensitivity to changes in the AO relationship, and an outcome devaluation test, which assesses sensitivity of action to changes in the value of a specific outcome. In animals with an intact PL, degrading the AO relationship or devaluing the outcome produced an appropriate change in action. In rats without a PL, performance was inflexible and the animals failed to adjust.

Similar effects have emerged in humans. When trained to press buttons for specific food outcomes, variations in the instrumental contingency modified performance and altered self-reported measures of the causal status of actions with respect to their consequences. When assessed using fMRI, goal-directed actions were found to activate regions of medial and ventromedial PFC (area 32) and anterior medial orbital cortex (area 14) (Liljeholm et al. 2011; Tanaka et al. 2008). Importantly, recent work suggests these areas mediate different functions: area 32 activity mediates the encoding of specific AO associations (Morris et al. 2022), whereas, in both humans (Morris et al. 2014) and rats (Bradfield et al. 2015), anterior medial orbital activity appears more essential for the performance of "action" based on the retrieval of a specific valued "outcome." With regard specifically to degradation of the instrumental contingency in humans, evidence suggests that, with contingency reduction, activity in vmPFC (particularly areas 32 and anterior 14) is modulated by dlPFC (BA9): the latter tracks concomitant changes in the value of the action (Morris et al. 2014), whereas changes in the value of the background as the action value declines is tracked by ACC (area 24). The covariance between action and background activity is tracked by caudate nucleus in humans (Morris et al. 2022), which is similar to findings in rodents in which activity in mPFC ultimately results in changes in posterior dorsomedial striatum associated with the long-term encoding of specific AO associations (for a review, see Balleine 2019).

Importantly, possibly similar effects have been reported in the marmoset in which lesions of both OFC and perigenual ACC (including areas 24 and 32) abolished sensitivity to contingency degradation in acquisition (Jackson et al. 2016). Subsequently, in a more extensive comparison of established instrumental performance using both pharmacological inactivation and overactivation, this effect was restricted to area 24 (Duan et al. 2021); this suggests that rodent area 32 (particularly its most dorsal aspects stretching into the ACC) may have some compatible functions with area 24 in the primate in controlling goal-directed action and its balance with habitual behavior (Figure 4.3a, p. 72). It is thus possible that area 24 does not directly control AO learning but other processes important to degradation of the instrumental contingency. This could fit with work in macaques and highlight a role of area 24 in sustaining responding after changes in contingency (Kennerley et al. 2006). An important aspect of the latter is the role of detecting changes in the background rate of reward. From an associative perspective, during instrumental acquisition, the action (A) is the best predictor of its specific outcome (O). However, in contingency degradation, during which O is presented unpaired with A, the background or context (C) becomes a better predictor. This is because, during initial conditioning, AC → O whereas C predicts no outcome (C → Ø) whereas, during degradation, C now predicts O: that is, AC→O plus C→O, causing A to lose predictive power to C. The question, with regard to Duan et al. (2021), is whether there is any evidence that area 24 in the marmoset mediates sensitivity to these changes in context conditioning? If so, then perhaps area 24 is not directly involved in A→O learning/performance but in the competing C→O learning. Unfortunately, at present, the evidence is not straightforward. Although marmosets can clearly show evidence of context conditioning (Duarte et al. 2014, 2015), no studies to date have evaluated the role of BA24 in this effect. There is some evidence, however, for BA32 and adjacent BA24 involvement in contextual conditioning (Lang et al. 2009) and, as mentioned above, for context associations during contingency degradation in humans (Morris et al. 2022) as well as for context conditioning in NHPs (Chien et al. 2023; Mansouri and Buckley 2018), although not in directly comparable situations. As such, this interpretation of Duan et al. (2021) awaits a more definitive test.

Motivational Control of Goal-Directed Action

Another source of functional PFC similarity across species emerges from consideration of the motivational control of goal-directed action. As mentioned, there is evidence that medial PFC circuits mediate sensitivity to changes in outcome value. Interestingly, these circuits do not mediate sensitivity to the control of action by specific predictions based on environmental stimuli. Our ability to extract predictive information from the environment to inform future actions is a critical component of decision making. This psychological process encapsulates the essential function of the cognitive control of action as being (a) fundamentally integrative, requiring the ability to integrate predictive information with action-related learning processes, but nevertheless (b) its function is not simply to acquire information but to do so in the service of future actions; that is, in a manner which allows the animal to use this information to choose between distinct (and sometimes competing) courses of action to achieve specific future goals.

To study this interaction in the laboratory, researchers have refined over a number of years a paradigm called *Pavlovian-instrumental transfer*. Here, subjects, whether rodents or humans, are first given the opportunity to learn various predictive relationships between stimuli (S) and specific outcomes (O) (e.g., S1O1, S2O2) as well as various goal-directed actions (e.g., A1O1, A2O2). These relationships are acquired across separate experimental phases before the effect of the stimulus predictions on action selection is assessed, usually in the absence of any outcomes, to ensure any changes in choice performance are determined by prior learning. Typically, the stimulus events (S1 and S2) strongly bias choice between the two actions (A1 and A2) toward the action that previously earned the predicted outcome. For example, given S1O1, S2O2 and A1O1, A2O2, S1 biases choice toward A1 (S1: A1 >A2) and

S2 toward A2 (S2: A1 <A2). This effect is referred to as *specific transfer* (for a review, see Cartoni et al. 2016).

We have learned quite a lot about the neural circuit that mediates this transfer effect, which implicates subcortical structures interacting with the PFC in a well-defined neural circuitry. In rodents, studies have found that during Pavlovian conditioning, the BLA is key to encoding specific SO associations and for coordinating conditioned responses based on these associations (Ostlund and Balleine 2007). However, to influence future actions, the BLA encodes these specific relations in the nucleus accumbens shell (NAc-S) via activity in a direct amygdalo-striatal pathway (Morse et al. 2020). This encoding is complex and is reviewed in detail elsewhere (e.g., Laurent and Balleine 2021). Briefly, during encoding, BLA inputs to NAc-S cause cellular changes in both the NAc-S and in its inputs from the IL cortex, which differ based on each specific SO. During retrieval in the transfer test, stimulus presentation produces activity in the IL NAc-S pathway, resulting in increased activation of specific targets of the NAc-S in ventral pallidum. The ventral pallidum output targets both the ventral tegmental area and mediodorsal thalamus, and it is this latter projection that has been found to be critical for the transfer effect (Leung and Balleine 2015). The ventral pallidum sends an inhibitory projection to the mediodorsal thalamus, which ultimately causes the activation of ventrolateral OFC, from which its targets in the dorsal striatum directly modulate action selection. As a consequence, this research establishes evidence for a PFC–striatal–pallidal–thalamic–PFC feedback network whose function is critical for the cognitive control of action.

A similar circuit has been implicated in human transfer effects. The initial studies using fMRI found evidence for activation in a ventral putamen/pallidal area (Bray et al. 2008) and in the BLA (Prévost et al. 2012), produced during the increased performance of an action when it was associated with the outcome predicted by the stimulus. More recently, dynamic causal modeling identified evidence of a circuit involving VS modulation of mediodorsal thalamus in this same effect (Balleine et al. 2015) and, in another study, for activation of lateral OFC in this effect and specifically when the action was associated with the same versus a different outcome to that predicted by the stimulus (Perkes et al. 2023). Interestingly, in this latter study, causal evidence for OFC activity was established with reference to transfer effects in adolescents with obsessive-compulsive disorder. This group was found not to express the specific transfer effect; instead, predictive stimuli were found to have no effect on action selection and, when assessed using fMRI, the lateral OFC was found to be hypoactive in these adolescents. These data provide clear evidence for functional similarity across this same circuit. This identification of neural circuitry in the motivational control of goal-directed behavior is highly relevant to the discussion of network organization of PFC circuitry below.

How Did the PFC Evolve and How Has This Evolution Led to Produce Higher-Order Cognition, Including Social Behavior and Language Elements in Humans?

One possible avenue to analyze the evolution of PFC is to use a phylogenetic comparative framework. Such studies do not focus on particular model species but rather on many species within the context of their phylogenetic relatives (Passingham 1975). Such analyses have huge potential but also clear limitations. The potential lies primarily in being able to capture patterns of cross-species differences that provide a detailed view on how brain regions have changed in response to evolutionary pressures. The informative nature of this type of variation is derived from the fact that present-day variation across species is the result of a series of natural experiments that have taken place over millions of years of evolution, across all continents, and in all species. The scope of these experiments is such that they can never be replicated in the lab. The results of these experiments provide an unmatched and largely untapped wealth of information on how genotypic changes can shape phenotypic changes in response to environmental changes. One of the primary limitations of the phylogenetic comparative approach is that there is a clear tradeoff between a higher comparative resolution (in terms of number of species) and the resolution of neuroanatomical specificity. Significant advances have, however, been made such that recent studies incorporate a variety of different measures (e.g., size, modularity, neuronal density, synapse density) across an increasingly wider variety of different brain regions and different species. The expectation is that the field of phylogenetic comparative analyses of the brain will continue to increase its neuroanatomical specificity and, as such, become increasingly relevant for understanding neurocircuitry, neurodevelopment, and neurogenetics.

The phylogenetic comparative approach can also be used to investigate the evolution of PFC. Because brain region sizes all scale with brain size, comparisons between the size of PFC with the size of brain regions with which PFC shares a type of neurobiological association are most informative (Passingham and Smaers 2014). For example, comparing PFC size with V1 uses first-order visual input as a baseline to assess volumetric investment in PFC's higher-order processing. Such comparisons reveal stepwise grade changes in great apes and humans, indicating a selective expansion of PFC size relative to V1 in these species (Smaers et al. 2017). In other words, great apes and humans have significantly more PFC size relative to V1 than expected for their brain sizes. The same pattern of evolution is observed when comparing PFC volume against the volume of frontal motor cortex, and when using either the Brodmann or Smaers datasets (Figure 4.2).

Because size is a good indicator of growth, the occurrence of such evolutionary grade shifts suggests that great apes and humans both indicate concordant shifts in the developmental body plan of prefrontal growth (Smaers et al. 2019). This evolutionary expectation aligns with evidence for a developmental heterochronic shift in human prefrontal growth (Somel et al. 2009; Somel et al. 2011). The recapitulation of evolutionary grade shifts in ontogenetic growth patterns provides a largely untapped source of information that may help elucidate the molecular pathways that underpin prefrontal growth.

Additionally, phylogenetic comparative analyses can also contribute to insights on which brain circuits have expanded the most throughout evolution. In primates, volumetric variation in brain regions involved in the cortico-cerebellar system have been found to explain almost all of variation in brain size across species (Smaers et al. 2019). This suggests that aspects of the same neural system may be selected across primates. In turn, this may suggest that primate brain evolution may emphasize domain general abilities. One concept that provides a powerful explanatory framework is that of relational learning (Genovesio et al. 2014). Part of the broadly defined prefrontal-parietal network, relational learning can be materialized across modalities and results in complex behavior across the social, motor, and affective domains. When considering putative behavioral evolutionary drivers of brain evolution, emphasizing cognitive processes that have interpretable roots in neural circuitry may be preferred over emphasizing particular behavioral outputs of such processes (e.g., sociality) (Passingham et al. 2017). In the case of human evolution, it is clear that any behavioral specializations were ultimately the result of early humans having to adapt to a changing environment when the formation of the Great Rift valley separated early Australopithecus from early Pan, confronting the species that ultimately lead to *Homo* with a changing climate and an environment that was more unpredictable than the jungle environment (King and Bailey 2006). Relational learning was hereby the likely key to the success of early Homo to adapt to this new, unpredictable, and seasonal environment.

As mentioned above, one of the key drivers of relational learning was likely sociality (Humphrey 1976) but living in an uncertain environment, where understanding the behavior of prey or availability of food, most likely contributed as well. Indeed, one way to improve foraging success in sparse and unpredictable environments is to forage or hunt with a group of conspecifics. The chances of finding food is increased if each member of a group alerts the others when sustenance is found, widening the search area. Such foraging, therefore, has a major social component to it. As further noted above, the ACC in humans and other primates has been identified as a brain area that plays a key role in both foraging and social aspects of behavior. For instance, humans choosing to change foraging locations show increased activity within the dorsal ACC (Kolling et al. 2012), neurons in macaque dorsal ACC ramp in anticipation of changing foraging locations (Hayden et al. 2011), and lesions (Kennerley et al. 2006), inactivations (Shima and Tanji 1998), or electrical stimulation (Sarafyazd and Jazayeri 2019) of dorsal ACC lead to a decrease in the rate of reward procurement during foraging. A similar pattern of effects

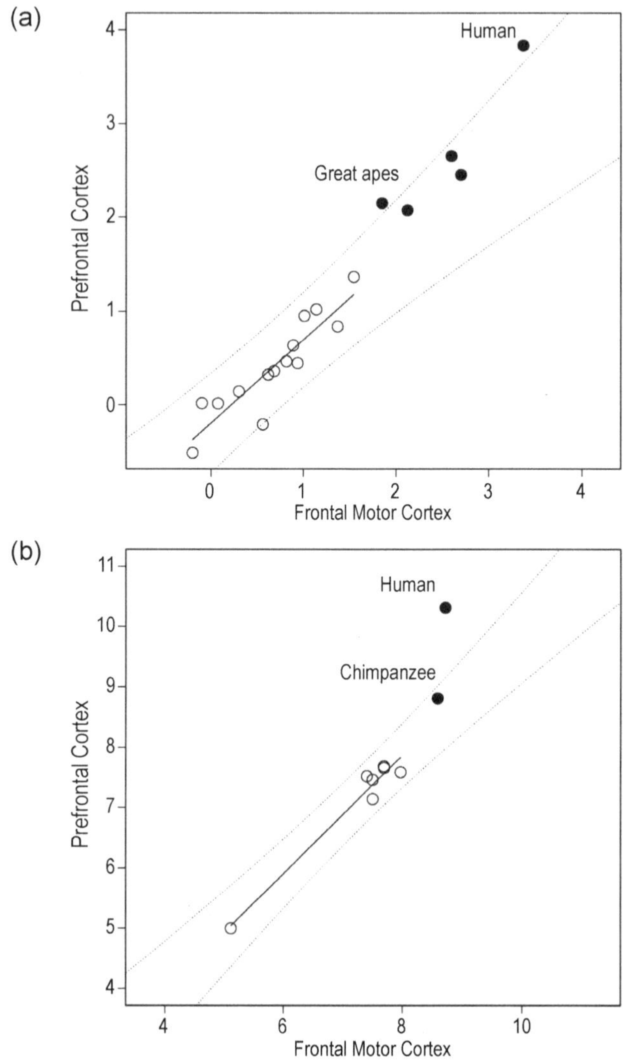

Figure 4.2 Phylogenetic regression of PFC volume against the volume of frontal motor cortex for the (a) Smaers and (b) Brodmann data. Prefrontal volume and volume of frontal motor cortex for all species in the sample, rank-ordered according to the size of PFC for (c) Smaers and (d) Brodmann data. Modified from Figures 2–4 in Smaers et al. (2017).

is also evident in the equivalent of primate dorsal ACC in rodents, CG1/CG1 subfields of medial frontal cortex (Lapish et al. 2008; Seamans et al. 2008). At the same time, medial frontal cortex, including dorsal ACC, has been shown to be essential for appropriately guiding social behaviors in humans, macaques, rats, and mice (Basile et al. 2020; Rudebeck et al. 2006; Rudebeck et al. 2007;

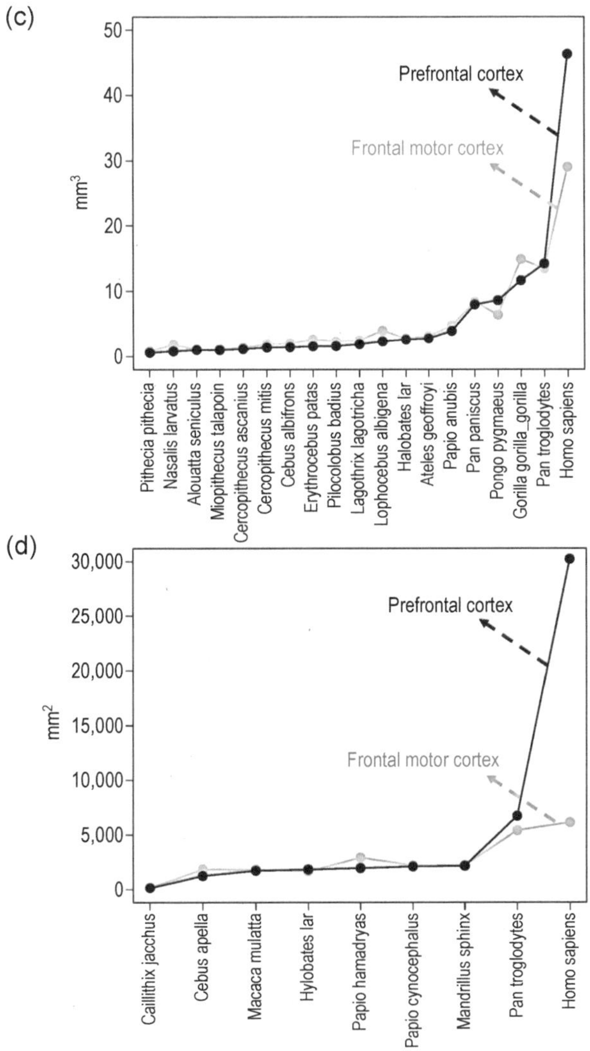

Figure 4.2 *(continued)*

Yizhar et al. 2011). This correspondence between species is notable. It indicates that the role of the dorsal ACC in both social behavior and foraging has a common origin (Apps et al. 2016), possibly in cognitive processes that are not specific to social behaviors (Humphrey 1976). The expansion of ACC in primates has likely led to these areas taking on additional functions to accommodate higher-level cognitive operations such as relational learning.

Altogether, we are far from understanding how and why new anatomical PFC areas arose throughout evolution. In addition to the ideas summarized in this section, several investigators have put forth ideas regarding the expansion of the PFC. For instance, consistent with the previously discussed ideas, it has been suggested that the selective pressures leading to the large brains of primates reflect the emergence of complex social systems (Dunbar and Shultz 2007). Others have suggested that because early primates were nocturnal, PFC expansion was likely related to foraging behaviors and diet (DeCasien et al. 2017). On this view, the earliest new PFC areas (e.g., granular OFC and FEF) provided adaptive advantages in the ability to identify, attend to, and plan grasping movements aimed at valuable nutrients in the fine branch niche in which they lived (Murray et al. 2017). Additional PFC areas that emerged in simian primates (e.g., vlPFC, dlPFC) have been proposed to improve foraging efficiency by reducing the frequency of poor foraging choices and reducing predation risks. Additional ideas are that expansion of visual cortex and frontal cortex in primates is tied to adaptive advantages related to predation and maternal investment, among others. It seems likely that no single driving force is responsible for the multiple stages of PFC expansion and that PFC expansion and the evolution of new areas within the PFC occurred in response to several selective factors: at different times and in different ancestral species.

What Are the Main Organizational Principles of PFC?

Definitively answering this question, of course, requires a textbook in and of itself and is above and beyond the week of discussion that we had together. Given this time constraint, we considered three features: (a) cortical folding, (b) network organization of the frontal lobe and its relationship to goal-directed action, and (c) hierarchies and gradients in PFC.

Cortical Folding

Our discussion considered how the structure and function of different aspects of PFC contributed to different aspects of behavior and cognition across many species that had either smooth, lissencephalic, or convoluted gyrencephalic brains (Miller et al. 2021b; Van Essen et al. 2013). For example, the cerebral cortices of mice and rats lack indentations, or sulci, whereas the cerebral cortices of macaques, chimpanzees, and humans have an extensive amount of sulci—in which human association cortices have sulci that are even absent in nonhuman hominoid hemispheres. Here, we focus on cortical folding features that are specific to the human cerebral cortex and address how those features relate to individual differences in functional organization with cognitive and clinical implications. Separately we consider lateral PFC, medial PFC, and OFC. As tertiary sulci are small in surface area and shallow in depth, we refer to newly identified small and shallow sulci as putative tertiary sulci. Future studies examining these sulci in lateral PFC, medial PFC, and OFC will determine if they are truly tertiary sulci based on their emergence in gestation, which is the classic definition (Armstrong et al. 1995; Chi et al. 1977; Welker 1990).

In human lateral PFC, there are several putative tertiary sulci that are (a) identifiable in every hemisphere (Petrides 2019) and (b) functionally (Miller et al. 2021a, b) and cognitively relevant (Voorhies et al. 2021; Willbrand et al. 2023d; Yao et al. 2022). In addition, some putative tertiary sulci in lateral PFC are not identifiable in every hemisphere, but their presence or absence is functionally and/or cognitively relevant. For instance, the presence of one such sulcus is related to a 20–34% improvement in reasoning ability in children, adolescents, and adults (Willbrand et al. 2023b). Further, this sulcus is absent in macaques and seldomly present in chimpanzees (Hathaway et al. 2023) and interestingly, the presence or absence of this sulcus is related to the functional architecture of lateral PFC (Willbrand et al. 2023a). Thus, future work should test the relationship between the presence/absence of these sulci relative to the functional and structural organization of lateral PFC in different clinical populations and species (Hathaway et al. 2023). While these studies focus on local structural-functional links, we emphasize that previous findings serve as a foundation for uncovering the infrastructure of a complex neural network linking aspects of brain structure and function to cognition in lateral PFC.

In medial PFC, perhaps the most widely studied and variable tertiary sulcus is the paracingulate sulcus across age groups, species, and in different clinical populations. The morphology of the paracingulate sulcus is related to individual differences in functional representations, cognitive performance, and the severity of clinical symptoms (Amiez et al. 2013, 2018; Amiez and Petrides 2014; Borst et al. 2014; Cachia et al. 2016; Crosson et al. 1999; Fornito et al. 2004, 2006; Garrison et al. 2015; Lopez-Persem et al. 2019; Rollins et al. 2020). The presence/absence of the paracingulate sulcus is also related to the boundaries of cytoarchitectonic areas in medial PFC (Amiez et al. 2021; Palomero-Gallagher et al. 2009a; Vogt et al. 1995). Recent research has shown that the paracingulate sulcus is present in nonhuman hominids but not NHPs such as baboons and macaques (Amiez et al. 2019, 2021; Miller et al. 2021a). Additional putative tertiary sulci have also been identified and related to different aspects of the functional organization of medial PFC (Amiez et al. 2013; Amiez and Petrides 2014; Lopez-Persem et al. 2019). Future research is needed to pinpoint whether individual differences in the morphology of these putative tertiary sulci in medial PFC are also related to individual differences in cognition.

In human OFC, sulcal morphology is related to the complexity of representations of value (Li et al. 2015). Different OFC sulcal patterns (or "types") are also related to the complexity of different clinical disorders (Cardenas et al. 2011; Drevets 2007; Eckart et al. 2011; Nakamura et al. 2020; Patti and Troiani 2017; Rogers and De Brito 2016). Recent findings also show that the local gyrification of specific parts of OFC are related to emotion-related impulsivity, which is a transdiagnostic feature of several different clinical disorders (Elliott 2022). Future research is needed to bridge the gap with the results in lateral and medial OFC by testing if the morphology of sulci, including putative tertiary sulci, in OFC is related to cognition and the severity of clinical symptoms.

Altogether, as in other cortical expanses—such as ventral temporal (Ammons et al. 2021; Chen et al. 2023; Parker et al. 2023; Weiner 2019; Weiner and Willbrand 2023), lateral parietal (Willbrand et al. 2023d), and medial parietal cortices (Aponik-Gremillion et al. 2022; Willbrand et al. 2023c; Willbrand et al. 2022)—putative tertiary sulci in lateral and medial PFC, as well as OFC in hominid brains seem to serve as a mesoscale infrastructure bridging between micro-architectonic and network features. This has cognitive and clinical implications, and awaits further elucidation through future research, especially as pertains to the hypothesis of fundal cognition (Weiner 2023).

Network Organization of the Frontal Lobe and its Relationship to Goal-Directed Action

Resting-state fMRI (rs-fMRI) has emerged as an important method for assessing neural networks and has enabled extensive connectivity analyses between multiple brain regions (Gratton et al. 2023; Lurie et al. 2020). Another interesting and important cross-specifies comparison should also be made between prominent network analyses of PFC, based on resting-state functional connectivity versus the circuits that have been implicated in goal-directed action by more conventional functional analyses. Although there are clear differences in brain complexity and function, reports of sensory, motor, and default networks in human, NHPs, and rodents suggest that common principles may underlie resting-state brain organization across species (Xu et al. 2020). This research has obvious implications for studying the evolution of brain function and connectivity as well as our understanding of the fundamental mechanisms underlying sensory perception, motor control, and cognitive processes. Sensory networks corresponding to visual, auditory, and somatosensory modalities, involving corresponding functional regions of the cortex, have been described and, in NHP and rodents, studies have shown similar resting-state networks associated to those described in humans, reflecting the spontaneous activity and functional connectivity of brain regions involved in sensory perception. Similarly, common resting-state motor networks have also been identified associated with motor planning, control, coordination, and execution during rest, suggesting a common role in the preparation and execution of motor tasks. Similarities in a default mode-like network have also been reported across species (summarized in Buckner and DiNicola 2019) and although those described in NHP and rodents may not be as complex as in humans, their presence also suggests a level of conservation, as well as differences, in brain organization related to cognitive functions (Garin et al. 2022).

Nevertheless, despite these impressive similarities, the relationship of these networks to those underlying goal-directed action is not at all clear. As described above, this form of action is strongly linked, across species, to the integration of cognitive and emotion processes, controlling both the learning process through which the relationship between specific actions and their consequences is encoded (Figure 4.3a) and integrated with goal values. As such, one might expect close similarities to the sensory, motor, and cognitive networks described using rs-fMRI. This should be particularly true of the default mode network (DMN), which is concerned primarily with "higher" cognitive processes (Raichle 2015). As commonly conceived, the DMN in humans and NHPs includes regions of ventromedial PFC (BA9, 10, 11), ACC (BA24 and 32), and, more posteriorly, the retrosplenial cortex, the precuneus, posterior parietal cortex, and medial temporal lobe. Many of the prefrontal structures implicated in the DMN are also involved in goal-directed action (Figure 4.2a); however, the more posterior structures have not been implicated (although activity in the caudate nucleus and posterior parietal cortex has been reported to track outcome identity covariance during changes in AO contingency (Morris et al. 2022). With increasing attention being paid to individual differences rather than group averaging, it appears likely that networks such as DMN may become further subdivided (DiNicola et al. 2023), thus accounting for this apparent discrepancy.

Perhaps more notably is the almost complete silence of the basal ganglia in rs-fMRI, given that interactions between prefrontal regions and the striatum have been heavily implicated in goal-directed control in rodents, NHPs, and humans. The same distinction can be drawn with sensory and motor resting-state functional connectivity networks. These identify converging regions of sensory and motor cortices (including the dlPFC, posterior cingulate, and cerebellum), respectively, but again, completely avoid the basal ganglia, most notably the ventral striatal networks identified with the stimulus control (Figure 4.3b) and value-based control (Figure 4.3c) of goal-directed performance. These general networks involving sensory, motor, and default modes, including the executive network, do not appear, therefore, to have much in common with any of the networks implicated in goal-directed action using cross-species functional analyses. However, this may not be as true of another resting-state network associated with more specialized sensory processing, often referred to as "the saliency network" (Menon and Uddin 2010). This network has been argued to involve strong interconnectivity of anterior insular cortex and ACC together with midline thalamus, ventral striatum, and central amygdala and could be argued, therefore, to have much in common with some features of the stimulus- and value-based control networks described by Balleine and O'Doherty (2010) and illustrated in Figure 4.3. However, the results of a meta review of this literature showed that the ACC and insular cortex respond to saliency independently of changes in value (Bartra et al. 2013), whether predicted or experienced, and appears more closely linked to autonomic feedback

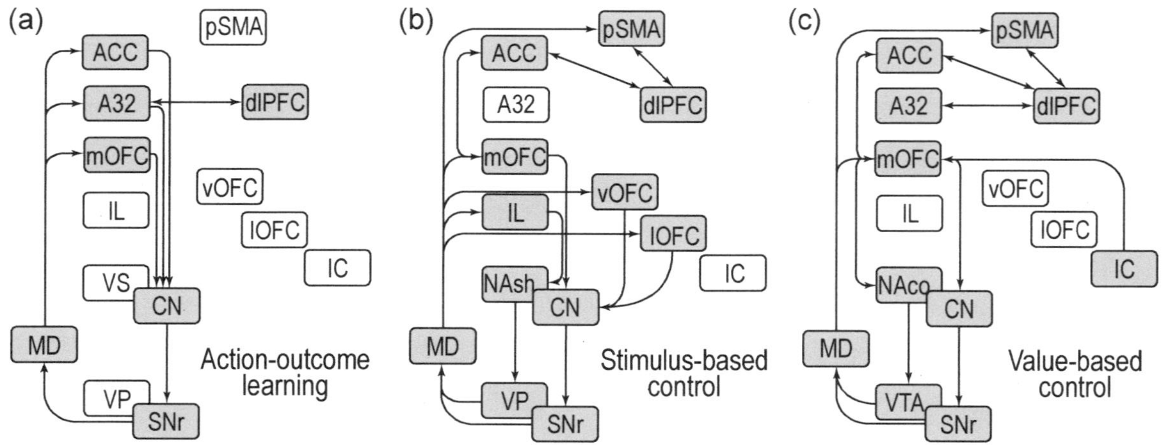

Figure 4.3 Circuit models of the neural structures involved in various aspects of goal-directed learning. (a) Action-outcome learning: Prefrontal regions involving dorsal and anterior regions of medial PFC (BA32, extending to BA10) and ACC (BA24) have been directly implicated in encoding the action-outcome contingency during goal-directed learning, with dlPFC (BA9) implicated in action value comparisons. These structures play a key role in striatal learning processes, particularly the caudate nucleus (CN), for long-term memory and together with basal ganglia feed back to medial OFC (BA14) to control action-outcome retrieval for subsequent performance. (b) Stimulus-based control: Predictive learning, particularly specific stimulus-outcome associations, have been found to exert control over goal-directed actions via a circuit involving mOFC and infralimbic (IL) cortical control of the nucleus accumbens. For specific outcome predictions, this circuit involves IL to accumbens shell (NAsh), ventral pallidum (VP) and feedback to lateral OFC (lOFC) via substantia nigra pars reticulata (SNr) mediodorsal thalamus (MD), and consequent activation of a cortical-basal ganglia feedback circuit to cingulate and presupplementary motor areas (pSMA) that elicit changes in performance. (c) Value-based control: The control of actions after changes in the reward value of their consequences is critical to maintaining value-based decision making. Such control is mediated by a limbic-cortical “incentive memory” involving amygdala and insular cortical (IC) connections with mOFC and its inputs to other medial wall structures including A32, and its output to accumbens core (NAco) and the parallel in retrieval of specific action-outcome associations via CN and basal ganglia feedback to motor regions.

or homeostatic demands (Seeley 2019). As such, it seems reasonable to remain agnostic on the relationship between activity in this network and functional networks mediating the motivational and emotional control of goal-directed action. It should also be noted that there may be technical reasons for the relative lack of evidence for basal ganglia network involvement, particularly the use of ultrafast (multiband) imaging protocols which tend to favor cortical structures (Srirangarajan et al. 2021).

Hierarchies and Gradients in PFC

A ubiquitous organizational principle in the portions of human PFC (and somewhat in the species discussed here) is different types of hierarchy. For example, in different portions of PFC, Burt et al. (2018) showed a tight coupling between transcriptomic expression and structural imaging correlated with myelin that contributes to an area's position in a cortical hierarchy, including PFC, in both human and macaque. These authors also considered position in the cortical hierarchy in macaque as determined based on the ratio of efferent to afferent projections (see Murray and Constantinidis, this volume), which further provides details of the microcircuitry contributing to the anterior-posterior PFC hierarchy.

In the medial PFC, there is also a gradient in both humans and macaques running along an anterior-posterior axis in which primary/sensory motor regions are situated more posteriorly and transmodal regions associated with the DMN are situated more anteriorly (Margulies et al. 2016). Consistent with this anterior-posterior hierarchy in medial PFC, there is also evidence of a hierarchy of concepts, again with simpler concepts represented more posteriorly and vice versa (Theves et al. 2021). Earlier in this chapter, we provided other examples of apparent hierarchical PFC organization, including neurochemical gradients (see Table 4.2 and Rapan et al. 2023).

Furthermore, Murray et al. (2014) showed a hierarchy of intrinsic timescales across primate cortex; for example, the intrinsic timescale was slowest in ACC compared to OFC and lateral PFC (Knudsen and Wallis 2022; Padoa-Schioppa 2009). These findings, which are based on measures such as spiking autocorrelations, fit nicely with task findings in macaques (Lin et al. 2020). Nevertheless, in recent years, several studies have explored timescale hierarchies in humans (Baldassano et al. 2017; Huntenburg et al. 2018) that are also consistent with this hierarchy.

Further consistent with this dorsal-ventral hierarchy, Hunt et al. (2018) recorded in macaque OFC, ACC, and dlPFC and found that (a) OFC performs a value comparison, (b) ACC integrates several features of individual values to a decision bound, and (c) dlPFC routes attention to salient features of the task, relevant for decision making. Single unit and population activity were largely consistent with this pattern, indicating an increasing level of complexity from ventral-to-dorsal (or dorsal-ventral control) of PFC function in macaque. A similar pattern holds true in rat OFC and ACC for value comparison and final actions, respectively. Rat OFC is involved in value computations of specific outcomes (Schoenbaum et al. 2011), whereas ACC is involved in *relative* value comparisons in the action or effort space (Akam et al. 2021; Hart et al. 2020; Mashhoori et al. 2018). Thus, ACC likely contains an integrated, multiplexed signal with information from OFC and more.

Though not discussed extensively during the Forum, hierarchies in lateral PFC should also be mentioned. Whereas previously it was thought that the most anterior regions of the frontal pole in humans were located at highest stages of the processing hierarchy (Badre, this volume, Badre 2008; Badre and D'Esposito 2009), recent findings support two separate hierarchical gradients: one related to temporal abstraction and the other to feature abstraction. They both converge in the mid-PFC (Nee and D'Esposito 2016, 2017), which would be considered at the "apex" (Badre, this volume) of the hierarchy. The findings of this modification of the anterior-posterior gradient in lateral PFC was also supported by rs-fMRI data (Margulies et al. 2016). Interestingly, this is consistent with connectivity data in macaque. As connectivity is commonly used to assess hierarchical positions in the brain—specifically a ratio of efferent to afferent connections—Goulas et al. (2014) explored this ratio in lateral PFC in which an anterior-posterior hierarchy predicts the highest ratio at the frontal pole in BA10 and identified the highest asymmetry within the middle portion of dlPFC. A recent meta-analysis by Abdallah et al. (2022) also shows that there is evidence for a dorsal-ventral hierarchy in dlPFC across over 14,000 studies. In a much smaller sample size, this is consistent with a recently proposed dorsal-ventral hierarchy within lateral PFC in which the mid-dlPFC was identified as critical for working memory, whereas the mid-vlPFC was proposed to be critical for active retrieval and encoding of information (Petrides 1994, 1996, 2005; Petrides et al. 2002).

In human OFC, previous work shows evidence of a hierarchy of value representations along an anterior-posterior axis: simpler reward representations were situated more posteriorly and more complicated reward representations were situated more anteriorly (Sescousse et al. 2010, 2013). These findings were consistent with a proposed hypothesis of an anterior-posterior functional gradient, reflecting the abstractness of reinforcers in OFC (Kringelbach 2005; Kringelbach and Rolls 2004). Interestingly, functional regions related to the complexity of reward also couple with sulcal morphology at the level of individual participants in human OFC (Li et al. 2015). Linking back to our discussion earlier, more anterior sulci emerge later in gestation in OFC; this indicates that the sulcal-functional coupling in anterior OFC may develop later in life than posterior OFC. Further, the posterior region is located in dysgranular cortex, while the anterior region is located in granular cortex (Henssen et al. 2016; Mackey and Petrides 2009; Öngür et al. 2003; Öngür and Price 2000; Price 2007). Because OFC also contains representations other than value and reward (Knudsen and Wallis

2022; Wallis and Miller 2003b), future research is needed to show if human OFC contains hierarchies for additional representations.

Mechanisms By Which Major Circuits Exert Control: Is There Anything Special about Neuronal Physiology of PFC?

Oscillations

In comparing meso- to macroscopic measurements across primates and rodents (e.g., oscillations), it is important to consider that primate dlPFC, for example, appears to have some clustering of neurons that show similar task-relevant tuning (e.g., Wallis et al. 2001). In rodents, these features are less observed (Rikhye et al. 2018; Schmitt et al. 2017). This is not dissimilar from the differences observed in visual areas of the two species: primate V1, for instance, shows clustering in the form of orientation columns (Hubel and Wiesel 1977), whereas rodent V1 shows a salt and pepper organization (Priebe and Ferster 2012).

Network-level oscillations are features of cognition and behavior, though ideas differ as to whether they are considered mechanisms or epiphenomena. Irrespective of the strong opinions, measuring oscillatory activity can capture information transfer across regions, across hemispheres, and many neuropsychiatric conditions like schizophrenia and bipolar disorder are characterized by aberrant oscillations. Certain frequency bands have been previously associated with certain functions of the PFC, including, for example, gamma oscillations (40–100 Hz) in working memory, as well as the theta frequency band (5–10 Hz) in reversal learning and value-based decision making across species (Amarante et al. 2017; Fatahi et al. 2020; Knudsen and Wallis 2020; Marquardt et al. 2017; Ye et al. 2023b). Furthermore, Sohal et al. (2009) first showed that mouse medial PFC parvalbumin+ neurons play an important role in generating synchronized rhythmic activity in the gamma frequency range. In more recent work, Cho et al. (2020, 2023) empirically showed that this synchrony was necessary for learning about rule shifts in an attentional set-shifting task and not required for learning initial associations between cues and rewards, or even in reversals of individual cue-reward associations. This type of specificity on behavior is an interesting and important extension to lesion experiments in rats, indicating medial frontal cortex is necessary for attentional set shifting (Birrell and Brown 2000). Thus, overall, there is good evidence that fronto-cortical gamma and theta oscillations could be studied as biomarkers across species, particularly as preclinical models of disorders in which one finds impaired reward learning and value-based decision making paired with aberrant oscillatory activity. Future work should combine measures of oscillations partnered with viral-mediated, cell type specific targeting.

While we have shown that temporal hierarchical, transcriptomic, and receptor architectural features differ in the main PFC regions that are the focus of this chapter, further details can be provided that lead to mechanistic insight into aspects of cognition associated with PFC, such as working memory. For example, NMDA, but not AMPA, receptors are prominent in subregions of PFC, which is meaningful as NMDA receptors have slow decay time constants and AMPA receptors have fast decay time constants (Constantinidis and Wang 2004; Wang 2001). Thus, the former have sustained firing rates associated with a delay period during a memory task, while the latter do not (Murray and Constantinidis, this volume).

Rapid Learning

A key domain in which PFC circuits may exert control over sensorimotor transformations is rapid learning. Human and other primates can adjust behavioral strategies within seconds, even following a single error (Thoroughman and Shadmehr 2000). This behavioral capacity is thought to rely on mechanisms that are faster than what synaptic plasticity supports. The notion of computation through dynamics has thus been suggested as a mechanism for this process (Sohn et al. 2021). Within this framework, a cortical area's population activity patterns are a function of its internal connectivity and external drive (Vyas et al. 2020). That is, changing the external drive alters the initial conditions of the neural dynamical system and, in turn, would change the quality or even the nature of the implemented computations (Gurnani and Cayco Gajic 2023). One example is derived from primate dorsomedial PFC of monkeys trained to generate a timed motor response based on a sensory measurement of a corresponding interval on single trials (Remington et al. 2018). Changing the sensorimotor context, or the relationship between the sensory measurement and motor output, generated dorsomedial PFC motor production neural dynamics consistent with changing their initial conditions. Computationally, this resulted in different speeds at which the population activity evolved, allowing monkeys to produce flexibly different time intervals within exceedingly short periods of time. Although not explicitly measured in this setup, the lack of synaptic-level adjustments in such rapid learning conditions was observed in a primate motor adaptation task, in which single trial adjustments did not result in any changes to the activity covariance structure within either premotor or motor cortex (Perich et al. 2018). These population-level activity mechanisms may also be relevant for more cognitive strategy adjustments as recently observed in changes of dlPFC neural geometry in macaques performing value-based decisions (Wang et al. 2023).

What Unique Contributions Does Work with Optogenetics, Chemogenetics, Large-Scale Electrophysiology, and Calcium Imaging Contribute to Understanding PFC Function?

There has been a steep increase in the use of high-channel, high-density probes for electrophysiological recordings in both rodents and primates, enabling the collection of an unprecedented amount of data from just a few animals (Juavinett et al. 2019; Jun et al. 2017; Luo et al. 2020). These methods are expected to offer unique insights into a functional dorsal-ventral "gradient" organization for PFC, as described by Rich and Averbeck (this volume). Viral-mediated technology also allows powerful correlate and causal approaches in both rodents and primates (see Izquierdo, this volume). The major advance associated with transgene targeting using specific promoters is the ability to identify selectively and track individual cells and cell types over time or over processes (i.e., learning). For example, targeting and manipulating pyramidal neurons in PFC is now possible with single-cell calcium imaging combined with opsin/optogenetic tagging. This level of resolution is commonplace in mice and rats (for a review, see Resendez et al. 2016) and has gained momentum over recent years in macaques (Jazayeri and Afraz 2017; Oguchi et al. 2021a; Seidemann et al. 2016). Though optogenetic techniques probing PFC circuits have demonstrated promise in NHPs, they have mostly been applied to the interrogation of sensorimotor systems, less to learning, decision making, or other functions of the PFC. An important factor to consider here is the duration of activation/inhibition, especially because PFC functions tend to unfold over longer timescales, whereas sensorimotor functions occur much more quickly. In addition, optogenetic perturbation relies on implantation of a fiber to deliver different wavelengths of light, making it less of a viable therapeutic option for human patients in the future (i.e., limiting its translational appeal), though progress is being made in delivering light to deep brain structures transcranially (Chen et al. 2021). Similarly, fiber photometry enables the measurement of bulk calcium signals (analogous to the relationship of local field potentials to single-unit activity measures in electrophysiology) and is often used to confirm causal manipulations in systems neuroscience experiments in rodents. This technology, however, has not been widely adopted in nonhuman or human primate studies.

Of particular promise for cross-species translation is the chemogenetic approach. Chemogenetic techniques work through viral introduction of mutant G-protein coupled receptors or designer receptors exclusively activated by designer drugs, DREADDs (Armbruster et al. 2007; Roth 2016). Though this technology does require invasive intracerebral surgery to introduce the mutant receptors, the timescale of this method during behavior is ideal, similar to traditional pharmacological approaches, requiring no chronic implant for activation. Similar to pathway-specific DREADD experiments conducted in rodents, the more refined double-virus method to introduce retrograde cre in the target region (i.e., terminals), cre-dependent DREADD at the origin (i.e., cell bodies), and similar approaches are now frequently being employed in NHPs (Oguchi et al. 2021b; Oyama et al. 2022; Vancraeyenest et al. 2020; Wood et al. 2023). Consequently, there should be a critical mass of studies in the near future to provide a thorough cross-species comparison of rodent and NHP studies on the function of PFC circuits.

Utility of Animal Models

Understanding the structure and function of PFC circuits in species other than humans is an important intellectual goal in its own right. However, this endeavor also has utility in various applications of our understanding and treatment of human mental disorders, even despite their evident complexity and heterogeneity. The optimal approach may be to model human symptoms or symptom clusters by explaining them in terms of theoretical constructs (at both functional neural and behavioral levels) derived from cross-species studies, as described earlier in this chapter. We anticipate that this would at least provide building blocks for understanding the greater complexities of human executive function. This approach may then, for example, identify a relatively finite number of neural systems or circuits, which in many cases (limited by homology) can be investigated using such methods as chemogenetics or optogenetics, combined with suitable behavioral measures having cross-species translational validity. Ideally, tests which show functional similarities (e.g., see earlier discussion on criteria for homology) across species should be employed rather than simple behavioral "readouts."

The second important component of any such model is to simulate a deficit in a particular neural circuitry that may mirror what has been discovered in studies of a human disorder. Of course, in most cases, the etiology of some human mental disorders is obscure and multifactorial, which makes single transgenic preparations and global manipulations of stress of limited use. However, given knowledge about neural systems involvement in human mental disorders, it may now be more feasible to make these simulations. For example, overactivation of subcallosal cingulate cortex in marmosets to mimic the overactivation of this region in depression has revealed both anxiety and anhedonia-like symptoms, which appear dependent upon separate pathways to the amygdala and different parts of the nucleus accumbens, respectively (Wood et al. 2023). Other excellent examples are provided by the use of optogenetics to provide excitatory or inhibitory drive, respectively, to the medial and lateral OFC in *Sapap* knockout mice to produce compulsive grooming behavior mediated by the striatum, as well as other behavioral signs, which may likely be relevant to human obsessive-compulsive disorder (Ahmari et al. 2013; Burguière et al. 2013).

The aim would be to develop interventions such as selective microinfusions of pharmacological agents, electrical stimulation (deep brain stimulation, DBS), or even behavioral interventions. For instance, in the work on marmoset subcallosal cingulate cortex, only anhedonia, but not anxiety, was ameliorated by the rapidly acting antidepressant ketamine administered systemically 24 hours earlier (Alexander et al. 2019). Further, in the work by Ahmari et al. (2013), the behavior was remediated by treatment with SSRIs as used (with limited success in the clinical population), which also further strengthens the validity of their model. The advent of newer technologies, however, also enables much more specific-circuit interventions, which have greater cellular specificity in the form of optogenetic and, more feasibly from the clinical therapeutic perspective, chemogenetic interventions via DREADDs receptors. This approach may also help us understand how some existing treatments (e.g., DBS) actually work at a mechanistic level.

It should not, of course, be underestimated just how ambitious such an undertaking actually is. It is inconceivable, for instance, that chemogenetics could be readily applied to human mental health disorders without monumental ethical groundwork. Perhaps an early tractable approach could be to use chemogenetics to reduce the side effects of existing and successful pharmacotherapies. Viral-mediated technology is also technically challenging to employ in NHPs, although considerable progress is being made. Nevertheless, parallel work with rodents validating, for example, noninvasive methods for implant-free deep brain transcranial photoactivation of deep brain circuits (Chen et al. 2021) should help to establish proof of principle, given the constraints on translation imposed by species differences.

Conclusion

As reflected in this chapter, our discussion at this Ernst Strüngmann Forum focused on key principles that underpin the determination of homologies and analogies of PFC. Our discussion built on previous work that dates back to the 1800s as well as highlights ongoing efforts to determine how the structure and function of PFC relates to similarities and differences across species with cognitive, developmental, and translational insights. Throughout, we highlighted areas for future research to motivate future experiments, both empirical and theoretical. In addition, we hope that this discussion will spur further discussion and reviews and eventually lead to a consensus regarding the ambitious goal of determining the homologies and analogies of PFC, as well as the cognitive, developmental, and translational insights gleaned from those homologies and analogies.

Acknowledgments

We thank other delegates for useful discussion and contributions to this chapter, especially Clay Curtis, Caterina Gratton, Betsy Murray, John O'Doherty and Angela Roberts.

5

Neurophysiology

Differentiating Functional Contributions across Prefrontal Cortex

Erin L. Rich and Bruno B. Averbeck

Abstract

A long history of research in neuropsychology has supported the idea that there is functional specialization within the prefrontal cortex (PFC). To better understand how a region subserves a specific function, neuron activity is often recorded from multiple areas as subjects engage in prefrontal-dependent cognitive tasks. Contrary to expectations, these studies have generally found that neurons across PFC encode all manner of task-relevant information, with relatively little difference among regions. These data are important because they demonstrate the vast representational capacity and flexibility of PFC, yet they have been less useful when trying to glean a mechanistic understanding of how regions differ and interact with each other. In this chapter, these data are first reviewed, then considerations are proposed that might better direct future studies. Discussion includes the anatomy and evolutionary origins of the primate PFC, which suggest a gradient organization, with a main division between dorsal and ventral trends, rather than a series of smaller discrete regions. These gradients are observable in neural recordings within and across regions and may provide insights into the functional organization of PFC. It is important to note that gradients are consistent with functional differentiation across PFC, but they suggest continuous rather than discrete changes in function. Second, recent advances in neural analysis are reviewed, which focus on representations and temporal dynamics in neural populations, as opposed to individual neurons. These population codes may reveal unique insights into local function and cross-regional interactions and help us understand the unique properties of the main divisions of PFC.

Introduction

The idea that the brain can be divided into functional regions dates back to the 19th century. While functions of motor and sensory regions were quickly discernable, there has been significantly more debate about parcellation of function in the prefrontal cortex (PFC). Early attempts to understand functions of the PFC led Penfield to believe the region was "uncommitted" at birth and specialized function was learned over a lifetime (Penfield 1965). However, investigations over the ensuing decades have supported the notion that functional specialization not only exists within PFC but is consistent across individuals and species. This is largely based on neuropsychology studies that find reproducible patterns of behavioral alterations following damage or dysfunction in subregions of the PFC. Logically, then, one would assume that the activity of neurons in a subregion should reflect its function. Indeed, there are many instances where neurophysiological correlates are found in the same region where lesions impair a particular function. For example, after finding that inactivation of the lateral PFC impaired performance in the delayed-response task, Fuster and colleagues recorded from this region to search for neural responses that underlie this dependence (Fuster and Alexander 1970). They found that neurons displayed elevated firing rates in the delay period of the task, which was interpreted as the neural mechanism maintaining information in mind to perform the delayed-response task (Fuster and Alexander 1971). This ability was later dubbed "working memory." Since then, however, elevated delay period activity has been reported in a wide variety of brain regions, including other frontal areas, such as the frontal eye fields, orbitofrontal and medial frontal cortices (Chafee and Goldman-Rakic 1998; Enel et al. 2020; Kamiński et al. 2017), as well as nonfrontal areas including parietal cortex, inferotemporal, medial temporal, auditory, and temporal pole regions (Chafee and Goldman-Rakic 1998; Fuster and Jervey 1982; Gnadt and Andersen 1988; Kamiński et al. 2017; Kornblith et al. 2017; Nakamura and Kubota 1995; Napoli et al. 2021). Therefore, elevated delay period activity is not a unique property of regions required for working memory tasks. To complicate matters further, there have been demonstrations of intact working memory in the absence of elevated delay period activity (Lundqvist et al. 2018). Although there could be many explanations for these discrepancies, working memory stands as an example of a pattern that has played out in many subfields focused on different cognitive functions putatively localized within subregions of the PFC. Neuropsychology studies implicate functional localization and initial recording studies find logical task correlates in the corresponding brain region, but these are followed by tempered enthusiasm when it is realized that the signals are neither unique to that region nor reliably found there in different task scenarios. Overall, it is now safe to say that functional localization is less apparent in neurophysiology than anticipated. This conclusion has led to a resurgence of the idea that, while some specialization is inherent in anatomical connectivity, the dominant regime is that of distributed, homogenous functionality across PFC.

Here, we propose that we should not dispense with the idea of functional localization at the level of neurophysiology. Instead, we highlight two considerations for future studies. First, we review anatomical evidence that PFC may be organized by gradients rather than discrete boundaries and consider how this might impact neural responses within and across regions. Gradient organization is consistent with function varying across PFC and therefore could be consistent with results from lesion studies. Most lesion results are interpreted, however, as evidence for functional localization within architectonic areas, which are circumscribed areas whose function is often thought to not depend on their two-dimensional location on the cortical sheet. Gradient organization, rather, suggests that function varies continuously across the cortical sheet with few clear areal boundaries. Second, we suggest that advances in large-scale recording and corresponding analysis techniques provide more valid measures of neural mechanisms and may ultimately help to differentiate functional regions of PFC. We limit our focus to nonhuman primates, where there is abundant neurophysiology data and reliable similarities to humans in prefrontal anatomy and function, but we note that there are a number of excellent reviews on functional organization of frontal regions in rodents (e.g., Heidbreder and Groenewegen 2003; Laubach et al. 2018).

Functional Localization from the View of Neuropsychology

Our strongest framework for understanding how functions localize in PFC has come from examining the consequences of circumscribed lesions or other manipulations that create loss of function. Striking contrasts are found between the lateral regions, particularly the areas surrounding the principal sulcus, compared to the ventral and ventromedial regions. In general, damage to the lateral PFC produces deficits in processes like working memory, attention, and planning, often grouped together as cognitive control or executive function. On the other hand, lesions to the ventral frontal cortex produce disturbances of emotional processing, including emotional regulation and social behavior, primarily dependent on the ventromedial regions, as well as evaluation and decision making, primarily dependent on the orbitofrontal regions. Data on the medial PFC, including anterior cingulate cortex (ACC), are more mixed, with proposed functions including linking goals to actions, signaling or adjusting to errors, or using contextual information to interpret outcomes (Kolling et al. 2016a).

Based on this evidence, it is widely held that, although the major divisions of PFC work together to orchestrate behavior, they each contribute a unique function. There are still many open questions relating to the precise nature of these functions, as well as the anatomical locations that produce certain effects on behavior. For instance, more localized lesions can sometimes parse effects further yet at other times result in no detectable deficits where a broad manipulation did. Moreover, the lack of behavioral effect following a lesion does not definitively indicate that the lesioned area is not involved in the task. Behavioral measures commonly obtained in these studies, such as percent correct or reaction time, are coarse and do not preclude the possibility that the contributions of the impaired neurons are simply not measurable at this level. Alternatively, another intact region may be able to compensate for the loss of neurons elsewhere, which is particularly important in the case of permanent lesion, when plasticity could take place over time. Despite these caveats, it is indisputable that reproducible patterns of behavioral alterations do occur following damage or interference to different regions of PFC, with clear parallels across species. This supports the widely accepted notion that there is, indeed, functional specialization in PFC. For further discussion on the degree and evidence for parcellation of function within frontal cortex, see Chapter 8 by Murray et al. (this volume).

Given this conclusion, one would expect neuron responses to differ across regions of PFC. In particular, neurons in different regions should be driven by, or *encode*, different factors related to ongoing behavior or cognitive processes. We use the term "encode" operationally, meaning that variance in a neuron's activity is explained by variance in an experimentally defined feature, such as stimulus identity, direction of a motor response, or current task rule. This premise has guided the design of neurophysiology studies in the PFC for decades. A typical approach is to record from a specific region during a task that is impaired by loss of function in that region. Such experiments commonly reveal neural correlates of the task being performed. For instance, neurons in dorsolateral regions (dlPFC) encode information held in working memory (Chafee and Goldman-Rakic 1998; Constantinidis et al. 2018; Funahashi et al. 1989; Fuster and Alexander 1971; Goldman-Rakic 1995; Kubota and Niki 1971; Lara and Wallis 2014; Niki 1974; Niki and Watanabe 1976; Watanabe et al. 2006) as well as rules or categories in cognitive tasks (Blackman et al. 2016; Freedman et al. 2001, 2002; Wallis et al. 2001; White and Wise 1999), and neurons in orbitofrontal cortex (OFC) encode the value of choice options as well as expected and received rewards in decision-making tasks (Critchley and Rolls 1996; Hosokawa et al. 2005, 2007; Kimmel et al. 2020; Morrison and Salzman 2009; O'Neill and Schultz 2010; Padoa-Schioppa and Assad 2006, 2008; Padoa-Schioppa and Conen 2017; Rich et al. 2018; Setogawa et al. 2019; Tremblay and Schultz 1999, 2000).

Although these results are consistent with neuropsychology data, further investigation has revealed a more complicated picture. If we expect that computations differ across regions of the PFC, and our behavioral tasks can uniquely tax these abilities, then this leads to a few concrete predictions, illustrated for OFC and dlPFC in Figure 5.1. First, if a task is impaired by inactivation of region A and not region B, then neurons in region A should carry more task-relevant information than region B, or at the very least, neural responses in the two regions should differ measurably (columns of Figure 5.1). Second, if a region is required for task X and not Y, then neurons in this region should encode more task-relevant information during task X than task Y, or at least they should differ measurably (rows of Figure 5.1). Dissociations of this sort have

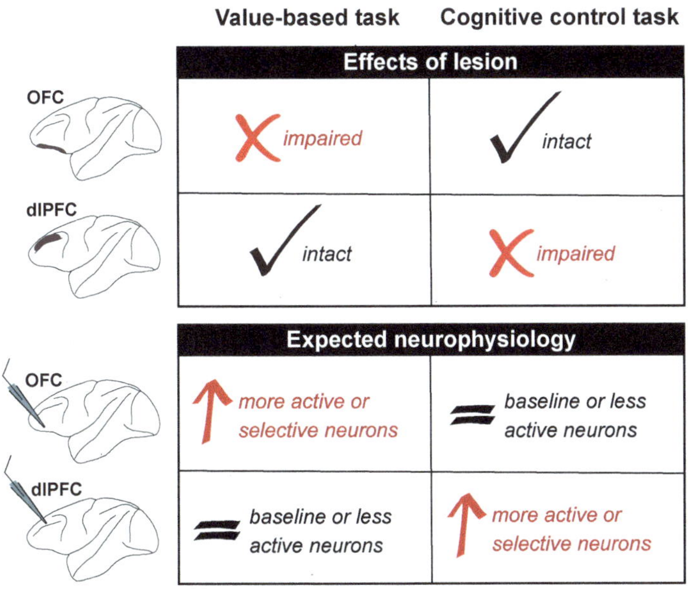

Figure 5.1 Conceptual comparisons between orbitofrontal cortex (OFC) and dorsolateral prefrontal cortex (dlPFC), from the perspective of neuropsychology and neurophysiology. The top half shows the general framework supported by loss of function studies, where OFC is important for performing value-based tasks and dlPFC is important for various cognitive control tasks. This leads to the prediction that neurophysiology should vary across regions and tasks in a similar manner (bottom half).

been sought in many instances to better understand the unique contributions of different regions of PFC. However, one of the striking outcomes has been that differences are much more limited than one might expect. Moreover, we see this equivocal outcome in published studies where there may be a bias toward identifying and reporting differences that align with each region's presumed function; this suggests that there could be a number of unpublished observations that are even more mixed. Below we briefly outline some of these data, summarized in Table 5.1, that have led to this impression. We emphasize comparisons of ventral and lateral PFC, primarily OFC and dlPFC as an example case, because there is strong neuropsychology data to support their unique and dissociable functions.

Contrasting Neuron Responses in OFC and dlPFC

The OFC and neighboring regions are important for emotional appraisals as they relate to decision making. However, decision-relevant information such as expected values are also strongly encoded by neurons in dlPFC (Leon and Shadlen 1999; Roesch and Olson 2003; Tsutsui et al. 2016b; Watanabe 1996), as well as supplementary and premotor regions of the dorsal and lateral frontal

Table 5.1 Percent of OFC or dlPFC neurons encoding task variables in value-based or cognitive control paradigms, from studies that recorded neurons in both regions in the same experiment. The most consistent difference is a tendency for more encoding of spatial information, such as location or response direction, in dlPFC compared to OFC. Proportions shown are percentage of all neurons recorded in an area; *percentages estimated from published figures.

Task Variable Encoded	% OFC Neurons	% dlPFC Neurons	Reference
Value-Related Tasks			
Any decision variable	56%	49%	Kennerley et al. (2009a)
Main effect of expected reward (by trial epoch)	5, 9, 5, 6%	7, 4, 2, 7%	Wallis and Miller (2003b)
Reward × Picture (by trial epoch)	10, 5, 8, 12%	7, 8, 9, 2%	
Reward × Location (by trial epoch)	5, 7, 4, 11%	7, 16, 11, 17%	
Stimulus	34.9%	rostral 46.0 mid 46.3 caudal 55.9%	Tang et al. (2022a)
Outcome	32.2%	rostral 39.5 mid 46.1 caudal 57.0%	
Actual payoff (i.e., reward)	45.3%	41.2%	Abe and Lee (2011)
Hypothetical payoff	16.9%	21.4%	
Juice type (by trial epoch)	13, 16, 21, 18%	10, 10, 11, 6%	Lara et al. (2009)
Receipt of reward	27%	32%	Kennerley and Wallis (2009b)
Response direction	6%	13%	
Probability of receiving reward	12%	8%	
Chosen (integrated) value	9%	14%	Hosokawa et al. (2013)
Decision type (category)	57%	68%	
Cue value	57.5%*	42%*	Hunt et al. (2018)
Action (right/left) value	6%*	17%*	
Attribute (magnitude/probability) value	23.5%*	12%*	
Spatial (location) value	7%*	16.5%*	
Cognitive Control Tasks			
Strategy	14%	12%	Tsujimoto et al. (2011)
Task rules (pre-cue epoch)	17%	29%	Yamada et al. (2010)
Abstract rules	32%	49%	Wallis et al. (2001)
Category	28%	8%	Tsutsui et al. (2016a)
Rule	26%	28%	
Contingency	48%	41%	
Strategy	12%	8%	Fascianelli et al. (2020)
Directionally-selective delay period activity	3.9%	29.9%	Ichihara-Takeda and Funahashi (2007)

cortex (Roesch and Olson 2003). These areas encode value during the delay period of working memory tasks (Leon and Shadlen 1999; Roesch and Olson 2003; Watanabe 1996), when dlPFC is believed to hold relevant cognitive information online, as well as in value-based decision making tasks (Cai and Padoa-Schioppa 2014). For example, in a task where monkeys had to weigh an amount of juice against either the delay or effort needed to obtain it, decision-relevant values were encoded by similar proportions of OFC and dlPFC neurons with only minor differences between regions (Kennerley et al. 2009). In this case, more dlPFC neurons encoded movement direction, consistent with the common finding that directional or spatial information is preferentially represented in lateral regions (Grattan and Glimcher 2014; Hunt et al. 2015; Kennerley and Wallis 2009a; Tang et al. 2022a; Wallis and Miller 2003b). Others, however, have emphasized that, although it is not as strongly encoded, spatial information is not absent from OFC (Strait et al. 2016; Yoo et al. 2018). Beyond spatial selectivity, there was very little that distinguished these regions in how they encoded decision-related information.

On the other hand, processing cognitive information, particularly rules and strategies that guide behavior, is believed to be the domain of dlPFC, yet OFC and ventrolateral PFC also robustly encode task rules (Fascianelli et al. 2020; Wallis et al. 2001; Yamada et al. 2010). In a variant of the classic Wisconsin Card Sorting Task adapted for monkeys, OFC encoded both abstract rules that define the relevant feature domain (e.g., shape or color), as well as concrete rules indicating the currently correct feature (e.g., choose red) (Sleezer et al. 2016). In addition, both OFC and dlPFC neurons encoded response strategies in a stay versus shift task, and OFC even encoded the strategy earlier (Tsujimoto et al. 2011). Taken together, the encoding properties of individual neurons tend to be primarily informed by the task the monkey is engaged in, rather than the prefrontal region where they were recorded.

It is less common to evaluate the same neurons in multiple tasks, in part because this involves training monkeys to perform tasks in interleaved fashion. Some blocked designs have been used and suggest that prefrontal neurons flexibly adapt to encode information about the current task, but do so fairly uniformly, without one particular region being uniquely engaged by one task. For instance, monkeys learned to select rewarding actions or objects in different trial blocks while large populations of neurons were recorded from the full rostro-caudal extent of principal sulcus. Across this region, neuron activity shifted between encoding the rewarded actions or objects, depending on which was relevant in the current trial block (Tang et al. 2021). Another study recorded from OFC and ACC while monkeys similarly chose a rewarding cue or rewarding action (Luk and Wallis 2013). In this case, slight differences were found in the choice phase of the task, where more ACC neurons encoded actions (16% versus 10% in OFC) and more OFC neurons encoded stimuli (20% versus 10% in ACC), but this occurred while actions, stimuli, and their associated outcomes were encoded in similar proportions during all other task phases. Again, although small degrees of difference can be found, there is an overwhelming pattern of similarity across tasks. Finally, a recent study approached this question by recording different populations of dlPFC neurons across days from the same monkeys as they performed four different tasks, only one of which was impaired by dlPFC lesions (Tremblay et al. 2023). In this case, no metrics of neuron responses were found to distinguish the tasks. Although the expectation that there would be measurable differences is as reasonable as the expectation that two regions should differentially encode information in a given task, the supporting evidence remains quite weak.

Reconciling Neuropsychology with Neurophysiology

The contradictions between lesion effects and neurophysiology data have led to different interpretations. To start, the tasks used to study prefrontal function are typically relatively simple, and it has been suggested that tasks with more complexity, that are designed to better tax prefrontal function, or those with better construct validity might find distinctions that are not found with simpler tasks. While this may be true and task design is of critical importance, it is often the case that uniform neural responses are found in tasks that are the same or highly related to those in which neuropsychology studies have demonstrated functional dissociations. This argues against the notion that more refined tasks are likely to reveal neurophysiological differences among prefrontal areas. Conversely, the impacts of prefrontal damage on human behavior are most evident in daily activities rather than highly structured laboratory tasks, suggesting that less constrained tasks may be better at tapping into the unique functions of different prefrontal regions. While this is an attractive hypothesis, there are a host of challenges in parsing and interpreting unconstrained behavior and concomitant neurophysiology. Advances in markerless-tracking algorithms, such as DeepLabCut (Mathis et al. 2018), have improved our ability to parse unconstrained behavior at the level of motor movements. Still, critical gaps between observable motor output and underlying cognitive processes have so far limited the degree to which computer vision tools have improved our understanding of PFC.

Another view notes that the dense interconnectivity of prefrontal subregions could suggest that information spreads easily, and this makes neuron responses relatively uniform. If this is the case, temporal analyses, such as latency to encode information, could reveal an origin and direction of spread, and in this way point toward specialization. For instance, similar proportions of neurons in OFC and dlPFC encode expected rewards, but encoding among OFC neurons begins about 80 ms earlier, which has been taken to suggest that reward information enters PFC via OFC and is then passed to dlPFC to influence behavior (Wallis and Miller 2003b). While this may be true, it does not explain why these signals are present in both areas. For instance, if dlPFC represents reward values because it is a node on the path to expressing reward-guided behavior as motor output, then dlPFC lesions should produce measurable changes in motivated behaviors such as value-based decision making. Alternatively, these signals could be only passively present. However, they are curiously prevalent, and potentially metabolically costly, to be just a byproduct. A related idea suggests that information becomes more shared across regions as a result of extensive practice or training, which is common in monkey studies, though this encounters the same problem in explaining why this is an efficient way the brain would operate.

An alternative proposal is the "content differentiation" model, in which different regions of PFC perform the same basic computations, but do so on different types of information, which depend on the anatomical inputs that they receive (Goldman-Rakic 1987; Zald 2007). From this view, neuron responses in different regions might appear similar because specialization arises from the large-scale circuits in which each region participates. While this is plausible, anatomical evidence has also been used to argue the opposite; namely, that different regions are specialized for fundamentally different computations, such as holding information online in working memory (Petrides 1994). Anatomically, lateral and orbital prefrontal regions differ in their cellular architecture, including granularity, density of neurons in superficial layers, type and density of interneurons, and lateral connectivity among pyramidal neurons (Zald 2007). These marked differences are hard to reconcile with the notion that the areas carry out the same fundamental computations.

Finally, other views have, to greater or lesser extents, rejected the notion of functional specialization within PFC and instead posit that information appears distributed because function is distributed (e.g., Sleezer et al. 2016). The most extreme version of this argument, in which there is no functional organization, is not commensurate with the extensive neuropsychology literature. A more nuanced suggestion is that there are discrete, localizable processes that each contribute to a broader, integrated function of PFC as a whole (e.g., Wilson et al. 2010). This offers more parsimony with the neuropsychology literature, by accounting for differential effects of lesions, but leaves the prevailing problem that the information encoded by single units varies so little across prefrontal areas, making it hard to discern the unique components of function that occur in one area versus another.

Rather than conclude that neurophysiology is homogenous across PFC or question the notion of specialization altogether, we propose two directions for reconciling the clear distinctions in neuropsychology with the relative homogeneity of neurophysiology. First, we consider the anatomical organization of PFC as it relates to larger brain circuits, where gradients of connectivity and cellular architecture are more prominent than discrete subregions. This suggests that physiological properties may also vary in a graded fashion, producing a source of variability that muddies the waters when trying to understand localization of function from the perspective of discrete regions. Second, our standard analytic approaches that investigate neural coding might be missing the forest for the trees, and new perspectives on information coding and dynamics in neural populations could help us understand how prefrontal regions are specialized for particular functions.

Anatomical Organization of Prefrontal Circuitry

The anatomical organization of PFC is an important guide to understanding its functional organization. A long history of anatomical work has fractionated PFC into discrete regions, each given a corresponding number or acronym (Brodmann 1909; Walker 1940). Early studies relied on the size and location of cells, and stains for myelin. More recent studies have used stains for increasingly complex sets of markers (Carmichael and Price 1994) or, when using imaging in humans, measures of functional connectivity (Van Essen and Glasser 2018). For the most part these studies assume that discrete regions exist, and then proceed to determine how many regions there should be and where the boundaries should be placed. Placing a boundary is always an inference process. For instance, modern techniques that use clustering algorithms use a free parameter to determine the number of clusters. Although the assumption that there are in fact discrete areas in PFC has been questioned multiple times (Kaas 1987; Lashley and Clark 1946; von Bonin and Bailey 1947), the dominant view is that discrete areas exist. Furthermore, it is assumed that each area subserves a unique function. Brodmann went so far as to assume that each area was a separate organ of the mind, with its own function. This notion of discrete areas is also reflected in the placement of lesions in neuropsychology studies.

Despite this tendency to parcellate anatomy, the balance of evidence seems to support a different interpretation. Consideration of both anatomical connectivity and comparative anatomy across species suggests that PFC can be better understood from the perspective of gradients than discrete areas with sharp boundaries. While there are some cases of clear distinctions between cortical regions (e.g., between primary and secondary sensory areas), this does not apply as well to association circuits, including PFC. The large-scale organization of PFC and related circuits has instead led to a model (Figure 5.2a) which suggests that, at the cortical level, parietal-frontal and temporal-frontal circuits are organized as nested loops, similar to an onion (Giarrocco and Averbeck 2023). At the core is primary somatosensory and motor cortex (M1/S1). At the next level there is a dorsal parietal to dorsal premotor circuit, and a ventral parietal to ventral premotor circuit. Beyond this there is a dorsal-medial parietal to dorsal prefrontal circuit, and a temporal to ventral prefrontal circuit. Although considerable anatomical complexity is not captured by this simplified model, it does capture the strongest trends in connectivity. In particular, the model articulates both hierarchical organization and specific connectivity that is likely to influence the organization of function. Furthermore, the

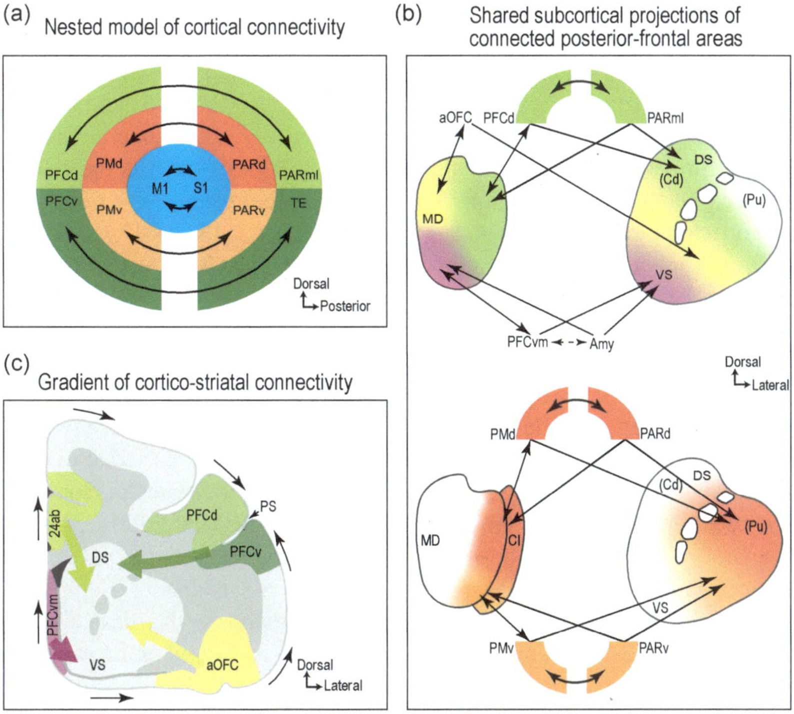

Figure 5.2 Gradient organization of cortical-striatal circuits (Giarrocco and Averbeck 2023). (a) Nested model of cortical-cortical connectivity. Connectivity in neocortex is organized in a nested architecture, with posterior and frontal areas connected in circuits that show ventral-dorsal, posterior-anterior structure. (b) Prefrontal cortex connections to the striatum are organized in a gradient, such that ventral-medial and caudal orbital areas are connected to the ventral striatum, dorsolateral areas are connected to the dorsal striatum, and intermediate areas are connected to intermediate portions of the striatum. (c) Connected posterior and anterior cortical areas, in the nested architecture, have overlapping projections into the striatum and thalamus. Additionally, circuits through the striatum to the pallidum also project to similar overlapping areas in the thalamus, forming closed loops. Anterior cingulate cortex (24ab), anterior orbitofrontal cortex (aOFC), mediodorsal (MD) thalamus, dorsal parietal (PARd), mediolateral parietal (PARml), ventral parietal (PARv), dorsal prefrontal (PFCd), ventromedial prefrontal (PFCvm), ventral prefrontal cortex (PFCv) dorsal premotor (PMd), ventral premotor (PMv), principal sulcus (PS), temporal cortex (TE), dorsal striatum (DS), ventral striatum (VS).

dominant white-matter tracts linking posterior (i.e., behind the central sulcus) and anterior cortical areas connect posterior and anterior areas at the same level of the hierarchy (Yeterian et al. 2012).

When the cortical-subcortical circuitry is examined, it can be shown that the posterior and anterior nodes of these nested circuits share subcortical projections in both the striatum and thalamus (Figure 5.2c). Thus, nodes in the dorsal parietal to dorsal premotor circuit project to overlapping regions in the dorsal putamen, the lateral mediodorsal nucleus, and the adjacent central-lateral nucleus of the thalamus. Nodes of the ventral parietal and ventral premotor circuit project to a corresponding ventral region in the same basal ganglia and thalamic nuclei. Similar overlapping projection targets can be shown for each of the connected posterior and anterior areas (Giarrocco and Averbeck 2021, 2023). Although not all connected cortical areas have overlapping subcortical projections (Selemon and Goldman-Rakic 1988), it is the case that connected areas that correspond to the nested architecture have overlapping subcortical targets.

Within this nested organization, the striatal projection target of prefrontal areas can be predicted using only the coronal and anterior-posterior locations of tracer injections (Averbeck et al. 2014), consistent with the idea that connectivity between PFC and the striatum follows a gradient. Here, the ventral-medial PFC and the caudal OFC project into the ventral striatum, the dlPFC (area 46) projects into the dorsal striatum, and areas between these two poles project to intermediate locations in the striatum on a ventral-medial to dorsal-lateral axis (Figure 5.2b). This is true whether one translates dorsomedially from ventromedial PFC or anterolaterally from OFC, toward dlPFC. A similar topography can be seen in downstream striatal projections to the pallidum and cortical and pallidal projections to the mediodorsal thalamus (Figure 5.2c). This gradient in frontal projections, combined with the overlapping subcortical projections of posterior and anterior areas, implies an overall gradient architecture in cortical-subcortical circuits.

Beyond anatomy, this model suggests an organizing principle for the functions of PFC, as well as their corresponding neurophysiological mechanisms. Specifically, there might be gradients of function within and across traditionally defined prefrontal regions. This would be important in comparisons across prefrontal regions because it would introduce a source of variability within each population, particularly if neuron sampling is wide and sparse. Indeed, when relatively large swaths of cortex are sampled at high density, graded trends are often found. For instance, spatial receptive fields in dlPFC broaden from posterior to anterior, and selectivity for objects and colors drops in a graded fashion (Riley et al. 2017; Tang et al. 2021). In contrast, moving from posterior to anterior in OFC, value encoding tends to increase (Rich and Wallis 2017). While we still lack a mechanistic picture of how these graded responses reflect an underlying function, recognizing heterogeneity can help form hypotheses of how function is organized and maps to neurophysiology.

Evolutionary Origins and Ventrodorsal Trends

Evolutionary perspectives help to integrate the concepts of anatomical and functional gradients with what is known about the main divisions of PFC. Anatomical gradients and nested organization have been identified using modern tract-tracing methods in macaques, and this is also consistent with human resting-state studies, but this architecture likely reflects the evolutionary expansion of neocortical areas. Early comparative work in reptiles identified two dominant nodes in the pallium (the vertebrate homolog of the mammalian cortex). The medial pallium is homologous to the hippocampus, and the lateral pallium is homologous to pyriform cortex. Between these nodes there are transition areas. This early work, therefore, established a tripartite model of the pallium (Abbie 1940, 1942; Dart 1934) with medial, lateral, and intermediate (possibly dorsal) areas.

Subsequent work based on developmental gene expression has extended and provided further support for this model and suggested that the pallium, and the mammalian cortex, can be divided into four regions (Puelles et al. 2017): a medial-hippocampal region, a dorsal neocortical (neopallial) region, a lateral region that develops into the claustrum and insula, and a ventral region that develops into pyriform cortex and the cortical or pallial amygdala. Whether fish, amphibians, and nonmammalian amniotes have a dorsal pallium that is homologous to mammalian neocortex is the subject of ongoing debate (Striedter and Northcutt 2020). Recent work using gene expression data from single cells has suggested that reptiles do have a neopallium, homologous to neocortex (Tosches et al. 2018). What is clear is that the neopallial region in fish, amphibians, and nonmammalian amniotes is relatively small when compared to the massive expansion of the neocortex, particularly in primates. While there has been considerable expansion in the mammalian cortex, the slope is steepest for neocortical areas (Finlay and Darlington 1995). Thus, the dorsal pallium is relatively small in nonmammalian vertebrates, relative to the medial and ventral pallium. Particularly in primates, however, the neocortex has become much larger than the medial (hippocampal) and ventral-lateral (pyriform) areas.

Sanides (1970) and subsequent authors suggested that the organization of PFC could be understood from the perspective of the tripartite model. In the dual-origin theory, prefrontal cortical areas expanded across evolution, as cortex expanded, starting from the medial-hippocampal and ventral-pyriform areas. The gradient anatomical organization of cortical-striatal-pallidal-thalamocortical circuits can, therefore, be understood as a topological expansion of this circuitry, from a Cambrian or possibly Precambrian ancestral vertebrate brain that was dominated by medial (hippocampal) and ventral (olfactory) circuits. As the dorsal pallium expanded, the anatomical connectivity between pallial, striatal, pallidal, and thalamic areas maintained their topological adjacency relationships as they also expanded, leading to the gradient of connectivity identifiable in primates.

The anatomical data suggests that the ancestral vertebrate brain was dominated by medial-hippocampal and ventral-pyriform pallial areas, and at most an incipient dorsal pallium. The medial and ventral pallial (cortical) allocortical areas in primates project to the ventral striatum, which projects to the ventral pallidum. The ventral pallidum is a single structure to which both the direct and indirect pathway neurons from the striatum project, similar to the pallidum in fish and amphibians. The neocortical areas, on the other hand, project more dorsally into the striatum, which then projects to the dorsal pallidum, which is divided into internal and external segments, with direct pathway neurons in the striatum projecting to the internal segment and indirect pathway neurons projecting to the external segment. The division between internal and external segments in the pallidum, in the circuitry connected to the neocortex, is prominent in the primate.

We have previously defined the areas connected to the ventral striatum as the ventral circuit and the areas connected to the dorsal striatum as the dorsal circuit. The ventral circuitry is dominated by conserved (i.e., present across all vertebrates) medial and ventral-lateral pallial circuits, whereas the dorsal circuitry is dominated by the recently expanded neopallial circuits. The medial pallial circuits correspond to the hippocampus and the ventral-lateral pallial circuits correspond to piriform cortex. At all levels, including the cortex, ventral circuitry, similar to the classically defined limbic system, has strong connections with the hypothalamus, whereas the dorsal circuitry has minimal connections with the hypothalamus (Figure 5.3) and instead projects, via the substantia nigra, to the mid-brain tectum (i.e., the colliculus). Because the hypothalamus plays an important role in physiological homeostasis, this suggests a model where the ventral circuitry is important for identifying internal needs, and matching these needs to objects in the environment that can satisfy these needs (Averbeck and Murray 2020). The dorsal circuitry, on the other hand, is situated to use egocentric spatial information to direct actions toward objects in the environment. The ventral circuitry, therefore, establishes goals and the dorsal circuitry implements actions to achieve those goals.

This organization aligns well with lesion data and shows clear distinctions between ventral regions such as the OFC (which is important for evaluation and emotion, processes that relate to internal needs) and dorsal regions like the dlPFC (which is implicated in cognitive control used to direct attention and action). It is also echoed in the tendency of neurons in dorsal areas to encode spatial or directional information. In addition, there is some indirect neurophysiology support for separation of large-scale dorsal and ventral circuitry. Specifically, during reinforcement learning tasks in which monkeys have to learn which objects are more frequently rewarded when they are chosen, ventral circuit areas (including the cortical amygdala, orbital frontal cortex, and the ventral striatum) maintain a representation of the values and identities of behavioral goals between trials and during baseline hold periods before choice options are presented (Tang et al. 2022a). Presumably, this value- and goal-related information (in the form of the representation of the to-be-chosen visual stimulus) reflects a match between mechanisms in the hypothalamus that code thirst or hunger depending on the unconditioned reinforcer used in the experiments and the visual stimulus on the screen. Further, it has been found that, when the choice options are presented, the value and identity information flows into dorsal circuits where it is used to identify and direct an action toward

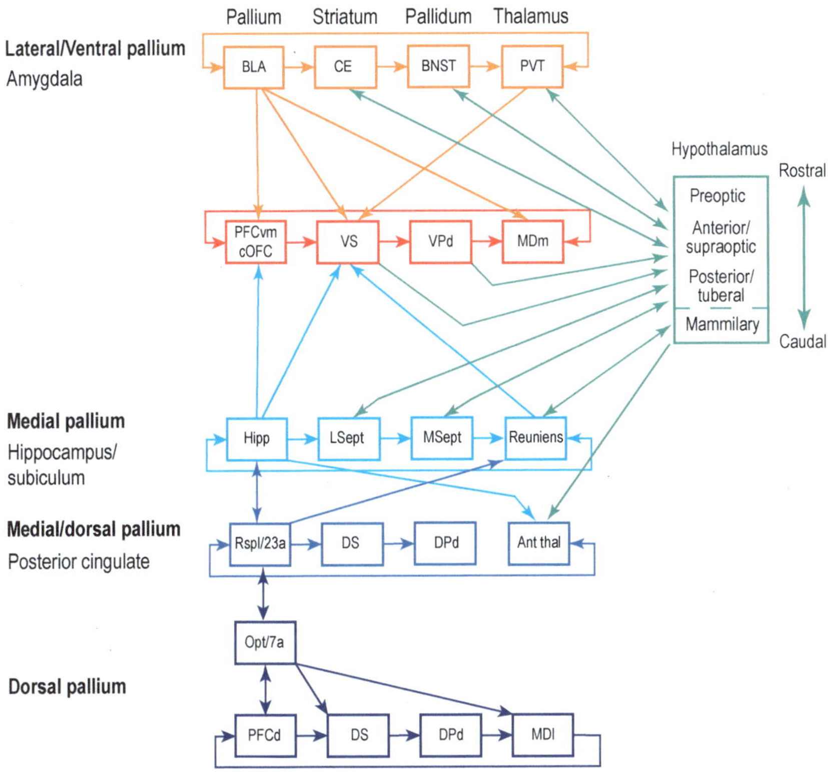

Figure 5.3 Organization of pallial-striatal-pallidal-thalamo-pallial circuits (Giarrocco and Averbeck 2023). Lateral and medial pallial areas are strongly connected to the hypothalamus, whereas recently evolved dorsal pallial areas have minimal connectivity with the hypothalamus. BLA: basal-lateral amygdala, CE: central nucleus of the amygdala, BNST: basal nucleus of the stria terminalis, PVT: paraventricular nucleus, PFCvm: ventromedial prefrontal cortex, cOFC: caudal orbital frontal cortex, VS: ventral striatum, VP: ventral pallidum, MDm: medial portion of the medial dorsal thalamus, Hipp: hippocampus, LSept: lateral septum, MSept: medial septum, Rspl: retrosplenial, DS: dorsal striatum, DPd: dorsal pallidum, Ant Thal: anterior thalamic nuclei, PFCd: dorsolateral prefrontal cortex, MDl: lateral portion of the medial dorsal thalamus.

the spatial location of the object (Tang et al. 2022a). The value and identity information is not, however, strongly represented in the dorsal circuit during the intertrial interval or other periods in the task when actions cannot be planned or directed to goal objects. This hypothesis was motivated by a consideration of the anatomical circuitry, and specifically by differential connectivity between forebrain circuits and the hypothalamus. The current neurophysiological data that supports the hypothesis is based on a stronger representation of the behavioral goal, which in this case is a visual stimulus, during the intertrial interval and initial fixation period in the ventral circuitry, and a stronger representation of the actions, at the time of choice, in the dorsal circuitry (O'Reilly 2010).

Population Coding and Dynamics

In addition to anatomical organization, functional dissociations may be obscured by the methods used to analyze neural responses. Historically, prefrontal neurophysiology has focused primarily on the activity of single neurons and identifying the experimental factors that change their firing rates. More recently, as it has become common to record many neurons simultaneously, efforts have increased to understand how information is represented at the population level and how computation is performed over such representations. Although information can be extracted from single neuron firing, these neurons are embedded in interconnected networks, both local and long range. Therefore, it may be more accurate to conceptualize neuron responses that we record as snapshots of activity in a larger dynamical system. If population perspectives have increased validity over single unit analyses, they may also be able to reconcile the disconnect between neuropsychology and neurophysiology in the search for functional specialization.

Similar to single units, encoding properties can be assessed in neural populations. One might analyze how activity varies across time or conditions, when the functional unit is not a single neuron but a population of neurons. This can be done by considering each neuron as an axis in a high-dimensional space. For instance, if we record the activity of 100 neurons, the population response can be considered as a 100-dimensional representation that evolves over time, with any time window characterized by a vector of 100 firing rates. By doing this, the response of any given neuron is necessarily considered in relation to others in the population, so that information is not represented in the activity of any one neuron, but as a pattern of activity over the population. From this starting point, one can take multiple approaches. If the population is sampled repeatedly under different conditions, classifiers can extract task information from the population vectors by differentially weighing elements (neurons). Similarly, population dynamics can be captured by the path the vector takes through the high-dimensional neural state space. Repeated samples of these paths define the region of neural space in which activity resides, referred to as a manifold.

Because neural activity is not random and includes shared variance, the population activity that defines a manifold usually exhibits structure and is lower in dimensionality than the theoretical potential of a sampled population (Gao et al. 2017). That is, a good deal of variance in our 100-neuron population might be captured by only a few dimensions. Dimensionality reduction finds dimensions of shared variance, allowing us to understand whether they correspond to task or cognitive variables. Heading direction, for example, is a two-dimensional variable. Thus, activity in circuits representing heading direction might reside on a two-dimensional manifold, perhaps nonlinear, in population coding space. Given multiple samples of a population under different conditions, shared variance across samples could be found agnostically with an approach like principal component analysis. Projecting the original samples onto the first principal component summarizes the original data in a single dimension, or "subspace," and allows us to ask whether activity in that subspace varies across conditions. In this case, subspace is a generic term referring to a lower-dimensional linear projection of population activity, defined by applying weights to each neuron in a population vector. These weights might be determined in a number of ways. While a principal component analysis captures the axes of maximum variance in population, they may be poorly aligned with the dimensions in which task conditions vary. Therefore, an alternative subspace might be defined by axes oriented to condition-wise variance. In any of these reduced-dimensionality spaces, one can assess how dynamics evolve and vary with experimentally defined conditions. Indeed, much of this decrease in dimensionality may have to do with the relative simplicity of tasks used to study neural activity (Gao et al. 2017). For example, if a population response is a (potentially nonlinear) mapping from task variables into population coding space, then low-dimensional tasks will necessarily lead to low-dimensional population activity. By extension, more naturalistic tasks that include many dimensions of variability are expected to increase the dimensionality of neural representations. However, the dimensionalities of populations in natural conditions are not yet clear, in part due to the challenges of interpreting behavior and neurophysiology in unconstrained tasks.

Because population approaches afford a different perspective on neural coding, they may provide unique insights into how representations, and the computations over these representations, vary across PFC. For instance, although task-relevant information tends to be encoded by single neurons throughout PFC, different features may be emphasized by different populations, such as expected rewards in OFC and cognitive variables in dlPFC. An example of this is population activity that creates a geometry where the relevant condition on a trial is clearly distinguished by a large separation of different conditions in state space, with other task-relevant information embedded in that structure (Chien et al. 2023). Large separations can lead to a form of abstraction, in which different instances that share a common feature occupy nearby or overlapping regions of the neural state space, which may allow the concept to generalize to new instances (Bernardi et al. 2020). Such possibilities can be investigated by evaluating the geometry of population representations.

In another domain, population dynamics traverse different landscapes, the features of which could vary in different PFC regions. For instance, dynamics in PFC often tend toward consistent dynamical trajectories, fixed points, or other attractor basins. These are believed to be stable points in the neural activity space that may be formed by patterns of synaptic weights within a network (Averbeck 2022). Therefore, attractor states could be influenced by both intrinsic architecture and experience-related plasticity, both of which could vary across PFC regions. Importantly, these dynamics arise from the collective activity of a group of neurons, so that any one unit might reflect some fragmented features but is unable to reveal the overall picture. For instance, in lateral prefrontal (prearcuate) cortex, population dynamics separate sensory inputs from the computation of an upcoming choice, even though these are intermixed at the single neuron level (Mante et al. 2013). In OFC, population dynamics reveal transient representations of two choice options that alternate during deliberation, where single units only revealed the chosen option (Rich and Wallis 2016). Although these studies each focused on one region at a time, cross-regional comparisons that use similar techniques could help us better understand the neural mechanisms that support unique functions within and across regions of PFC.

Summary and Open Questions

It is widely believed that there is functional specialization within PFC, so it is natural to expect neurophysiology to provide clarity on the nature of the unique function of a region. To date, however, this clarity has not emerged. Instead, single neurons tend to represent "everything everywhere." Although these data demonstrate the flexibility of prefrontal neurons, they have so far failed to reveal major differences between neurons recorded from different regions. In light of this, we have highlighted two considerations for future studies. First, evolutionary and anatomical data suggest two dominant trends within PFC, each with gradient-like organization that is more prominent than discrete boundaries. Investigating neural coding with respect to this anatomy may be fruitful for understanding local and global organization of function. Second, examining the representations and temporal dynamics that emerge from neural populations may provide unique insights into local function and cross-regional interactions. One recent study has taken steps in both of these directions, by using population representations to assess the flow of information across lateral PFC. Here, information flowed in the caudorostral direction when the location of a valuable object needed to be identified, and in the dorsoventral direction when preparing an eye movement to that location (Tang et al. 2021). Approaches such as these hold promise in revealing how populations represent and communicate information.

In addition to the approaches highlighted here, there are others that could provide important insights. In particular, a defining feature of different prefrontal regions is their unique patterns of connectivity, and approaches aimed at understanding interactions among interconnected regions could reveal key differences. One way to accomplish this is to combine perturbation studies with neural recording. A study that did this found that neurons in both OFC and ACC encode reward values, but only OFC neurons showed altered value coding following amygdala lesion (Rudebeck et al. 2013a). Similarly, studies that quantify functional connectivity between regions can determine how PFC interacts with targets elsewhere. To the extent that these interactions differ between PFC subareas, these approaches may also shed light on functional specialization.

Although we have suggested avenues for future investigation, there are still many open questions. Population approaches are increasingly popular in neurophysiology, yet it remains to be determined whether they will ultimately provide unique insights into functional specialization. To this end, we need to know which anatomical or functional properties define a population. This may be particularly challenging to address if functions are graded, meaning discrete boundaries do not apply. Populations in nonhuman primate recordings are often samples of opportunity, defined by the access the researcher achieved and limited by current recording methods. However, if high-density recordings are collected along the entire anterior-posterior length of the principal sulcus, should they be analyzed as one population or many, and if the latter, where should divisions be drawn? The rapid advance of technology, in terms of the scale and type of recordings we can collect, presents new opportunities to address these questions. In addition, population approaches are usually agnostic to neuron type, connectivity, or laminar location, none of which are typically known when neurons are recorded from nonhuman primates. Yet methods for identifying subtypes of neurons or their projections or recording across cortical laminae are becoming more prevalent, which means that we should soon be able to evaluate some of these questions rigorously. Taken together, although the neurophysiological distinctions among prefrontal regions are not obvious and neural encoding appears superficially similar, there is reason to be optimistic that pursuing in-depth understanding of anatomical organization and neural coding may help parse the neurophysiological mechanisms that distinguish the fundamental functions of different regions of prefrontal cortex.

6

The Position of the Prefrontal Cortex in the Cortical Hierarchy

John D. Murray and Christos Constantinidis

Abstract

The prefrontal cortex (PFC) exerts control on the flow of sensory information in cortical circuits, integrates current stimulus streams with stored memories, and plans motor action. Prefrontal neurons exhibit quantitatively distinct firing properties relative to its afferent inputs. These can be traced to unique anatomical morphology, neurotransmitter receptor composition, and relative distribution of different interneuron types. This evidence suggests a position of PFC on the top of the cortical hierarchy that processes sensory information and controls behavior. A functional specialization is also present within the PFC, as it comprises multiple areas that are hierarchically organized. Other brain structures exert influence on PFC activity critical for the control of behavior, including the thalamus and neuromodulator systems. In that sense, PFC is a critical node of the broader circuit that instantiates intelligent behavior.

Introduction

Our understanding of cortical function has been shaped by the hierarchical processing of sensory systems. The visual cortex provides a prototype of this organization (Felleman and Van Essen 1991): neurons in the primary visual cortex with small receptive fields respond to elementary properties of the visual scene. Inputs from multiple V1 neurons are integrated into higher cortical areas to extract progressively more complex properties of images, over larger parts of the visual field, ultimately allowing objects to be segmented, identified, and categorized. Other sensory systems (auditory, somatosensory) perform similar transformations along their own hierarchies. The output of all sensory systems is ultimately propagated to the prefrontal cortex (PFC) and in this sense the PFC sits atop the processing hierarchies. PFC is also connected with a number of subcortical structures, most importantly the thalamus (via the medio-dorsal nucleus) and the basal ganglia (Giguere and Goldman-Rakic 1988; Middleton and Strick 2002). Cortico-thalamic and cortico-striatal loops are essential for the maintenance of information in working memory, one of the critical functions of PFC (Jaffe and Constantinidis 2021). In turn, PFC broadcasts top-down signals to the rest of the brain influencing sensory processing, integrating current events with stored memories, and prompting motor action (Badre and D'Esposito 2009), commonly referred to as executive function.

The Baddeley and Hitch model of working memory, which has been tremendously influential, encompasses a central executive and three subsidiary systems: the phonological loop, the visuospatial sketchpad, and episodic buffer (Baddeley 2012). Although introduced as a conceptual model, parallels between the function of the PFC and the central executive, the module responsible for the control and regulation of the other components and the ability to switch between tasks, have been highlighted. In this view, PFC sits atop of the cortical hierarchy, whereas the subsidiary systems maintain the contents of working memory at the sensory cortices (D'Esposito and Postle 2015). This division of function, however, is tenuous, as strong evidence exists for sensory information being maintained within the PFC by the same neurons that implement top-down control, thus supporting the idea that PFC is the anatomical seat of both executive and subsidiary systems of working memory (Riley and Constantinidis 2016). The concept of "mixed selectivity," which has been popularized recently, provides a vivid illustration of how deeply maintenance and executive functions are entwined: individual neurons exhibit selectivity for stimuli that differ depending on what cognitive task a subject executes (Rigotti et al. 2013). At the population level, the "representational geometry" of stimuli also changes with task demands (Bernardi et al. 2020; Cueva et al. 2020; Minxha et al. 2020; Okazawa et al. 2021). This finding also suggests that neurons within the PFC instantiate executive function by flexibly altering the type of information they represent.

In this chapter, we consider experimental evidence of the PFC performing integrative functions, the neural substrates that allow the PFC to play such a role, and the organization of the PFC itself. We will emphasize experimental evidence primarily from nonhuman primate models, as these have allowed the most detailed experimental data pertinent to this question. However, when available, parallels with human neuroanatomy and imaging will be integrated.

Prefrontal Specializations

The position of a brain area in the cortical hierarchy can be assessed based on objective criteria related to anatomical circuits and properties of neuronal firing. Prefrontal neurons do not represent additional sensory attributes relative to those already represented at the top levels of the sensory pathways; instead, they exhibit distinct properties during the maintenance of information in working memory and modulation of neuronal activity during execution of different tasks. The generation of persistent activity is thus a critical property of PFC, which differs at least quantitively from areas connected to it (Leavitt et al. 2017), though some controversy exists around this point (Christophel et al. 2017). Understanding which underlying specializations produce these unique prefrontal properties will be instructive as to the position of the PFC in the cortical hierarchy. Here, we review elements of cortical morphology and circuitry that differ systematically in the PFC and other areas, and specifically the anatomical morphology of pyramidal neurons, myelination of axonal fibers traversing the cortex, types of interneurons, and receptor composition for different neurotransmitters and neuromodulators.

The classical view of pyramidal neurons has been that they are essentially uniform across the cortex. This idea has been challenged by experimental findings that demonstrate a systematic difference across the cortical hierarchy, with the prefrontal pyramidal neurons exhibiting the most extensive dendritic trees and the largest number of spines among cortical neurons (Elston 2000, 2003). Functional correlates of this anatomical specialization are also reflected in the patterns of neuronal discharges at different areas. Prefrontal neurons receive a greater proportion of distal synaptic inputs compared to the neurons at other brain areas, with a substantial proportion of these inputs originating at distances greater than 1 mm. By contrast, the majority of inputs to posterior parietal neurons appear to originate from neurons at shorter distances, in the order of 0.2–0.5 mm (Hart and Huk 2020; Katsuki et al. 2014). It has long been speculated that prefrontal neurons with similar memory fields are grouped in clusters with reciprocal connections, often visualized in anatomical tracer studies (Goldman-Rakic 1984; Kritzer and Goldman-Rakic 1995; Levitt et al. 1993; Pucak et al. 1996). Modeling studies suggest that more extensive networks of interconnected neurons in the PFC account for the improved stability of prefrontal persistent activity during working memory and its ability to resist distracting stimuli, compared to equivalent neural circuits of other areas (Mejias and Wang 2022).

Independent evidence of prefrontal anatomical specialization has also been provided by anatomical studies of cortical myelin content. The MRI-based T1-weighted/T2-weighted ratio (T1w/T2w) is indicative of the extent of myelin presence within gray matter (Glasser and Van Essen 2011; Huntenburg et al. 2017). The cortical map of T1w/T2w exhibits a large-scale hierarchical gradient, with high values in primary sensory cortex and low values in association areas of the cortex. In the monkey, this was found to correlate with a measure of cortical hierarchy, based on the laminar patterns of feedforward and feedback inter-areal projections (Burt et al. 2018). As a proxy measure of cortical hierarchy applied to the human brain, the T1w/T2w map was found to align with the dominant spatial pattern of transcriptomic variation in human cortex, derived from the Allen Human Brain Atlas, which reflects multiple aspects of cellular specialization across cortex (Burt et al. 2018).

Another specialization that reveals the prefrontal position in the cortical hierarchy is the concentration and composition of NMDA receptors. These are glutamate-gated cation channels, critical for the generation of persistent activity, as they are capable of extending the duration of the postsynaptic depolarization by virtue of their slow decay time constant (Constantinidis and Wang 2004; Wang 2001). Thus, a circuit containing exclusively AMPA receptors, which produce synaptic currents with very fast decay time constants, would require unrealistically high firing rates to sustain neural activity during the delay period of a memory task (Wang 1999). Experimental results further support the role of NMDA receptors in the generation of persistent activity, as the systemic administration of ketamine, a nonspecific NMDA antagonist, seems to decrease the effective connectivity between prefrontal neurons, demonstrated by a decrease in the synchronous spiking between simultaneously recorded neurons (Zick et al. 2018). The area-specific expression of NMDA further underlies its role in facilitating the prevalence of persistent activity in the PFC. For example, GluN2B (the NMDA receptor subunit with the slowest decay time constant) is expressed in a gradient across the primate brain, with highest levels of expression observed in the PFC. A hierarchical cortical gradient of increasing expression of the GRIN2B gene is observed in human transcriptomics (Burt et al. 2018), and a higher ratio of NR2B/NR2A in pyramidal neuron EPSPs has been observed in rat medial frontal cortex compared to primary visual cortex (Wang et al. 2008).

A further physiological signature of prefrontal cortical specialization has to do with the intrinsic timescales of neural activity. At the level of single-neuron recordings, the autocorrelation of spontaneous neuronal firing exhibits a characteristic timescale, which increases across the cortical hierarchy, with faster dynamics in sensory areas and slower dynamics in prefrontal areas (Murray et al. 2014). These timescale differences presumably reflect different properties of the local microcircuit operating regime across areas, with contributions from cellular and synaptic properties (e.g., NR2B/NR2A ratio due to different time constants associated with those NMDA subunits) and differences in net recurrent strength (e.g., stronger recurrence in PFC, as hypothesized to play a role in generation of persistent activity for working memory).

Another line of evidence suggestive of prefrontal specialization is related to inhibitory interneurons. Prefrontal interneurons exhibit persistent activity with higher baseline firing rates and broader tuning than pyramidal neurons (Constantinidis and Goldman-Rakic 2002). Their action thus serves to "sculpt" the spatial and temporal tuning of prefrontal neurons (Constantinidis et al. 2002), without which stimulus-specific persistent activity is much less viable in computational models (Compte et al. 2000). Multiple types of cortical interneurons are hypothesized to form specialized networks for the purpose of facilitating stimulus-specific persistent activity (Wang et al. 2004b). Three broad types account for the vast majority of interneurons in the cortex: those that express (a) parvalbumin (PV), (b) vasoactive intestinal peptide (VIP), which tends to co-localize with calretinin (Gabbott and Bacon 1997), and (c) somatostatin (SST), which tends to co-localize with calbindin. PV interneurons target the cell bodies of pyramidal neurons and, when activated by their preferred stimulus, they would tend to suppress the activation of pyramidal neurons with different spatial tuning than their own and sharpen the tuning function of those with similar tuning (Li et al. 2020b). Without feedback inhibition, recurrent excitation may shift the excitatory/inhibitory balance and bring the network into an unstable, hyperexcited state, which would be deleterious for the maintenance of working memory (Constantinidis and Wang 2004).

VIP/calretinin interneurons are thought to inhibit other types of interneurons, including SST/calbindin ones (Fish et al. 2018; Melchitzky and Lewis 2008; Meskenaite 1997). Furthermore, interneuron-targeting cells are more abundant in association cortices, particularly in the PFC, compared to the sensory cortex whereas soma-targeting PV interneurons exhibit the opposite trend (Defelipe et al. 1999; Elston and Gonzalez-Albo 2003). SST interneurons, on the other hand, are peridendritic-targeting cells and are thought to exhibit high spontaneous firing rates that may tonically inhibit all pyramidal neurons during baseline, prior to any stimulus presentation. SST neurons would lift inhibition on the pyramidal neurons that are excited by a stimulus maintained in working memory, whereas other SST neurons, not recruited by the maintained stimulus, would continue to inhibit nonactivated pyramidal neurons, thus suppressing both background noise and any potential activation by subsequent, distracting stimuli (Wang et al. 2004b).

Anatomical and physiological evidence supports the greater prominence of the disinhibiting circuit in PFC compared to other areas. Calretinin-positive interneurons are more numerous in PFC compared to the medial temporal and medial superior temporal visual cortical areas (Torres-Gomez et al. 2020). Moreover, interneurons with high baseline firing rate and inverted tuning (consistent with the profile of disinhibiting neurons) have also been found to be more numerous in the PFC than in the posterior parietal cortex (Zhou et al. 2012). The importance of inhibitory-to-inhibitory connections has been confirmed by neural network modeling studies (Kim and Sejnowski 2021). Such connections emerge in the network as training of synaptic weights progresses, and they play a critical role in maintaining working memory activity. Thus, these circuits underlie the prefrontal specialization toward persistent activity.

The final specialization informative about the prefrontal position in the cortical hierarchy has to do with dopamine. Dopamine innervation is concentrated in the frontal lobe (Levitt et al. 1984), and the D1 receptor has, in particular, been implicated in the generation of persistent activity. Iontophoretic application of D1 receptor antagonists, at least in large doses, compromise working memory function and erode persistent activity in the oculomotor delayed response task (Sawaguchi and Goldman-Rakic 1994; Williams and Goldman-Rakic 1995). In contrast, D1 agonists increase activity for preferred stimuli and suppress nonpreferred responses (Ott et al. 2014; Vijayraghavan et al. 2007). However, the effects of dopamine receptors are complex and depend on dosage (Vijayraghavan et al. 2007; Williams and Goldman-Rakic 1995), with differential effects on pyramidal neurons and interneurons (Jacob et al. 2016). D2/D3 antagonists also suppress persistent activity, though their action primarily modulates motor responses of prefrontal neurons (Wang et al. 2004a). Computational and experimental studies suggest that the overall effect of dopamine is to improve the signal-to-noise ratio of persistent activity (Chen et al. 2004; Durstewitz et al. 2000; Seamans et al. 2001; Yang and Seamans 1996). Thus, dopamine innervation in the frontal lobe endows PFC with properties that are distinct from its afferent pathways.

Prefrontal Intrinsic Organization

Thus far we have referred to the PFC as a single brain region; however, embedded within this collective term is considerable heterogeneity of structure and functional specialization. The PFC can be subdivided into a medial, a lateral, and an orbital aspect, each consisting of several cytoarchitectonic areas. By itself, the lateral PFC comprises at least 12 functionally distinct areas, defined by unique cytoarchitectonic patterns and interconnected with a different set of brain areas (Petrides 2005). Many of the aspects of prefrontal specialization, delineating differences relative to other non-prefrontal areas in the cortical hierarchy, are also present in different areas *within* PFC: feedforward/feedback projection patterns, intracortical myelination variation, transcriptomic and cell-type gradients. An anterior-posterior hierarchical specialization has been suggested within the lateral PFC based on anatomical and imaging studies, with more abstract operations localized anteriorly in the prefrontal surface (Baird et al. 2013; Cole et al. 2015a; Koechlin et al. 2003; Ramnani and Owen 2004; Strange et al. 2001). A newer, more nuanced view suggests that the organization of function along the rostro-caudal axis is not an absolute gradient serving a unitary model of frontal control function; instead, separate frontal networks interact within an overall hierarchical structure to support task demands (Badre and Nee 2018). This idea is further developed by Badre (this volume). Neurophysiological evidence from animal models supports the idea of a rostro-caudal hierarchy in two respects: more neurons are selective for stimulus properties in posterior PFC, and more abstract qualities (e.g., task variables and rules) in anterior areas (Figure 6.1). Depending on task demands, plasticity of responses to the same stimuli is also more prominent in anterior areas.

Neuronal selectivity for the spatial location of stimuli was thus found to decrease along the anterior-posterior axis, so that the most highly selective neurons for stimulus properties were located more posteriorly in the PFC (Riley et al. 2017). Conversely, neurons in more anterior areas exhibited little selectivity to stimuli per se but were more likely to represent task variables. Neuronal selectivity for nonspatial stimulus attributes, such as shape and color, was also found to decrease along the anterior-posterior axis. The most highly

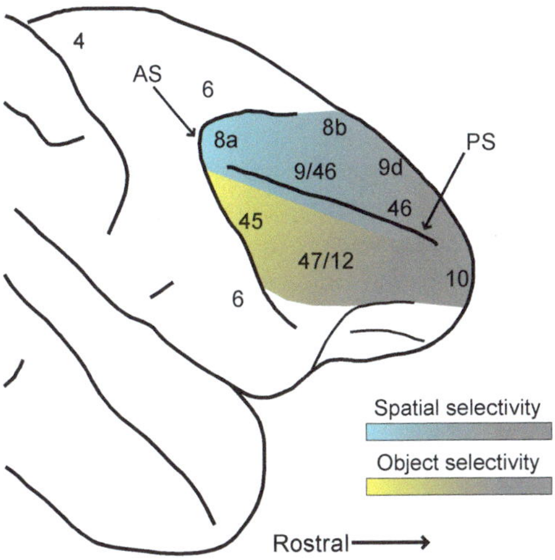

Figure 6.1 Diagram of the macaque monkey brain, with the lateral PFC highlighted. Labels denote anatomical areas: AS: arcuate sulcus; PS: principal sulcus. From Constantinidis and Qi (2018).

selective neurons for stimulus properties were located more posteriorly in the PFC (Riley et al. 2017).

Similar to the anterior-posterior axis of specialization, ventral areas are also more sensitive to task variables and cognitive factors rather than stimulus properties, so that robust selectivity to the location of stimuli may emerge as a result of training in task that requires tracking of reward. Ventral PFC has greater sensitivity to the learning of new, rewarded conditions, and this is likely due to the action of dopamine D1R receptors (Puig and Miller 2012).

A hierarchical prefrontal organization with respect to abstract functions implies that areas at the apex of this hierarchy are not activated solely by sensory stimuli, but by certain stimuli only in a specific context. In turn, such a context-depending property implies a greater capacity for plasticity of neural responses according to task demands. Direct evidence of systematic variation of plasticity markers between limbic and eulaminate areas, roughly organized across the anterior-posterior axis of the PFC, has been documented (Garcia-Cabezas et al. 2017). Calcium/calmodulin-dependent protein kinase II (CaMKII), which is essential for plasticity, is more impoverished in area 46d compared to anterior limbic areas. By contrast, makers of cortical stability, including intracortical myelin, perineuronal nets, and parvalbumin, show the reverse pattern. Changes in neuronal morphology, molecular profiles of the synaptic apparatus, and the influence of neuromodulator systems have also been implicated in long-term prefrontal plasticity (Laroche et al. 2000; McEwen and Morrison 2013), and these differ as well from posterior to anterior PFC.

Conclusions and Future Directions

Over the past decade, substantial progress has been made in our understanding of the unique functional properties of the PFC and the neural circuit substrates responsible for this specialization. This work has made it possible to account for the role of PFC, at least in some tasks, in terms of elemental neural circuits. A great deal of progress has been made in uncovering the functional organization of the PFC itself, as well as the hierarchy within its borders.

Nonetheless, some questions and controversies remain. Although we emphasize the hierarchical organization of areas within the PFC, comparative studies that would provide a comparison of their properties are lacking, particularly for the most anterior ones (e.g., area 10). These are critical regions for understanding the function of PFC, and additional experimental data will be valuable to inform our understanding.

The concept of executive control of other circuits is still poorly understood, though several chapters in this volume provide a good overview of the state of knowledge. To what extent functions of the PFC, such as working memory maintenance, are centralized in the PFC or more distributed is also a matter of controversy. Finally, how generalizable the conclusions drawn from animal models are on the human brain is still relatively unexplored. Attending to these questions over the upcoming decade will allow us to determine definitively not only the position of the PFC in the cortical hierarchy but also the mechanisms of executive cognitive functions.

Acknowledgments

This work was supported by NIH grant award R01 EY017077 (CC). We wish to thank Mark D'Esposito for helpful comments on an earlier version of this manuscript.

7

What Is the Nature of the Hierarchical Organization of Lateral Prefrontal Cortex?

David Badre

Abstract

An influential view of lateral prefrontal cortex (lPFC) is that it is organized hierarchically to support cognitive control function. Specifically, regions more rostrally are hypothesized to engage in more abstract control processing than those caudally. Further, rostral regions are proposed to asymmetrically influence those caudal to them. This chapter provides an updated background on this view of lPFC organization and reviews evidence for two theoretical commitments of lPFC hierarchy: (a) functional differentiation along the rostro-caudal dimension of the lPFC and (b) super-to-subordinate hierarchical interactions within the lPFC. It will be seen that the standard view has undergone important revisions. In particular, what makes control more or less abstract along the rostro-caudal axis has been defined and redefined. The original assumption of a rostro-caudal gradient has been revised in favor of a hierarchy of interacting networks, which include association cortex outside of lPFC and subcortical structures. In addition, the apex of the hierarchy has shifted from rostro-lateral prefrontal cortex at the most anterior extent of the PFC to the mid-dorsolateral prefrontal cortex (mid-dlPFC) that lies just caudal to it. This discussion speaks directly to the topic of the functional organization of the PFC.

Introduction

The lateral prefrontal cortex (lPFC) has an established association with higher cognitive function, including cognitive control or executive function (Badre 2020; Devinsky and D'Esposito 2004; Duncan 2013; Miller and Cohen 2001; Stuss and Benson 1987). Broadly speaking, these functions control and organize our behavior, flexibly mapping sensory input to action outputs based on internal representations of goals, plans, and our behavioral context. They allow us to perform a wide range of different behaviors in the open-ended complexity of our everyday world.

However, beyond this broad functional association, there is little agreement regarding the functional organization of lPFC. We do not understand whether or how the various systems and networks that are encompassed within lPFC are distinguished from one another in terms of their computational nature or how they interact to support the complex cognitive control functions we attribute to them. Indeed, some theorists have proposed that lPFC is functionally homogenous and without a systematic organization, at least in the portion of lPFC that supports cognitive control (Assem et al. 2020; Duncan 2010, 2013). The strong version of this perspective proposes that lPFC is part of a multiple demand network that supports performance of demanding tasks in a domain general way, but that no particular part of PFC is devoted to a particular task. Thus, what functional differentiation lPFC might exhibit from task to task is not governed by consistent organizing principles that would generalize across task-independent cognitive demands or other computational-level factors.

In this chapter, I review the state of an alternative, influential class of theory regarding lPFC organization that does not ascribe particular tasks to localized regions of the lPFC, but which does assume functional organizing principles at a computational level that are systematic and generalizable. Specifically, I will consider the proposal that the lPFC is organized as a functional hierarchy along its rostro-caudal axis in the service of control function (Badre and Nee 2018; Christoff et al. 2009; Fuster 2001; Koechlin et al. 2003; Nee 2021; Soltani and Koechlin 2022).

It is important to clarify what is meant by an organizing hierarchy with regard to lPFC, as there are at least two ways hierarchy is used when discussing lPFC. The first is mostly uncontroversial. Most theorists accept the proposal that lPFC holds a hierarchical relationship to the rest of the brain. Theories of cognitive control propose that lPFC broadcasts top-down signals to influence processing in other areas or networks of the brain that support basic cognitive processes like perception and memory (Badre and D'Esposito 2009; Badre and Nee 2018; Cole et al. 2015b; Duncan 2013; Hazy et al. 2006; Miller and Cohen 2001). These control signals modulate ongoing processing in these networks so that they are coordinated toward particular behavioral goals. This is a hierarchical organization of the brain, and theories of lPFC organization are situated within a commitment to this larger architecture.

Nonetheless, the hierarchy we will primarily be concerned with in this chapter—and the one that remains controversial in the cognitive neuroscience literature—describes the intrinsic functional organization of lPFC itself. Specifically, this refers to the proposal that functionally distinct regions, networks, or gradients within the lPFC form a representational and/or processing hierarchy, with higher-order areas controlling and influencing the activity in lower-order areas.

Though specifics differ, this architecture is generally hypothesized to support the control of behavior at multiple levels of abstraction and/or over multiple timescales that array along the rostro-caudal axis of the frontal lobe. In particular, the most caudal areas of frontal cortex—those participating in motor, premotor, and attention networks—are associated with control over specific, concrete movements and externally directed attention. Moving rostrally in lPFC, progressively complex control functions are proposed to be supported, serving goals or rule structures that are more abstract, multiply contingent, internally generated, counterfactual, and/or that prevail over longer timescales.

Though this axis of organization has sometimes been used to describe both dorsal and ventral aspects of lateral frontal lobe organization, debate about the hierarchical organization of lPFC has mostly concerned the networks related to cognitive control and adaptive task performance. These include the frontoparietal control networks (Fedorenko et al. 2013; Gordon et al. 2017; Gratton et al. 2018b; Ji et al. 2019; Power et al. 2011; Yeo et al. 2011) as well as sensorimotor and attention networks. As such, our discussion of hierarchy in lPFC will primarily concern the motor and dorsal premotor areas in caudal frontal cortex, pre-premotor (prePM) and mid-dorsolateral prefrontal cortex (mid-dlPFC around the inferior frontal sulcus) more rostrally, and the lateral frontal pole or rostro-lateral prefrontal cortex (rlPFC) at the most rostral extent.

Over the last two decades, this hierarchical view of lPFC organization has been supported in some ways, and in others, has undergone important revisions. Here, I will provide a brief update and background on the current state of the literature on the hierarchical organization of the lPFC.

This discussion is of direct relevance to the central topic of this Forum on the organization of PFC. Not only does it provide one putative answer to the question of how part of the frontal lobe is organized, it also has implications for the theory of the organization of PFC more generally. First, hierarchical theories are examples of theories of functional organization that do not localize particular task-related functions or executive skills, like task switching, to particular areas of the frontal cortex, per se. Rather, a hierarchical organization defines a processing architecture that describes how general functions like cognitive control emerge from the interactions among regional or network computations. Second, as the lPFC has direct interactions with other parts of the frontal lobe, the organization of lPFC will have implications for the organization of those other areas that interact with it. Thus, the debates, challenges, and discoveries arising from investigation of hierarchy in lPFC are of relevance to understanding frontal lobe organization more generally.

Functional Hierarchy in Rostro-Caudal Lateral PFC

Functional Differentiation Along the Rostro-Caudal Dimension of lPFC

A hierarchy along the rostro-caudal axis of the lPFC takes as its premise that there exists functional differentiation along that axis that can be related to generalizable task demands. A persistent obstacle to finding such evidence is that experiments testing particular task contrasts, while being well-controlled and thus amenable to mechanistic theory, will also feature many idiosyncratic choices, both at the level of the task and study implementation. Any of these choices could drive observed effects in a region like lPFC that is adapted to shape task performance. Thus, inconsistencies in the literature might arise from overgeneralizing these idiosyncratic effects.

Meta-analyses are one way of testing the premise of a rostro-caudal organizing axis in lPFC that can overcome the limitations of particular idiosyncratic tasks (see also de la Vega et al. 2018). Recently, Abdallah et al. (2022) studied meta-analytic connectivity in fMRI activation across 14,371 studies from the NeuroSynth database, spanning a wide range of tasks and contrasts. Meta-analytic connectivity refers to a pattern of co-activation across study contrasts. More concretely, for any point in lPFC, one can compute the probabilities that (a) activity is reported in other brain regions given activity in that lPFC location and (b) activity is reported in other brain regions when there is no activity in that lPFC location. Meta-analytic connectivity, then, is the odds ratio computed from these two probabilities. At the scale of thousands of task contrasts, it provides an estimate of systematic co-activity across diverse differences in tasks and study implementations.

Importantly, Abdallah et al. (2022), tested whether the high-dimensional variance in these meta-analytic connectivity values across lPFC locations could be reduced to lower-dimensional components. They found that a rostro-caudal dimension of organization emerged from this analysis to account for the most variance in meta-analytic connectivity of lPFC regions (around 40%), followed next by the dorsoventral dimension of organization (around 20%) (Figure 7.1a). To clarify, these dimensions of organization are in reference to variance in meta-analytic connectivity. They do not speak to organization of function along these gradients. In other words, across many different tasks and studies, the variance in co-activation between regions of lPFC and other regions in the brain is systematically related to its position along a rostro-caudal axis. This observation was robust across several controls and ways of doing the analysis; it was also evident at the single subject level using an independent dataset in which people were scanned doing many different tasks.

Two further insights were evident from this study. First, this rostro-caudal lPFC gradient was situated within a broader hierarchy of brain networks defined by their distance from unimodal sensorimotor regions (Huntenburg et al. 2018). Breaking the lPFC into quintiles from caudal to rostral, the meta-analytic connectivity of the most caudal lPFC quintile was found to overlap with visual networks and external attention networks more so than the rostral portions of the gradient. The connectivity of the most rostral lPFC quintile overlapped with cognitive control and default mode networks, with a gradual transition from visual and attention networks to cognitive control and default networks across the quintiles. Notably, this pattern of network overlap is also consistent with observations by Choi et al. (2018), who directly compared activations from a task of hierarchical control (Badre and D'Esposito 2007) to the Yeo et al. (2011) parcellation.

Importantly, this network connectivity pattern situates the local lPFC rostro-caudal gradient within a global hierarchical organizing principle in the brain. Specifically, Margulies et al. (2016) analyzed whole brain connectivity data in both human and macaque monkey and found a principal gradient spanning networks in terms of distance from sensorimotor regions, with cognitive control and default mode networks at the furthest extreme. Directly comparing their local gradient of meta-analytic connectivity in lPFC with that macroscale hierarchy of networks identified by Margulies et al. (2016) in resting state connectivity, Abdallah et al. found that the network profile of the local lPFC gradient of meta-analytic connectivity correlated with this macroscale hierarchical organization.

The second insight to emerge from this meta-analysis is that the gradient is differentially related to distinct task demands. Abdallah et al. used a topics analysis (Poldrack et al. 2012) of terms, describing studies in the NeuroSynth database to cluster experiments into 38 topics, ranging from attention and decision making to lexical semantics and memory encoding. They then tested how these topics were associated with activation along the gradient.

This analysis found a systematic pattern of associations across tasks that are consistent with an abstraction gradient (Figure 7.1b). Roughly, the most caudal zone was primarily associated with tasks involving basic sensorimotor functions such as eye movements and attention. The middle caudal zone was most associated with more complex controlled tasks, such as those involving response selection or task switching. The middle rostral zone was chiefly associated with mentalizing and decision making. The furthest rostral zone was most associated with topics related to subjective experience, empathy, declarative memory, and feedback/error processing. Notably, this set of task associations is roughly in line with the gradient of function inferred from the network connectivity profile, ranging from more concrete, external, sensorimotor function to more internal, abstract cognitive function.

Thus, while the lPFC is clearly associated with multiple demands, it is also not uniformly or arbitrarily associated with all tasks without an organization that can be described at a cognitive or computational level of analysis. Rather, a rostro-caudal abstraction gradient is supported across thousands of contrast measurements from fMRI studies of the human brain. Another recent meta-analyses (de la Vega et al. 2018), while coarser in resolution, reached similar conclusions to Abdallah et al. regarding functional differences in lPFC. Thus, whether hierarchical in nature or not, a rostro-caudal functional gradient in lPFC appears evident and robust.

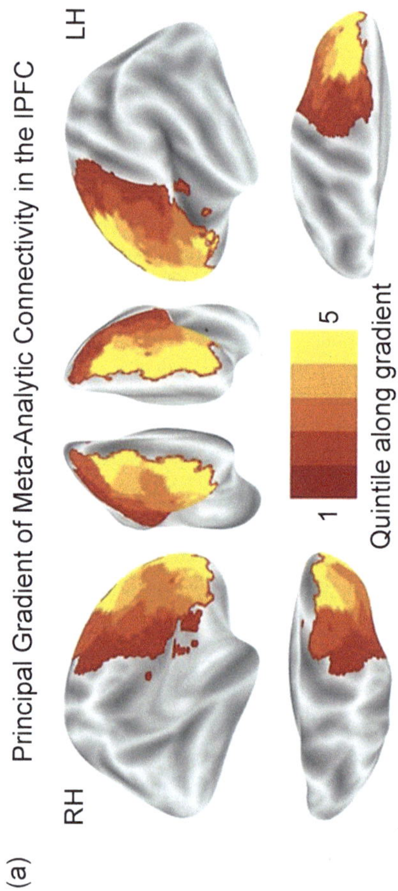
(a)
Principal Gradient of Meta-Analytic Connectivity in the lPFC
RH
LH
1
5
Quintile along gradient

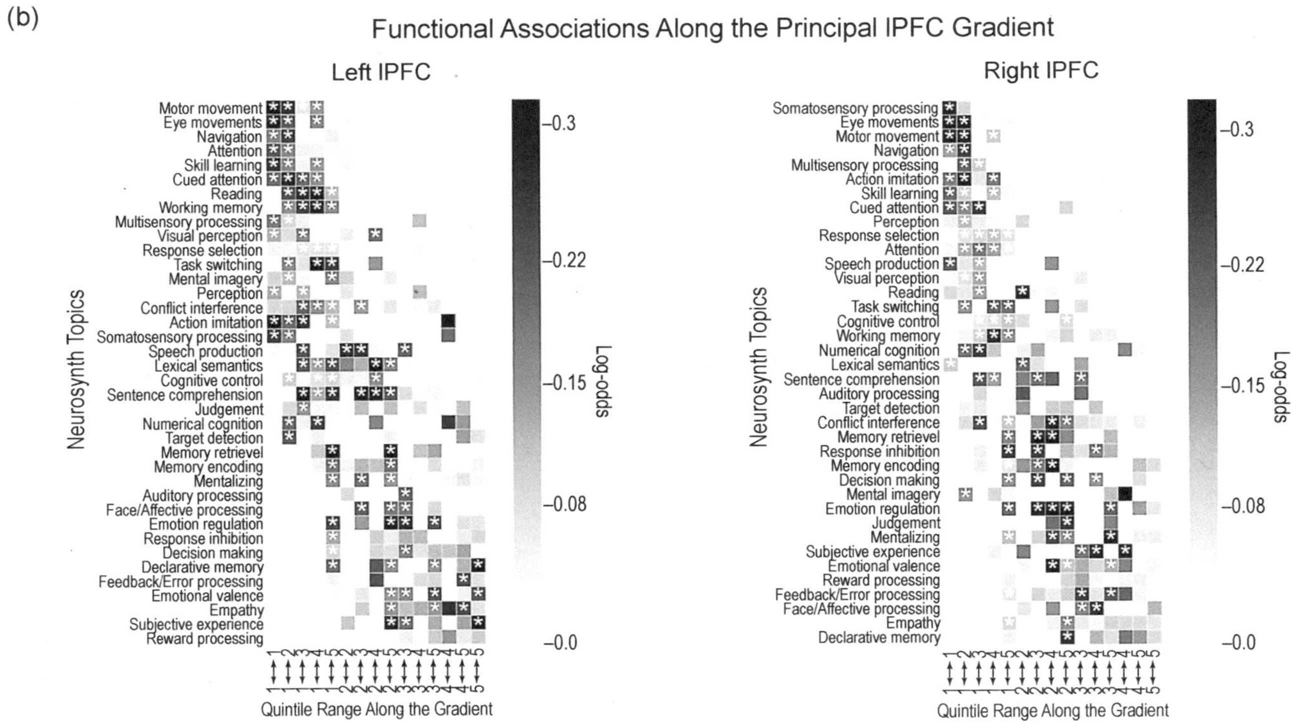

Figure 7.1 Meta-analysis results from Abdallah et al. (2022). (a) Surfaces showing the rostro-caudal gradient in meta-analytic connectivity. (b) Results from the topics analysis associating experiment topics with quintiles of the meta-analytic connectivity gradient going from 1 caudally to 5 rostrally. Box shading shows log-odds and an asterisk indicates statistical significance. Adapted from Abdallah et al. (2022, Figures 1 and 4).

What Defines the Functional Hierarchy in lPFC?

Meta-analysis provides evidence that lPFC has a rostro-caudal organization based on cognitive demands that generalize across tasks. Nonetheless, the specifics of these demands, as can be described at a computational or mechanistic level, remain vague when using this large-scale approach. Even in the Abdallah et al. analysis, for example, it is not clear why particular topics, such as "cognitive control," are distributed more caudally than others, such as "response inhibition." More specific task analyses and computational accounts are needed for this level of interpretation.

Carefully controlled laboratory experiments have been useful for testing hypotheses at this more specific computational and functional level. However, several proposals and tests (reviewed in Badre 2008; Badre and Nee 2018) have failed to produce a consistent interpretation. Here, we will consider three that have received recent attention: policy abstraction, internal versus external attention, and present versus future focus.

Policy Abstraction

The earliest establishing fMRI experiments that tested rostro-caudal differences in lPFC manipulated rule complexity in terms of the number of contingencies needed to make a response (Badre and D'Esposito 2007; Koechlin et al. 2003; Yao and Hsieh 2022). These studies found that as choice conflict was manipulated among responses, stimulus-to-response mappings, context-to-stimulus-to-response mappings, and episode-to-context-to-stimulus-to-response mappings, the focus of activation related to that choice demand moved rostrally from premotor cortex to prePM to mid-dlPFC to rlPFC.

One interpretation of this observation is that it reflects a change in policy abstraction needed for hierarchical control. Cognitive control generally refers to cases where a higher-order context is needed in order to distinguish which response to make (Badre and Nee 2018; Botvinick 2008). Hierarchical control refers to cases in which these contextual signals are themselves conditioned on superordinate contexts, which can be conceptualized as a hierarchical or branching rule tree.

The concept of policy abstraction is closely tied to this definition of hierarchical control. Policy, as a concept, comes from machine learning and reinforcement learning, where it refers to a relationship between a particular context, an action, and the expected outcome this will produce (Botvinick et al. 2009). Policy abstraction refers to learning abstractions over these relationships, wherein classes of lower-order policy are captured within a particular higher-order policy. For example, a sequence of specific policies that enact individual movements of an artificial agent around a grid could be abstracted over in terms of a policy that defines the start and end position. Abstraction of not only states and actions, but also their combination, is increasingly influential in AI and robotics, where these agents face similar dilemmas in planning and decision making in complex settings as humans do (Konidaris 2019).

In the context of hierarchical control, policy takes a similar definition and its abstraction is defined in terms of contingency. Thus, a simple S-R mapping, such as pressing a key when a particular color is presented in order to succeed at a trial, represents a concrete form of policy; it determines what action to take based on the stimulus context that yields a desired outcome. Policy can be more abstract by adding contingencies. Thus, task switching involves grouping sets of S-R mappings to follow based on a task cue (e.g., following all the color response rules when a stimulus appears in the upper half of the screen). In this sense, task-set-level policy is more abstract because the task context does not specify what specific response to make (as occurs in simple S-R mapping), but rather specifies how to interpret a set of stimulus contexts with regard to selecting a response. Further contingencies could be added, for example, specifying how to interpret screen position with regard to the task to perform. As contingencies are added, the policy becomes higher order.

People's performance suggests that they break complex tasks down and represent them efficiently as hierarchically structured, abstract policies. In task-switching manipulations, switches of higher-order policy, as defined above, show larger switch costs than lower-order switches, consistent with traversing different choice points on a policy tree (Kleinsorge and Heuer 1999; Rac-Lubashevsky and Frank 2021; Ranti et al. 2015). An advantage to hierarchies is that decisions can be made at each level separately, and the status of lower-level decisions need not affect upper-level ones. Consistent with this, people will solve hierarchical rules in parallel, in line with decisions being made partly independently at each level (Rac-Lubashevsky and Frank 2021; Ranti et al. 2015). Hierarchies also permit generalization, such as over lower-order S-R mappings and transfer to new tasks with the same contingency structure. Indeed, people learn hierarchical rules more rapidly through reinforcement than rule sets that are not structured hierarchically (Badre and Frank 2012; Badre et al. 2010; Eichenbaum et al. 2020; Frank and Badre 2015). Further, when possible, they impose a hierarchical structure on tasks which builds a more abstract structure that can support transfer to new tasks (Collins and Frank 2013; Collins et al. 2014). Indeed, people structure tasks hierarchically, even when doing so conveys no immediate behavioral advantage and potentially comes with a cost in mental effort (Sayali et al. 2023).

It is evident, then, that people control their behavior during complex tasks in hierarchically structured ways based on more abstract policy. Thus, one account of abstraction differences along the rostro-caudal axis is that they reflect the neural processing needed to behave according to increasing levels of policy abstraction. The results from the fMRI experiments described above are consistent with this interpretation, in that manipulating selection demands at higher levels of policy abstraction resulted in more rostral lPFC activation. More recent experiments using different tasks (Nee 2021; Nee and D'Esposito

2016), but which can again be analyzed in terms of progressive increases in policy abstraction (Sayali et al. 2023), show a similar rostro-caudal pattern of activation differences. Further, complementary support for the hierarchical assumptions made in these experiments has come from studies in patients with lesions in lPFC (Azuar et al. 2014; Badre et al. 2009) and transcranial magnetic stimulation (Nee and D'Esposito 2017), and it supports differences in the necessity of rostral versus caudal PFC for following complex versus simpler rules, respectively.

Observations from other experiments, however, are difficult to explain within the simplest policy-abstraction account. Hierarchical manipulations of the 12AX-CPT task—in which a higher-order number context (1 or 2) determines which context (A or B) determines whether lower-order items (Xs or Ys) are targets—have produced inconsistent results with regard to the rostro-caudal gradient, despite the clear policy manipulation. One fMRI study using a blocked design failed to locate differences along this axis when comparing these different levels of contingency (Reynolds et al. 2012). However, other designs using this task did locate rostro-caudal differences in fMRI activity associated with higher- and lower-order contexts, though in different locations along the rostro-caudal axis than would be predicted by prior studies (Nee and Brown 2012, 2013).

It is conceivable that these differences might relate to specific aspects of the experimental protocols, such as the serial versus simultaneous nature of presentation (see Badre and Nee 2018). Nonetheless, if that is the case, it also indicates that factors beyond policy abstraction are relevant to the lPFC organization. Similarly, a recent study of the multiple demand network found that while there was consistent functional differentiation rostro-caudally in this network, it was due to factors like reward and time pressure that were not clearly attributable to a policy-abstraction hierarchy (Crittenden and Duncan 2014; Shashidhara et al. 2019). Thus, while policy abstraction may be important for the rostro-caudal organization of lPFC, it is evidently not the only relevant factor.

The function of rlPFC presents another problem for a single policy-abstraction gradient. While there is ample evidence from neurophysiological recording of abstract rule processing in mid-dlPFC (e.g., Mansouri et al. 2020; Wallis and Miller 2003a; Wallis et al. 2001), the few recording studies of the anterior frontal pole in the monkey have not found more abstract rule coding or, indeed, rule coding at all (Tsujimoto et al. 2010). Neuroimaging studies of hierarchical sequence control in humans have found a strong association of rlPFC with superordinate or sequence-level effects that could be interpreted as higher-order policy (Desrochers et al. 2015a, b, 2019). The specifics, though, do not fit with a simple policy gradient. The activity in rlPFC ramped toward the end of sequences and brain stimulation with TMS also had disruptive effects at the end of the sequence. How this ramping dynamic relates to simple policy-abstraction demands is unclear. Further, in terms of its functional relationship to other regions of PFC, rlPFC is unlikely to be the apex of the control hierarchy and so should not necessarily be expected to represent the highest levels of policy abstraction.

Thus, while our ability to behave according to higher-order policy is closely related to lPFC and its rostro-caudal gradient, it is unlikely that policy abstraction ranks the entire functional gradient in lPFC. Other ideas have continued to be pursued in recent years regarding the organization of lPFC along its rostro-caudal axis that relate to policy abstraction, but which also differ in important ways.

External- versus Internal-Oriented Cognition

Returning to the ordering of tasks in Figure 7.1b, another hypothesized organizing dimension of the lateral gradient can be recognized. Specifically, one can informally describe the tasks as shifting from those primarily requiring externally oriented processing, such as tasks of multisensory perception and cued attention, to those requiring an internal orientation, such as tasks of declarative memory or emotion regulation.

In this view, going from sensorimotor networks to internal processing of the default mode network, the gradient in lPFC is not different qualitatively from the larger hierarchical network organization of the brain (Margulies et al. 2016). According to this hypothesis, mid-dlPFC has a multi-demand and integrative nature due to its interposition between areas or networks involved primarily in internal (rlPFC and default mode network) versus externally oriented (sensorimotor and dorsal/ventral attention networks) control. The mid-dlPFC may, therefore, play a crucial role in linking the internal control of thought with its externalization in behavior.

As with policy abstraction, the first neuroimaging experiments that provided evidence of a rostro-caudal organization of lPFC could be interpreted in terms of a progression from internally to externally oriented control (Badre and D'Esposito 2007; Koechlin et al. 2003). In particular, the highest levels of both tasks—those associated with the most rostral lPFC activation in each study, though differing in their specific location of activation—placed a demand on "episodic control," which meant that the particular task episode acted as a context for selecting the appropriate set of S-R mappings. There was no external cue for this episode, so it had to be tracked internally.

A similar observation was made in a series of studies by Nee and D'Esposito (2016, 2017) using the Comprehensive Control Task (CCT), which manipulated simple S-R selection (sensorimotor control), contextual S-R selection (contextual control), and temporally extended selection based on items held in memory (temporal control). These manipulations and their network associations fit with a shift from external to internal processing and have consistently associated these three demands with progressively rostral areas of lPFC. Hence, there is some empirical support for defining the rostro-caudal organization of lPFC in terms of an external to internal processing dimension.

Nevertheless, there are also limitations. First, at a more mechanistic level, outside of proximity to sensory versus default networks, it is not fully clear what defines internal versus external processing. For example, most models of cognitive control assume that information must be held in working memory by PFC for it to affect current processing, even of contextual information available in the external world. This requirement is important as not all information we sense should serve as a control signal. So, deciding to represent something in working memory constitutes a decision about whether to allow it to influence behavior. As such, this information must be internally represented, even for simple sensorimotor control. Indeed, studies of selection from within working memory, based on an internally maintained context, have located activation in prePM, caudal to mid-dlPFC (Chatham et al. 2014). This type of selection from within working memory, also termed output gating in the context of hierarchical control, is an internally oriented cognitive control demand, yet, it is not selectively associated with rostral PFC.

Second, the direct experimental support distinguishing episodic control from other types of control demands is, at present, weak. Pitts and Nee (2022) modified the CCT in a way that manipulated episodic control demands factorially, relative to other demands. The contrast of low versus high episodic control resulted in a pattern of overlapping activation, which for contextual control was based on a stimulus cue and caudal to the original "temporal control" manipulation. Thus, at present, conceptualizing the rostro-caudal hierarchy along a strict gradient of external to internal processing dimension has mixed theoretical and empirical support.

Present- versus Future-Oriented Control

A related alternative to the internal-external gradient distinguishes between present- versus future-oriented control (Badre and Nee 2018; Nee 2021; Soltani and Koechlin 2022). Most experimental manipulations of cognitive control are in the moment. Whether one is naming the ink color of a word or switching between tasks, one is selecting a particular response or task set to perform right now as well as in expectation of whatever outcomes follow from that behavior. However, we are also able to monitor cues and encode information that is relevant to our future or hypothetical behavior, rather than what we are doing right now. Pitt and Nee (2022) have pointed out that the external versus internal focus of control is often confounded in experiments with this temporal focus on present versus future.

The hypothesis that the rostro-caudal axis of lPFC is differentiated by a present to future orientation is broadly consistent with several observations from the neuroimaging literature regarding, in particular, rlPFC. FMRI studies have associated with rlPFC cognitive branching (Koechlin et al. 1999; Koechlin and Hyafil 2007), with monitored conditions driving exploration of future alternative behaviors over exploitation of current behaviors that are leading to present rewards (Badre et al. 2012; Boorman et al. 2009; Culbreth et al. 2023; Daw et al. 2006), and with counterfactual task set control, as in what task was performed versus what could have been performed (Donoso et al. 2014a; Soltani and Koechlin 2022). In common, these tasks require some monitoring of which alternative paths to take. This form of monitoring may not affect behavior on the present trial, but to the degree that one successfully tracks future or possible rewards, this tracked information could affect behavior on future trials. Consistent with the present/future characterization of lPFC organization, Nee (2021) found that activation in more caudal areas of lPFC correlated with response times on current trials of the CCT, whereas activation in the more rostral areas was associated with response times on future trials in these experiments. In sum, there is some convergence across different tasks and experiments that a shift of focus from present to future orientation might characterize processing along rostro-caudal lPFC.

The distinction between future- and present-oriented control may also fit with the connectivity of these regions of lPFC and at a computational level, a distinction between representations suitable for action planning versus action execution. Vaidya and Badre (2022) recently observed that independent lines of research have separately implicated medial temporal lobe and orbitomedial PFC (MTL-OMPFC) versus frontoparietal networks in the same function, the representation of abstract task set information. Why do distinct networks represent the same abstract information? Vaidya and Badre (2022) proposed that this might reflect differences in the format or use of these representations rather than their content (Lovett and Anderson 2005), with the MTL-OMPFC network representing abstract tasks to plan what to do versus frontoparietal networks that format abstract task information for efficient task performance (Figure 7.2). Specifically, MTL-OMPFC representations may organize task information in a map-like format that allows multiple relationships among task states to be represented, and new relations to be inferred. By contrast, frontoparietal networks may represent task information as productions that specify what action to take in a given a state. Productions are unidirectional, and so while not as useful for planning, they can yield efficient and controlled action selection given a set of states.

Several convergent lines of research support this hypothesis. Results across species implicate a network involving ventromedial and orbitofrontal cortex along with the medial temporal lobe (MTL-OMPFC network) with the representation of abstract task information, such as latent contextual states, and with drawing inferences based on these representations (e.g., Bradfield et al. 2015; Chan et al. 2016; Coutureau et al. 2002; Iordanova et al. 2007; Jones et al. 2012; Schuck et al. 2016; Wilson et al. 2014; Zhou et al. 2021a, b). These results have specified that this MTL-OMPFC network encodes a cognitive map of task-space as a way of efficiently representing structured task relationships useful for planning and inference.

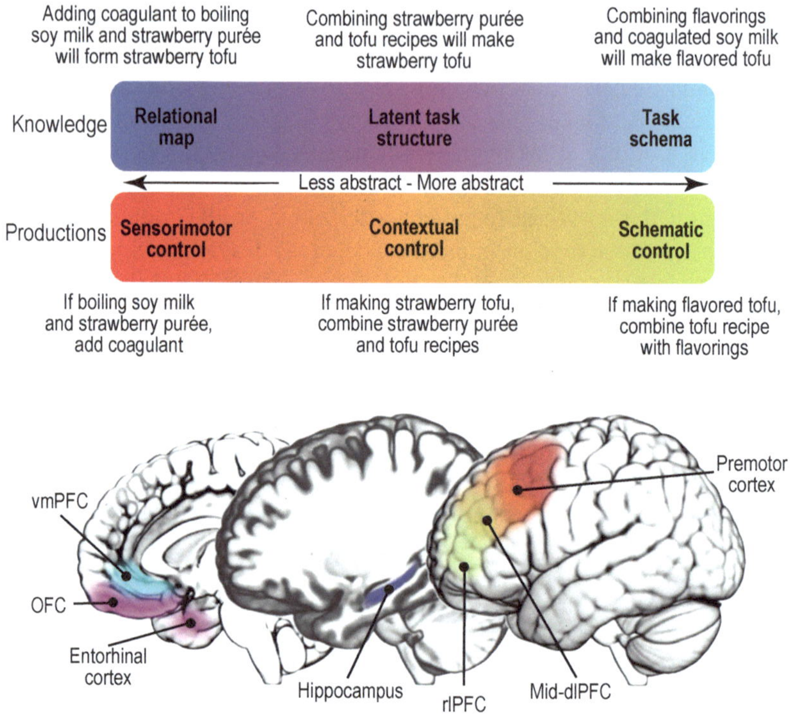

Figure 7.2 Schematic from Vaidya and Badre (2022) summarizing the hypothesized relationship between the lPFC control network and the MTL-OMPFC network in representing abstract task knowledge. It shows two abstraction gradients of organization task planning and inference (blue to cyan) and task production (red to yellow). The gradient in lPFC follows the distinction from sensory to cognitive to schematic control proposed by Badre and Nee (2018). The examples in black are intended to illustrate the shared content at each level, but differences in format. Reproduced with permission from Vaidya and Badre (2022).

On the other hand, there is evidence that the lateral frontoparietal network represents abstract task sets and can leverage inferred relationships in order to behave. For example, we recently observed that the frontoparietal network, particularly the mid- and anterior portions, represented inferred abstract task relationships while performing an acquired equivalence task. (Vaidya et al. 2021). Specifically, across learning phases, participants learned a latent, abstract task set that they could use to generalize behavior to new cases they had not previously encountered through inference. People were able to do this generalization and perform the task. However, possibly because learning occurred during performance of the task rather than the period of inference, decoding results found the latent task set information to be maintained in mid and anterior lPFC, with limited activity in MTL-OMPFC.

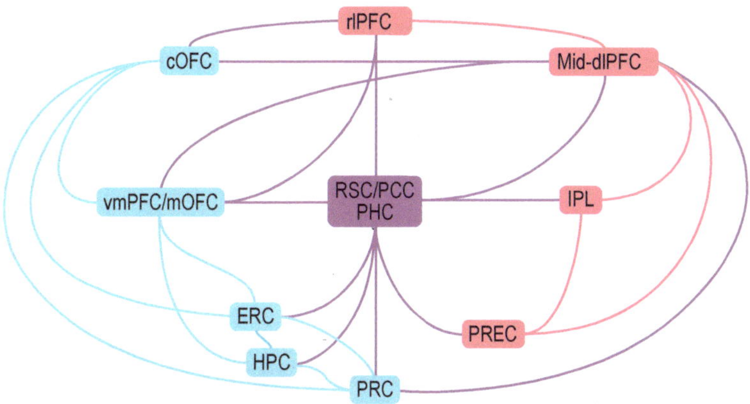

Figure 7.3 Schematic of connections between regions along the rostro-caudal axis of lPFC and those in the MTL-OMPFC network, based on a review of anatomical studies in the nonhuman primate. Light red shows regions and connections in the frontoparietal network. Light blue shows regions and connections in the MTL-OMPFC network. Purple marks the connections between the networks. Notably, rlPFC and mid-dlPFC, in the rostral lPFC, hold direct connections with regions in the MTL-OMPFC network. Reproduced with permission from Vaidya and Badre (2022).

Importantly, to connect our plans with their realization in our behavior, there must be an interface between these systems for planning and for execution of tasks. Figure 7.3 summarizes connections between these networks, based on a review of anatomical studies in nonhuman primates (Vaidya and Badre 2022). In the lPFC, it is notable that rlPFC and mid-dlPFC in the lPFC share connections with orbitofrontal cortex and the ventromedial PFC, perhaps positioning them as the interface between the MTL-OMPFC network and the frontoparietal control network. These ideas elaborate what Badre and Nee (2018) distinguished as "schematic control" (as distinct from sensory and cognitive control; see Figure 7.2) in their review of the literature around hierarchical control. Nevertheless, direct testing of these ideas in experiments designed to distinguish planning from execution is needed.

Hierarchical Interactions Within the Rostro-Caudal Organization of lPFC

Our discussion to this point has concerned the functional attributes that might characterize processing or representations along the rostro-caudal gradient of lPFC. An important implication of hierarchy in lPFC is, however, that it is not merely a description of function going from more concrete sensorimotor control to more abstract cognitive and then schematic control, but that it reflects a processing architecture. Specifically, a hierarchy of processing within the lPFC would mean that superordinate regions of lPFC asymmetrically influence the processing of subordinate regions of lPFC.

Several anatomical gradients have been noted across species in lPFC that are consistent with a hierarchical organization, including the transitions from granular to agranular cortex and changes in connectivity across areas from caudal to rostral (Badre and D'Esposito 2009; Jacobs et al. 2001; Phillips et al. 2021; Sanides and Sanides 1972; Thiebaut de Schotten et al. 2016; Yeterian et al. 2012). However, over the last several years, more directed anatomical and functional investigations have found evidence for a processing hierarchy in lPFC, as defined above, that emphasizes two main points. First, the rostral mid-dlPFC is the top of the hierarchy, in that it exerts an asymmetric influence over other lPFC regions, including rlPFC which is rostral to it. Second, cortico-striatal interactions appear integral to hierarchical processing within lPFC.

Mid-dlPFC As the Apex of the Hierarchy

One way to index position within a processing hierarchy is in terms of input versus output connectivity. As influence should be asymmetrically super- to subordinate, regions higher in the hierarchy should exert broader influence than those which are lower. It follows that superordinate regions would have a higher ratio of output to input than subordinate regions (Badre and D'Esposito 2009).

Goulas et al. (2014) used the CoCoMac database of anatomy (Kotter 2004) in the macaque prefrontal cortex to test this ratio of efferent to afferent connections across regions of lPFC. A strictly rostro-caudal hierarchy predicts that the rlPFC should show the highest ratio. However, Goulas et al. did not find evidence that area 10 in the monkey was at the top of the hierarchy. Rather, they observed that mid-dlPFC showed the highest output asymmetry and so should be the apex of the hierarchy by this definition.

Evidence from effective connectivity studies in humans provided converging support. Across multiple studies using dynamic causal modeling of fMRI data from the CCT (Nee 2021; Nee and D'Esposito 2016; Pitts and Nee 2022; Wood and Nee 2023), a consistent pattern of effective connectivity within lPFC has emerged that places mid-dlPFC on top (Figure 7.4). In particular, mid-dlPFC exerts influence over both caudal sensorimotor control areas of lPFC, as well as rlPFC and schematic control areas. Further, stimulation of regions in this network using TMS produced hierarchical effects on behavior that were consistent with this pattern of super/subordinate relationships (Nee and D'Esposito 2017).

Taken together with the functional attributions we have discussed along the rostro-caudal axis of lPFC, the picture emerging from these studies is one of multiple zones of integration within lPFC, hierarchically ordered with respect to each other. In particular, caudal regions of lPFC sit at the interface of sensory input and movement. The most rostral regions are at the interface between planning and inference networks. In the middle, at the apex, the mid-dlPFC

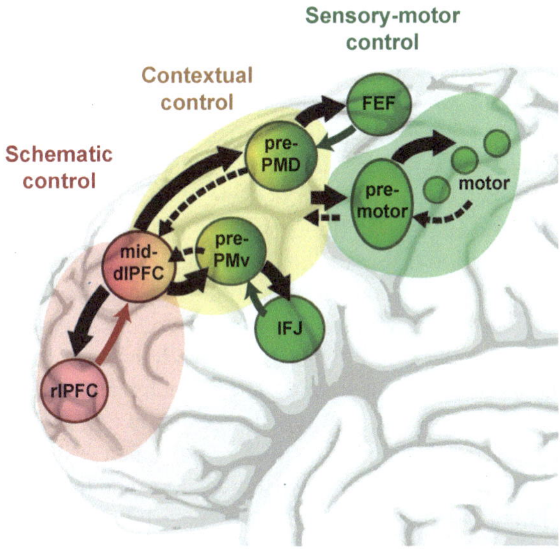

Figure 7.4 Schematic showing summarizing interactions between regions along the rostro-caudal axis of lPFC. The three zones defined by Badre and Nee (2018) of schematic, contextual, and sensorimotor control are distinguished by colored shading with regions labeled in each. Large arrows show the consistent primary direction of influence. Broken or colored arrows are weak or task-dependent influences. Mid-dlPFC shows the strongest outgoing influences both rostrally to schematic control regions, and caudally, to cognitive and sensory control regions. Reproduced with permission from Badre and Nee (2018).

is the link between these two systems, allowing control to be influenced by multiple forms of information from both the world around us, as well as our internal planning, memory and affective systems.

Cortico-Striatal Circuits and Hierarchical Control

Evidence from effective connectivity analysis of fMRI data in humans is consistent with the asymmetric anatomical connections found in nonhuman primates. However, effective connectivity measured in fMRI is not necessarily due to direct cortico-cortical interactions. Functional and effective connectivity can also reflect complex, polysynaptic network interactions. Indeed, one such contributor to hierarchical interactions among regions of lPFC may be the basal ganglia and its interactions with lPFC through the thalamus.

A recent study using Granger causality analysis of fNIRS data at rest (Schumacher et al. 2019) reproduced the asymmetric cortical pattern of functional connectivity observed using fMRI, in that there was an overall rostro-to-caudal pattern of influence, but with mid-dlPFC showing the strongest influence on other regions. Interestingly, a second study (Schumacher et al.

2021) of patients diagnosed with Parkinson disease undergoing deep brain stimulation of the subthalamic nucleus replicated this observation, but then found that turning on the stimulator enhanced this pattern of rostro-to-caudal lPFC asymmetry in patients early in the progression of the disease. Though limited by sample size and sample heterogeneity, and not linked directly to behavior, these observations provide evidence that interactions with basal ganglia structures, like the subthalamic nucleus, might play a causal role in the asymmetry of influence among lPFC regions.

What is the functional role of the basal ganglia with regard to the rostro-caudal hierarchy? One hypothesis is that the basal ganglia supports working-memory gating in the service of hierarchical cognitive control (Frank and Badre 2012; Frank and O'Reilly 2006). As already introduced, working memory plays a central role by maintaining information that can serve as a control signal. The lPFC, in particular, is hypothesized to maintain the control representations needed for this function (Miller and Cohen 2001). Computational models of this mechanism have demonstrated, however, that working-memory gating is required to select what information to hold as a control signal in working memory, "input gating" and when to allow it to influence behavioral choice, "output gating" (Frank et al. 2001; Hochreiter and Schmidhuber 1997; O'Reilly and Frank 2006). Further, tasks often require selectively input gating and/or output gating one representation, while holding others in mind. This is particularly important for hierarchical control tasks in which superordinate contexts are held steady, while subordinate goals are frequently updated.

Frank and O'Reilly (2006) proposed the prefrontal cortex basal ganglia working memory (PBWM) model, which implements selective working-memory gating using a cortico-striatal-thalamic mechanism that is assumed to perform the same core computations as more established mechanisms of motor control. In motor control, candidate actions are represented in premotor cortex but not executed because cortico-thalamic drive is under inhibition from the globus pallidus. However, topographically organized loops through striatum can disinhibit or gate particular responses. Moreover, through dopaminergic signaling, reinforcement learning can modulate synaptic weights in the striatum so that its gating responses reflect a cost-benefit decision about the candidate actions.

PBWM proposes that parallel cortico-striatal-thalamic loops perform analogous computations on cortical representations in lPFC, thereby acting as an adaptive gate on working memory. Further, because of its topographic organization, subcircuits can modulate specific cortical populations thereby allowing selective gating. The relationship between cortico-striatal interactions, working memory, and dopamine signaling predicted by the model has been supported by multiple lines of evidence in humans and animals (Chatham et al. 2014; Cools et al. 2006; Dagher and Robbins 2009; Frank et al. 2004; Jin et al. 2014; Lee et al. 2015; McNab and Klingberg 2008; Schmitt et al. 2017; Schonberg et al. 2010; Stollstorff et al. 2010; Tai et al. 2012; Voon et al. 2010).

PBWM also proposes dynamics among multiple cortico-striatal loops that serve as the basis of hierarchical cognitive control (Badre and Frank 2012; Frank and Badre 2012). There is a well-established organization of anatomical connections rostro-caudally between striatum and lPFC. This organization has been observed in detailed tracing studies in animals (Haber 2003; Haber et al. 2020), as well as in human studies using diffusion-weighted tractography (Verstynen et al. 2012) and functional connectivity (Choi et al. 2012, 2018). PBWM suggests that different loops can update contextual representations in working memory at different levels of a task hierarchy. For example, one loop might gate working-memory superordinate goals (e.g., making a sandwich), while another loop is gating working memory for the subordinate goal (e.g., slicing bread). Importantly, asymmetric diagonal connections from the superordinate loops higher in the rostro-caudal organization, to the striatal region gating the subordinate loops, allow these higher-order contexts to influence differentially the gating decisions made at the lower level. Note, this hierarchical gating system could operate over whatever factors are functionally differentiating regions of lPFC along its rostro-caudal axis. They would allow control to be made compositional and executable through productions to match the complex tasks we confront (Bhandari and Badre 2018).

Some initial evidence supports this hypothesized nested looping architecture in the context of hierarchical control. Evidence from model-based fMRI, which correlates parametric functions estimated from a computational model of learning with fMRI BOLD signal change, indicates that specific cortical and striatal sites, aligned along the rostro-caudal dimension, were sensitive to reward prediction errors at specific levels of policy abstraction (Badre and Frank 2012). A study of artificial grammar learning observed three separate pairs of lPFC-striatal foci associated with different levels of task complexity (Jeon et al. 2014). These sites were connected based on diffusion tractography. Further, it has been observed in high-fidelity diffusion tractography that lPFC-striatum connections are not only ordered rostro-caudally; when connections deviate from this pattern, they are more likely to do so from rostral lPFC to caudal striatum, than vice versa, consistent with asymmetric diagonal connections that imply a hierarchy (Verstynen et al. 2012). Nonetheless, more evidence is needed to connect selective gating at multiple levels during cognitive control to interacting cortico-striatal loops.

In sum, there is evidence that both cortico-cortical and cortico-striatal connections in lPFC may support a hierarchical architecture with mid-dlPFC at its apex. Taken together with the functional divisions described in the preceding section, these interactions may describe how information from the sensory and planning systems are not only integrated but used as contextual signals for control.

Conclusions and Future Directions

Over the last two decades, our understanding of the hierarchical organization of the lPFC has progressed in a number of ways. There is now strong evidence that an axis of functional differentiation exists from caudal to rostral. Further, it seems clear that this lPFC organization is situated within a larger organization of brain networks that bridge from concrete sensorimotor function at one end, to more abstract cognitive function at the other. There is also consistent evidence that this functional organization is hierarchical in its processing character, with the mid-dlPFC at the apex of this hierarchy and cortico-striatal interactions playing an important role.

In addition to continuing efforts to test and revise these hypotheses, there are a number of future directions and open questions to be addressed. For example, mechanistic investigation of the ways that this lPFC hierarchical organization interacts with other areas of PFC—or the brain more generally— such as with medial PFC (Shenhav et al. 2018; Venkatraman et al. 2009b; Wood and Nee 2023) or cerebellum (D'Mello et al. 2020), is needed to add specificity regarding the lPFC role as a controller. This line of investigation should be combined with lesion or other manipulations that allow us to understand the causal influence that PFC has on other frontal lobe regions and processing in the brain more generally. Understanding how this architecture changes and contributes to the development of cognitive control will provide important insights (e.g., Amso et al. 2019; Freier et al. 2021; Unger et al. 2016).

Further, while considerable evidence for a functional hierarchy in PFC has been reported from humans studies using fMRI, lesion, and TMS approaches, this organization has not been thoroughly examined and tested in animals models. What physiological evidence we do have suggests that neural populations in lPFC encode most task information, and while there are some gradients (see Rich and Averbeck, this volume), there are not large qualitative differences in cell coding along the rostro-caudal axis. Thus, reconciling these literatures will require direct study of hierarchical control and tasks thought to engage this axis in animal models, as well as complementary approaches in humans, such as those using intracranial recordings (e.g., Johnson et al. 2023).

A related open question concerns the organization of the neural representations themselves that occur in different regions along the rostro-caudal axis. There is a growing focus in the broader field on the geometry and dynamics of neural representations in terms of how similarly neural populations represent their inputs during a task (e.g., stimuli, contexts, responses, task sets) in their patterns of neural activity, and how these patterns of similarity change over short and long timescales (Badre et al. 2021). The geometry of neural population coding is known to affect computation (Fusi et al. 2016). For example, whether neural populations encode their inputs as a small set of abstract low-dimensional components or as high-dimensional mixtures balances a trade-off between generalizability versus separability. That trade-off might affect behavior, if one is engaged in learning versus interference resolution (Badre et al. 2021; Fusi et al. 2016). Thus, how these features of neural representational geometry interact with the rostro-caudal organizing gradient of lPFC will be important to understand. For example, it may be that lower-dimensional representations that comprise the components of a hierarchical task structure are represented separately from more integrated conjunctive representations that represent specific instances of a task. From studies with human EEG, we now know that such high-dimensional, conjunctive representations are important for determining performance on a trial-to-trial basis (Kikumoto and Mayr 2020) as well as for maintaining and prioritizing action plans in working memory (Kikumoto et al. 2022). These neural representations are involved when performing hierarchically structured tasks; thus, it is important to understand their relationship to the rostro-caudal organization of the lPFC.

Just as repeated experimentation and testing of ideas around the rostro-caudal organization of lPFC has changed our view of this organization over the last several years, experiments in these domains promise to continue to do so. Study of this problem across levels of analysis and using a range of approaches will give us a clearer picture of the functional significance of this dimension of PFC organization.

Acknowledgments

I am grateful to current and former Badre Lab members, as well as to Mark D'Esposito, Derek Nee, John Duncan, Theresa Desrochers, Michael Frank, and many other colleagues for helpful discussions on these topics over the years.

8

Functional Fractionation and Integration

Physiology, Networks, and Behaviors

Elisabeth A. Murray, Bruno B. Averbeck, David Badre, Christos Constantinidis, Roshan Cools, Clayton E. Curtis, Lesley K. Fellows, Anna S. Mitchell, John D. Murray, Erin L. Rich, and Mark D'Esposito

Abstract

From the early 1900s onward, anatomists have parcellated the cerebral cortex, including the frontal cortex. Initial approaches were based on both the features of stained cell bodies and the pattern of myelinated fibers, together called architectonics. The labels provided by these architectonic investigations are still widely used today. This chapter considers the extant evidence for functional fractionation of the frontal lobes, and whether the organization of the frontal lobes should be conceptualized in terms of functional and anatomical gradients, instead of discrete areas with well-delineated boundaries. Discussion includes how the frontal lobes interact with other parts of the brain to influence behavior as well as the identification of critical gaps in knowledge. The authors conclude that a greater understanding of frontal lobe function would emerge from advances in theory that connects different levels of explanation, that take into account evolutionary perspectives, and that lead to the development of a common cognitive-behavioral ontological framework.

General Introduction

The frontal lobes remain a formidable frontier in neuroscientific study, both literally and figuratively. Frontal cortex forms the furthest extent of the brain,

Group photos (top left to bottom right) Elisabeth Murray, Mark D'Esposito, Lesley Fellows, David Badre, John Murray, Clayton Curtis, Anna Mitchell, Christos Constantinidis, Roshan Cools, Bruno Averbeck, Elisabeth Murray, Mark D'Esposito, Erin Rich, Roshan Cools, Clayton Curtis, David Badre, Bruno Averbeck, Erin Rich, Lesley Fellows, Anna Mitchell, and John Murray

anteriorly, providing guidance in decisions ranging from the mundane—like what to eat for breakfast—to the profound—like the selection of a life partner. Frontal cortex is also very much an outer limit in the field of neuroscientific study, one in which the opportunities for research and development, and the promise of understanding and treating maladaptive behavior—whether arising from brain injury or dysfunctional neural circuits—have not been fully realized. Until we have identified and modeled the functions of frontal areas and their circuit interactions, we cannot fulfill one of the key objectives of translational neuroscience: effective treatments of neurological and psychiatric disorders.

Over the last few decades, the field has made substantial progress in defining the functional neuroanatomy of the frontal lobes. The underlying premise of this work is that localization of function arises in part because each frontal cortex region has a unique pattern of afferent and efferent connections. Here we discuss progress toward understanding frontal lobe function not only from identifying functions of single areas, but also in identifying the functions and computations of the networks in which those areas are embedded. We first address the evidence for functional specializations within the frontal lobe and whether the identified functions align with identified anatomical subdivisions. We then explore organizational principles of frontal cortex and how the frontal lobes influence behavior. Finally, we discuss what approaches might unravel the nature of circuit interactions involving the frontal lobe and how we might address gaps in our knowledge.

Anatomical Subdivisions in the Frontal Cortex

From the early 1900s onward, anatomists have parcellated the cerebral cortex, including the frontal cortex. Initial approaches were based on both the features of stained cell bodies (cytoarchitecture) and the pattern of myelinated fibers (myeloarchitecture), together called *architectonics*. Although the number of parcellations in frontal cortex has varied across investigators, as do the locations of boundaries, the labels provided by these architectonic investigations are still widely used today. This is in large part because the architectonic labels provide a common framework for presenting findings across experimental approaches. Recently, chemoarchitectonics has been added to the roster of methods, based on histochemical stains or patterns of receptors. Where cell types are similar across cortical areas, it is also possible that the relative distribution of those cell types could help delineate functionally distinct cortical fields. These new methods can refine classical cortical maps and offer an additional basis for generating hypotheses regarding the functions of these regions.

There is no consensus on whether anatomically identified regions in the frontal cortex correspond to meaningful functional zones. Neuropsychological evidence in humans and animals has generally pointed to a division of labor;

it seems likely that, in the frontal cortex, as in other parts of the brain, there is functional specialization, though the granularity of this evidence for regional specificity tends to be coarser than the fine cytoarchitectural parcellations suggested by anatomy. Recently, some have questioned whether architectonic fields have relevance to function at all (Hayden 2023), harkening back to similar arguments by Lashley and his colleagues in the 1940s. However, such challenges to orthodox frontal lobe maps have yet to offer precise and testable alternatives for defining the organization of function within the frontal lobes. A related consideration is whether the organization of frontal cortex should be conceptualized in terms of functional and anatomical gradients, instead of discrete areas with well-delineated boundaries. Whereas early sensory areas often have physiological properties that allow one to define clear areal boundaries, the extent to which this can be extrapolated to frontal areas (or indeed, other regions of association cortex) is unclear. Here, we revisit these questions and consider the extant evidence, as well as critical gaps in knowledge.

If There Is Functional Fractionation of the Frontal Lobes, What Would It Look Like?

The underlying premise of frontal cortex neurobiology is that localization of function arises from the unique pattern of afferent and efferent connections within each region, as well as local differences in connectivity and cellular properties, and that this spatial variation supports unique cognitive operations. If this premise holds, we should be able to gain insights into the functional organization of the frontal lobes from the convergence of anatomical and functional methods. Tract-tracing studies in macaques have provided anatomical data to support this idea. In identifying the major connections of individual areas, investigators have observed different patterns of connections across architectonic fields. Modern anatomical methods based on structural and functional magnetic resonance imaging (fMRI) have also contributed to our understanding of the anatomical organization of the frontal cortex. Diffusion-weighted imaging, which allows mapping of the patterns of diffusion of (mainly) water molecules in brain tissue, has been used to study white matter connectivity and integrity. Although this method was initially thought to hold promise for mapping the connections of the human brain, its accuracy is known to be limited by technical factors that are unlikely to be overcome by improved data acquisition or analysis methods (Thomas et al. 2014). Another approach has been to examine "connectional fingerprints" of different frontal lobe regions based on resting-state covariance of activations acquired during fMRI (Mars et al. 2016). This method is particularly useful because it can be applied in both macaques and humans and used to infer homology across frontal cortex regions. The downside of this approach is that covariation in physiological signals between areas does not necessarily reflect actual anatomical connectivity. However, using this approach, the architectonic delineations of Price and colleagues (Öngür et al. 2003) and Petrides and Pandya (1999, 2002) have been largely supported by the resting-state functional connectivity studies in macaques and humans (Mars et al. 2016; Sallet et al. 2013). The one exception is the lack of a rostral lateral region in macaques with a connectional fingerprint matching that of the lateral frontal polar cortex of humans (Balsters et al. 2020; Neubert et al. 2015). Thus, macaques and presumably other simians most likely lack a homologue of human lateral frontal polar cortex. An alternative is that its homologue is relatively small in macaques and related simians. Figure 8.1 illustrates the frontal architectonic subdivisions in humans and macaques.

If frontal areas perform specialized functions, then the anatomical maps should align with functional data. Outside of the motor and premotor areas (see below), no single method has provided a reliable index of functional boundaries, so this information is often inferred from converging techniques. Arguably, within the frontal cortex, the strongest evidence for functional specialization comes from studies of people or animals with brain damage. Whether brain damage is accidental or experimental, it provides unique insight into whether a given bit of the brain is essential for a given behavior (Murray and Baxter 2006; Vaidya et al. 2019). Other loss-of-function experimental methods include reversible inactivations of cells with GABA agonists like muscimol, locally applied pharmacological agents that selectively increase or decrease cell activity, and recently developed chemogenetic methods that use virally delivered constructs in combination with systemically administered activators to shut down processing. Temporary disruption of function in humans is accomplished noninvasively with transcranial magnetic stimulation (TMS). TMS can be used to alter regional cortical function in frontal areas, exhibiting a higher degree of anatomical precision than in patients who have experienced accidental brain injuries. For example, Blumenfeld et al. (2014) observed distinguishable effects on memory function when applying TMS to neighboring sites within the middle versus inferior frontal gyrus. A relatively recently developed noninvasive method for producing regional inactivation of neural tissue is focused transcranial ultrasound stimulation (Folloni et al. 2019; Tufail et al. 2011; Yoo et al. 2011). Unlike TMS, focused transcranial ultrasound stimulation can be applied to deep structures in the brain (Folloni et al. 2019). Like experimentally induced cortical ablations, these methods have provided valuable insights into structure-function relationships within the frontal lobe. Another method for gaining causal insights into structure-function relationships involves applying electrical microstimulation to targeted regions of the brain, which can be performed in animals as well as in patients with depth electrodes placed for neurological disorders (e.g., epilepsy and Parkinson disease). Using logic analogous to that used for revealing topographic maps (such as primary visual cortex, V1), electrical stimulation can reveal systematic body maps in motor and premotor cortex areas, as well as somatic sensory areas (Halley et al. 2020), and might also provide evidence regarding the localization of function

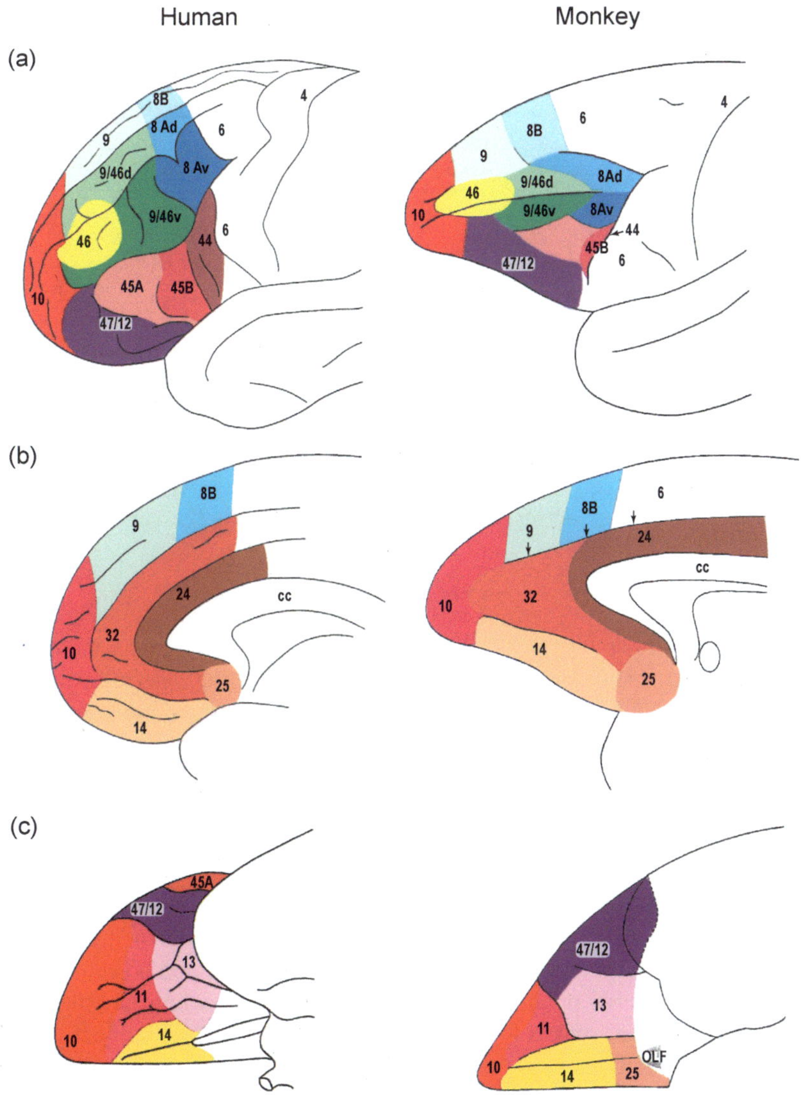

Figure 8.1 Schematic diagram of anatomically defined areas of human (left) and macaque monkey (right) frontal cortex, depicted on lateral, medial, and ventral surface views of the frontal lobe (top to bottom, respectively). Numerals refer to different architectonic fields; rostral is to the left. Adapted from Petrides and Pandya (1999, 2002).

in frontal areas outside the motor areas. Beyond manipulations, electrophysiological recording of the activity of neurons is commonly used to understand brain-behavior relationships. This is most common in awake-behaving monkeys, usually macaques, though there are also some opportunities to record neural signals directly from the brains of neurosurgical patients. Much more widely used, although far less regionally precise methods that also provide evidence for functional specialization include scalp EEG and functional MRI. These methods reveal correlations of neural activity with behavior and can establish whether neural activity patterns differ across areas under particular behavioral conditions (e.g., evidence for encoding of distinct variables in different PFC areas).

Evidence for Functional Fractionation of the Frontal Lobe in Primates

In this section we discuss the evidence for functional fractionation within the frontal cortex of human and nonhuman primates. Detailed reviews are available elsewhere, as are book-length treatments of the topic (e.g., Passingham 2021). Here we evaluate the strength and consistency of the evidence within and across methods as well as the extent to which functional dissociations respect anatomical boundaries, illustrated with some examples.

Electrical Stimulation

Although no single approach is definitive, brain stimulation maps and neuropsychological studies have provided the most compelling data regarding the fractionation of function in the frontal lobes. Electrical stimulation of primary motor cortex, M1, and supplementary motor cortex (SMA), or M2, reveal body maps (Graziano et al. 2002; Halley et al. 2020; Mitz and Wise 1987; Penfield 1954; Woolsey 1963; Woolsey et al. 1952). Within these cortical areas in primates there are well-characterized and consistent stimulation-elicited movements arranged systematically according to body part. These data provide clear evidence of modularity of function within the anatomically defined areas M1 and SMA. In nonhuman primates, in addition to the SMA, five additional premotor areas have been identified (Dum and Strick 2002; Luppino et al. 1991): the dorsal and ventral premotor areas (PMd and PMv) and three cingulate motor areas. Like M1, each of these premotor areas has substantial direct projections to the spinal cord. It is possible to evoke movements of the distal and proximal forelimb using intracortical stimulation at relatively low currents in all six of the established premotor areas.

In addition, stimulation of the frontal eye fields (FEF), which reside in the arcuate sulcus of the macaque, yields a systematic map of the contralateral visual field in monkeys (Bruce and Goldberg 1985) as does stimulation of its presumed homologue in humans (Blanke et al. 1999). Electrical stimulation of FEF reliably induces saccades of a particular direction and amplitude (Bruce et al. 1985; Robinson and Fuchs 1969), providing evidence for modularity of function within FEF that is based on saccade direction relative to the current eye position.

More recently, investigators have used modified population receptive field modeling of fMRI measurements to define visual areas across individuals and species. Using this technique, two visual field maps of contralateral space have been identified along the superior and inferior portions of the precentral sulcus in humans (Mackey et al. 2017). The map in the superior precentral sulcus is thought to be the homologue of the macaque FEF (Vernet et al. 2014); alternatively, this region could be one of several premotor oculomotor representations (Passingham 2021; Schall et al. 2020). At least some evidence suggests the macaque FEF might also contain two topographic maps (Savaki et al. 2015). Critically, these visual maps in frontal cortex, together with the body movement and eye movement maps evoked by electrical stimulation, constitute well-defined anatomical units that researchers can reliably target for study; they serve not only as a basis for alignment of maps generated by fMRI and other methods, but also offer a view of the flow of information from prefrontal areas to output effectors (e.g., eyes, head, forelimbs, hindlimbs) in humans and macaques.

Neuropsychological Studies

Human neuropsychological studies based on patients with accidental brain injury, such as traumatic penetrating head injury, brain damage incurred by stroke, tumor removal, or ruptured aneurysms, have been pivotal in identifying functional zones within the frontal lobes. These cases have provided causal evidence for functional fractionation of the human frontal cortex, with a level of explanation that has immediate relevance to the clinic. This method also has constraints: lesions incurred in humans most often are moderate in size, with varying degrees of overlap at regional/subregional levels, as well as involvement of underlying white matter which may lead to dysfunction beyond the anatomically defined lesion boundaries. Voxel-based lesion-symptom mapping can, in principle, improve the spatial resolution of functional inferences, but in practice is limited by available sample size and non-independence of how patterns of damage relate to the etiology of the damage. Lesion etiology limits what can be tested: patterns of injury typically segregate in lateral frontal (LF), dorsomedial prefrontal (DMF), and ventromedial-orbitofrontal lobes (VMF/OF). Damage to VMF/OF and DMF is often bilateral (to varying degrees); even unilateral lesions likely disrupt callosal integrity, thereby introducing the possibility of disconnecting regions in the other hemisphere. LF damage, in contrast, is rarely bilateral. The level of anatomical resolution that can be tested in human lesion studies is typically no finer than 3–6 regions: motor/premotor, DMF, VMF/OF, LF, and frontal pole. Studies variably consider the effects of damage to the left versus right hemisphere versus both.

Given the possibility of nonspecific or "off target" effects of brain injury, or illness more generally, double dissociation provides the strongest evidence for regional specialization. In such studies, two cohorts with lesions affecting different frontal cortex areas are compared on two or more functional assessments, administered as tasks. If one cohort demonstrates impairment on task A but not B, while the second cohort is impaired on task B but not A, this double dissociation is strong evidence that the functions assessed by tasks A and B are independent and depend critically on different neural substrates. Single dissociations (e.g., when a lesion of a particular area causes impairment in task A but not B) are also relevant. Compared to double dissociations, however, single dissociations are more open to alternative, nonspecific explanations such as general differences in task difficulty or reliability (Vaidya et al. 2019).

There are multiple examples of double dissociation in humans with regionally specific focal damage within the frontal lobes. For example, patients with VMF/OF damage are impaired in probabilistic stimulus-reward reversal learning, i.e., choices between two "objects" (decks of cards) yielding different monetary outcomes. They perform similarly to healthy controls in a task with the same dynamic reward structure, but where a reward is associated with one of two actions. The opposite pattern of results was obtained in humans with damage centered in the dorsal anterior cingulate cortex (ACC), part of the DMF (Camille et al. 2011b). Adding further assurance (and allowing more anatomical specificity), this finding replicates a similar observation in macaques with experimentally induced lesions to either orbitofrontal cortex (OFC) or ACC (Rudebeck et al. 2008b). There is a larger literature showing regionally specific lesion effects (single dissociations) across prefrontal cortex (PFC), tested with either multiple regions of interest or voxel-based lesion-symptom mapping (e.g., Gläscher et al. 2009; Tsuchida and Fellows 2013).

Lesion studies can also help to dissect distinct structure-function mappings that separately contribute to overall task performance. For example, damage to DMF and left LF, but not VMF/OF, disrupts different aspects of working memory performance in a two-back task (Tsuchida and Fellows 2009). Lesion studies can also fail to find task dissociations, which could be considered evidence that two tasks are drawing on the same component process. For example, performance on both Stroop and task-switching tasks is impaired after left LF lesions (Tsuchida and Fellows 2013), suggesting that these tasks tap into a common underlying function carried out by LF. Given the inherent heterogeneity of lesions, however, together with individual differences in structure-function relationships, studies of this type provide relatively weak evidence. Stronger conclusions can only be drawn by considering multiple lines of evidence.

As indicated above, brain damage in humans rarely respects anatomical boundaries and may affect underlying white matter pathways. As a result, based on human neuropsychological studies alone, it has been difficult to refine frontal cortex function beyond the broad anatomical regions outlined above. TMS in humans can produce more localized effects, but its influence is limited to frontal areas at or near the surface of the cranium. More fine-grained causal tests for structure-function relationships, respecting areal boundaries, require experimentally controlled lesions or other causal regional manipulations in nonhuman animals. Because new PFC areas emerged in early primates (Preuss and Wise 2022), and because extant nonhuman primates like macaques (Old World monkeys), marmosets (New World monkeys), and humans inherited these new PFC regions from a common ancestor, nonhuman primates have been indispensable for unraveling the function of the PFC.

Neuropsychological studies in macaques and marmosets have identified specialized functions for several parts of granular PFC. Here we focus on a few studies and paradigms that yield key findings. For example, within the ventral PFC, multiple studies have found doubly dissociable effects of lesions of OFC (areas 11/13/14) versus ventrolateral PFC (area 12/47) (Baxter et al. 2009; Dias et al. 1996a; Rudebeck et al. 2017b). In one study, monkeys performed two different tasks requiring them to take into account stimulus-reward-value associations while performing object choices (Rudebeck et al. 2017a). Both tasks manipulated reward value and in each, task performance indexed the ability to update rapidly object-reward-value associations. However, one task required updating of the desirability of food based on internal state, whereas the other required updating of the availability of food based on external contingencies. Selective lesions of OFC (areas 11/13/14) led to severe impairments on the task requiring value updating based on internal state but no impairment on the task requiring updating based on external contingencies, whereas selective lesions of ventrolateral PFC (area 12/47) led to the opposite pattern of results. This result supports the idea of fractionation of function in the ventral frontal cortex and is consistent with the idea that different types of specialized representations reside in OFC versus ventrolateral PFC (for review, see Murray and Rudebeck 2018; Rudebeck et al. 2017a).

An even finer fractionation of function within these areas was achieved using reversible inactivation. In one study, temporary inactivation of caudal OFC area 13 but not rostral OFC area 11 led to impairments in updating values based on changes in internal state (i.e., satiety). By contrast, temporary inactivation of rostral OFC but not caudal OFC led to a selective impairment in choosing between visually presented objects based on that updated value (Murray et al. 2015). In this case, the increased temporal specificity of pharmacological infusions over permanent lesions was critical to revealing a finer dissociation of processes involved in reward updating; inactivations delivered to different regions at different points in the task (before versus after satiation) produced distinct effects. Consistent with this finding, a human fMRI study found dissociable activations within OFC in a stimulus-reward task employing multiple foods. Participants in the experiment first learned a variety of arbitrary image-food associations. An important aspect of the design was that multiple images mapped onto individual foods. Then, using a repetition-suppression design, the investigators showed that rewards (i.e., specific foods) activated caudal OFC area 13 whereas stimulus-reward associations led to activation of rostral OFC area 11 (Klein-Flugge et al. 2013). This finding supports the idea that these OFC subregions have analogous functions in macaques and humans.

Additional fMRI studies in humans yield findings consistent with those in macaques; OFC activations reflect changes in food value that accompany object choices (Howard and Kahnt 2017). Finally, damage to the VMF/OF in humans, like damage to OFC in macaques, results in impairments in choice behavior in humans that resemble what is observed in macaques (Reber et al. 2017); they are impaired at switching their choices from the objects leading to a sated food to those leading to a nondevalued food. There is, however, an obvious difference between the human and macaque studies: We can ask humans why they made their choice. Humans with damage to VMF/OF explicitly indicate that after selective satiety, they no longer want the sated food, despite the fact that they usually make the choice that leads to getting that food. This points to a disconnection of knowledge and action that is evident in the choice behavior of humans with damage to VMF/OF sectors of the frontal lobe, and which resembles the choices of macaques with selective inactivations within OFC (Murray et al. 2015; Reber et al. 2017).

Notably, there are dissociable functions *within* areas consequent to selective neurotransmitter depletions. Using tasks known to be dependent on OFC—in this example, stimulus discrimination extinction—it has been shown that depletions of either serotonin or dopamine produce different patterns of behavioral impairment. For example, marmosets with OFC serotonin depletion showed an inability to overcome their bias toward responding to the previously rewarded stimulus, whereas those with OFC dopamine depletion were not biased toward the previously rewarded stimulus but nevertheless persisted in responding in the absence of reward (Walker et al. 2009). These and related results point to ways in which monoamine neurotransmitters can influence PFC-dependent behavior in regionally specific ways (Clarke et al. 2004, 2007; Walker et al. 2009).

There is also evidence for fractionation of function at the resolution of architectonically defined regions of medial frontal cortex and OFC with respect to threat reactivity. Activation of marmoset subgenual cingulate area 25 appears to induce an overall negative state, biasing basal cardiovascular activity toward sympathetic control, increasing reactivity to predictable as well as unpredictable threats, and enhancing avoidance of threats in an approach-avoidance task (Alexander et al. 2019, 2020; Wallis et al. 2019). In contrast, activation or inactivation of areas 14, 11, or 13 has no impact on basal cardiovascular activity and has more selective effects on threat responsivity. Specifically, whereas activation of area 14 produces little reactivity to predictable, certain threat, it increases reactivity to uncertain threat (Stawicka et al. 2020). Enhanced reactivity to uncertain threats is also seen in relation to areas 13 and 11, but in contrast to area 14 and 25, it is inactivation of these regions rather than their activation that heightens reactivity to uncertain threats (Stawicka et al. 2022). There are also clear distinctions between the effects of inactivation of area 11 and area 12/47 with respect to negative biasing as a consequence of threats in approach avoidance. Whereas inactivation of area 12/47 biases responding away from threats at the time of threat exposure, inactivation of area 11 had no such effect; instead its effects, which involve the enhancement of negative bias in responding, are only observed the next day, likely the result of an altered threat memory (Clarke et al. 2015). This overall pattern of results whereby granular PFC regions, as distinct from agranular cingulate cortex, have a greater role in contexts of uncertain threat is consistent with the findings from Mobbs and colleagues in humans. According to these investigators, prefrontal regions are only engaged when a threat is distal (in time, space or probability) and there is time to engage PFC mechanisms (e.g., in OFC and ventrolateral PFC) compared to when the threat is proximal and rapid response selection is required (Mobbs et al. 2020).

Within dorsal PFC regions there is also evidence for functional specializations that map onto anatomical subdivisions. Here, the parcellation of Petrides and Pandya shows three distinct subdivisions: area 9, area 46, and area 9/46. As shown in Figure 8.1, area 46 occupies the banks of the rostral half of the principal sulcus in macaques, area 9/46 occupies the caudal half, and area 9 lies above (dorsal and medial to) the principal sulcus. It has long been known that areas 46 and 9/46, typically referred to collectively as dorsolateral PFC, are essential for performance of delayed response tasks, including delayed response and delayed alternation. Although the original reports were based on aspiration lesions (Goldman et al. 1971), more recently the result has been confirmed using a more selective method: chemogenetic inactivation (Upright et al. 2018). There may be even further fractionation of function within this region. It has been suggested that the two regions have specialized functions. Passingham points out that based on anatomical projections, the more rostral of these dorsal PFC regions, area 46, is likely involved in identifying goals for saccadic eye movements (Passingham 2021:226). These regions are active when monkeys need to learn to perform sequences of movements (Averbeck et al. 2006) and are essential when monkeys make judgments about temporal order (Petrides 1991).

A key point related to the foregoing discussion is that the circuitry dedicated to reaching movements and eye movements is not only dissociable, but has different functions in choice behavior. The breakthrough concept is that saccades are not a movement so much as a mechanism for orienting attention (overt attention in this case), whereas reaching is not a mechanism for orienting attention. Selective pressures would have operated differently on circuits for eye movements and those for arm movements, for the simple reason that no primate ever "grasped" anything with an eye movement.

The findings reviewed above—based on effects of lesions, temporary inactivations, and pharmacological manipulations—provide strong evidence for functional fractionation within the frontal lobes. Next we consider the evidence for task-based regionally specific patterns of activity, first in humans (fMRI), then in nonhuman primates (neurophysiology).

Functional Magnetic Resonance Imaging

In general, fMRI in humans takes the approach of investigating differences in the blood-oxygen-level-dependent (BOLD) response across experimental conditions as a means of testing functional fractionation. Such differences can be tested as a univariate difference in overall voxel activity, differences in decoding or similarity matrices (i.e., representational similarity analyses), or patterns of functional connectivity. Fractionation is considered evident in region by effect interactions, such that the differences between conditions can be shown to change as a function of region. Although these interactions are sometimes interpreted as dissociations, it is rare to find full cross-over double dissociations (Chatham and Badre 2012, 2020; Fletcher and Henson 2001). Evidence of single dissociations distinguishing regions of the frontal lobe is much more common from fMRI. Although the latter suggest functional differentiation, they come with limitations to inference.

An advantage of fMRI is that it allows for measurement of activity while humans are performing a wide range of tasks. This is particularly important for studying the human frontal lobe, as it permits the study of the kinds of complex and higher-order tasks for which the frontal lobes are thought to be crucial. It follows that fMRI is one of the primary sources of evidence for functional differences across regions of the frontal lobe. Studies using fMRI have located evidence for the coarser frontal lobe distinctions that are supported by multiple sources of evidence, such as between VMF/OF, LF, and DMF. However, it is also a source of evidence for finer grained differences in function. For example, differences in fMRI activity within the LF, specifically between dorsal premotor cortex and dorsolateral PFC, have been observed based on demands for sensorimotor versus cognitive control (Badre and D'Esposito 2009; Nee and D'Esposito 2016; Badre, this volume). In rarer cases, these finer regional differences observed in fMRI have been supported by convergent evidence, such as from lesions in human patients. For example, patients with lesions in regions overlapping the zones activated in the aforementioned fMRI studies exhibited a pattern of behavioral deficits consistent with a hierarchical relationship between sensorimotor and cognitive control (Azuar et al. 2014; Badre et al. 2009). Likewise, TMS of these subregions, guided by fMRI, yielded a similar pattern of deficits (Nee and D'Esposito 2017).

More routinely, however, observations of regional differentiation with fMRI at a finer scale have not seen convergent evidence from other methods. For example, several fMRI studies have consistently reported activation in the LF polar cortex (i.e., the most rostral portion of the LF cortex), when processing abstract functions such as exploration over exploitation (Badre et al. 2012; Daw et al. 2006) or counterfactual predictive task-set processing (Koechlin and Hyafil 2007). However, corresponding lesion evidence in humans or lesion or physiological evidence in animals has not been reported.

Crucially, in this case, this failure may be due to methodological limitations, such as the ability to train animals in tasks hypothesized to involve frontal pole or the lack of homologous areas across species. As indicated earlier, it appears that macaques lack a homologue of human LF polar cortex (Neubert et al. 2015). Nonetheless, at finer granularity, fractionation of function in PFC is often supported only by fMRI evidence.

Of course, evidence from fMRI is limited in several ways. These include the indirect and correlative nature of the signal, its lack of temporal resolution, and in some cases, smaller effect sizes with limited samples. With regard to understanding the fractionation of frontal lobe, outside of the premotor and motor regions discussed above (see section on Electrical Stimulation) it is challenging to localize activity acquired from fMRI with reference to a map that aligns with anatomical features and allows for comparison across individuals and species. Thus, findings of functional differences in patterns of activation across regions observed using fMRI, even ones found repeatedly and reliably, are only a starting point. Investigations using causal methods and detailed physiological analysis are crucial. Such work would benefit from a more refined anatomical framework to align findings across individuals.

One recent advance has come from anatomically aligning fMRI data with tertiary sulci, which are small, shallow sulci that show a good deal of variation in presence and location across individuals (Weiner 2023). In one example, investigators aligned data from individual subjects to their (variably located) paraintermediate frontal sulcus and found that the sulcus marked a transition in function within the LF cortex (Willbrand et al. 2023a). Thus, anchoring data to tertiary sulci may be a way to overcome at least some individual differences in brain shape and structure. Rather than averaging data across brains, which tends to blur the pattern of activations, anchoring activations to a tertiary sulcus before averaging allows finer structure-function mapping.

Neurophysiology

The suggestion from methodologies such as neuropsychology or fMRI that a particular region of PFC subserves a given function has, in many cases, led researchers to seek neurophysiological correlates of those functions in the same area to understand underlying mechanisms. These approaches have led to a wealth of data showing that the activity of neurons in PFC can encode or represent a wide range of information, from external stimuli or motor responses to reward expectations to abstract concepts and rules. Based on the findings of double and single dissociations, investigators have expected to observe neuronal activity that not only reflects the differences in function, but also serves as the origin of it. To date, distinctions in the neurophysiology of different frontal regions have, however, been much less clear cut than many would have expected. As reviewed by Rich and Averbeck (this volume), this is true even at a coarse level of anatomical parcellation, where evidence for functional dissociations using other methods is quite strong. For instance, the lateral PFC is strongly implicated in cognitive control functions, including the use of rules and strategies; however encoding of rules and strategies is found not only in lateral PFC, but also in other areas such as OFC (Fascianelli et al. 2020; Wallis et al. 2001; Yamada et al. 2010). Conversely, OFC and neighboring PFC regions are involved in evaluation and value-based decision making, but decision-relevant information is strongly represented not only by OFC neurons but by those in lateral PFC (Leon and Shadlen 1999; Roesch and Olson 2003; Tsutsui et al. 2016b; Watanabe 1996) as well as medial PFC (Cai and Padoa-Schioppa 2012; Chien et al. 2023; Kennerley and Wallis 2009a; Matsumoto et al. 2003).

Despite the encoding of many variables across multiple frontal regions, there are some counter examples. Some reports show clear-cut differences in the activity of neurons across regions. For example, in a study by Tsujimoto and colleagues, who recorded neurons in three frontal cortex regions while monkeys performed a cued strategy task, only neurons in frontal polar cortex signaled responses that were correct according to the cued strategy (before feedback); only OFC neurons signaled the response that had been made (after feedback), whether correct or incorrect; and dorsolateral PFC encoded responses in a modality specific way. These signals support a role for dorsolateral PFC in generating responses, a role for OFC in assigning outcomes to choices, and a role for frontal polar cortex in assigning outcomes to cognitive processes (Tsujimoto et al. 2012). In addition, a few consistent trends across studies can be found. Perhaps most clearly, dorsolateral regions tend strongly to represent factors related to space, including action or attention that is directed in space, and these variables are typically poorly represented by ventral regions such as the OFC (reviewed by Rich and Averbeck, this volume). This is generally consistent with the idea that LF cortex plays a role in translating goals to actions (Averbeck and Murray 2020; Cai and Padoa-Schioppa 2014). In addition, there have been many reports of small but significant distinctions in the proportion of neurons encoding different types of information. For instance, when monkeys chose a rewarding cue or rewarding action, more neurons in the dorsal ACC, compared to OFC, tended to encode actions, whereas more OFC neurons encoded stimuli (Luk and Wallis 2013). This is consistent with the human and monkey neuropsychology data reviewed in the section above, where damage to OFC and ACC disrupt the assignment of value to stimuli or actions respectively. However, in the neurophysiology study, the magnitudes of the encoding biases were small and only found briefly, in one phase of the task. Most studies focus on the small differences because the differences are consistent with the hypothesis of functional localization. However, the strongest patterns in the data indicate widespread encoding in PFC of most variables at roughly comparable levels.

In another example, analysis of neural activity during a baseline "hold period" in a reinforcement learning task, rather than during the trial itself, revealed that OFC neurons maintain a representation of values and target stimuli, whereas lateral PFC regions had only a weak representation of these variables (Tang et al. 2022a). Once the choice options were presented, directional analyses indicated that value and identity information flowed to dorsal circuits. This is consistent with other cases, where differences in the timing of responses can suggest a flow of information from one region to another. For instance, similar proportions of neurons in OFC and dorsolateral PFC encode rewards, but encoding begins about 80 ms earlier in OFC, again suggesting that this information is passed from OFC to dorsolateral PFC to influence behavior (Wallis and Miller 2003b). Taken together, there are small differences in neural encoding across prefrontal areas, and these support the idea of functional fractionation. However, these differences are embedded in a predominant pattern of similarity across regions that to date has made neurophysiology one of the less useful methodologies for distinguishing functional regions of PFC or establishing finer grained parcellations of functional areas. Viewed from another perspective, perhaps differences in encoding have more to do with differences in the areas to which each subregion of PFC projects, or to the distributed nature of representations, as opposed to functional parcellations. Nonetheless, neurophysiology is the most direct method of investigating mechanisms that produce complex cognition and behavior and is, therefore, a critical component of understanding the functional organization of PFC.

Taken together, convergent findings from multiple methods have led to a widespread consensus that there are distinct functional specializations within frontal cortex. What remains to be elucidated, however, are the degree and particulars of finer parcellations, the computation that each region contributes, how it participates in larger networks, and how behavior emerges from interactions of those distributed networks.

What Are the Organizational Principles of the Frontal Lobe?

Although much remains to be learned about the organization of function in the frontal lobes, several organizational principles are evident. Here we consider these general principles in the hope they will inform theory and thereby speed progress toward a more thorough understanding of frontal lobe function.

As mentioned above, the functional fractionation of PFC regions emerges in part from specialized, topographically defined inputs and outputs. In this regard, one can consider connectivity between frontal lobe regions, cortico-cortical connectivity more broadly, and cortico-subcortical circuits. After addressing potential hemispheric specialization of function, we discuss cortico-subcortical connectivity, in part because these subcortical inputs and outputs substantially influence frontal lobe neuronal activity and behavior.

Hemispheric Specialization

All mammals have a hemispheric specialization of premotor and motor areas that control the movement of the contralateral limbs and eye movements that direct gaze into the contralateral visual field. Beyond this specialization for motor control, there is abundant evidence for hemispheric specialization of function in humans but little or no evidence for hemispheric specialization in macaques. In humans, one of the strongest specializations involves speech and language processing in the left hemisphere. Even here, however, specialization is relative; with few exceptions, both hemispheres can process most types of information.

In both humans and macaques, evidence suggests that visual working memory in the PFC operates largely independently within each hemisphere, with each processing information in the contralateral visual hemifield. For instance, working memory capacity limitations depend on the number of memoranda per visual hemifield and are generally unaffected by stimuli presented in the opposite (unattended) field (Buschman et al. 2011; Delvenne 2005; Umemoto et al. 2010). Similarly, neurons tend to show stronger encoding of contralaterally presented cues (Brincat et al. 2021; Funahashi et al. 1990; Kastner et al. 2007; Kornblith et al. 2016; Luria et al. 2016; Rainer et al. 1998). However, beyond spatial specificity, the processes carried out in each hemisphere appear similar, and under natural viewing conditions, information is likely to be transferred rapidly from one hemisphere to the other (Brincat et al. 2021).

Although the organizing principles of PFC lateralization remain unclear, there are some examples of lateralized structure-function lesion effects in the PFC, beyond language and motor processes, where lateralization is very well established. In one human lesion study (Geddes et al. 2014), effective interference resolution was found to require either right or left lateral PFC, depending on the nature of the task. In another study (Stuss and Alexander 2007), the left lateral PFC was found to play a pivotal role in task-setting—a function that entails the establishment of a stimulus-response relationship—whereas the right lateral PFC was engaged in monitoring processes involving the continuous assessment of task performance for quality control and the implementation of required behavioral adjustments.

Another interesting domain of specialization involves affect. In humans, evidence suggests that posterior regions of the right hemisphere are specialized for the interpretation of emotional information, including information contained in tone of voice and facial expressions. In addition, anterior regions of the right hemisphere are specialized for the production of emotional cues (e.g., facial expressions) that serve a communicative function. Correlates of mood states, while represented bilaterally, show some asymmetry. For example, fMRI studies suggest that greater activation of left than right frontal regions is associated with positive mood and approach behaviors. In contrast, greater activation of the right than left frontal regions is associated with negative mood and avoidance behavior (Davidson 1992). It has been proposed that this asymmetry is due to the asymmetric autonomic innervation of the heart (Craig 2009). In addition, there is an asymmetry in the effects on autonomic output following electrical stimulation of the insular cortex where, in humans, there appears to be right-sided dominance for sympathetic effects (Oppenheimer et al. 1992).

Cortico-Basal Ganglia-Thalamic Loops

All cortical areas, including the frontal lobes, participate in cortical-basal ganglia-thalamocortical "loops" (Alexander et al. 1986). This fundamental loop architecture involves a series of projections from cortex to striatum, striatum to pallidum, pallidum to thalamus, and, finally, thalamus back to cortex. Importantly, the loops project back to the same regions of cortex from which they originated. Because this feature of cortical organization is well known, and discussed extensively elsewhere, we will not repeat it here. That said, we note that cortico-striatal connections are more complex and less segregated than stated above, and that interactions between functional territories are extensive. Thus, within the striatum, there appears to be integration of information across what are classically considered reward, cognitive, and motor territories of the frontal cortex (Haber 2016). In addition, the existence of "focal" and "diffuse" cortical projections to the striatum opens the possibility that these two termination patterns serve different functions (Haber et al. 2006; Watakabe et al. 2023).

Cortico-Thalamo-Cortical Connectivity

In mammals, the frontal cortex and thalamus are anatomically interconnected and share a common developmental trajectory. Several thalamic nuclei connect directly with the frontal lobes including the mediodorsal (MD) thalamus, motor thalamus, anterior thalamus, pulvinar, intralaminar nuclei, and the nucleus reuniens. Different thalamic neurons provide either targeted, or more diffuse, frontal inputs, replicating patterns of thalamocortical connectivity across different thalamic nuclei, now referred to as thalamocortical motifs (Halassa and Sherman 2019). Each thalamic nucleus also has reciprocal modulation with Layer VI of the frontal lobes via the reticular thalamic nucleus (Halassa and Sherman 2019). The entire frontal cortical mantle is reciprocally interconnected to different MD subdivisions. These MD thalamocortical projections in primates target deep Layer III and Layer IV, while Layer V projects directly back to each of these MD subdivisions, or indirectly via cortico-striatal-thalamic loops (Barbas et al. 1991; Giguere and Goldman-Rakic 1988; Goldman-Rakic and Porrino 1985; Porrino et al. 1981; Ray and Price 1993; Saunders et al. 2005; Schwartz et al. 1991; Timbie and Barbas 2015; Xiao et al. 2009). In human neuroimaging, multi-domain thalamic network hubs have now been identified

(Hwang et al. 2021; Shine et al. 2023). These cortico-thalamo-cortical circuits are consistent with the idea that frontal cortico-thalamic interactions are essential to cognitive function (e.g., for review, see Mitchell 2015; Perry et al. 2021).

Gradients

Within PFC, at least two spatially organized gradients of anatomical circuitry can be discerned. First, two large-scale anatomical circuits are evident: dorsal and ventral. The ventral PFC regions, including OFC (areas 11, 13, 14) and ventrolateral PFC (areas 12/47), are part of a larger network that has prominent connections with the amygdala and inferior temporal visual cortex, the ventral striatum, the medial portion of MD, and the hypothalamus. The dorsal PFC is part of a network that has prominent connections with parietal cortex, the dorsal striatum, the lateral portion of MD, and few connections with the hypothalamus (Averbeck and Murray 2020). This pattern of connections suggests that ventral and dorsal PFC regions have distinct functions. Specifically, it has been proposed that, operating in the networks in which they are embedded, the ventral and dorsal PFC define behavioral goals and orchestrate behavior to achieve behavioral goals, respectively (Averbeck and Murray 2020; Giarrocco and Averbeck 2023; Marquand et al. 2017; O'Reilly 2010).

The dorsal-ventral dichotomy outlined above can be viewed as part of a larger medial versus lateral pattern that is evident in all mammals. Comparative neuroanatomical studies have revealed that cerebral cortical organization can be viewed as a set of concentric rings around a core of eulaminate cortex, with the core containing, among other things, primary sensory areas S1, A1 and V1. Medial to the core is cortex with one developmental origin, and lateral to the core is cortex with a different developmental origin. Thus, the mammalian neocortex can be described as two sheets (Cisek 2022). As indicated earlier, after the divergence of rodent and primate lineages—roughly 80 million years ago—additional frontal cortex regions emerged in primates. We note that calling this pattern a "gradient" is a convenient label; there is no evidence that the frontal neocortex evolved in an ordered sequence (Murray et al. 2017). The emergence of new frontal and parietal cortex areas eventually led to the long-range frontoparietal connectional networks described next (Figure 8.2).

A second spatially organized gradient of circuitry in frontal cortex involves rostro-caudal patterns of connections (Murray and Constantinidis, this volume). Setting aside the details of point-to-point projections makes it easier to see this organization, which essentially looks like a series of reciprocally related concentric bands. Specifically, frontoparietal circuits are topographically organized such that primary somatosensory cortex and primary motor cortex are reciprocally related, the posterior parietal and premotor areas are reciprocally related, the inferior parietal and ventral premotor areas are reciprocally related, and, finally, the medial parietal and adjacent areas in the posterior intraparietal sulcus and dorsolateral prefrontal regions are interconnected

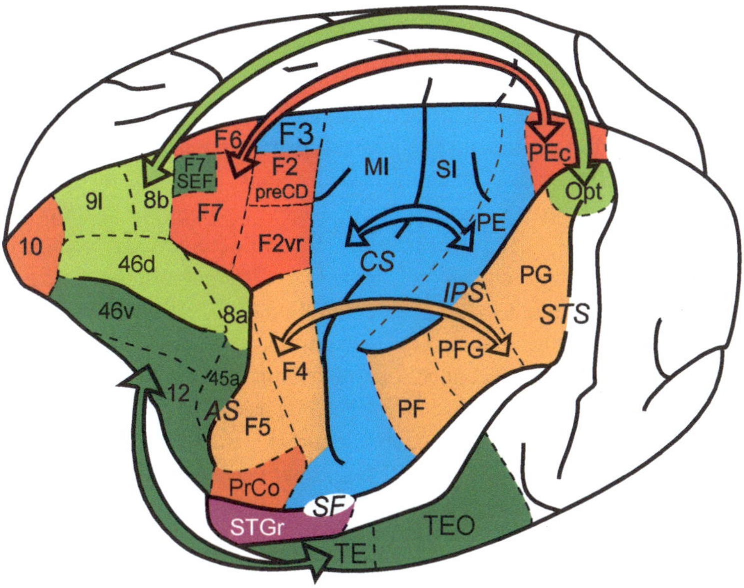

Figure 8.2 Patterns of frontoparietal and frontotemporal connections in macaques. Arrows indicate reciprocal anatomical projections between regions. Adapted from Giarrocco and Averbeck (2021). Sulcal abbreviations: AS, arcuate sulcus; CS, central sulcus; IPS, intraparietal sulcus; SF, Sylvian fissure; STS, superior temporal sulcus. F2-F7, areal abbreviations for premotor cortex regions identified by Matelli et al. (1985, 1991). PF, PG, PFG, Opt, PE, PEc, areal abbreviations for parietal cortex areas identified by Pandya and Seltzer (1982). Areas 8a, 8b, 9, 45a, 46v, 46d, 10, 12, areal abbreviations for areas identified by Petrides and Pandya (2007). PrCo, precentral opercular area; STGr, rostral superior temporal gyrus; TE, rostral inferior temporal cortex; TEO, caudal inferior temporal cortex.

(Cavada and Goldman-Rakic 1989; for review, see Giarrocco and Averbeck 2021).

Additional gradients may be evident in neurotransmitters and their respective receptor distributions within frontal cortex. For example, a recent report revealed frontal cortex regional differences in receptor densities, based on analysis of 14 distinct receptor types. In general, rostral frontal areas were characterized by higher receptor densities, whereas more caudal areas had lower receptor densities. This information was combined with information about MR-based connectional fingerprints and cytoarchitecture to suggest novel cortical subdivisions. Thus, the rich information about laminar and regional receptor distributions may provide additional insight into the molecular structure underlying the fractionation of function within the frontal cortex (Rapan et al. 2023). Similarly, a consideration of the combined morphological, electrophysiological and transcriptomic properties of neurons may yield insight into the functional organization of the frontal lobes (Gouwens et al. 2020).

Networks

Functional imaging studies have led to the description of several large-scale systems or networks of functionally interconnected brain regions. As discussed earlier, these functionally connected networks emerge from observations of covariation in fMRI activations (i.e., the temporal association between the patterns of activations in two or more brain regions) and have been proposed to be important for particular aspects of brain function. As summarized by Gratton (this volume), the PFC possesses several networks, which has led to the idea that different networks—as opposed to different architectonic areas—might carry out different aspects of PFC function (e.g., specific types of executive function or cognitive control). The network approach has also identified "hubs," regions that have connections distributed across multiple networks. Although there is as yet no consensus regarding assignment of particular functions to specific networks, the network approach may offer insights into regional interactions both within the PFC and between the PFC and other brain regions at a systems level (Menon and D'Esposito 2022).

Hierarchy

Anatomically, the laminar origin and termination of projections have been used to classify a connection as "feedforward," "feedback," or "lateral." This classification was initially used in the visual system, although it applies to other sensory systems as well. For example, projections identified as feedforward have their origin in deep layer 3, whereas projections identified as feedback originate in (typically) layers 5 or 6.

The feedforward and feedback architecture has been used to infer cortical hierarchy: feedforward connections are efferents from regions lower in the hierarchy toward regions higher in the hierarchy, and feedback are the inverse. Another idea is that hierarchy can be based on the asymmetry of connections: regions higher in the hierarchy exhibit more efferent connections to regions lower in the hierarchy than to those higher in the hierarchy. These ideas have important implications for computational models of cortical and network interactions.

In general, the frontal cortex exhibits a feedforward pattern of projections from rostral eulaminate to caudal dysgranular and agranular regions. For example, according to the laminar-based hierarchy, rostral OFC area 11 exhibits a feedforward projection to caudal OFC area 13 (i.e., mainly layer 3 neurons in area 11 give rise to projections to the deep layers in area 13), which in turn feeds forward to the caudal orbital agranular insular areas (Barbas 2000; Carmichael and Price 1996). On the lateral surface, this model suggests that area 10 is located at a higher level than more posterior regions, namely areas 45, 46 and 8A.

Unfortunately, the laminar- and asymmetry-based definitions of hierarchy do not always agree. For example, the asymmetry-based model reveals that area 10 does not sit atop the hierarchy as would be predicted (Goulas et al. 2014). In addition, both laminar- and asymmetry-based classifications may mask other fine-grained differences in connectivity that inform modes of communication (Rockland 2022).

As noted above, an anterior-posterior hierarchical specialization has been suggested within the lateral PFC based on anatomical and imaging studies, with more abstract operations localized anteriorly on the prefrontal surface (Badre, this volume; Badre et al. 2009; Koechlin et al. 2003). Neurophysiological evidence supports this idea: neurons with shorter response latencies, smaller receptive fields, and greater selectivity for stimulus properties are encountered in posterior regions of the PFC, but neurons responsive to more abstract qualities are more frequent in anterior areas (Riley et al. 2017). Plasticity of responses, dictated by task demands, is also more prominent in anterior areas (Riley et al. 2018).

Direct evidence of systematic variation of plasticity markers between eulaminate and agranular areas has been documented in the PFC. For example, the expression of calcium/calmodulin-dependent protein kinase II (CaMKII), which is essential for plasticity, is greater in medial frontal areas 25 and 32 relative to polar PFC area 10 and dorsolateral PFC area 46. By contrast, markers of cortical stability, including intracortical myelin, perineuronal nets, and parvalbumin show the reverse pattern (Garcia-Cabezas et al. 2017). Changes in neuronal morphology, molecular profiles of the synaptic apparatus, and the influence of neuromodulator systems have also been implicated in long-term prefrontal plasticity (Laroche et al. 2000; McEwen and Morrison 2013) and may differ between areas. Finally, short-term synaptic plasticity, depression, or facilitation has been documented in the PFC, and this too may be critical, particularly for task-related plasticity (Hempel et al. 2000).

How Do the Frontal Lobes Influence Behavior?

There is an extensive literature on the role of the PFC in executive function. This summary term provides a succinct way to discuss the planning and control of behavior (sometimes called cognitive control), the withholding of behaviors, and the pursuit of both immediate and long-term goals over hours, days, months, or years into the future, including embedded, intermediate, and nested goals and strategies for achieving such goals (for review, see Friedman and Robbins 2022). We have discussed frontal lobe function in other terms, but for readers interested in a consideration of executive function, we recommend discussions offered by Shenhav et al. (this volume) and Duncan and Friedman (this volume). Here we focus on just a few of the many ways in which the frontal lobes interact with other regions to influence behavior.

Cortico-Striatal-Thalamocortical Interactions

As indicated in the prior section, the PFC is embedded in a larger network of areas including cortical-cortical and cortical-subcortical projections. Efforts to understand the functional organization of PFC, therefore, need to take these larger networks into account. As discussed by Rich and Averbeck (this volume), there is a topographic organization of the cortical-cortical and cortical-subcortical circuits. At the highest level, ventral-medial PFC (e.g., area 25) and caudal OFC (area 13) are connected to the ventral striatum, ventral pallidum, and medial, magnocellular mediodorsal thalamic nucleus. The lateral PFC (e.g., area 46) is connected to the dorsal striatum, dorsal GPi, and lateral, parvocellular MD.

Consistent with this network organization, lesions of the caudate produce deficits in patients that closely resemble those that follow damage to dorsolateral PFC (Sandson et al. 1991). This finding mirrors early lesion work in monkeys which suggested similar behavioral impairments following lesions to either structure. Experiments in monkeys that have simultaneously recorded in area 46 and the caudate, to which area 46 projects, have shown similar responses in tasks that require spatial learning, although the caudate did have stronger correlations with the values of specific actions (Seo et al. 2012). Similar activity has also been seen across OFC and the ventral striatum in tasks in which monkeys have to learn the values of images (Costa et al. 2019; Tang et al. 2022a). In related work, lesions to medial MD thalamus, the part of MD most prominently connected to OFC, have shown deficits similar to those seen following lesions to ventrolateral PFC areas 12/47 (Chakraborty et al. 2016; Rudebeck et al. 2017b). Thus, a consistent set of findings, across human lesion, animal lesion, and neurophysiology in monkeys, have demonstrated that connected prefrontal cortical and subcortical areas show similar neurophysiological responses, as well as similar effects of lesions.

Cortico-Cortical Interactions

The extensive anatomical connections of the PFC place it in a privileged position to send feedback signals to the rest of the brain. Empirical support for the existence of such signals was obtained by Joaquin Fuster. In one study, a cooling probe was used to disrupt PFC function while neural activity was simultaneously recorded in the visual association cortex of monkeys performing a delayed match-to-sample task (Fuster 1985). When PFC was cooled, there was a reduction in delay-related neural activity in the temporal cortex. This finding indicated that PFC modulated the activity of the temporal cortex. In addition, PFC cooling affected the selectivity of neural responses in the temporal cortex. For example, neurons in the temporal cortex that originally coded for distinct color attributes displayed reduced selectivity for color following PFC cooling, consistent with similar recent studies. These findings have been replicated in human studies, both through fMRI investigations in healthy individuals utilizing TMS to perturb PFC function and through scanning patients with focal PFC lesions (Buschman et al. 2011; Lee and D'Esposito 2012).

Additionally, frontal-temporal interaction in macaques is essential for rapid acquisition of visual stimulus-reward, stimulus-stimulus, and stimulus-action associations (Bussey et al. 2002; Clark et al. 2013; Eacott and Gaffan 1992) and for the retrieval of stimulus-stimulus associations (Tomita et al. 1999). Lesions that surgically disconnect frontal and temporal cortex disrupt associative learning while leaving intact basic visual sensory, motor, and reward processing. Thus, cortico-cortical interactions involving the frontal lobe appear to be involved in a variety of functions, including top-down modulation and/or attention, learning, and retrieval.

Cortico-Amygdala Interactions

The ventral (areas 12, 11, 13 and 14) and medial (areas 25, 32 and 24) frontal cortex areas have extensive reciprocal connections with the basolateral portion of the amygdala (for review, see Murray and Fellows 2022; Aggleton et al. 2015). In macaques, studies using crossed disconnection surgeries have examined the consequences of functional disconnection of OFC and ACC from the amygdala. Crossed lesions of the amygdala and frontal cortex regions (i.e., involving removal of the amygdala in one hemisphere and a frontal cortex region in the other hemisphere) produce a functional disconnection because the amygdala projections to frontal cortex are ipsilateral. Using this approach, Murray and colleagues found that a network composed of the OFC, amygdala, and medial MD thalamus is critical for performing the devaluation task, a task in which changes in food value need to be taken into account before making object choices; this circuit is essential for linking objects in the environment with food value and adjusting those valuations in real time based on current biological needs (Murray and Rudebeck 2013).

Crossed surgical disconnection of the amygdala from ACC yields a different impairment, one involving loss of social interest/and/or social signaling (Pujara et al. 2022). Consistent with this finding, simultaneous recording in amygdala and ACC during a social reward allocation task reveals neural signatures (e.g., based on coherence between ACCg spikes and BLA local field potentials) of prosocial behavior and of vicarious versus experienced rewards (Dal Monte et al. 2020; Putnam et al. 2023). These data point to a role for ACC in social evaluation and, together with the information about the effects of OFC disconnection from amygdala mentioned above, provide a neural framework for distinct value assignment processes in the PFC.

Little information is available about the mechanisms underlying amygdala interactions with frontal cortex. Studies combining electrophysiology with causal manipulations indicate that amygdala inputs are important for acquiring as well as maintaining representations of the value of liquid rewards (both anticipated and received) in OFC but not ACC (Rudebeck et al. 2013a, 2017a). It is known that neurons in both OFC and amygdala of macaques signal the value of anticipated and received foods and fluids, as well as types of fluid, during choice tasks and appetitive Pavlovian conditioning (Morrison and Salzman 2009; Padoa-Schioppa 2011; Paton et al. 2006). Thus, these data suggest the possibility that the basolateral amygdala plays a general role in maintaining representations in frontal cortex, according to the types of representations stored in that area. Consistent with this idea, neurons in rat gustatory cortex lose representations of taste palatability but not identity after temporary inactivation of amygdala (Piette et al. 2012).

Although the amygdala is often considered responsible for processing "emotion" and neocortex for processing "cognition," this division of labor is almost certainly incorrect. Instead, it seems likely that emotional and cognitive parameters are inextricably linked and represented in dynamic neural circuits within amygdalo-frontal circuits, among other regions (Salzman and Fusi 2010).

What Are the Key Knowledge Gaps?

A lack of synthesis and coordination among disparate fields of research hampers progress in understanding the PFC. Research from the cellular and molecular level up to the systems level would benefit from closer integration, as would translational work in animal models and human subjects. Evolutionary, cross-species comparative research could provide the broader perspective needed for progress along these lines. There is also a large gap in our understanding of how subregions of frontal cortex interact with each other and within larger networks. Theoretical work, both conceptual and computational, could play an important role in bridging these gaps.

Theory That Aims to Connect Different Levels of Explanation

While there are exceptions, an ongoing gap in the field's approach to understanding functional fractionation of the frontal lobe has been a failure to explicitly bridge across levels of analysis and to integrate work done in cells and circuits with that at the systems and functional level. Computational neuroscience is an indispensable part of any strategy to overcome these obstacles and draw these links in a formal way (Badre et al. 2015). In neuroscience, computational modeling is often pursued at individual levels of analysis, from biophysically realistic models at the cellular level to abstract mathematical descriptions of behavior. Over the last decade, however, there has been fruitful progress in developing computational models that bridge levels of analysis (for detailed discussion, see Frank 2015 as well as Koechlin and Wang, this volume). These approaches allow modeling frontal function at one level to inform questions and models of other adjacent levels (e.g., Frank and Badre 2012;

Moolchand et al. 2022; Neymotin et al. 2020; Shin et al. 2017). Thus, theory provides a principled way of linking levels of analysis, going from molecules and cell types to systems-level function to complex behavior.

Theory and Experiments on Multiregional Interactions

Another major gap to be bridged concerns how regions of the frontal lobe interact, and how they interact with the cortical-cortical and subcortical networks in which they are embedded. For example, given that lesions to connected areas of cortex, thalamus, and striatum can lead to similar deficits, what is specific about the contribution of frontal cortex? Progress on these questions will require development and synthesis of both theoretical and empirical research. Some work has looked at computational implementations of PFC cognitive processes (Hart and Huk 2020; Wimmer et al. 2014), but these models lack biophysical detail and are confined to a single prefrontal subregion.

Ideally, the development of models that represent computations distributed across multiple areas would be a coordinated effort between theorists and experimentalists, and there would be consensus about how to assess the usefulness of such models. Such models should be able to perform a set of core PFC-dependent tasks and serve as a platform to integrate a wide range of experimental findings to achieve a cross-level mechanistic understanding of frontal lobe function. In practice, it would be beneficial to see studies of multiregional systems (e.g., cortico-cortical, cortico-striatal, or cortico-thalamocortical) in which high-level cognitive processes can be mapped onto specific regions or circuits, together with a computational model of how behavior is implemented in neurons, networks, and systems.

There has been progress on several theoretical questions in PFC related to representation and processing, mostly within specific regions. For example, there has been a shift in focus toward understanding representation and computation at the level of neuron populations, with the perspective that the fundamental unit of the brain's computation is a collection of interacting neurons that create dynamical activity. As discussed by Rich and Averbeck (this volume), the activity of large neural populations can often be captured by a low-dimensional manifold exhibiting a particular geometry (Chung and Abbott 2021). Examining these geometries can reveal dimensions that emphasize certain types of information over others, or that maintain information in orthogonal subspaces. For example, a recent study in mice found that within a high-dimensional space of neural activity, different subspaces were functionally connected with different networks of brain regions (MacDowell et al. 2023). This allowed a single area to interact simultaneously with multiple circuits. In addition, changing the alignment of the subspace (i.e., changing the geometry of neural responses) switched communication among networks, suggesting a mechanism that could support cognitive flexibility. These types of analyses are based on the notion that properties of neural populations are fundamental to brain computation, and that those computations cannot be studied by examining single neurons in isolation. The conceptual step from single neurons to populations is part of a larger step from one area to a network of interconnected areas. If our level of analysis is mismatched to the level of the PFC computation, we will fail to understand the link between neural activity and behavior. An alternative possibility is that the homogeneity of neural responses across the frontal lobes could reflect computation that is truly distributed. In this case, an open challenge will be to determine how modular functions arise from distributed systems. Intersections with theory and modeling will be critical to further these ideas.

Another series of theoretical and experimental studies has shown that a unique property of PFC is that it houses complex, high-dimensional representations (Fusi et al. 2016; Rigotti et al. 2013). These high-dimensional representations encode combinations of actions, stimuli, contexts, and outcomes in a way that allows downstream areas to decode any combination of the information. One view suggests that, from these high-dimensional representations, the striatum learns to select actions or action sequences specific to those stimuli in a given context that led to advantageous behavior (Parker et al. 2022). To further connect this idea to the discussion about functional organization, different prefrontal-basal ganglia circuits may be specialized for selection of different types of information. For instance, OFC-ventral striatal circuits may specialize in identifying or selecting goals based on combinations of stimuli and outcomes. By contrast, circuits involving medial and lateral PFC may combine actions or hierarchically organized action sequences and/or cognitive mechanisms (e.g., working memory or response inhibition) that allow goals to be achieved. If so, the striatum would play an integral part in learning to link the stimulus and context representations to the actions and cognitive mechanisms required to reach a given goal. The thalamus would be relevant to relaying the associations formed by the basal ganglia back to the cortex, perhaps for action execution through descending motor systems or for updating frontal representations. Consistent with an updating process, recent evidence has shown that the high-dimensional representations in PFC become lower-dimensional with learning, presumably because they become more sculpted to task demands (Mack et al. 2017; Wojcik et al. 2023). Testing the role of thalamocortical-basal ganglia circuits in this model is not straightforward, however, as disrupting any single element (e.g., frontal representations, striatal learning mechanisms, or thalamic relay mechanisms) would be expected to lead to deficits in behavior. Additional assumptions would have to be made about how each of these processes is implemented. Notably, some studies that have examined cortical and striatal representations of actions, selected to achieve specific goals, have shown that the cortex represents the chosen actions before the striatum, at least under some conditions (Seo et al. 2012; cf. Pasupathy and Miller 2005).

Although computational models have been developed that account for these specialized coding properties, these models have not taken into account the dynamical properties of PFC. Thus, the previous models have only formally described these processes, without embedding them in dynamical systems, or more ambitiously, multi-area dynamical systems models. More work is needed to continue to develop these models. Increasingly sophisticated computational models that respect the functional organization of the networks within which specific prefrontal areas are embedded will allow the development of quantitative predictions that can be tested empirically, using, for example, simultaneous recordings across multiple areas. Currently, few such predictions exist.

Theory That Aims to Connect PFC Function with Evolutionary Perspectives

In general terms, the frontal cortex is thought to store knowledge about behavioral goals and actions that could achieve them, along with outcomes that should result from such actions (Miller and Cohen 2001; Passingham 2021; Passingham and Wise 2012). As mentioned in the prior section, high-dimensional representations in PFC combine this information and confer several adaptive advantages, such as empowering individuals to learn from specific (less averaged) events (Massi et al. 2018). For example, when a new type of representation brings together previously unassociated stimuli, contexts, goals, and action sequences, selective forces can favor such representations to generate the cortical maps characteristic of each species (Murray et al. 2017). A consideration of representations and their adaptive advantages could provide the theoretical perspective to bridge several current gaps in knowledge, such as why cortical areas in the PFC are so much more difficult to define than in sensory areas of cortex.

It is well established that new PFC areas appeared in early primates and more emerged later in anthropoid primates. Accordingly, as discussed by Weiner et al. (this volume), there are many more frontal areas in primates than in rodents. However, a collection of sensory-cortex-like areas might not be the best way to think about the PFC. Although it is appealing to think that PFC areas are organized like early sensory areas, which have a well-defined function and discrete boundaries, that is not the only way representations can be distributed in the cortex. Biologically significant representations may be more widely distributed within the PFC. In these instances, it will be difficult or impossible to discover area-function relationships that look like maps of visual areas. The variation among published architectonic maps of the PFC, the lack of agreement about the precise number of areas or their boundaries, and the distributed encoding of variables observed in neurophysiological studies strongly suggest that there is some other organizing principle underlying PFC function. Accordingly, future work may benefit from a renewed focus on neural representations that smaller units of cortex, such as individual columns, generate and store, as well as the advantages such representations confer on animals in their natural habitats.

Consideration of the adaptive advantages provided by specific representations could lead to the development of new tasks that are more closely related to ethologically relevant behaviors. For example, recent work in freely moving macaques has set the stage for bringing together behavior (e.g., pose estimation), wireless electrophysiology, and autonomic measures in social contexts (Hayden et al. 2022; Maisson et al. 2023; Milton et al. 2020). Further work along such lines promises to bridge another key gap in knowledge: that between laboratory or clinical settings and the natural habitats of species that serve as animal models.

Conceptual Theory

A general challenge for studying structure-function relationships within the frontal lobes has been to develop a common cognitive-behavioral ontological framework, especially one that crosses human and animal models. A first problem is that typical experimental paradigms systematically remove elements that likely require the frontal lobes. That is, tasks meant to assess frontal cortex function often provide simple, salient stimuli, strongly constrain possible responses, use instruction to set explicit expectations about the task, provide practice to refine performance before the "real" task starts, and provide trial-by-trial feedback which, with enough repetitions, may eliminate the need for the frontal lobes entirely by converting the behavior to a well-learned "habit." Despite that limitation, there are many tasks that rely on the frontal lobes, but little standardization of what may be the crucial details of instruction, practice, timing, and trial number. The standardized human tasks (i.e., from clinical neuropsychology) tend to be grounded in a more classic theoretical framework, largely aiming to tap lateral PFC-mediated attentional and set-shifting abilities, which may not be readily mapped to current conceptual or computational models that include decision making, social behaviors, and flexible learning from reward. It would be helpful to develop a set of more standardized tasks grounded in current theories and useful for modeling. Tasks that address how we navigate novel or changing environments, select and pursue ecologically relevant motivational goals, and learn rapidly from real or vicarious experience may be especially useful, particularly for linking across humans and other primates. Further validation could come from considering the correspondence of such tasks and the clinical phenomenology that we think may relate to fractionated frontal lobe function, setting a direction that could connect with the clinic.

One measure of our understanding of frontal lobe fractionation is our ability to predict how prefrontal regions would be activated by cognitive tasks. This can be quantitatively formalized as the challenge of predicting cortical maps of activation across conditions of a given arbitrary task design (e.g., via an encoding model). Encoding models in fMRI develop voxel-level tuning functions that identify the features of a task that drive activation of a given voxel. (These are similar to tuning functions, for example, in visual cortex, that describe the properties of a visual stimulus that activate a neuron.) Meta-analytic approaches that combine task fMRI data across many different tasks provides the current state of the art. For example, NeuroQuery (Dockès et al. 2020) predicts brain-wide maps associated with various neuroscience terms (e.g., "working memory" or "reward"), from compiling activation coordinates and extracting text terms, across many fMRI studies. The resulting maps tend to be much coarser than the group-level activation contrast maps of any particular fMRI study, revealing finer regional differentiation than can currently be predicted *a priori* for a novel task. These meta-analytic approaches use neuroscientific terms from publications rather than a standardized description of the task itself. Thus, these approaches will be limited in their ability to capture neural effects of task manipulations.

Prediction of task fMRI maps from task description is limited by our ability to represent computationally a novel task in relation to other tasks. An embedding of tasks in some latent space could potentially capture how tasks differentially engage cognitive processes in a way that enables a mapping into the space of neural activations. This encoding model approach has been fruitful in the study of naturalistic perception. For instance, encoding models can predict voxel-wise cortical map activations by natural visual images (via receptive field models) and by spoken text (via semantic category labels) (Huth et al. 2016; Kay et al. 2008). A challenge for the study of frontal cortex function is to apply similar approaches to cognitive tasks.

9

How the Brain Creates Unity and Diversity of Executive Functions

John Duncan and Naomi P. Friedman

Abstract

Different executive functions, such as response inhibition, working memory updating, and mental set shifting, are correlated but separable. The focus of this chapter is the neural substrates of this "unity and diversity," with particular reference to the "multiple demand" (MD) system, a set of well-localized frontal, parietal, and posterior temporal brain regions that are active in tasks with diverse cognitive demands. After evidence for unity and diversity in behavioral studies is reviewed, the anatomy and function of the MD system is described and its potential mapping to unity and diversity discussed. Unity is evident in strong patterns of activation in core MD regions across tasks with different demands. Diversity is evident in differential activation of adjacent, more domain-specific regions, with strongest activation sometimes at the boundary between the MD core and these adjacent regions, suggesting communication between the two. It is suggested that the MD core serves to combine information from many brain regions and networks, integrating the diverse contents of an attended cognitive operation. Overlaps of the MD system and executive function unity with general cognitive ability are discussed, as are difficulties in integrating studies focusing on group-mean contrasts with individual-differences results. Understanding how behavior arises from the brain will involve understanding how information is represented, communicated, and transformed within and between brain networks, with the MD system likely contributing a core, integrative role.

Introduction

The terms executive functions and cognitive control generally refer to the cognitive processes used to regulate thoughts and action in the course of goal-directed behavior (Friedman and Miyake 2017; Friedman and Robbins 2022). Although these processes involve a large network of frontal, parietal, and other regions, they are often called "frontal lobe" functions due to their notable impairment after frontal lobe damage. Commonly examined executive functions include, but are not limited to, response inhibition, interference control, working memory updating, and mental set shifting. The relationships among these functions have been described by several researchers with the phrase "unity and diversity" (Duncan et al. 1997; Miyake et al. 2000; Teuber 1972): Executive functions are correlated with one another (show some unity), but those correlations are often only moderate, indicating that particular executive functions have unique variance (diversity). Here we discuss the neural substrates of this unity and diversity, with particular reference to the "multiple demand" (MD) system, a set of well-localized brain regions that are active in tasks with diverse cognitive demands (Duncan 2010).

Unity and Diversity of Executive Functions

Teuber (1972) first used the term "unity and diversity" to capture a "bewildering variety" (p. 637) of behaviors observed after frontal lobe damage, which shared elements of "compulsiveness" or "abnormally stimulus-bound behavior" (p. 640). Duncan et al. (1997) echoed this "unity and diversity" phrase in their study of deficits after frontal lobe injuries, noting that despite low correlations among scores on so-called frontal lobe tests, these tests shared a common element of goal neglect and association with general fluid ability. Later, Miyake et al. (2000) invoked this "unity and diversity" phrase once again to describe the pattern of correlations among executive functions and so-called frontal lobe tasks in neurologically intact young adults: Using confirmatory factor analysis, they found evidence for separable factors of response inhibition (Inhibiting), working memory updating (Updating), and mental set shifting (Shifting) tasks (i.e., diversity), yet there were also moderate correlations between these factors (i.e., unity), as shown in Figure 9.1a. This figure also shows the range of correlations observed across subsequent studies (reviewed by Friedman and Robbins 2022).

Although the model examined by Miyake et al. (2000) is sometimes called the "three-factor model" of executive functions by others, the focus on these three functions was purely practical (only so many latent factors can be assessed in one study) and was not intended to imply that these three functions are the only executive functions that exist or are necessarily "core" executive functions (Friedman and Miyake 2017; Miyake et al. 2000). Indeed, Miyake et al. (2000:90) noted that "although our choice of the three target functions in this study seemed a reasonable one, it is certainly not exhaustive and there are other important relatively basic functions that need to be added to the current list." Subsequent research has examined how other executive functions (e.g., dual tasking, verbal fluency, verbal and spatial working memory capacity, interference control) relate to these three functions (for further discussion, see Friedman and Miyake 2017). In addition, Miyake et al. discussed the possibility that there might be more complex functions (e.g., planning, problem solving) which draw on these three functions, as well as the possibility that these three functions might "be decomposed into more basic component processes" (Miyake et al. 2000:90), such as the monitoring, energizing, inhibiting, and adjustment of contention scheduling processes proposed by Stuss et al. (1995). Nevertheless, as the most widely studied executive functions, the three examined by Miyake et al. (2000) are a useful set with which to consider the key question of whether executive functioning is a unitary construct.

In Figure 9.1a, unity and diversity live in the correlations between the factors, specifically in the magnitudes of these correlations (the fact that they are greater than zero, but less than one). Alternative model parameterizations can be used to capture unity and diversity with latent factors. The statistically equivalent hierarchical model shown in Figure 9.1b illustrates that these correlations can be described with a higher-order "Common Executive Function (EF)" factor, which does not explain all the variance in updating or shifting abilities. Hence, in this model, unity is captured by the Common EF factor, and diversity is evident in the residual variances for the Updating and Shifting factors. The lack of significant residual variance for Inhibiting in several independent samples using similar batteries of EF tasks indicates that the Common EF factor captures all of the covariance among the response inhibition/interference control tasks (Friedman and Miyake 2017). It is important to note that this lack of Inhibiting-specific variance does not mean that there is no Inhibiting factor, just that individual differences in the Inhibiting factor are closely related to what is shared across many EFs.

More recently, Miyake and colleagues have used an alternative "bifactor" or "nested factor" model to capture EF unity and diversity in latent factors, rather than in their intercorrelations (Friedman and Miyake 2017). As shown in Figure 9.1c, the bifactor model has a Common EF factor that predicts all executive tasks directly, and orthogonal Updating-specific and Shifting-specific factors that capture remaining correlations among the updating and shifting tasks after the variance captured by the common factor is removed. Although the models shown in Figure 9.1 look a bit different, they all capture the data well; rather than being pitted against each other, they should be considered as complementary ways of carving up the variance among EF tasks. For example, Friedman et al. (2008) adopted the bifactor model because it enabled them to examine how other variables (such as speed and intelligence) relate to the EF unity and diversity factors directly, whereas the correlations of these other variables with the correlated factors shown in Figure 9.1a could reflect

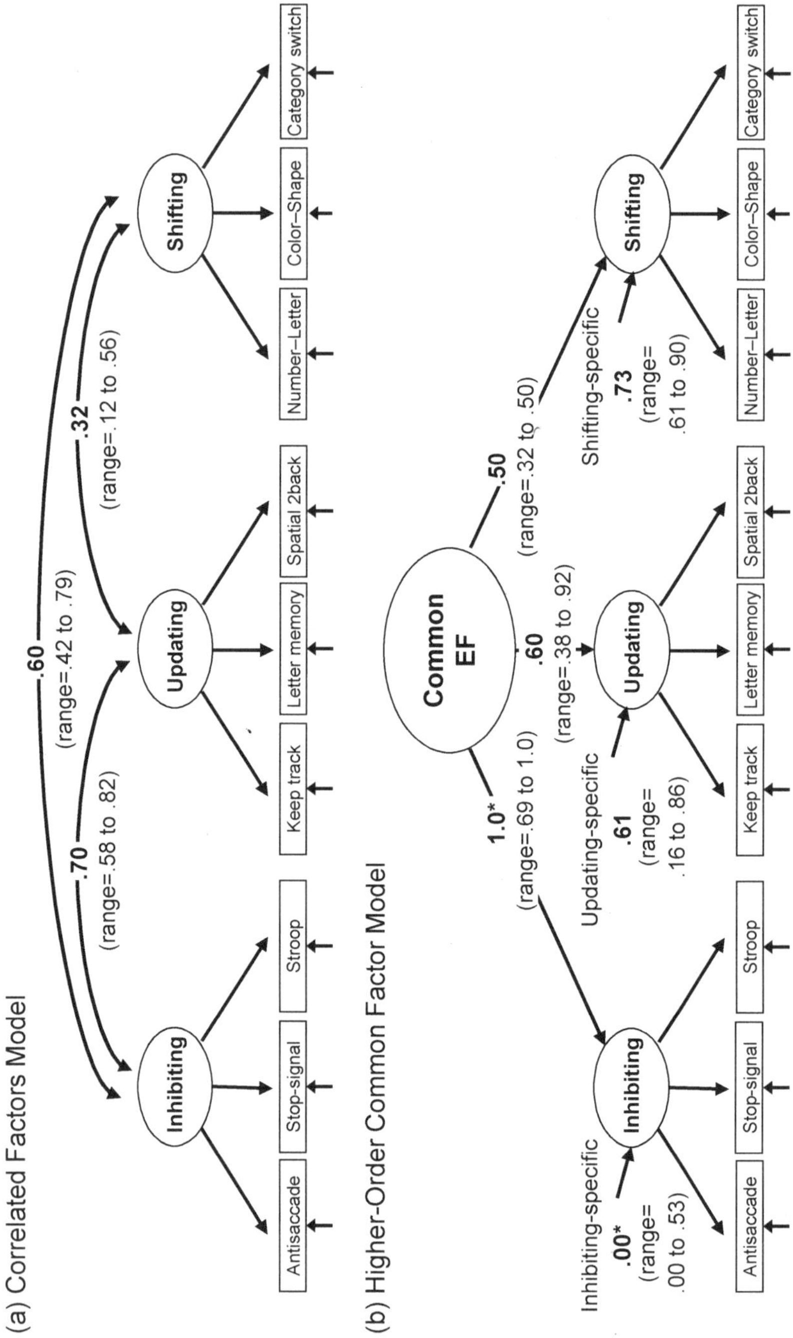
(a) Correlated Factors Model
Inhibiting
Updating
Shifting
Antisaccade
Stop-signal
Stroop
Keep track
Letter memory
Spatial 2back
Number–Letter
Color–Shape
Category switch
.70
(range=.58 to .82)
.60
(range=.42 to .79)
.32
(range=.12 to .56)
(b) Higher-Order Common Factor Model
Common EF
1.0*
(range=.69 to 1.0)
.60
(range=.38 to .92)
.50
(range=.32 to .50)
Inhibiting-specific
.00*
(range=
.00 to .53)
Updating-specific
.61
(range=
.16 to .86)
Shifting-specific
.73
(range=
.61 to .90)
Inhibiting
Updating
Shifting
Antisaccade
Stop-signal
Stroop
Keep track
Letter memory
Spatial 2back
Number–Letter
Color–Shape
Category switch

(c) Bifactor Model

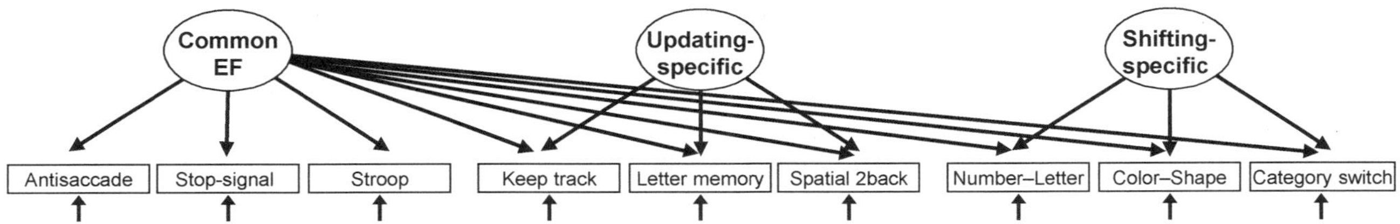

Figure 9.1 Latent variable models of three executive functions (EFs). Latent variables are depicted with ovals, whereas the nine observed tasks that define them are depicted with rectangles. Single-headed arrows indicate factor loadings and double-headed arrows indicate correlations (numbers shown are average correlation estimates from six studies using a similar battery, *N*s = 137–786). (a) In the correlated factors model, unity is represented by the fact that the correlations among the factors are greater than zero, and diversity is represented by the fact that the correlations are less than one. (b) In the higher-order common factor model, unity is captured by a higher-order Common EF factor that predicts the lower-order Inhibiting, Updating, and Shifting factors; diversity is captured by the significant Updating and Shifting residual (specific) variances. (c) In the bifactor model, unity is captured with a Common EF factor that directly predicts all nine tasks; diversity is captured by orthogonal Updating- and Shifting-specific factors, which capture remaining variance in the Updating and Shifting tasks, respectively, once Common EF variance is removed.

correlations with unity variance, diversity variance, or a combination of both unity and diversity variance.[1]

This general pattern of unity and diversity has been observed in numerous studies of individuals at different ages (Friedman and Miyake 2017), although specific patterns (e.g., which functions have been examined and how strongly they correlate) vary from study to study (Karr et al. 2018). This general unity and diversity pattern has also been observed at different levels of analysis (e.g., at genetic and environmental levels). For example, twin studies suggest that unity and diversity are due to common and specific genetic and sometimes environmental influences on different EFs (Engelhardt et al. 2015; Freis et al. 2022; Friedman et al. 2016, 2020; Gustavson et al. 2018, 2022). That is, the Common EF factor shows both genetic and environmental influences that are shared by all the executive function tasks examined in these studies, but there also appear to be different sets of genetic and environmental influences (i.e., reflecting different genes and environments than those that influence Common EF) that only affect the Updating- and Shifting-specific factors. Taken together, a wealth of evidence suggests that EFs can be distinguished but also share something in common, prompting the search for cognitive mechanisms and neural correlates that underlie this pattern.

With respect to cognitive mechanisms, the Common EF component is hypothesized to capture the ability to form, actively maintain, and use goals to bias ongoing processing (Friedman and Miyake 2017), consistent with classic models of cognitive control and frontal lobe function (Duncan 1986; Luria 1966a; Miller and Cohen 2001; Norman and Shallice 1986). In laboratory experiments, goals are set by experimental instructions. In the real world, candidates for goals must be derived by working forward from the current state (including stimulus input), suggesting goals that are currently achievable, and working backward from active overarching goals to suggest subgoals that are desirable. Candidates must then be weighted by some measure of importance, allowing one to be selected (Duncan 1990). This broad characterization of Common EF shares many similarities with other models of executive function/cognitive control (for a review, see Friedman and Miyake 2017). The mechanisms underlying diversity are less investigated. The variance specific to the Updating factor might reflect processes related to working memory gating,

---

1 For example, Friedman et al. (2008) reported that intelligence correlated .53, .70, and .19 with the Inhibiting, Updating, and Shifting factors shown in Figure 9.1a, but it is not clear from those correlations to what extent intelligence correlated with the Common EF factor and whether the nominally higher correlation with Updating could be attributed to a correlation between intelligence and the variance unique to updating abilities. Those questions are answered in the same dataset by examining the correlations of intelligence with the bifactor parameterization shown in Figure 9.1c, which were .50 ($p<.001$), .47 ($p<.001$), and –.17 ($p>.05$) with the Common EF, Updating-specific, and Shifting-specific factors, respectively. These correlations reveal that intelligence showed a higher correlation with the Updating factor in Figure 9.1a because it was correlated approximately equally with both the Common EF and the Updating-specific variance in Updating.

perhaps enabled by updating signals from the basal ganglia to the prefrontal cortex (Chatham et al. 2011), as well as item maintenance and potentially episodic retrieval (Friedman and Miyake 2017). These mechanisms include processes that are more general to working memory, as opposed to specifically related to the updating process. Although the Updating factor is so named because it is based on tasks that all involve continuous updating (e.g., the *n*-back task), the dependent measures do not subtract conditions that require working memory but no updating. Therefore, this factor necessarily reflects individual differences in working memory capacity as well as updating. Indeed, individual differences as measured by this Updating factor are closely related to those tapped by a working memory capacity factor based on complex span tasks (Schmiedek et al. 2009). Finally, according to one model, the variance specific to the Shifting factor may relate to differences in persistence of no-longer-relevant goal representations, which could be linked to multiple sources including local GABAergic inhibition within cortical regions (Herd et al. 2014).

Although EF unity and diversity at the behavioral level was discovered in conjunction with neuropsychological studies of brain damage, its neural substrates have remained unclear—or, rather, the substrates of diversity remain unclear. With respect to unity, brain imaging studies robustly show that different EF tasks recruit a similar network of frontal, parietal, and sometimes posterior temporal areas (Collette et al. 2005; Fedorenko et al. 2013; Niendam et al. 2012). This network has been dubbed the MD system due to its response during a wide variety of demanding cognitive tasks (Duncan 2010).

The Multiple Demand System: Anatomy and Function

This common pattern of brain activity associated with diverse cognitive demands has been known since the early days of brain imaging (Duncan 2006; Duncan and Owen 2000). A state-of-the-art version of this MD pattern is shown in Figure 9.2a, on inflated views of lateral and medial cortical surfaces (left) as well as on a cortical flatmap (right). To obtain this version, Assem et al. (2020) used data from 449 participants in the Human Connectome Project (HCP), averaging activations across three different demands (working memory, reasoning, arithmetic processing). The MD pattern consists of nine distinct patches in each hemisphere, widely distributed across lateral frontal (regions 1–4), insular (5), dorsomedial frontal (6), lateral and medial parietal (7, 8) and posterior temporal (9) cortex. Dividing these nine patches at a finer scale, using the cortical parcellation of Glasser et al. (2016a), Assem et al. (2020) identified a set of 27 individual MD regions, defined by the conjunction of significant activation for all three contrasts. Core MD regions, defined by the strongest common activity, are individually shown and labeled in Figure 9.2b (bright green), with additional MD regions ("penumbra") in darker green. At the higher resolution afforded by HCP methods (Glasser et al. 2016b), the MD

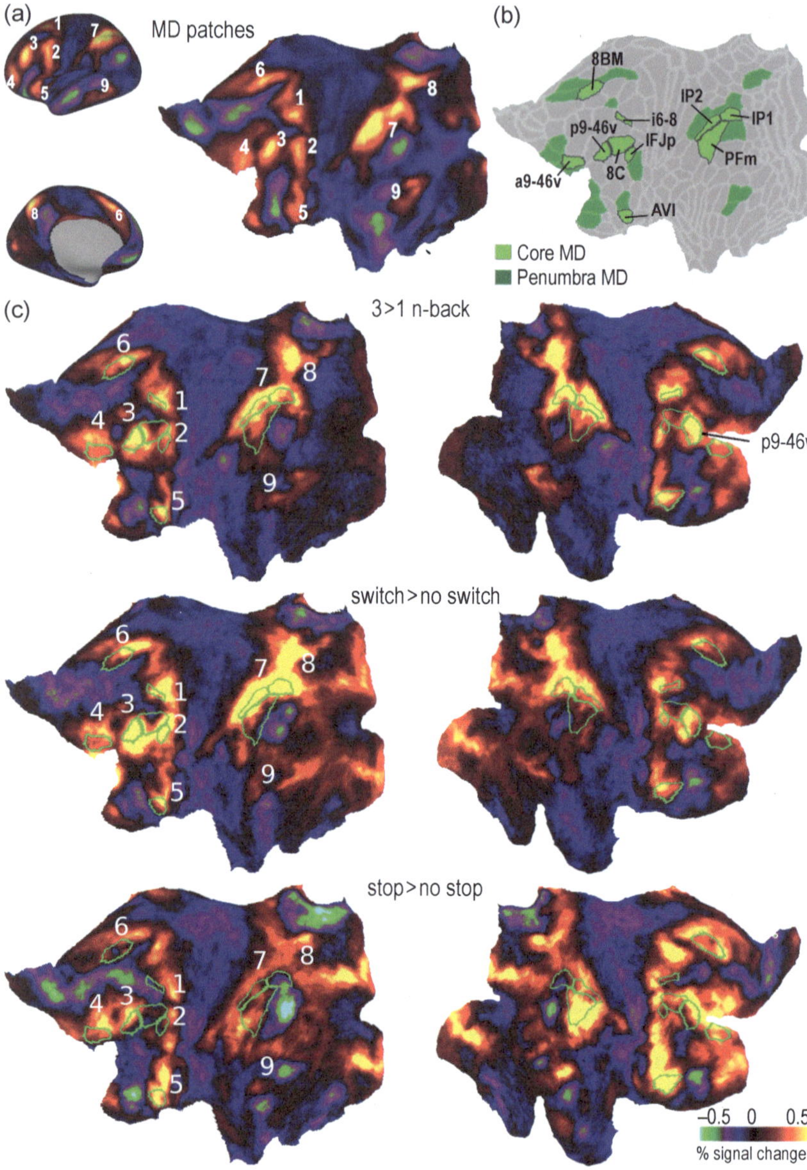

Figure 9.2 Multiple demand (MD) pattern. (a) MD regions from Assem et al. (2020) shown on inflated brain (left) and cortical flatmap (right). MD activity was largely symmetrical in the two hemispheres; just left hemisphere data are shown. Numbers indicate the nine individual MD patches identified in each hemisphere. (b) Fine-scale MD parcels from the Glasser et al. (2016a) parcellation. Bright green indicates the core; dark green, the penumbra. (c) Activations for three executive contrasts in Assem et al. (2024). Results for each contrast are shown on cortical flatmaps, with core MD regions from Assem et al. (2020) marked in green.

system identified here strongly resembles previous versions (e.g., Fedorenko et al. 2013; Niendam et al. 2012). Noteworthy is the tight definition of each patch: On the lateral frontal surface, for example, MD patches are immediately adjacent to patches with quite different functional properties, such as involvement in language (Fedorenko et al. 2013) or preferential response to visual versus auditory stimuli (Assem et al. 2022).

The suggestion of "unity" in these findings is obvious, but what of the "diversity" of EFs? Figure 9.2c shows results from a recent follow-up study (Assem et al. 2024), again using HCP methods on a new group of 37 participants. To examine the unity-diversity framework, the study used three contrasts: one for updating (3- vs. 1-back working memory), one for switching (blocks with or without cued task switches), and one for inhibition (blocks with or without stop signals). Figure 9.2c shows results for each contrast, again on cortical flatmaps, with core MD regions from Assem et al. (2020) marked in green.

On one hand, results show unity, with much the same nine MD patches visible in each contrast and each hemisphere. In individual participants, many individual surface vertices showed significant activation for all three contrasts. On the other hand, there is diversity, with the detailed pattern of activity for each contrast different from the other two. With the high-quality data obtained using HCP methods, many of these differences were significant. At the coarse level, for example, switch activations were stronger in the left hemisphere than the right (see also Crone et al. 2006; Tsumura et al. 2021), whereas stop activations were stronger on the right (see, e.g., Apšvalka et al. 2022; Aron and Poldrack 2006).

Many differences can also be seen at a much finer level; for example, in core region p9-46v, activation was most dorsal for stop, firmly within this region for *n*-back, and spreading more ventral for switch, whereas around the core MD regions of lateral parietal cortex, activation was most anterior/ventral for stop, especially visible in the right hemisphere, intermediate for *n*-back, and most posterior/dorsal for switch. In part, these shifts relate to additional networks preferentially involved in each contrast. To make this point, Assem et al. (2024) used networks defined by Ji et al. (2019) using HCP resting-state data. Of the 27 MD regions from Assem et al. (2020), the ten core MD regions all belonged to the Ji et al. (2019) frontoparietal control network, while penumbra regions were distributed between frontoparietal control and several other networks. In their EF data, Assem et al. (2024) found differences between the three contrasts in several of the original penumbra regions, plus others in their associated networks. Stop, for example, preferentially activated the cingulo-opercular network, and even in core MD regions, activations shifted somewhat in the direction of adjacent cingulo-opercular regions. Switch preferentially activated the dorsal attention network, and even in core MD regions, activations were somewhat shifted in the direction of adjacent dorsal attention regions. Such shifts were unique for each contrast, different from a simple expansion or contraction of the activated region. Often activation was strongest at the border of core MD and adjacent regions. A reasonable interpretation is that each type of executive control involves activation at the point of communication between core MD and adjacent networks, including those contributing to MD penumbra.

Core MD regions have strong functional connectivity with each other (Assem et al. 2020), and multivoxel pattern analysis shows extensive MD coding of task stimuli, rules, and responses (Woolgar et al. 2016). Widespread coding of task-relevant content is also well known in potential macaque homologues of human MD regions, including regions of lateral frontal and inferior parietal cortex (Goodwin et al. 2012; Miller and Cohen 2001). These properties suggest a potential function for the MD system with its multiple, strongly interacting parts. For any cognitive operation, multiple components (e.g., stimuli, rules, responses, context) must be integrated into a computational structure that reflects their required roles and relationships and allows the operation to be executed. A simple example would be integrating the parts of an attended object, and the actions directed toward it, but a similar integration is required for components of any cognitive operation. With parts widely distributed through the cortex, strongly interconnected with one another, the core MD system is well placed to take in and integrate representations of many kinds and flexibly feed out the results for selective cognitive control. Duncan et al. (2020) call this process "attentional integration," suggesting that the combined elements of a current cognitive operation correspond to the momentary contents of "attention."

This proposal fits well with the unity-diversity findings outlined above. As local communication is strong in the cortex, different kinds of cognitive content will enter the core MD system through different routes, perhaps reflected in coactivation of other adjacent networks, including penumbra regions, and strong activation at borders. Wherever information enters the core MD system, however, strong functional connectivity allows information to be widely distributed throughout this system, in turn giving local access to many potential outputs. On this view, core MD regions play a central role in integrating the components of any cognitive operation, explaining activation in tasks of many different kinds and bringing an element of unity to cognitive control. Regions of the network differ, however, in the information to which they have the most immediate local access, perhaps reflected in the relative functional preferences shown by different MD regions (Figure 9.2c) (Assem et al. 2020). As core MD regions exchange information so freely, clear functional dissociations will be hard to discern in the slow fMRI signal. At the same time, this interaction of core MD regions with different, more domain-specific regions brings an element of diversity to different executive tasks.

In Figure 9.2c, unity and diversity are illustrated with three canonical executive tasks from the unity-diversity framework, but on the above interpretation, the MD core brings together the diverse components of any task.

Are there a relatively small number of separate patterns of MD recruitment, reflecting a relatively restricted set of core EFs, or is every task likely to have its own, unique pattern of activity? With the high resolution of HCP methods, we are now in a position to examine exact patterns of MD activity for tasks of many different kinds, dependent on different combinations of cortical networks and regions.

Common EF Versus Spearman's *g*

The issue of whether there is a restricted number of separate patterns of MD recruitment, reflecting a core number of EFs, raises another question: Is there a restricted set of functions that we might call "executive," or does any task require its specific content to be drawn together, giving it an element of al integration reflected in its unique pattern of MD recruitment? While the Common EF factor is defined from a test battery including just inhibition, updating, and shifting tasks, there is some similarity to the concept of general intelligence or Spearman's *g*. In any broad cognitive test battery, correlations tend to be universally positive. To explain this result, Spearman proposed that some *g* factor contributes to success in all kinds of mental activities. If this explanation is correct, a measurement of *g* can be obtained as the first principal component extracted from the task battery, explaining the largest amount of shared variance between all the tasks it contains. This similarity naturally raises the question of whether Common EF is just a recapitulation of *g*; that is, a common element that contributes to success in all manner of tasks, including but not restricted to those commonly called "executive," and perhaps closely linked to the attentional integration functions of the MD system.

Certainly, the common element extracted for a typical executive battery (Common EF) is not identical to the *g* extracted from a typical IQ test battery. For example, Friedman et al. (2011) examined how the models shown in Figure 9.1 related to full-scale intelligence scores based on eleven subtests of the Weschler Adult Intelligence Scale III; Friedman et al. (2008) found similar results when examining a latent *g* factor for the eleven IQ subtests using an earlier subset of the same dataset. Using the bifactor parameterization shown in Figure 9.1c, IQ correlated .51, .49, and –.24 with the Common EF, Updating-, and Shifting-specific factors, respectively (Friedman et al. 2011). These correlations could not be constrained to 1, suggesting separability of IQ and Common EF. Moreover, the genetic correlations of IQ with the Common EF (*r*G = .57) and Updating-specific (*r*G = .56) factors were also only moderate, indicating that IQ and Common EF share some overlapping genetic influences; they are, however, not equivalent even at the genetic level (see also Gustavson et al. 2022).

Similar results have been obtained in other twin studies using different EF batteries and different aged samples. In a sample of middle-aged male twins,

Gustavson et al. (2018) found that a Common EF factor based on neuropsychological tests genetically correlated (*r*G = .59) with the Armed Forces Qualifying Test, a measure of *g*, as did a Working Memory-specific factor (*r*G = .24). Interestingly, studies of younger twin samples have found higher genetic correlations between Common EF factors and IQ or *g* (Engelhardt et al. 2016; Freis et al. 2022), raising the possibility that the association between Common EF and *g* may be stronger at younger ages; there is also some evidence that executive functions show more unity at younger ages (e.g., Brydges et al. 2012).

Neuroimaging studies also suggest a stronger association of IQ with updating ability compared to inhibition and shifting abilities. Specifically, a recent meta-analytic analysis of fMRI and PET data (Santarnecchi et al. 2021) suggested that an activation map for fluid intelligence overlapped 80% with a map of activation for updating tasks, but only 34% and 17% with maps for inhibition and shifting tasks, respectively.

Further evidence for this genetic separability of IQ and Common EF comes from a recent genome-wide association study of a Common EF factor estimated in the middle-aged UK Biobank sample (Hatoum et al. 2023). Common EF and IQ were genetically separable (*r*G = .74); moreover, the two constructs showed discriminant validity in their associations with other measures, with Common EF more strongly associated genetically with psychopathology factors compared to and controlling for IQ, but IQ more strongly associated genetically with educational attainment compared to and controlling for Common EF.

Several scenarios would be consistent with the observation that IQ is correlated with Common EF but not perfectly so, and that Common EF and IQ sometimes show different patterns of associations with outcomes like psychopathology and educational attainment. A common interpretation of a less than perfect correlation between two factors is that each factor captures some unique variance: In this case, it may be that Common EF and IQ both capture the same cognitive function(s), but that each also captures some cognitive function(s) that the other does not. A less than perfect correlation between two factors can also arise when one factor captures some unique variance in addition to the variance captured by the second factor, but the second factor does not capture anything unique from the first factor. Thus, it might be that Common EF captures something extra that is not captured by *g*, or that *g* captures something extra that is not captured by Common EF. The latter possibility is consistent with the findings that measures of IQ are related to both the Common EF and Updating-specific factors. That is, *g* seems to capture both Common EF and Updating-specific ability and may capture other abilities as well.[2]

---

[2] Friedman et al. (2008) found that the correlation (*r* = .70) of WAIS IQ with the full Updating factor (which does not separate Updating-specific variance from Common EF variance) was significantly lower than 1, suggesting that individual differences in IQ are not fully explained by working memory Updating.

Results might also depend on the task batteries that have been used to extract Common EF and *g* factors. In general, a common factor extracted from any set of tasks, explaining the most shared variance between them, varies from quite specific to quite general, depending on the breadth of the task battery. If the battery contains only verbal working memory tasks, for instance, then its first principal component will reflect specific strengths in verbal working memory. If batteries are very broad, their first principal component corresponds to a measure of *g*, and with enough breadth, the content of the individual battery ceases to affect the result. Thus, future studies could target this question with broader EF task batteries. What would happen if, in addition to inhibition, updating and shifting, an executive test battery was extended to include a wide range of other, putatively "executive" functions? If the first principal component resembled the Common EF extracted for just inhibition, updating, and shifting, this would be strong evidence that executive functions are indeed a natural kind, with a shared element not also common to non-executive tasks. If instead this first principal component resembled any other measure of *g*, the implication would be that EFs share little that is not also shared by any kind of task. In this case, the Common EF extracted from a more restricted executive battery would in part reflect the specific content of the particular executive tasks employed.

We also note that the universal positive correlations that underlie *g* may not be entirely explained by a single common element, such as a common cognitive or brain function, shared by all tests. One alternative is the idea of mutualism: Through development, acquisition of one cognitive strength (e.g., reading) promotes development of others, producing a pattern of universal positive correlations but no real *g* or shared brain function (Kievit et al. 2017). Another possibility is that positive correlations reflect process overlap, with many different overlaps underlying different correlations (Kovacs and Conway 2016). It may be that the cognitive process(es) captured by a Common EF factor or an Updating/Working Memory factor—as opposed to those captured by a broader *g* factor—are most closely related to a specific aspect of cognitive/brain function.

Returning to neural substrates, there is evidence relating both Common EF and Spearman's *g* to the MD system. In line with the results in Figure 9.2, meta-analyses of fMRI studies examining inhibition, updating, and shifting tasks all produce similar findings, with a strong MD pattern (e.g., Niendam et al. 2012). The same is true of fMRI studies that employ standard problem-solving tests, often termed tests of fluid intelligence, and widely used to measure *g* (Mitchell et al. 2023). In addition, some evidence shows that after brain damage, fluid intelligence deficits are predicted by volume of damage to the MD system (Barbey et al. 2012; Woolgar et al. 2010; but see Cipolotti et al. 2023). Given the strong relationship of IQ to EF, this neural overlap is not surprising. However, this overlap does not necessarily suggest that *g* and Common EF are equivalent; we still do not know whether the Common EF factor measures something more constrained than a broader *g* contributing to a wider range of tasks.

In the above studies, it is striking that *g* is often more related to Updating than to Inhibition and Shifting. Outside research specifically aimed at EFs, the tests with the highest *g* loadings tend to be complex reasoning tasks, such as Raven's Progressive Matrices, as well as measures of vocabulary. Duncan et al. (2020) argued that problem-solving tests, such as matrices, put an especially strong load on attentional integration and the MD system. Typically, such tasks are solved in a series of component steps, each requiring a new set of fragments to be assembled into the right cognitive operation. Attentional integration might also be important for building crystallized knowledge like vocabulary (e.g., to learn the meanings of unfamiliar words based on context). At the same time, a long history of research has linked performance on these tasks to individual differences in working memory, particularly the "central executive" or "controlled attention" component of working memory (e.g., Carpenter et al. 1990; Engle et al. 1999; Kyllonen and Christal 1990). There is much overlap between the concepts of attentional integration and working memory, particularly those aspects of working memory that go beyond simple storage capacity. Specifically, working memory tasks that require individuals to process and/or mentally manipulate information (e.g., to rearrange a list of items in order of size), in addition to remembering items, show stronger relationships with intelligence and reasoning tasks compared to tasks that only require individuals to remember a list of items (e.g., Engle et al 1999). Similarly, in line with the proposal of Friedman and Miyake (2017), there is much in common between the idea of attentional integration and the ability to form, maintain, and use a current goal. These overlapping concepts of Common EF, executive working memory, and attentional integration provide some explanation for the shared variance tapped by *g*, but the data suggest that they do not fully explain *g*. As we indicated above, *g* is likely to reflect sources of shared variance across tasks additional to any one shared cognitive function or brain system. While the attentional integration functions of the MD system may be linked to both Common EF and *g*, there is quite likely no simple one-to-one correspondence.

Linking Cognition to Brain: Methods and Levels

Our characterization of EFs here has relied on two quite different methodological approaches. Evidence for behavioral unity and diversity of EFs was based on analyses of individual differences in performance, whereas evidence for the MD system was based on analyses of group-mean activations for imaging tasks. We have attempted to link these two lines of research yet acknowledge that that there is no necessary correspondence between brain regions or functions and factors derived from individual-differences studies. For example, individual differences could be heavily influenced by genetic or developmental factors that are common to several or even all brain regions, leading to behavioral correlations with no link to specific cognitive functions or brain regions. Alternatively, as noted above, correlations in individual-differences data could reflect factors, such as mutualism, not captured in imaging data.

In general, different methods and levels of analysis produce different links of behavior to brain, making it challenging to integrate across these levels. In particular, studies that focus on which brain areas are active during EF tasks versus those looking for brain areas associated with individual differences in performance yield different results. For example, in a large fMRI study ($N=546$) with a design similar to that used by Assem et al. (2024), Reineberg et al. (2022) evaluated how individual differences in Common EF scores (based on six EF tasks) were associated with brain activity in three tasks selected to tap response inhibition, working memory updating, and mental set shifting. At the group-mean level, all three tasks robustly activated the MD system. However, individual differences in Common EF were not uniformly related to the degree of MD activation in each task, nor even to the degree of activation in the same areas outside the MD system across tasks. Only when the constraint that Common EF be related in the same direction (e.g., to greater activation) across the three tasks was discarded were there significant areas of conjunction, which included the bilateral middle frontal gyrus, medial superior frontal gyrus, left angular/superior parietal cortex, and cerebellum. Some of these areas are similar to some MD regions identified by Assem et al. (2024) but Reineberg et al. (2022) noted that these areas were also all at the anatomical borders between major functional networks. They concluded that their results were inconsistent with simple models in which EF performance is associated with higher or lower MD activation across tasks. They suggested, however, that the results could be consistent with a model in which performance is related to the activation of task-specific targets of executive control. Reineberg et al. (2022) also found some evidence that individual differences in Common EF were related to task-based connectivity of lateral PFC to these task-specific targets, which they interpreted as reflecting prefrontal biasing toward task-relevant information; however, these connectivity results were considered "preliminary" as they did not survive whole-brain correction for multiple testing.

Results such as these suggest that the brain regions that show mean activation differences in executively demanding conditions, compared to less demanding conditions across participants, do not necessarily predict individual differences in performance. Several factors might contribute to this null finding. Individual differences could be related to activation of different task-specific areas that are the targets of control. For example, in the Stroop task, individuals who show higher activation in brain regions associated with color representations, when the task is to ignore words and name the colors in which they are printed, might perform better on the task than those who show lower activation of these regions. Such activations may not simply be related to the activation levels of the control regions across individuals. In addition, even if a cognitive activity depends on some specific brain system X, it is far from clear whether individuals with stronger function should show more or less activity in X. One plausible scenario is that in each individual, the link of activity to demand follows an inverted U, initially increasing as demand goes up, then declining once the task appears impossible (Mattay et al. 2006). For relatively low demand, activation of X may be stronger in the low-ability individual, reflecting a greater struggle to satisfy task demands. For higher demands, activation is already declining for the low-ability individual, whereas for the higher-ability person, this downturn has yet to be reached. Such explanations could also account for small effect sizes in brain associations more generally (Marek et al. 2022), although many other factors could also explain small effect sizes (e.g., low reliability of measures and brain activation, and distributed associations across networks rather than specific associations with particular regions).

Similarly, the view of a "multiple demand" system will vary depending on the level of analysis. At the level of univariate fMRI, even in single participants, there is strong evidence for similar MD activity across many cognitive demands. A region of univariate activity, however, is made up of many millions of neurons, and it cannot be that each neuron responds similarly to all these demands. Neurons contribute to information processing to the extent that they respond differently to different things. Correspondingly, in any one individual, higher-resolution methods (e.g., multivoxel pattern analysis) reveal that MD regions carry information about many kinds of task content, such as discriminating a task's stimuli, rules, and responses; exact patterns of neural recruitment differ for different task events (Woolgar et al. 2016). The same, of course, follows from single neuron recordings in potential monkey homologues of MD regions, with neurons showing a very large number of different, idiosyncratic patterns of selective activity (Miller and Cohen 2001). A "multiple demand" region of cortex in fMRI suggests a body of neurons that are differentiated and have the flexibility to carry information of many different kinds in different task contexts (Rigotti et al. 2010), integrating the components of each individual cognitive operation.

Conclusions and Future Directions: Beyond Classical Functional Localization

In this chapter, we have reviewed evidence for unity and diversity of EFs at a behavioral level in an attempt to understand how unity and diversity emerges from neural activity during various tasks. Two key take-home points are as follows: First, EFs show unity and diversity at the level of individual differences in task performance. This unity and diversity is most clearly seen with latent variable models of multiple executive functions, which show that multiple executive function factors are correlated (show unity, captured with a Common

EF factor) but are also separable (show diversity, captured with factors such as Updating-specific and Shifting-specific factors). Second, when people complete different EF tasks and cognitive tasks more generally, they activate a similar network of frontal, parietal, and posterior temporal regions (the MD system). We characterized the spatial distribution of MD activation across EF tasks and discussed differences in patterns across tasks, which might relate to the observed diversity of EFs. We also discussed a potential cognitive mechanism for the MD system, attentional integration, which could explain why this network is active across diverse cognitive tasks.

In addition to these key messages, we discussed two related questions. First, we outlined our view on whether the Common EF factor recapitulates the *g* factor or IQ, a question that has preoccupied the literature for decades. Though future analyses of different batteries of executive function tasks may be useful to consider, the current literature suggests that the Common EF factor is not equivalent to IQ. Importantly, IQ is related to individual differences in working memory capacity and updating, over and above the Common EF factor. These patterns suggest that the *g* factor/IQ may capture more cognitive processes than the Common EF factor captures. Second, we pointed out the difficulty with integrating across different methods or levels of analysis, particularly findings related to population-level effects (i.e., everyone activates the MD system during cognitively demanding tasks) as well as those related to individual differences (i.e., individuals who perform better on demanding tasks may not necessarily consistently activate the MD system more strongly or weakly than individuals who do not perform as well). We cautioned that there may be no simple mapping between brain regions or functions and factors derived from individual-differences studies and provided potential reasons for discrepancies across methods. Although we do not have an answer on how best to integrate across methods and levels, such integration is needed to produce a comprehensive view of EFs and their associated brain networks.

For more than a century, ideas of how brain function may relate to behavior have been limited by the methods available. Animal and human lesion studies as well as many early results from human brain imaging invite a link between coarse regions of the brain, perhaps "dorsolateral frontal cortex," and some specific aspect of cognition or behavior. In this context, the "unity" element of unity and diversity has sat uneasily, in tension with the enterprise of linking specific brain regions to specific cognitive operations. We now know substantial limits to this conceptualization of the problem. Brain functions must be understood not in terms of the coarse regions that might be studied through lesions, but in terms of distributed, strongly interacting cortical and subcortical networks. We need to understand how information is represented, communicated, and transformed within and between such networks, with the MD system likely contributing a core, integrative role. Closely adjacent cortical regions can belong to different networks, with quite different patterns of resting-state activity and functional activation. As demonstrated in Figure 9.2, functional specializations can reflect extremely fine-scale gradients of activation and between-network communication. Accordingly, the enterprise of linking brain to behavior is not a search for simple regional mappings. Instead, perhaps not surprisingly, it is a matter of asking how whole-system function is assembled from the detailed dynamics of many interacting parts.

10

Computational Models of Prefrontal Cortex

Two Complementary Approaches

Etienne Koechlin and Xiao-Jing Wang[1]

Abstract

The frontal lobe cortex is among the brain regions that evolve the most across mammals. In rodents, the prefrontal cortex (PFC) comprises the orbitofrontal cortex, the anterior cingulate complex (ACC), as well as the prelimbic and infralimbic areas in the medial wall. In primates, the PFC has evolved with the addition of the lateral PFC. In humans, the PFC features the further development of its most anterior part, especially in the lateral sector, and is often named the frontopolar cortex. Human patients with PFC lesions exhibit little impairments in basic sensorimotor, memory, learning, and language functions. Thus, the PFC function fulfills additional, more abstract functional demands. Its characterization has long remained elusive through the use of poorly defined notions such as executive/cognitive control, working memory, or cognitive flexibility. Here, computational models are shown to overcome these theoretical shortcomings by providing more precise accounts, predictions, and simulations of PFC function at the neuronal and behavioral levels. Two approaches have been developed in neurobiology and cognitive neuroscience, respectively. Time is ripe to integrate the two for a cross-level understanding of PFC function.

Introduction

Computational approaches of prefrontal cortex (PFC) function may start from a simple postulate: PFC function has evolved to enhance animal adaptive behavior. From that respect, computational models of PFC function should address two key overarching issues: (a) how PFC basic cognitive operations emerge from the neural networks that have evolved in the PFC and (b) which

---

[1] Alphabetical listing: both authors contributed equally to this work.

are the key functional limitations of basic adaptive processes external to the PFC that PFC processes overcome in the service of enhanced adaptive behavior. The first issue is addressed through neural network models of PFC operations, primarily based on neurophysiology of single cells from animals performing a task and neural circuit dissection. The second issue is addressed through computational cognitive models of PFC function, primarily informed by behavior and brain imaging like fMRI, in the tradition of cognitive psychology. The interactions between these perspectives are necessary to achieve an understanding of the PFC (Miller and Cohen 2001; Wang 2013). Because PFC is implicated in many psychiatric disorders, progress in this area has spurred translational research that gave rise to the nascent field of computational psychiatry (Wang and Krystal 2014).

Neural Network Models of PFC Operations

Fundamental Cognitive Processes

Biologically based neural circuit modeling strives to build mathematical models across levels, from molecules and cell types to collective neural circuit dynamics to functions. In frontal cortex research, this approach was initially developed for working memory, the brain's ability to internally hold and manipulate information that is essential to enable mental processes separate from direct sensory stimulation. Working memory is commonly studied in the laboratory using delay-dependent tasks, where information about a sensory stimulus must be held internally across a delay period to guide a behavioral response later. Since the discovery of stimulus-selective persistent neural activity during a mnemonic delay period (Fuster and Alexander 1971), its circuit mechanism was investigated experimentally by Patricia Goldman-Rakic (1995) and others, as well as through computational models (Amit and Brunel 1997; Brunel and Wang 2001; Compte et al. 2000a). The main idea is that working memory in the absence of any external input can be actively sustained by recurrent synaptic excitation. Modeling work found that recurrent excitation must be slow and depend on NMDA receptors (Wang 1999), a theoretical prediction that was supported by monkey experiments (Figure 10.1a) (Wang et al. 2013). Thus, slow reverberation is now considered as a characteristic of PFC. This finding is of clinical interest, because NMDA receptor hypofunction is implicated in PFC deficits associated with schizophrenia (Coyle et al. 2003).

In the cortex, excitation is balanced by inhibition, which is mediated by multiple subtypes of GABAergic cells. Motivated by the need for a working memory system to "gate out" behaviorally irrelevant stimuli, Wang et al. (2004c) proposed a disinhibitory motif (Figure 10.1b) composed of three interneuron subclasses. While parvalbumin-positive interneurons control spiking output of pyramidal neurons, interneurons that express somatostatin or calbindin target dendrites are well positioned to gate inputs to pyramidal cells. When pyramidal cells are inhibited by interneurons that express calretinin or vasoactive intestinal peptide, the "gate" would be open, allowing for inputs to enter the circuit. This theoretically predicted disinhibitory motif has now been well-established experimentally (Tremblay et al. 2016). It is noteworthy that, compared to primary sensory areas, the ratio of input-controlling and output-controlling interneurons is much higher in the PFC, presumably tailored to its functional requirements (Wang 2020).

Furthermore, the recurrent neural circuit model initially proposed for working memory turned out to be suitable to account for key computational processes in decision making, which depends on the PFC, posterior parietal cortex and other associated brain regions. Experiments revealed that quasi-linear ramping neural activity over time underlies accumulated information in perceptual decision making (Roitman and Shadlen 2002), which in the model

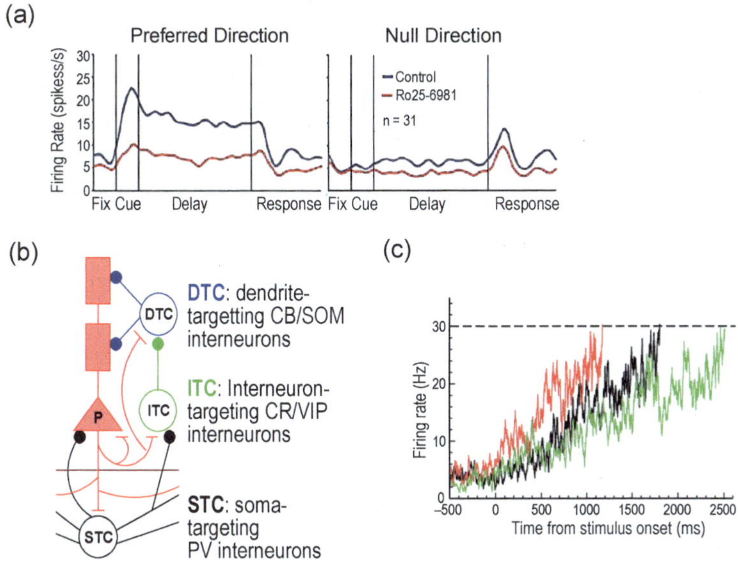

Figure 10.1 Intrinsic circuit properties and dynamics in the prefrontal cortex. (a) Dependence of delay period persistent activity on the NMDA receptors in a monkey experiment using the ODR task. Average response showing the mean firing patterns of 31 dlPFC delay cells for their preferred (left panel) versus nonpreferred directions (right panel) under control conditions (blue) and after iontophoresis of Ro25-6981, a selective antagonist of NR2B-containing NMDA receptors (red). Ro25-6981 markedly decreased task-related firing, especially for the neurons' preferred direction. Reproduced from Wang (2013). (b). The model scheme from Wang et al. (2004) with three inhibitory cell subclasses in addition to pyramidal (Pyr) cells: perisoma-targeting (parvalbumin-containing, PV), interneurons express somatostatin (SST) or calbindin (CB), VIP or calretinin (CR)-containing interneurons. (c) Ramping activity of a recurrent neural circuit model for working memory and decision making (Wang 2002).

is realized by slow reverberation (Figure 10.1c). Attractor dynamics underlying selective persistent activity during working memory produces a categorical choice in a decision process (Wang 2002). These results led to the proposal of "cognitive-type" cortical microcircuit (Wang 2013). Mathematically, the strength of recurrent excitation must exceed a threshold level, when a sudden transition called bifurcation takes place, leading to the functional capability to subserve working memory and decision making.

In summary, neural circuit modeling across levels has yielded several surprises: the idea of slow reverberation mediated by the NMDA receptors, the disinhibitory motif, and a common circuit mechanism for working memory and decision making.

Behavioral Flexibility

The PFC plays a central role in behavioral flexibility, illustrated by the Wisconsin Card Sorting Test as a clinical assessment of frontal lobe function. Can the attractor network model be generalized to rule-guided flexible behavior? Consider a simplified version of the Wisconsin Card Sorting Task. Given a sensory cue (a colored shape, e.g., red circle), a subject selects one of two test stimuli that matches the cue either in color or shape, depending on the task rule (color or shape) (Mansouri et al. 2006). Presumably, the rule that is currently valid, say color, is represented internally by persistent activity of "color rule cells," which must be maintained across trials, but switched off when the rule has changed (e.g., from color to shape), signaled by a negative feedback. To illustrate the problem (Figure 10.2a), assume that the neural activity (high or low, H or L) is determined by two types of inputs: recurrent drive which is high or low depending on whether the neuron is active or not (i.e., the internally maintained rule is color or shape), and feedback signal which can be positive (in which case the activity should stay) or negative (in which case the activity should switch). The required input-output mapping amounts to the exclusive OR operation.

The key to solving this problem is to introduce neurons that show conditional responses; for instance, having firing that is selectively high for a particular stimulus only when rule 1 but not rule 2 is currently valid. This reasoning led Rigotti et al. (2010) to propose the concept of mixed selectivity, by adding to a decision-making circuit a large "reservoir" of randomly connected neurons (RCNs) (Figure 10.2b). The basic idea is that by virtue of random connections, RCNs are naturally activated by a combination of synaptic inputs from external stimuli as well as rule-coding neurons (e.g., the color rule is currently in play *and* the network receives a negative feedback signal), and such mixed selectivity is exactly what is needed to solve the task. This model provides a general framework for describing context- or rule-dependent tasks (Rigotti et al. 2010). Figure 10.2c–d shows such a network model for the simplified Wisconsin Card Sorting Test. Notable is the high degree of variability

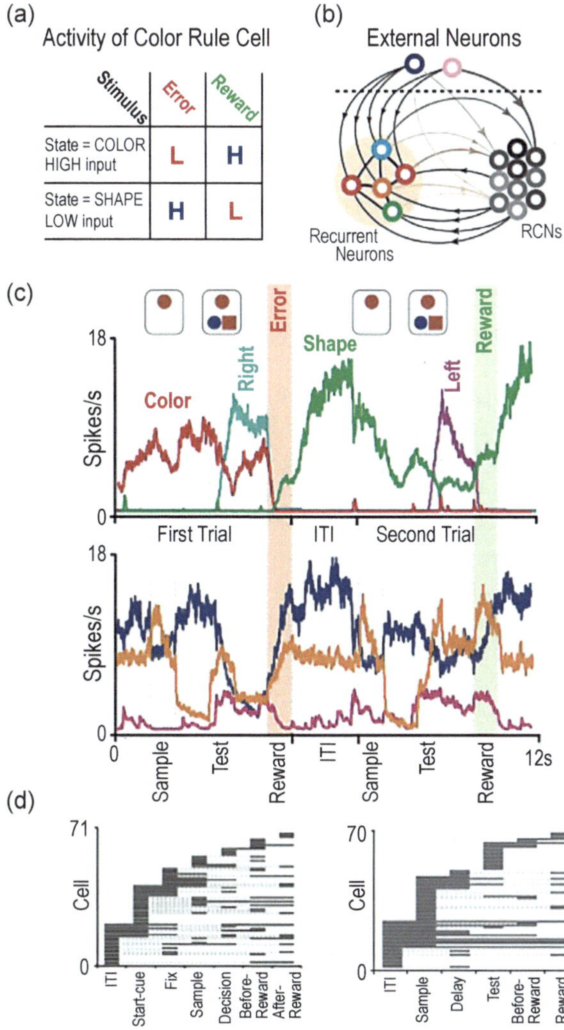

Figure 10.2 A network for rule-based behavior. (a) Exclusive or (XOR) computation by a cell that encodes the rule "color" in a simple variant of the Wisconsin Card Sorting Task; see text for further details. (b) Neural network architecture: randomly connected neurons (RCNs) naturally display mixed selectivity. (c) Firing activity time course for five sample neurons. Light pink vertical line: rule switch; light green line: rule stay. Top: two rule selective neurons; bottom: three RCNs. (d) Rule selectivity pattern is heterogeneous over time and across neurons. Left: rule selectivity for 70 simulated cells in the model. For every trial epoch (x-axis) a black bar is shown when the neuron had a significantly different activity in shape and in color rule blocks. Neurons are sorted according to the first trial epoch in which they show rule selectivity. Right: rule selectivity for spiking activity of single units recorded in prefrontal cortex of monkeys performing an analog of the Wisconsin Card Sort Task (Mansouri et al. 2006). Adopted from Rigotti et al. (2010).

of firing activity, across cells as well as for a single neuron across task epochs. Heterogeneity and mixed selectivity are salient yet puzzling characteristics of frontal cortical neurons recorded from behaving animals. Our model suggests that mixed selectivity is computationally desirable as it allows the network to encode a large number of facts, memories, events, and importantly their combinations, the latter being critically important for enabling the PFC to subserve context- and rule-dependent flexible behavior. The theoretical proposed concept of mixed selectivity has been supported by analysis of PFC single-neuron activity in behaving monkeys, establishing another principle for understanding how the PFC works.

More recent work investigated how a single brain area like the PFC may subserve many cognitive tasks. With the help of machine learning (Figure 10.3a), Yang et al. (2019) built a recurrent neural network capable of performing 20 cognitive tasks that are commonly used in monkey physiological experiments and which engage various core cognitive functions, including working memory, rule-based decision making, categorization, and inhibitory control of responses. This model made it possible to examine subportions of the model that represent neural clusters engaged in different types of cognitive building blocks. Concretely, the extent of engagement in a task by each model neuron is measured by a quantity called normalized task variance (Figure 10.3b). The task variances of each unit form a vector in the 20-dimensional space of tasks, and relationships between units can be assessed using clustering algorithms. Units were self-organized into distinct clusters through learning; those belonging to the same cluster are mainly selective in the same subset of tasks. For instance, inhibitory control is often studied using an anti-response task paradigm where a salient stimulus is shown, orienting toward it is prepotent but must be suppressed; instead, the correct action is a more deliberate response diametrically opposite to the stimulus. Three anti-response tasks (Anti-, reaction time-Anti, and delayed-Anti) primarily engage a distinct cluster #3 (purple), and computationally inactivating units in that cluster impairs only anti-response tasks but not the others.

This model needs to be biologically elaborated to provide insights into the brain mechanism of rule-guided behavior. First, it should obey Dale's law, which states that a given neuron contains and releases only one type of neurotransmitter, so that circuit wiring diagram can be identified with separate excitatory and inhibitory neurons. Second, as discussed above, gating may involve different inhibitory cell types which can be incorporated into the model. Third, the model can be extended to multiple modules that differentially encode task representation and task rule, guided by sensorimotor mapping.

Distributed Process with Functional Specificity

While neural correlates of a cognitive function, such as working memory, are commonly observed in PFC, they are also present in other parts of the cortex,

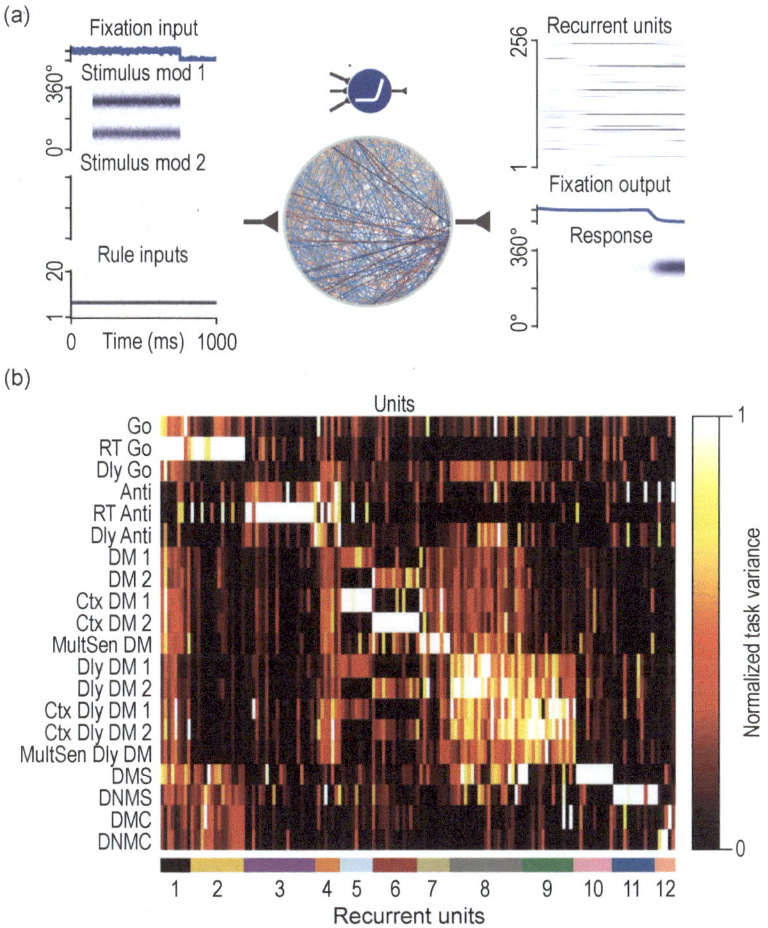

Figure 10.3 A recurrent neural network (RNN) trained to perform 20 cognitive tasks. (a) Schematic of model setting. Left: in a trial, the RNN receives a rule cue, sensory stimuli, and a fixation signal when the network should not produce a motor response. In this example of motion direction discrimination, the stimulus is shown in mod 1 pathway. Right: network dynamics with RNN units (top), fixation unit (middle), motor response unit (bottom). (b) Task variances across all tasks and active units, normalized by the peak value across tasks for each unit. Units form distinct clusters identified based on normalized task variances. Each cluster is specialized for a subset of tasks, such as those that involve a mnemonic delay (Dly). A task can involve units from several clusters; for example, delayed match-to-sample (DMS) engages clusters #1, 2, 8, and 10. Units are sorted by their cluster membership, indicated by colored lines at the bottom. Adapted from (Rigotti et al. 2010).

including the posterior parietal cortex (Leavitt et al. 2017). Because neurons are recorded from the intact brain where areas are interconnected, it is not a given that neural firing in an area related to working memory, even PFC, is generated locally or depends on interactions between multiple areas. As a matter of fact, studies using modern tools for neurophysiological recording and calcium imaging appear to show widespread neural correlates of behaviorally relevant attributes, thus raising the question of how distributed representation can be reconciled with functional localization.

Mejias and Wang (2022) developed a large-scale model of distributed working memory using a directed and weighted connectivity for macaque monkey cortex of Markov et al. (2014). Figure 10.4a–b show model simulation of a visual delayed response task. Notably, responses to an input during stimulus presentation occur in the portion of the model that simulates posterior parts of the cortex, whereas persistent activity during the delay period displays a spatial pattern involving frontal, parietal, and temporal areas of the model. Persistent activity of each area plotted as a function of its hierarchical position exhibits a gap in the firing rate that separates the areas that exhibit mnemonic activity

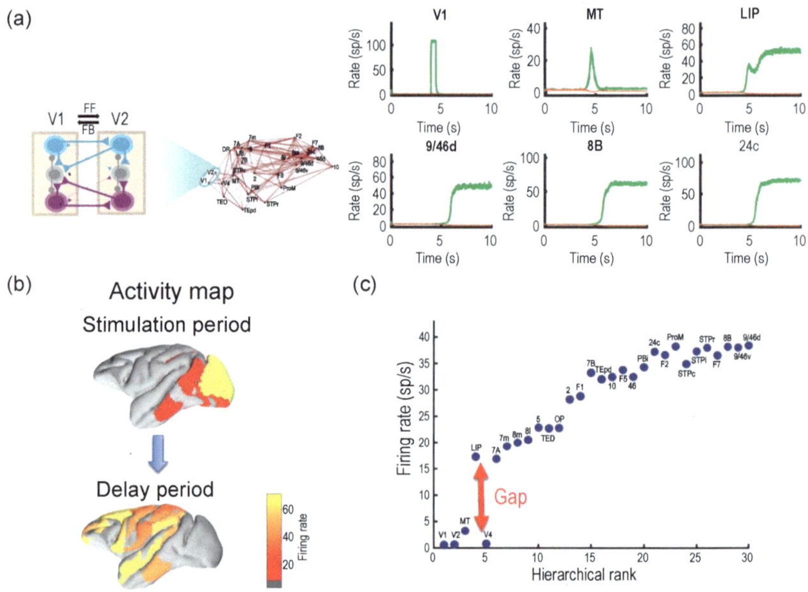

Figure 10.4 Distributed working memory representation in a large-scale monkey cortex model when none of isolated areas is capable of generating persistent activity. (a) Model schema is shown on the left; a model simulation of a visual delayed response task is shown on the right, where activities of the two excitatory neural populations are given for six sample areas. Green: preferred to the shown stimulus; red: nonpreferred to it. (b) Activity map is confined to the posterior part of the cortex during stimulus presentation. By contrast, it is distributed in the frontal, parietal, and temporal areas during the delay period after stimulus withdrawal. Firing rate is shown in color. (c) Mnemonic firing rate of the selective neural pool in each area during the delay period is plotted as a function of its hierarchical position. Those areas displaying persistent activity are separated from those that do not, by a gap in the firing rate. Reproduced from Wang (2020) with original data from Mejias and Wang (2022).

from those that do not (Figure 10.4c). This is reminiscent of a bifurcation, but it occurs in space rather than as a function of a parameter. The transition is robust: changes of network parameters would alter the location of cortical tissue where the transition occurs and show precisely which areas exhibit mnemonic persistent activity but would not abolish the transition itself.

This "bifurcation in space" phenomenon represents a mechanism for the emergence of functional modularity. In the model, parcellated areas follow an identical canonical local circuit organization, but certain properties, like the strength of synaptic excitation, vary systematically in the form of macroscopic gradients calibrated experimentally (Wang 2020). Interareal cortical interactions quantified by the connectomic analysis involve long-range connections, which makes it all the more remarkable that the sudden transition can occur locally in a multiregional cortex. Thus, working memory is distributed, yet depends on a specific subset of areas, in contrast to the absence of modularity manifest by merely graded variations of engagement across the entire cortical mantle. Furthermore, some areas show mnemonic activity as a result of sustained inputs from other core areas, including the PFC. These findings suggest a general principle for understanding functional specificity compatible with distributed cognitive processes.

Summary

In close interplay with experiments, theory has produced new concepts like slow reverberation, disinhibitory motif, cognitive-type microcircuits capable of working memory and decision making, mixed selectivity, and bifurcation in space as a mechanism for the emergence of functional modularity in a large-scale cortex endowed with a canonical circuit architecture. These concepts, derived from biologically based cross-level neural circuit modeling, have furthered our understanding of PFC function. Looking ahead, with the prospect of new availability of big data (ranging from genomic analysis to connectome to large-scale recordings), theory and mathematical modeling are poised to play an indispensable role in elucidating the complex inner working of the frontal lobe at the core of cognition and intelligence.

Computational Cognitive Models of PFC Function

Reinforcement learning (RL) is commonly viewed as describing animal (including human) basic adaptive behavior. Empirical evidence indicates that the basal ganglia interacts with the premotor cortex and lateral orbitofrontal cortex (OFC) to contribute to RL (and likely along with the insular cortex for punishments). RL is a simple, robust, and efficient adaptive process. RL, notably its temporal-difference algorithmic implementation (Sutton and Barto 1998), assumes the brain encodes stimulus-action and stimulus-reward associations that reflect experienced rewards (or punishments). These associations adjust online according to the discrepancy between actual and expected rewards/punishments encoded in these associations and gradually guide action selection toward the most valuable course of actions. We refer to such courses of action as *selective models*. Computational models using RL can learn complex selective models to adapt to complex situations. In particular, when action outcomes depend only on current external states and actions, RL potentially converges toward the behavioral strategy maximizing rewards, regardless of the situation complexity (Sutton and Barto 1998). RL is robust to uncertainty and contingency changes. While more efficient adaptive processes exist and adjust faster to changing situations, their gains relative to RL performances are often weak compared to their increased computational complexity and are obtained at the cost of decreased versatility. For instance, adaptive processes based on Bayesian inferences regarding the external contingency volatility and its variations across time (e.g., Behrens et al. 2007), adjust relatively faster than RL in varying volatile environments but perform much less efficiently than RL in stable environments with sparse environment feedbacks (Findling et al. 2021).

Still, RL exhibits key adaptive limitations. First, RL algorithms learn from reward subjective values, which vary according to animals' internal states or needs. For instance, a thirsty animal may learn through RL an efficient course of actions to obtain water. When the animal becomes hungry, however, this course of action becomes ineffective to get food, and the animal is forced to relearn from scratch a new course of action to acquire food. More generally, the problem arises because RL algorithms learn from the value rather than identity of action outcomes. Overcoming this adaptive weakness requires learning world models, which we refer to as *predictive models*, that link stimuli, actions, and outcomes, irrespective of rewarding values. Such predictive models enable RL algorithms to operate covertly (a process named model-based RL), according to current animal internal states/needs, to build effective selective models on demand and subsequently act in an efficient manner (Gershman et al. 2014; Liu et al. 2021). Learning predictive models remains, in principle, a basic process that corresponds to register the environment statistics. Future research is needed to understand how the animal is driven to learn such predictive models, which appear critical for responding to the ever-changing internal states and needs of an animal.

Second, learning and adjusting selective and predictive models in RL is achieved by erasing previously learned information. This naturally allows these models to adapt to new situations but it also requires the animal to relearn entirely these models when situations encountered in the past reoccur at a later time. Our natural environment actually features a constant mixture of new and recurrent situations: for instance, access to water sources may periodically change according to seasons but also suddenly when unique events occur like forest fires. New and recurrent situations are potentially unlimited; that is, external contingencies form a potentially infinite-dimension space, which prevents animals and actually any physical device from learning and parametrically adjusting only one comprehensive predictive model of the world. Thus, efficient adaptations require animals to gradually learn multiple predictive models as discrete entities, ideally as much as the number of encountered distinct situations: learned predictive models form a repertoire in long-term memory, thus defining a finite but expanding behavioral space, whose dimensions correspond to the number of situations encountered and perceived as distinct. This adaptive process, however, raises complex computational issues in terms of how animals identify situational changes or recurrent versus new situations as well as how they retrieve previously learned selective/predictive models or learn new models when facing recurrent versus new situations (Koechlin 2014).

These two RL adaptive limitations are tightly linked as learning predictive models unfold over time and rely on the assumption that the ongoing situation is identified as remaining unchanged. These limitations appear to be so fundamental for efficient adaptive behavior that we can reasonably assume that the PFC has evolved primarily to overcome them. Another RL functional limitation identified is the lack of learning rate adjustments according to the change frequencies of external contingencies, often referred to as volatility (Behrens et al. 2007). Indeed, efficient adaptive behavior requires learning rates to increase when volatility increases so as to discount previously learned information. Complex probabilistic inference models involving the PFC have been proposed to estimate volatility to make such learning rate adjustments (Behrens et al. 2007; Payzan-LeNestour and Bossaerts 2011). However, a recent computational study (Findling et al. 2021) shows that counterintuitively, such adjustments are likely to derive merely from neural computational imprecisions conforming to Weber's law (the more internal representations change, the more imprecise are representation updates) rather than from additional volatility estimate processes.

Medial OFC and ACC Overcome RL Adaptive Limitations

Empirical evidence suggests that through RL mechanisms, lateral OFC encodes/stores the *experienced* reward value of stimuli, irrespective of associated actions (i.e., in a Pavlovian fashion) (O'Doherty 2007; Rouault et al. 2019), while the premotor cortex encodes/stores stimulus-action associations. Accordingly, lateral OFC provides subjective values of actual action outcomes which enables the learning of stimulus-action associations in the premotor cortex likely via the basal ganglia. Lateral OFC, premotor cortex, and the basal ganglia thus form a basic functional network (possibly along with the insular cortex for punishments) that subserves the RL of *selective models* guiding behavior (Soltani and Koechlin 2022).

In contrast, empirical evidence suggests that medial OFC encodes/stores the identity of action outcomes (i.e., their probability of occurrences following action selection), along with the associated *prospective* reward values (Jones et al. 2012; Rouault et al. 2019). Based on experienced reward values encoded in the lateral OFC, prospective reward values correspond to the current valuation of prospective action outcomes likely based on animals' internal states or needs. This indicates that the medial OFC is likely to encode *predictive models* subserving model-based RL (Chan et al. 2016) to enable RL to operate covertly, and through the basal ganglia to readjust selective models encoded in the premotor cortex to guide behavior according to animals' current internal states/needs.

Computational models introduced the notion of *state beliefs*; namely, probability distributions over predictive models measuring to which extend they apply to the current situation or equivalently, their posterior ability to predict actual action outcomes through standard probabilistic inference processes (Chan et al. 2016). Computational models further introduced the notion of actor reliability—the belief that the predictive model guiding ongoing behavior applies to the current situation relative to any *known* or *unknown* alternative predictive models based on the maximum entropy principle (predictions in unknown situations are at chance level) (Collins and Koechlin 2012). Critically, actor reliability assesses whether the current situation is likely to remain the same or has changed; that is, whether the current predictive/selective model guiding ongoing behavior (referred to as the actor task set) remains reliable or not. In the former case, the corresponding predictive model continues to guide behavior and to improve through online learning. In the latter case, this task set is inhibited and replaced by another one (see below). Empirical studies provide evidence that the medial OFC indeed monitors online actor reliability based on actual action outcomes (Domenech et al. 2020; Donoso et al. 2014a).

Thus, the OFC encodes several signals, including experienced stimulus values in the lateral OFC, prospective outcome values, outcome probabilities, and actor reliability in the medial OFC. All these signals and possibly others may potentially guide behavior. A classical view originating in the rational decision theory states that to achieve action selection, the signals encoded in medial OFC are combined together to compute the subjective expected utility of each related behavioral option—a common currency used as a decision variable to arbitrate between the options. Recent computational studies, however, suggest that instead, these different signals independently compete and concurrently contribute to action selection within the ACC, after each signal type is normalized across available actions (Cao and Tsetsos 2022; Farashahi et al. 2019; Rouault et al. 2019). These studies show that these contributions are not weighted equally at choice time with the predominance of medial OFC signals related to predictive models. The weighting also varies depending upon the environment characteristics. For instance, the more volatile the environment, the less outcome probability signals were shown to predominate, in agreement with the fact that volatile environments prevent an organism from forming precise predictive models (Farashahi et al. 2019). Exactly how the weighting is determined remains an open issue. A possible hypothesis is that the weighting naturally arises from neural reciprocal interactions between the ACC and OFC regions rather than deriving from additional higher-order computational processes. For instance, less precise predictive models are likely encoded with increased neural variability in medial OFC, which in turn should weaken their remote influence on the ACC.

It is also likely that following action outcomes, actor reliability signals play a predominant role. Indeed, the medial OFC was observed to signal proactively that the situation might have changed right before experiencing action outcomes (i.e., the actor reliability is deemed as uncertain), leading the ACC to process actual action outcomes as confirming or denying this medial OFC prediction. ACC was observed to process actual action outcomes as a trigger to inhibit and switch away from the ongoing predictive model or to stay with the same predictive model to guide subsequent behavior (Domenech et al. 2020). Thus, the ACC is modeled as inducing behavior to switch to *undirected exploration* corresponding to the formation and learning of a new predictive and selective model (i.e., a new actor task set for guiding subsequent behavior). Computational models further propose that this new actor is first built from mixing previously learned predictive and selective models stored in long-term memory according to actual action outcomes and then adjusts subsequently to actual external contingencies (Collins and Koechlin 2012; Koechlin 2014) (see Figure 10.5). As the medial OFC monitors actor reliability, this new actor may eventually be deemed as reliable, in which case its selective and predictive model are consolidated in long-term memory to contribute to creating new actors in the future. Neural correlates of such covert confirmation events based on actor reliability were observed within the basal ganglia in the ventral striatum, which receives direct projections from medial OFC (Donoso et al. 2014a). In addition, once the new actor is deemed reliable, it will likely become unreliable at some point, in which case a new actor creation process will be triggered again. Although the neural mechanisms involved in the actor creation and consolidation processes remain poorly specified, we presume that these processes which rely on long-term memory involve a large network of brain regions, notably outside the PFC, which along basal ganglia certainly comprises the hippocampus, known for its central role in memory retrieval and world model constructs (e.g., Whittington et al. 2020). In this view, the medial OFC and ACC control only *when* to create and consolidate new actors whereas the creation and consolidation processes per se appear to unfold outside PFC control. Importantly, the computational model combining actor reliability monitoring, actor creation, and consolidation shows that the repertoire of task sets that comprise joint selective and predictive models, stored in long-term memory, extends in a way that associates more recurrent situations with task set replicas in long-term memory. As a result, actor creation relies more extensively on task sets associated with more recurrent situations. This computational model forms the optimal adaptive process with the constraint

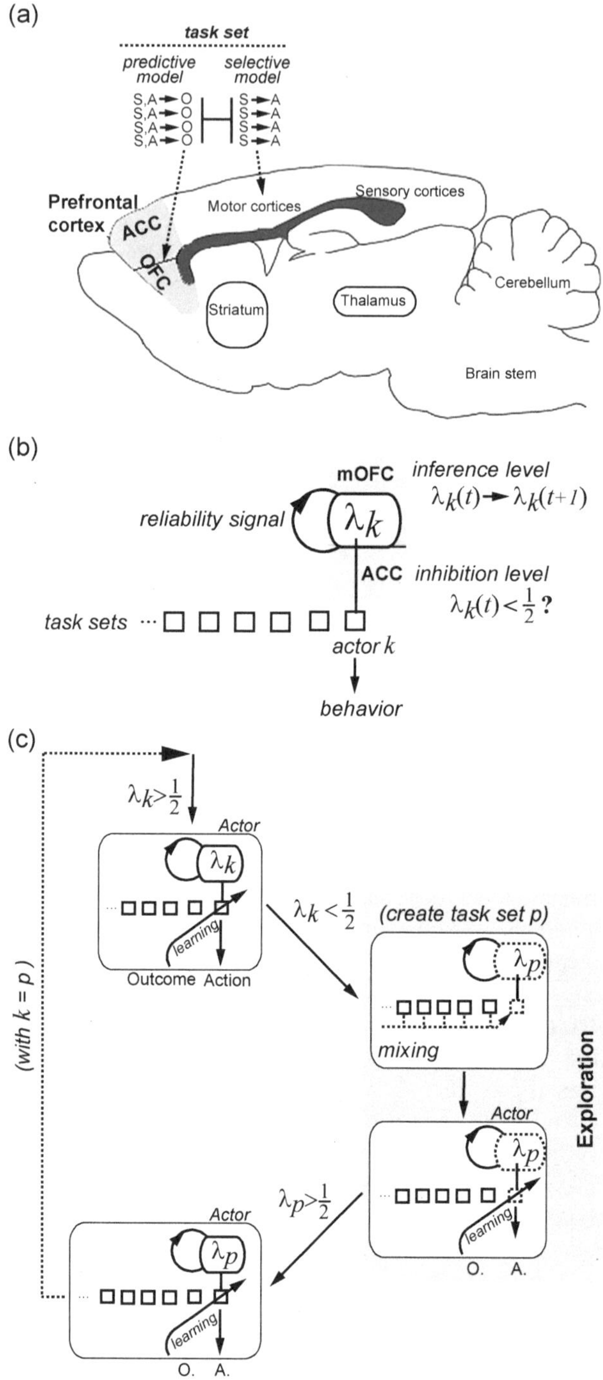
(a)
task set
predictive model
selective model
S,A→O
S,A→O
S,A→O
S,A→O
S→A
S→A
S→A
S→A
Prefrontal cortex
ACC
OFC
Motor cortices
Sensory cortices
Striatum
Thalamus
Cerebellum
Brain stem
(b)
mOFC
inference level
λk(t)→λk(t+1)
reliability signal
λk
ACC
inhibition level
λk(t)<1/2 ?
task sets
actor k
behavior
(c)
λk>1/2
Actor
λk
learning
Outcome Action
λk<1/2
(create task set p)
λp
mixing
Actor
λp
learning
O. A.
Exploration
λp>1/2
Actor
λp
learning
O. A.
(with k = p)

that only actor reliability is inferred online from action outcomes (Collins and Koechlin 2012).

Lateral and Frontopolar PFC Overcome OFC and ACC Adaptive Control Limitations

As described above, the medial OFC and ACC, in association with basal ganglia, form a consistent and efficient system that controls adaptive behavior in uncertain, changing, and open-ended environments beyond basic RL processes. Nonetheless, this medial PFC system exhibits three key functional limitations:

1. Actor reliability is inferred only from actual action outcomes, so that switching away from the current actor occurs only after experiencing actual action outcomes. This might be especially detrimental in case of adverse action outcomes.
2. Actor creation ignores the context in which selective/predictive models were learned, which may lead actor creation to start guiding behavior using maladaptive task sets (i.e., selective/predictive models) stored in long-term memory. For instance, the selective/predictive models I learned when interacting with people at work might not be well adapted when I interact with my roommates and vice versa.
3. By monitoring only actor reliability, the system is constrained to make irreversible decisions, when switching away from the current actor and creating new actors (actor creation cannot be reversed to re-instantiate a new actor creation).

In other words, the medial OFC-ACC system lacks flexibility, which is especially detrimental when dealing with discrete entities such as task sets (i.e., in non-parametric inferences). We have proposed that the evolution of lateral

Figure 10.5 Model of rodent PFC function. (a) Schematic representation of the rodent brain; PFC includes the anterior cingulate cortex (ACC) and orbitofrontal cortex (OFC) to manage the creation of actor task sets to guide behavior. Task sets comprise selective and predictive models (i.e., stimulus-action associations and stimulus-action-outcome associations, respectively). Selective models are encoded in motor/premotor cortices; the OFC encodes predictive models. (b) Diagram showing inferential and creative processes composing the rodent PFC function. OFC infers actor reliability λ (i.e., to predict action outcomes and monitor when the situation changes). While the actor remains reliable ($\lambda > 1 - \lambda$), the actor drives behavior and adjusts its internal models (learning, exploitation periods). ACC detects when the actor becomes unreliable ($\lambda < 1 - \lambda$) and the situation has presumably changed. ACC inhibits the unreliable actor and triggers the creation of a new actor. Actor creation results from mixing task sets stored in long-term memory (square) yielding to forming an unreliable actor. While this newly created actor remains unreliable, it drives behavior and learns external contingencies (exploration period). Once it becomes reliable, it is consolidated in long-term memory, and a new exploitation period starts to create new actors from long-term memory. Reproduced from Koechlin (2020).

PFC in primates overcomes the first two limitations, while further evolution of the frontopolar PFC in humans overcomes the third (Koechlin 2014).

There is ample empirical evidence that lateral PFC is involved in switching between sensorimotor mappings according to contextual cues. Computational cognitive studies in humans further show that stimulus-action associations are spontaneously learned and aggregated into clusters/chunks indexed by contextual cues (Collins and Frank 2013). Such clustering processes, which lead to hierarchical selective models, were shown to occur in posterior lateral PFC (Badre et al. 2010). These results also indicate that the actor task set guiding behavior comprises an additional internal model—the *contextual model*—that links the actor to contextual cues. Accordingly, actor contextual models can be modeled as learning to which extent external cues are predictors of actor reliability (Collins and Koechlin 2012; Koechlin 2014). Thus, lateral PFC enables actor reliability to be inferred from contextual cues and to switch away from the current actor proactively when such cues occur before acting and experiencing action outcomes (see Figure 10.6). Moreover, contextual models enable the contribution of task sets stored in long-term memory to actor creation to be weighted according to the current context of action. As a result, actor creation relies mostly on task sets which were possibly learned previously in similar contexts. In particular, when current contextual cues were previously associated with specific task sets, actor creation operates as if directly retrieving such task sets from long-term memory. Indirect empirical evidence from cognitive control and memory retrieval studies suggests that such cue-based inferences about the reliability of actors and actor creation involve mid-lateral PFC (Koechlin et al. 2003; Nee and D'Esposito 2016, 2017). The resulting executive system that spans the medial PFC (comprising medial OFC and ACC) and lateral PFC form an optimal adaptive system with the constraint that only actor reliability is inferred online (Koechlin 2014).

As noted above, this constraint implies irreversible decisions and yields a system that lacks flexibility to switch back and forth between multiple potential actors guiding behavior. Computational models indicate that overcoming this limitation requires inferring online the reliability of potential alternative task sets in addition to the current actor task sets guiding ongoing behavior (Collins and Koechlin 2012). For clarity, we refer to such potential alternative actors as counterfactual task sets, which as the current actor, consist of selective/predictive/contextual models forming consistent, discrete executive entities. Inferring in parallel the online reliability of multiple task sets is beneficial in many respects:

1. Reliability inference is improved as each task set now measures to which extent its predictive model applies to the current situation relative to the other task set predictive models along with any additional, unknown/unmonitored predictive models.

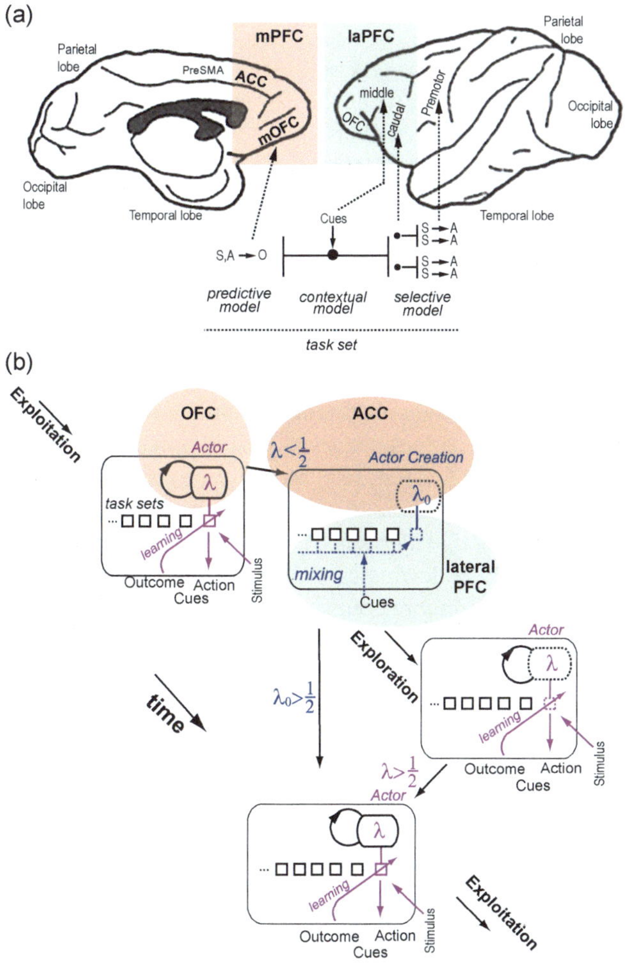

Figure 10.6 Model of monkey PFC function. (a) Schematic representations of the monkey medial and lateral cerebral cortex. Compared to rodents, monkey PFC has an additional lateral prefrontal cortex (laPFC) comprising a middle and caudal sector. In monkeys, task sets are assumed to comprise contextual models (associating task set to external cues) encoded in the laPFC. Contextual models indexing task sets are represented in the middle laPFC and allow chunking processes in caudal laPFC to operate within task sets (see text). (b) Illustration of the inferential and creative processes in the monkey PFC function. Inferential processes are similar to those in rodents (see Figure 10.5), except that contextual models enable the updating of actor reliability according to the occurrences of external cues (in addition to action outcomes). Actor creation may thus occur proactively. Contextual models also have a major role in refining actor creation: the mixture of task sets in long-term memory is now weighted by current external cues according to contextual models. As a result, new actors may be created as immediately reliable ($(\lambda_p > 1-(\lambda_p)$). In that event, the exploration period is skipped, leading to the ability to recreate new actors much more rapidly. Reproduced from Koechlin (2020).

2. When a counterfactual task set becomes reliable (implying that the current actor task set is deemed unreliable), the system simply switches to this counterfactual task set to replace the current actor and guide subsequent behavior, while the current actor guiding ongoing behavior becomes a counterfactual task set.
3. When no task sets are deemed as reliable, actor creation from long-term memory occurs for guiding subsequent behavior, while the current actor again become a counterfactual task set.
4. Actor creation may be rejected later on, whenever a counterfactual task set becomes reliable, while the newly created actor is still not deemed as reliable, preventing the newly created actor task set from being consolidated in long-term memory.

The resulting actor creation process thus resembles hypothesis testing: a new hypothesis (the newly created actor task set) is tested against alternative hypotheses (the counterfactual task sets) based on the acquisition of additional information from action outcomes or contextual cues. This computational model forms an optimal adaptive algorithm in uncertain, changing, and open-ended environments with the following constraint: only a limited number of counterfactual actors can be monitored online in parallel (see Figure 10.7). This computational model was shown to account well for human performances in such environments and performed better than several alternative models. Model fitting to human performances further suggests that humans monitor online no more than three counterfactual task sets in parallel (Collins and Koechlin 2012). When this capacity limit is reached, the least recently used counterfactual task set is simply discarded from online monitoring, while remaining stored in long-term memory. Furthermore, empirical evidence shows that the human frontopolar cortex monitors the reliability of counterfactual task sets (Donoso et al. 2014a; Mansouri et al. 2017), and as mentioned before, the current actor reliability is monitored in medial OFC. Rejecting actor creation to select a counterfactor task set monitored in the frontopolar cortex and deemed

Figure 10.7 Model of human PFC function. (a) Schematic representations of the human cerebral cortex. Compared to monkeys, human PFC comprises a frontopolar region (poPFC) in the lateral forefront of the PFC with no known homologues in monkeys. In humans compared to monkeys, task sets are likely to comprise two nested, abstract levels of chunking, involving BA 44 and 45, and may play a major role in language (see text). (b) Inferential, selective, and creative processes forming the human PFC function. Compared to monkeys (see Figure 10.6), the human poPFC forms an inferential buffer to infer and monitor the reliability of additional task sets (counterfactual task sets) in addition to the actor task set monitored in the medial OFC. This additional inferential capability endows humans with the ability to retrieve a counterfactual task set directly to drive behavior when it becomes reliable, in both exploitation and exploration periods. During exploration, this ability yields newly created actors to be rejected and disbanded and corresponds to hypothesis testing bearing upon task set creation. Reproduced from Koechlin (2020).

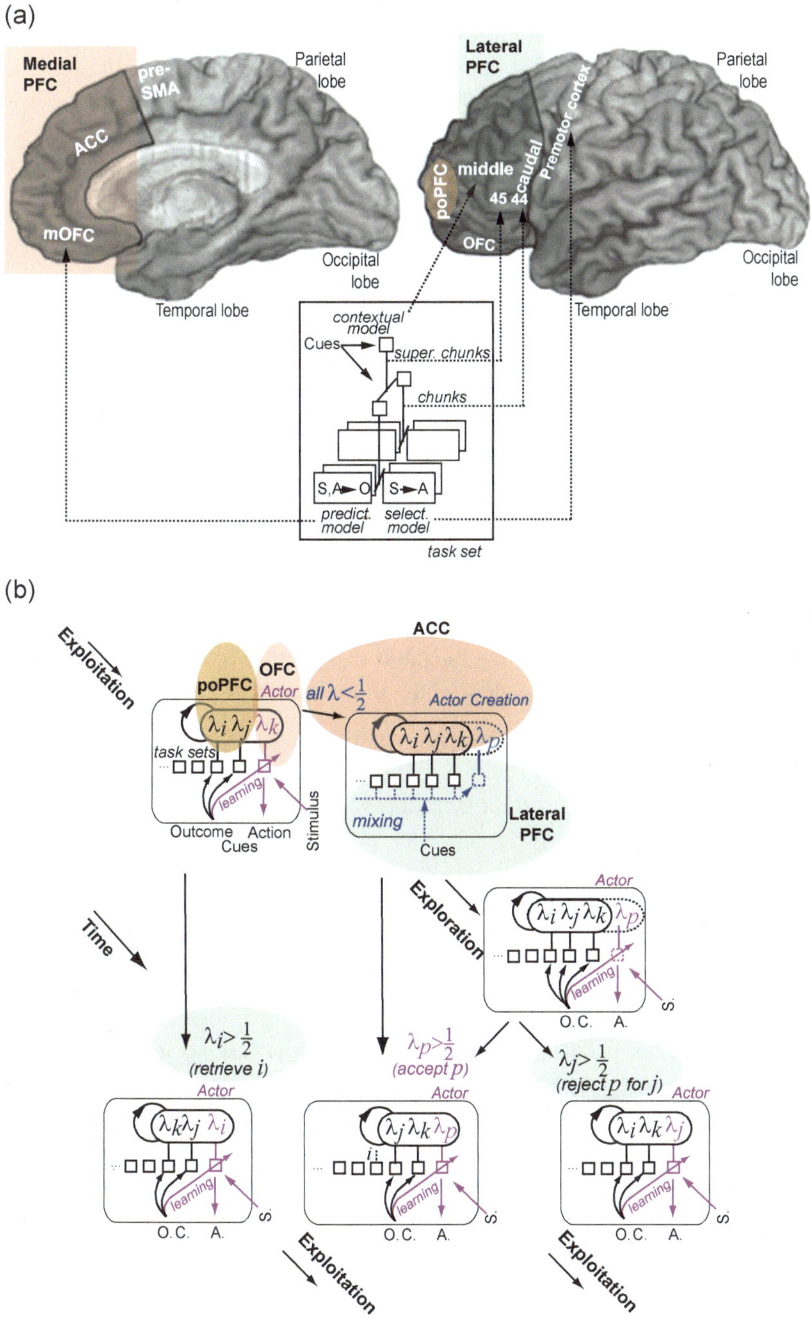
(a)
Medial PFC
pre-SMA
ACC
mOFC
Parietal lobe
Occipital lobe
Temporal lobe
Lateral PFC
poPFC
middle
caudal
Premotor cortex
45 44
OFC
Parietal lobe
Occipital lobe
Temporal lobe
contextual model
Cues
super. chunks
chunks
S,A→O
S→A
predict. model
select. model
task set
(b)
Exploitation
poPFC
OFC
Actor
ACC
all $\lambda < \frac{1}{2}$
Actor Creation
$\lambda_i \lambda_j \lambda_k$
task sets
learning
Outcome Action
Cues
Stimulus
$\lambda_i \lambda_j \lambda_k \lambda_p$
mixing
Cues
Lateral PFC
Time
Exploration
Actor
$\lambda_i \lambda_j \lambda_k \lambda_p$
learning
O.C. A. S.
$\lambda_i > \frac{1}{2}$
(retrieve i)
$\lambda_p > \frac{1}{2}$
(accept p)
$\lambda_j > \frac{1}{2}$
(reject p for j)
Actor
$\lambda_k \lambda_j \lambda_i$
learning
O.C. A. S.
Actor
$\lambda_j \lambda_k \lambda_p$
learning
O.C. A. S.
Actor
$\lambda_i \lambda_k \lambda_j$
learning
O.C. A. S.
Exploitation
Exploitation

as reliable to guide subsequent behavior was further found to involve mid-lateral PFC (Donoso et al. 2014a). Thus, this computational model suggests that the human frontopolar cortex forms a capacity-limited online monitoring buffer that allows switching back and forth across several potential task sets to guide behavior and regulate the online creation and storage of task sets in long-term memory through hypothesis-testing processes.

It is worth noting that *the* true optimal adaptive model requires additional features:

- A monitoring buffer with unlimited capacity, so that the reliability of all created task sets is inferred in parallel.
- Reliability inferences are not limited to online forward inferences but also involve critically offline backward inferences to enable constant online revision of actor creation.
- Actor guiding behavior is the continual parametric mixture of all created task sets weighted by their reliability.

More precisely, the optimal adaptive model involves mixtures of Dirichlet processes that generalize Bayesian inferences to open-ended environments (Doshi-Velez 2009; Gershman et al. 2010; Teh et al. 2006), but whose computational costs are exorbitant and even intractable, thereby hindering its optimality in practice. Accordingly, we have reviewed evidence that the human PFC has evolved as capturing tractable algorithmic approximations of key computational components underlying optimal adaptive behavior:

- Monitoring (a limited number of) multiple potential task sets,
- A minimal form of backward inferences through hypothesis testing involved in actor creation (creating new actors may be revised later on), and
- Mixing all created task sets weighted by contextual models when actor creation occurs to guide behavior.

Newly created actors are thus parametric mixtures of previously learned selective, predictive, and contextual models. Note that in contrast, mixing the task sets monitored in a *capacity-limited* buffer according to their relative reliability to guide behavior is detrimental, because the proper task set might actually be stored in long-term memory without being monitored.

Higher cognition comprising planning, reasoning, and language production might simply reflect the functioning of this whole computational PFC architecture (Koechlin 2020). As noted above, planning amounts to covertly navigating within the current actor predictive model through model-free RL using the actor task set. Reasoning can amount to combining reliability inferences about several potential task sets viewed as multiple behavioral hypotheses with hypothesis-testing regarding actor creation viewed as hypothesis generation. Language production may amount to actor creation viewed as generating linguistic sentences in reciprocal interactions with the superior temporal sulcus through the arcuate fasciculus (Rouault and Koechlin 2018).

What Drives Learning in Predictive Models

The preceding sections outline the key role of predictive models for efficient adaptive behavior. Predictive models as "world models" predict potential action outcomes, enable adjustment of selective models to internal states/needs, and have the ability to carry out planning covertly, and to detect situational changes that may result in actor changes through actor reliability inferences. We indicated above that learning predictive models is based simply on registering the experienced environmental contingencies. This could happen on the fly while other incentives, such as rewards, are driving animals' behavior. Given the critical role of predictive models, we reason that learning predictive models might also be an intrinsic motivation driving animals' behavior.

The classical theory is that animals/humans' behavior is primarily driven through the maximization of subjective rewards (e.g., Schultz 2015). To be efficient, reward maximization requires deviating episodically from what was learned as the most rewarding course of action and to explore alternative courses of action so as to avoid being trapped in local reward maxima. In this view, predictive models are learned on the fly; there are no specific incentives to learn them.

Another theory proposes that animals/humans' behavior is primarily driven by minimizing expected free-energy or "expected surprise" (Friston 2010): behavior aims at producing outcomes expected from predictive models, and predictive models are adjusted according to actual action outcomes. Under this view, potential subjective rewards are absorbed as highly expected outcomes in predictive models. The theory offers a general, principled view of adaptive behavior, revealing that behavior is centered on learning adequate predictive models and acting accordingly. The theory has, however, two key limitations. First, it assumes that agents have an exhaustive representation of all potential situations (latent states) they may encounter, corresponding to as many task sets that they monitor in parallel to form beliefs about their occurrences. This assumption is unrealistic in real-life environments that feature unlimited potential situations. As noted above, biological systems and physical devices are limited inasmuch as they only monitor a small fraction of potential situations/task sets. Discussion in the preceding sections actually outlines the optimal adaptive system, when the monitoring/inferential capacity is assumed to be limited and suggests that the evolution of PFC implements this capacity-limited adaptive system. Second, and more problematically, the theory relies on an *arbitrary* parametrization of potential subjective rewards aimed at transforming them into outcome expectations to absorb them into predictive models. This is problematic because parametrization actually determines the critical balance between reward- and information-seeking behavior;

that is, between exploitation and exploration. Accordingly, the theory appears to define this balance arbitrarily with no accounts of how it is determined and possibly controlled.

To address this issue, we proposed an alternative theory at FENS 2022, based on the fundamental principle of statistical physics. In contrast to Friston's free-energy theory, it distinguishes between the notion of energy and entropy which translate here to the notion of reward as energetic resource and information as negative entropy (Vaillant-Tenzer and Koechlin 2023). The general idea is that behavior aims at primarily maximizing expected information gain *with the homeostatic constraint* to maintain enough energetic resources (i.e., to get enough rewards compensating resource consumption) to pursue this information quest. (Note that unlike biological systems, physical systems "behave" in the converse way as maximizing their entropy with the constraint of maintaining their energy constant.) Within the computational framework outlined in the preceding sections, maximizing expected information gain means selecting actions where the outcomes are expected from the current actor predictive model to best improve predictive power or equivalently to best reduce its predictive entropy/uncertainty. The "statistical physics" formalization of this principle leads to the hypothesis that behavior aims at maximizing the weighted sum of expected subjective rewards and expected information gain within the current actor predictive model. Critically, the weighting of expected subjective rewards relative to information gain is fully determined by the Lagrangian multiplier relative to the homeostatic constraint. This Lagrangian multiplier is not computable in closed form but varies approximately as the inverse of the total amount of agent's energetic resources and consequently as the inverse of accumulated rewards over time. Accordingly, the more an agent is deprived, the more it will exhibit reward-seeking behavior. The more an agent accumulates rewards, the more it will exhibit information-seeking behavior. The more an agent acquires predictive knowledge of the current situation (i.e., expected information gain will vanish), the more it will exhibit reward-seeking behavior. Thus, the hypothesis predicts a complex dynamic balance between reward- and information-seeking behavior. For instance, when an agent faces a new, unknown situation, information-seeking behavior will first dominate as expected information gains within the current actor predictive model are initially at a maximum: thereafter, reward-seeking behavior will begin to dominate as expected information gains start to decline. Next, when received rewards start accumulating, information-seeking behavior will emerge again. And so on. The hypothesis thus predicts that the balance between reward- and information-seeking depends on the agent's homeostatic states and is likely mediated by brain regions monitoring such homeostatic states. A possible candidate where this occurs is the anterior insular cortex, which has been recently associated with homeostasis monitoring and which widely projects to medial PFC regions (Livneh et al. 2020).

Concluding Remarks

In this chapter, we have described the modeling of neural networks and cognitive computations subserving PFC function in two distinct sections. This division is unrelated to any distinctions between the classical Marr's levels of brain analysis—namely the physical, representational, and functional level—whereby the functional level describes the function of one system, the representational level how this function is achieved, and the physical level the material device realizing this function (Marr 1982). Both sections independently entail descriptions at all of the three levels. For instance, the first section describes the working memory function, whereas the second addresses the reliability monitoring function. The first section describes different classes of inhibitory neurons, whereas the second section describes different cortical areas in the PFC and so on. Instead, the first and second section address functional, representational, and physical issues at two distinct *scales* of brain organization: at the neuronal and cortical scale, respectively. The two sections reflect the idea that the functional, representational, and physical concepts differ between these two scales of analysis.

We view these conceptual differences across scales as similar to those present in physics. For instance, pressure makes sense at the scale of gas volumes but not at the level of gas molecular constituents. This does not imply that there are no connections between the elements describing the different levels of brain organization. On the contrary, quantitative models are especially useful, if not necessary, to understand how the different organization levels are connected and interact with each other. To date, however, there is little modeling work that aims to link the different neuronal and cortical levels in the PFC, in the way as in the visual system, models of cortical maps, and hierarchical visual processing have been developed. Filling this gap will certainly be an important future avenue in developing models and understanding frontal lobe function.

Acknowledgments

This work was partly supported by an NIH grant R01# MH062349.

11

What Do Network Approaches Add to Our Understanding of Prefrontal Cortex and Executive Function?

Caterina Gratton, Zach Ladwig, and Diana C. Perez

Abstract

Regions in the human frontal lobe form distributed large-scale brain networks, with connections to one another and other locations in the cortex, striatum, thalamus, and cerebellum. Here, evidence is reviewed that multiple networks lie side by side in the frontal lobe, and these networks are largely (but not entirely) parallel, or separate, from one another. These network findings improve our understanding of frontal lobe organization and help constrain theories of executive function and the impact of brain disorders. Ongoing challenges in the study of frontal lobe networks are discussed related to tracking functional associations of brain networks, individual differences, and changes in networks over time.

Introduction

Many approaches used to study the frontal lobe focus on the characteristics of isolated regions. In contrast, another class of approaches examines the frontal lobe through a distributed processing lens, characterizing how regions relate to one another and to other parts of the brain. In this chapter, we discuss the value of this "network" perspective.

Large-scale networks or "systems" of the human brain can be defined in various ways but most often refer to sets of brain regions that are interconnected anatomically or exhibit covarying activity patterns (referred to as structural and functional connectivity, respectively) (see Sporns 2016 and Appendix 11.1). These properties are taken as evidence that neurons in these regions frequently interact to complete different aspects of brain function (Petersen and Sporns 2015). Large-scale networks represent an important scale of brain organization (Churchland and Sejnowski 1988), and connectivity is often considered a defining feature for delineating brain areas, together with cytoarchitectonics, function, and topography (Van Essen and Glasser 2018).

We argue here that network approaches are crucial to the study of the frontal lobe, in particular the prefrontal cortex (PFC). The frontal lobe contains many distinct but closely juxtaposed networks with stereotyped patterns of connections across the cortex and subcortex. Several of these networks are linked to executive functions—the set of functions that allow one to control thoughts and actions in the pursuit of a goal, overriding automatic behavior. Thus, network characteristics can constrain theories of both function and dysfunction of the frontal lobes.

What Do Network Approaches Add to Our Understanding of the Frontal Lobe?

In this section, we review four insights gained from using network approaches to study the frontal lobe. We describe (a) principles of frontal lobe network integration and segregation: different regions in the frontal lobe are reciprocally connected within networks with a stereotyped spatial topography, but these networks are largely distinct, or parallel, to one another. These findings provide a platform (b) to examine how executive function is supported by multiple distributed networks, (c) to situate PFC networks within a whole-brain complex system, and (d) to understand the consequences of PFC disorders from the lens of network connectivity.

Regions of the Frontal Lobe Form Multiple Distributed and Parallel Networks

Early observations of large-scale networks in the frontal lobe were grounded in tract-tracing work in macaques (e.g., Goldman-Rakic 1988; Mesulam 1981, 1990; Selemon and Goldman-Rakic 1988). These influential studies mapped the afferent and efferent connections of PFC subregions and found evidence for parallel segregated networks, which involved interconnected parietal, frontal, temporal, and subcortical regions (Figure 11.1). For example, Mesulam (1981) proposed that spatial attention is not supported by a single region, as had been previously hypothesized, but is instead supported by a distributed system of reciprocally connected brain regions, including the posterior parietal cortex, frontal cortex, cingulate cortex, and subcortical areas in the intralaminar thalamic nucleus, brainstem raphe nuclei, and locus coeruleus. He proposed that while different components of that system may support different components of behavior (thus lesions to different areas within the network may create slightly different deficits), attention emerges as a property of the network as a whole. Goldman-Rakic (1988) furthered this view, showing that when areas of the parietal and PFC are anatomically connected, they often sent convergent connections to the same temporal, midline, and subcortical regions, forming a densely interconnected system. However, while regions in these networks have strong connectivity to each other, they have minimal connectivity outside of their network. This led Goldman-Rakic (1988) to propose that a fundamental property of these distributed networks is their largely parallel or independent nature from one another.

Modern tract-tracing work has characterized connectivity of large-scale networks further, partly thanks to the creation of large databases of tract-tracing results (Giarrocco and Averbeck 2023; Stephan et al. 2001; Wang et al. 2020). This work has extended earlier observations of structurally distinct networks (e.g., Hilgetag et al. 2000) and led to an understanding of how these networks function using a complex systems lens; for example, by identifying hub regions of the brain with distributed network connections (Markov et al. 2013; Sporns et al. 2007), map-like topographic patterns of organization that are mirrored across regions (Averbeck et al. 2014; Haber et al. 2006), and nested models of cortical connectivity, with subnetworks breaking up larger network systems (Giarrocco and Averbeck 2021). As the collection and analysis of macaque connectomes grows, this work will continue to influence the analysis of human connectomes based on fMRI that is the focus of the remainder of this chapter.

We focus on networks measured with fMRI functional connectivity, which measures patterns of covariation in BOLD activity between different brain

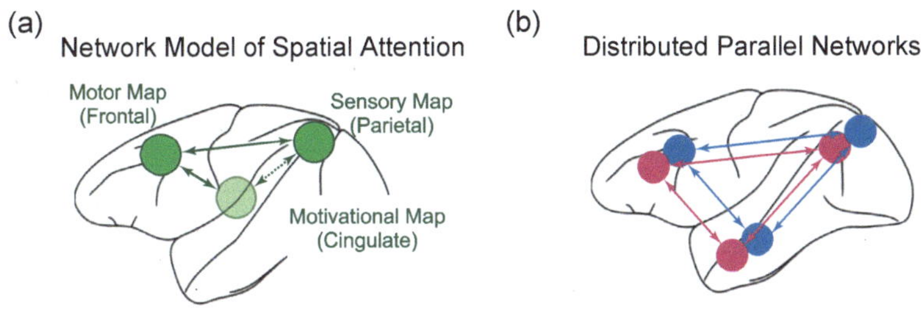

Figure 11.1 Early observations of large-scale networks were made using tract-tracing techniques in macaques and lesion studies in humans. (a) On the basis of axonal tracing and analysis of patients with focal cortical damage, Mesulam (1981) proposed that spatial attention is emergent from a network of distributed regions including frontal, parietal, and cingulate cortex. He proposed each node of the network supports a slightly different representation (sensory, motor, and motivational) of spatial attention. (b) Using axonal tracing, Goldman-Rakic (1988) identified that parietal area 7A and frontal area 46 in the macaque sent projections to many of the same regions, defining a large-scale connected network. Further, she found that regions which lie side by side in parietal cortex often project to regions that lie side by side in other parts of the brain, thus supporting a model of segregated, parallel networks. Figures were drawn by the authors based on work from Goldman-Rakic (1988) and Mesulam (1981).

regions, because it has emerged as a dominant approach to track brain networks in humans. Our working hypothesis is that large-scale networks measured with functional connectivity represent relatively stable organizational elements of the brain and, as such, they should converge with other methods of identifying brain systems, although it is important to remember that they are not the same.[1] For further discussion, see Appendix 11.1.

Using functional connectivity methods, substantial evidence has provided an extended account of distinct networks, components of which are in the human frontal lobe. This includes the delineation of the default, cingulo-opercular, and frontoparietal networks, among others (Figure 11.2). The existence and topography of these networks has been consistently found in several different datasets in different populations, different scanners, and using different network techniques (e.g., Gordon et al. 2017; Power et al. 2011; Smith et al. 2009; Sporns 2016; Yeo et al. 2011). In contrast with the more spatially proximal sensorimotor systems, association networks of the PFC are linked to cortical parietal, temporal, and midline regions, as well as with distinct subregions of the basal ganglia, thalamus, and cerebellum (Figure 11.2c) (Seitzman et al. 2020), forming a distributed pattern.

Figure 11.2 Using functional connectivity to define large-scale human brain networks. (a) Functional connectivity is a measure of the temporal association between the patterns of activity of two brain regions. If the two regions are functionally related, such as the left (orange) and right (blue) motor cortex, the correlation between their BOLD timeseries will be high (top box). In contrast, if the two regions are not functionally related, like the left motor cortex and the left visual cortex (green), then their activity correlations will be low (bottom box). See Appendix 11.1 for an extended description of functional connectivity measures from resting-state fMRI. (b) Functional connectivity across all regions in the brain can be displayed in a correlation matrix, where each cell represents the relationship between a pair of regions. In these matrices, we see a characteristic structure where the within-network correlations (on-diagonal) are high, whereas the between-network correlations (off-diagonal) are lower. These patterns can be used to group regions into networks with data-driven clustering methods (marked by lines in the matrix). (c) Functional connectivity can be used to map the network organization of the cerebral cortex, subcortex, and the cerebellum; different colors represent different networks and mapping onto rows in (b). A description of network terminology is provided in Table 11.1. Figures were drawn using conventions from Power et al. (2011), Seitzman et al. (2020), and Van Dijk et al. (2010).

[1] For example, functional connectivity from fMRI is moderately correlated to anatomical connectivity measured with diffusion MRI (Honey et al. 2009). While this relationship suggests a link between structure and function, a number of differences are also present. Some differences may have functional significance, but differences can also arise for methodological reasons (Mnih et al. 2015). For example, correlational measures from resting state measure indirect as well as direct connections (Petersen and Sporns 2015), and signals may be unreliable without sufficient data (Gordon et al. 2017); diffusion based measures of anatomical connectivity have difficulty tracking branching, turning, and crossing fibers (Grisot et al. 2021); head motion can bias both measures (Baum et al. 2018; Power et al. 2015). An avenue for continued research will be to study the relationships across these techniques and the merits of joining information across modalities.

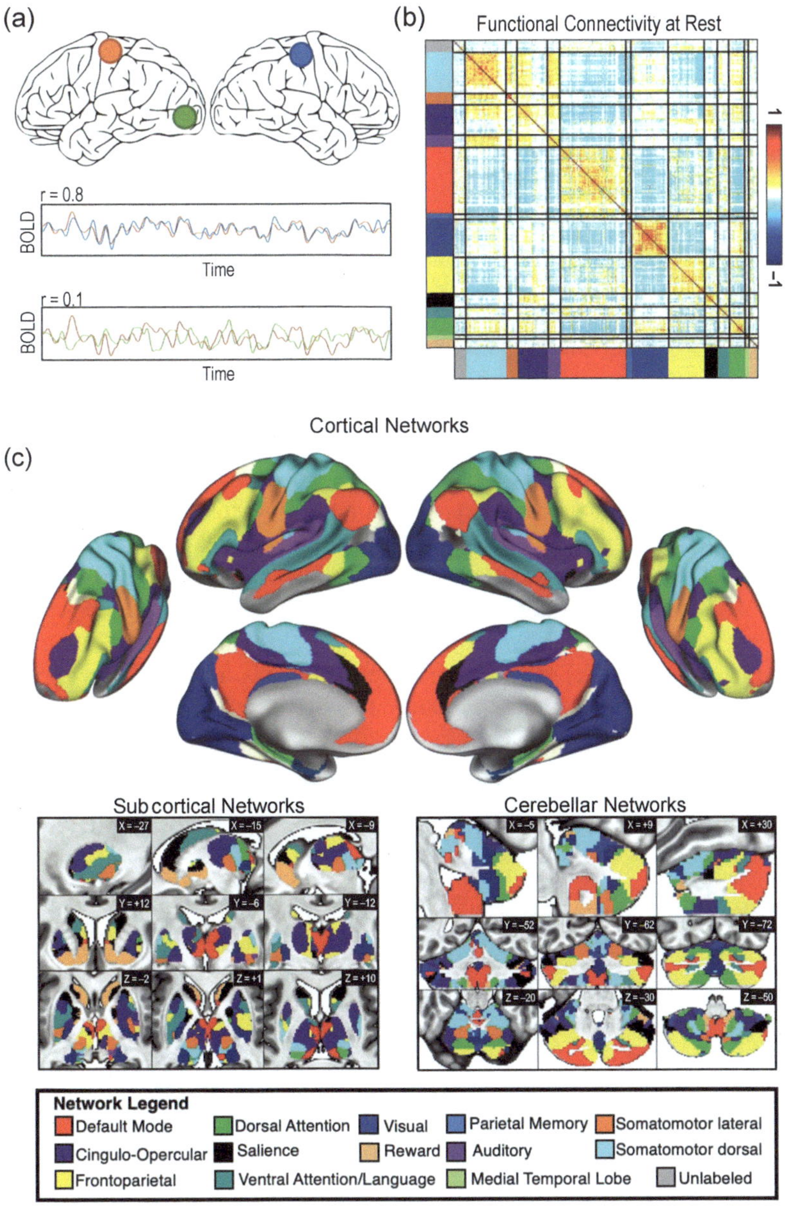
(a)
r = 0.8
BOLD
Time
r = 0.1
BOLD
Time
(b)
Functional Connectivity at Rest
1
Correlation (r)
−1
Cortical Networks
(c)
Sub cortical Networks
Cerebellar Networks
Network Legend
Default Mode
Dorsal Attention
Visual
Parietal Memory
Somatomotor lateral
Cingulo-Opercular
Salience
Reward
Auditory
Somatomotor dorsal
Frontoparietal
Ventral Attention/Language
Medial Temporal Lobe
Unlabeled

Notably, a source of confusion in the literature on human brain networks is that there is limited consensus in the field on the proper terminology (and taxonomy) with which to refer to these networks (Uddin et al. 2022). Some of the disagreement stems from differences in resolution: an apparent network at one resolution may divide into two or more separate components when examined at another. Different definition methods may also give rise to distinctions: e.g., single task contrasts may not correspond to full resting-state networks or may join multiple networks together. Finally, lack of anatomical specificity in published works, as well as anatomical variation in functional network locations across individuals, adds to the ambiguity of separating closely interposed systems. As the literature currently stands, it can be difficult to determine which network someone is referring to by name unless it is accompanied by an anatomical map (Uddin et al. 2019, 2022). Recent efforts aim to address these issues by providing quantitative representations of the variability in functional network definitions, delineating regions that commonly fall into these networks across most individuals as compared to brain regions that show variability in the network they belong to across individuals (Dworetsky et al. 2021).

Moving forward, these tools will aid researchers in addressing some of the ambiguity that has plagued past reports. For clarity, we provide a taxonomy for the large-scale networks of the PFC that we follow, see Table 11.1, based on Gordon et al. (2017), Power et al. (2011), and Seitzman et al. (2020), together with some common variations (e.g., Yeo et al. 2011). We also provide a visual representation of these networks and their typical (group average) anatomical patterns (Figure 11.2c) and point the interested reader to materials associated with the following references for full downloadable maps (Dworetsky et al. 2021; Power et al. 2011; Seitzman et al. 2020; Yeo et al. 2011).

Table 11.1 Large-scale networks of the PFC and their nomenclature.

Network	Terminological description
Frontoparietal network	Sometimes called the "central executive" or "cognitive control" network, it corresponds with the Yeo et al. (2011) "frontoparietal" network (7-network parcellation) and Control A from Kong et al. (2019). It is sometimes joined with the dorsal attention or cingulo-opercular networks (e.g., Fox et al. 2005). However, the cingulo-opercular and frontoparietal networks actually have very low (near zero, and often negative) intercorrelations, suggesting that they are unlikely to be closely related brain systems. Relative to the dorsal attention network, the frontoparietal network is more positively correlated with the default, and less to visual and somatomotor networks. Adding to confusion in descriptions of this network, it is consistently identified as highly variable across individuals (Gordon et al. 2017; Kong et al. 2019; Seitzman et al. 2019). This suggests that individual-level mapping is necessary to distinguish accurately the frontoparietal from other interdigitated systems (see Challenge 2, below).

Network	Terminological description
Cingulo-opercular	Many studies refer to networks with this distribution as the "salience," but we separate this network from another nearby system with that name but slightly differing anatomical distribution (Gordon et al. 2017; Power et al. 2011). While intercorrelated, the network we call cingulo-opercular has stronger relationships to somatomotor systems and weaker relationships to the default network than does the salience (Figure 11.2). In Yeo et al. (2011), this network is called "ventral attention" (7-network parcellation). As described, some studies join the cingulo-opercular and frontoparietal network into a single system, despite their low intercorrelations.
Default	This network is relatively consistently named and identified (Power et al. 2011; Uddin et al. 2022; Yeo et al. 2011) but can vary substantially in extent across papers; sometimes it encompasses regions of the ventral attention, language, and salience networks (note that these three networks are smaller and more variable in position across people, which may contribute to these differences). Recent evidence suggests that the default network is composed of at least two separable subnetworks (Andrews-Hanna et al. 2010). Due to anatomical variability, these most clearly differentiate in individuals (Braga and Buckner 2017).
Dorsal attention	This network was first described by Corbetta and Shulman (2002) on the basis of common task activations and monkey electrophysiological responses. In functional connectivity, it is sometimes joined with the frontoparietal, but differs in its connectivity to the default, visual, and somatomotor systems. In Yeo et al. (2011), it is also referred to as the dorsal attention network (separate from the frontoparietal network).
Language/ventral attention	Originally, we termed this network "ventral attention" (Power et al. 2011) in relation to the influential work by Corbetta and Shulman (2002). However, this network shows considerable overlap in distribution with language localizers (Braga et al. 2020) and has been labeled as "language" in more recent work. In general, this network is variable across individuals and not always consistently identified in group maps with data-driven methods. A network with this distribution is not identified in the 7-network parcellation (Yeo et al. 2011), although components emerge in the 17-network parcellation. Kong et al. (2019) name this the "temporal parietal" network.
Salience network	We identify a small network, "salience," which has a similar but distinct anatomical distribution from the cingulo-opercular network. This network appears in ventral regions of the anterior insula, and in more rostral components of the anterior cingulate, often extending further along the anterior cingulate gyrus in patterns that differ across individuals (Gordon et al. 2017). This network shows relatively higher correlations with the default and lower correlations with the somatomotor systems than the cingulo-opercular network. A network with this distribution was not identified by Yeo et al. (2011), but a similar network in Kong et al. (2019) is labeled "Control C."

As seen in Figure 11.2c, multiple networks have components in the PFC, including the frontoparietal (yellow), cingulo-opercular (purple), salience (black), default mode (red), dorsal attention (green), and language/ventral attention (teal). These networks are consistently separable across datasets and methodologies, and appear largely parallel to one another, with limited exceptions (e.g., see Figure 11.2b, off diagonals). These observations of a distributed but primarily parallel organization to PFC brain networks have implications for their function.

Executive Function Is Supported By Multiple Distributed Networks

Until recently, efforts to divide the frontal lobe into specialized components were dominated by a localization of function view, in which individual brain regions were the foci for specific functions. These studies linked the frontal lobe with executive functions (e.g., Banich 2009; Botvinick et al. 2001; Corbetta and Shulman 2002; D'Esposito et al. 1998a; Duncan and Owen 2000; Koechlin et al. 2003; Stuss and Alexander 2000). It has been difficult, however, to identify specialization within the frontal lobe, partly because these areas are less likely to show a one-to-one association with specific tasks. Many regions of lateral PFC are activated by a range of tasks tapping working memory, attention, inhibition, task set, and novelty (Duncan and Owen 2000). This led these regions to be labeled as part of a single "multiple demand" system (Duncan and Owen 2000). Network approaches provide new insights into this organization.

First, these approaches suggest that, rather than a single multiple demand system, there are at least[2] two networks central to executive functions (Dosenbach et al. 2008): the frontoparietal network and the cingulo-opercular network, sometimes called salience (Table 11.1). These networks were first segregated based on resting-state fMRI (Dosenbach et al. 2008; Seeley et al. 2007). The networks are activated in many tasks, especially tasks with executive function demands (Dosenbach et al. 2008; Nelson et al. 2010) (Figure 11.3). However, detailed analyses suggest they differ in their specific activation patterns associated with cues, error and ambiguity, task set maintenance, and decision making (reviewed in Gratton et al. 2018b). Notably, resting-state functional connectivity between the two networks is near zero, suggesting they function largely separately. These correlations are raised slightly during controlled task periods, but still remain low relative to correlations within each network (Cohen et al. 2014; Gratton et al. 2016). Focal lesion studies provide

[2] Additional networks (e.g., default, dorsal attention, ventral attention, salience; see Table 11.1) in the frontal lobe add further complexities to this view. At least a portion of these (dorsal and ventral attention) have been reported to show signals consistent with a role in sustained attention and shifts in attention (Corbetta and Shulman 2002), and others (e.g., default network) show signals that scale inversely with executive function demands (McKiernan et al. 2003). Executive function performance is likely supported by the cumulative processes of these large-scale systems.

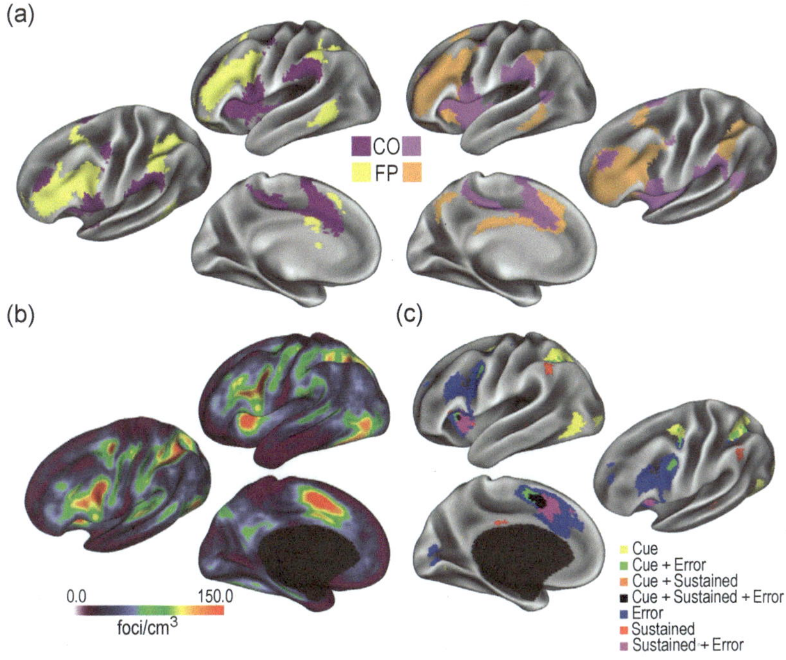

Figure 11.3 The cingulo-opercular and frontoparietal networks have been linked to executive function. (a) The frontoparietal (FP, yellow/orange) and cingulo-opercular (CO, purple/pink) networks are reproducibly identified across studies of large-scale networks of the human brain, shown here based on Power et al. (2011) and Yeo et al. (2011). (b) These networks are activated across many tasks; color bar shows the frequency of activation in a meta-analysis of 1000 task contrasts, based on Nelson et al. (2010). (c) These networks show activations related to task set, including for task cues, errors, and sustained across task periods. Figure reproduced with permission from Gratton et al. (2018b).

evidence for the independence of the networks: individuals with damage to the cingulo-opercular network have decreased functional connectivity extending throughout that system, but leave the frontoparietal network largely intact, and vice versa (Nomura et al. 2010).

This dissociation revealed by network approaches implies that there are subdivisions of executive function associated with spatially distinct neuroanatomical units. The function of each subdivision and how it supports goal-directed and context-dependent implementations of executive function remains an area of debate (see Challenge 1, below), with multiple proposed theories. For example, some suggest the cingulo-opercular (or sometimes labeled “salience”) network acts as a “switch” that engages default or frontoparietal networks, depending on current goals and the import of incoming information (Menon and Uddin 2010). We have suggested that the cingulo-opercular network is involved in maintaining task sets while the frontoparietal network rapidly adjusts control as needed, based on differences in task set responses (Dosenbach et al. 2008; Gratton et al. 2018b). A related, but distinct theory suggests that the cingulo-opercular and frontoparietal networks are involved in the control of tonic and phasic awareness, respectively, based again on differences in task activations and linked EEG signatures (Sadaghiani and D'Esposito 2015; Sadaghiani et al. 2012). These ideas also connect with other models of goal-directed functions, which propose that separate regions monitor for the need for control and implement top-down control biases (Banich 2019; Botvinick et al. 2001). An ongoing challenge in the field is to design experiments to tease apart these different interpretations. Regardless, network studies add to prior work by emphasizing that multiple distributed networks are associated with executive functions, each likely subserving distinct roles.

Second, brain networks are linked to individual differences in cognition. In an influential study, Finn et al. (2015) examined "fingerprints" of brain networks that were characteristic of different individuals. They found that these fingerprints were particularly unique in the frontoparietal network and could be used to predict fluid intelligence. The observation that brain network variability—especially of frontal-associated networks—can predict a range of cognitive abilities has been replicated by several groups (e.g., Cui et al. 2020; Kong et al. 2019; Marek et al. 2022). In a developmental sample, Cui et al. (2020) demonstrated that variability in brain network layout (especially in the two control networks noted above) could be used to predict executive function performance. While the levels of prediction were modest (~15–20% variance explained), they tend to be higher than those seen from anatomical measures and similar to task fMRI (Marek et al. 2022). Further improvements are likely to come from techniques addressing the precision of individual brain measures (Challenge 2, below).

Situating PFC Networks within a Whole-Brain Complex System

Early work on networks of the frontal lobe studied these networks in relative isolation, leaving open the question of how networks are interrelated. A complex systems approach provides a means to characterize network connections throughout the brain simultaneously and to understand properties of the brain's communication architecture (Sporns 2016). At a basic level, graph theoretical approaches reformulate brain network data as a graph, in which brain regions are represented as nodes of the graph and connections between brain regions are represented as links or "edges" in the graph (Figure 11.4). Graphs can be analyzed and contrasted with one another to reveal different properties of brain network organization. Graphs can help quantify properties of brain network structure through diverse metrics (e.g., path length, global and local efficiency, modularity, segregation) developed from the study of other complex networks such as the internet, social networks, and biological systems. These metrics

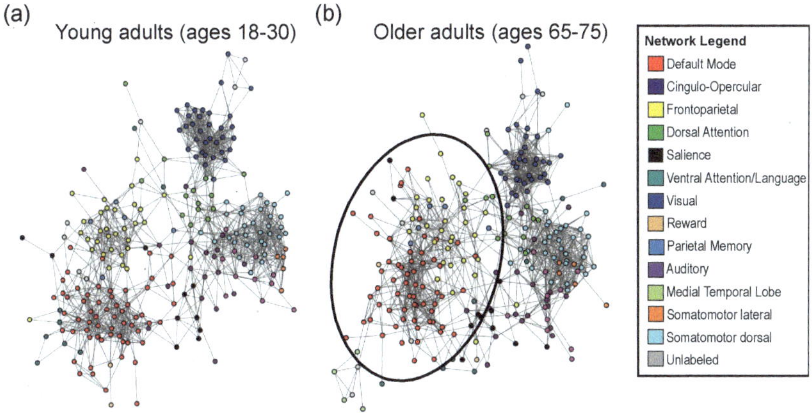

Figure 11.4 A complex systems representation of networks of the human brain. In these graphs, called spring embedding plots, each brain region (or node) is represented by a dot; the lines (or edges) represent high functional connectivity between pairs of brain regions. (a) Nodes that belong to the same network (same color in the plot) cluster together and lie separate from other networks. (b) In older adults, this segregation decreases, especially in association systems, such as the default and frontoparietal networks (black circle). This effect is known as desegregation. This figure illustrates unpublished results from our research group, reproducing similar findings from Chan et al. (2014).

can quantify global properties of the brain (e.g., identifying brain organization layouts that are more or less efficiently organized for transferring information), properties specific to particular networks (e.g., the degree of association, or closeness, between a pair of networks) and the roles of specific brain regions (e.g., identifying hubs, either based on their number of connections or the distribution of connections across different brain networks).

When considered within this context, a number of observations can be made about whole-brain systems organization, the positionality of particular networks within this organization, and the roles of individual brain regions. First, the complex systems approach highlights properties of whole-brain organization, including how efficiently the organization supports the transfer of local and/or global information, and the degree of modularity within a system. Modularity—or the extent to which the system (brain) can be separated into separable subunits (networks)—has been well characterized within the brain. This separability can be observed within graph theoretical depictions of brain connectivity (Figure 11.4), where nodes from a given network (marked with a distinct color) are clustered closely with one another and separate from other networks. This is true for the default (red), frontoparietal (yellow), and cingulo-opercular (purple) networks along with roughly a dozen others. This property underlies the parallel network identification findings cited earlier in this chapter. Using graph theory, modularity can be systematically quantified and compared across systems with statistics like Newman's Q (Newman 2006), which provides a unitary statistic that quantifies the extent to which a given complex system (brain) divides into modules, with higher values (closer to 1) indicating higher modularity, and values close to 0 indicating a nonmodular organization similar to that seen in randomly interconnected systems. Interestingly, a decrease in the segregation between brain networks (Chan et al. 2014) is one of the best replicated changes seen in brain networks with aging, and is particularly prominent for the default, frontoparietal, and cingulo-opercular networks (Figure 11.4b). Moreover, modularity (and its reconfiguration) has been linked to executive function (Eichenbaum 2017) and complex task performance (Bertolero et al. 2015, 2018; Braun et al. 2015; Cohen and D'Esposito 2016). Thus, graph metrics provide a means to quantify the macro-level architecture of brain organization, how it may differ across populations, and how these differences are linked to executive function.

Second, we can use these methods to assess relationships between specific networks. For example, despite their general parallel nature, some networks lie "closer together" in graph space, with more interconnections. This information constrains theories about how large-scale networks interact. For example, as introduced in the previous section, several theories posit that control networks are in competition, and that "switches" between their activity are important to aspects of executive function; deficits in the ability of these networks to switch modes is thought to underlie psychiatric disorders ranging from anxiety to autism (e.g., Dosenbach et al. 2008; Menon and Uddin 2010; Seeley et al. 2007). The network interactions seen in Figure 11.4 suggests that the frontoparietal network is well positioned to mediate between the default and cingulo-opercular system, while the cingulo-opercular network is well positioned to mediate interactions between somatomotor and default systems (for a different viewpoint on the relative positions of these networks, see Menon and Uddin 2010).

Third, in addition to considering entire networks, these approaches can be used to highlight the roles of specific regions within these networks (Sporns 2016). These measures can improve our understanding of executive functions. For example, we and others have proposed that connector "hub" regions of the brain are important for controlled behavior (Gratton et al. 2018b). Connector hub regions (by definition) have connections distributed across multiple networks that can be quantified with the participation coefficient statistic (e.g., nodes with connections across networks in Figure 11.4). This property makes connector hubs particularly well suited to regulating interactions between systems, as is likely needed for flexible, goal-driven behavior. Perhaps not surprisingly, connector hubs are found with high density in frontal and parietal brain regions (Gratton et al. 2018b).

Supporting the importance of connector hub regions for executive function, lesions to hub regions have particularly widespread consequences on modularity throughout the brain, even in regions remote from the sites of damage (Gratton et al. 2012). Disruption of hub regions occurs across a range of neurological and psychiatric disorders (Crossley et al. 2014), and lesions to hubs are associated with widespread deficits on neuropsychological tasks (Warren et al. 2014). Bertolero et al. (2015) demonstrated that these regions are activated in many task contrasts and we have shown that they also exhibit altered functional connectivity during tasks, that differs systematically relative to non-hub regions (Gratton et al. 2016). These pieces of evidence support the idea that connector hubs are important sites for coordinating effective complex behavior, an essential element of executive function.

Understanding the Consequences of Brain Disorders from the Lens of Network Connectivity

Many advances in cognitive neuroscience have come from linking specific behavioral deficits to localized damage to the brain, and cases like Phineas Gage, Tan, and H. M. permeate introductory textbooks. While lesion studies provide insights into the function of the frontal lobe (e.g., Stuss and Alexander 2000), deficits from damage to frontal regions can often be diffuse and hard to characterize. Indeed, it has also long been recognized that damage to connections between brain regions can also cause behavioral impairments and that localized damage can cause disruptions in the function of remaining intact regions (i.e., diaschisis) (Geschwind 1974).

Recent work has used network approaches as a way of characterizing these distributed effects of damage to the brain. These observations have borrowed from insights gained in the study of other complex systems, such as air transportation networks. When travel at one airport is disrupted due to bad weather, for example, this disruption can spread to other connected airports. If the airport is relatively isolated, the effects will be minimal, but if it is well connected to others, especially on an international level, then the effects can be particularly detrimental (e.g., as occurred in 2010 after volcanic eruptions in Iceland impacted European airport hubs). So, too, can we view damage in the brain: certain regions will have more extensive effects than others by virtue of their position within the network structure.

For example, we have shown that focal lesions which caused damage to (non-hub) nodes of the cingulo-opercular network were related to functional connectivity disruptions throughout the network but did not influence connectivity of the frontoparietal network (Nomura et al. 2010). Similar effects were seen in reverse after frontoparietal network damage. In contrast, damage to connector hubs (Gratton et al. 2012) produced more extensive disruptions that affected widespread multi-network organization. These findings suggest that network approaches can provide a way to contextualize and understand nonlocalized, but still selective, effects of brain lesions, in terms of how they extend across interconnected complex systems.

Network models are also relevant to the study of neurodegenerative diseases. Seeley et al. (2009) found that neurodegenerative disorders, including Alzheimer disease and frontotemporal dementia, exhibited distinct atrophy patterns and that these patterns corresponded to specific functional brain networks present in healthy individuals (Figure 11.5). Atrophy in Alzheimer disease tracked with the default network, while frontotemporal dementia shows a profile that overlaps substantially with the cingulo-opercular or salience networks. Seeley et al. (2009) proposed that this pattern arises because neurodegenerative diseases target and spread through specific large-scale networks. Just as with lesion studies, disruptions from neurodegeneration at key nodes (hubs) of these networks, with disproportionately numerous and long-distance

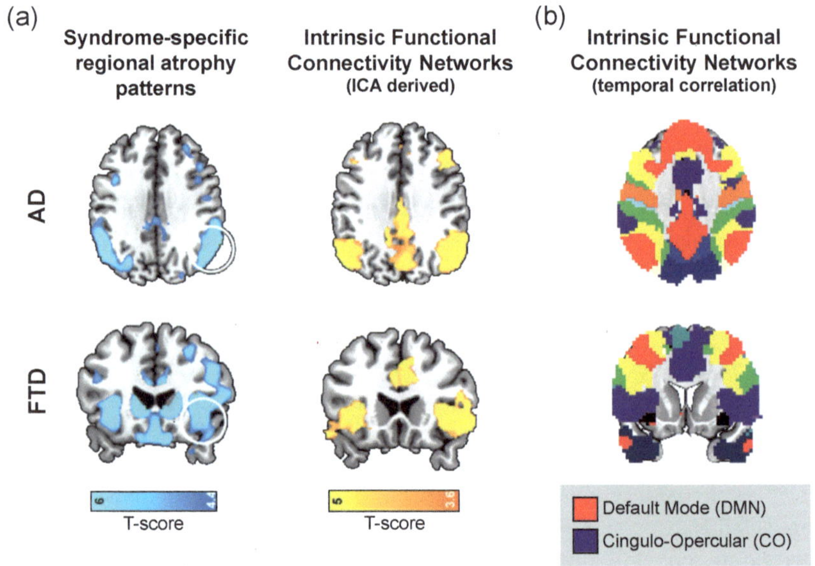

Figure 11.5 Neurodegenerative disorders affect specific networks. (a) Work from Seeley et al. (2009) demonstrated that subtypes of neurodegenerative disorders, like Alzheimer disease (AD) and frontotemporal dementia (FTD), exhibit syndrome-specific atrophy patterns (left column) that correspond spatially with intrinsic functional connectivity networks defined using independent component analysis (ICA). (b) The two networks shown in (a) overlap spatially with networks defined using temporal correlation. Namely, the atrophy pattern associated with AD overlaps with the default network (in red) and the pattern associated with FTD overlaps with the cingulo-opercular network (in purple). Results were reproduced with permission (Seeley et al. 2009) or redrawn.

connections, may produce a cascade of deleterious effects resulting in a weakening of functional circuits.

In addition to neurological disorders, psychiatric disorders ranging from depression to schizophrenia, ADHD, and autism have also been linked to disruptions in large-scale brain networks. To take depression as an example, a recent meta-analysis of lesions identified a distributed set of brain regions, rather than a single location, which contribute to depression when damaged (Padmanabhan et al. 2019). Similarly, depression symptoms are linked with abnormal functional connectivity between the default mode network, frontoparietal network, dorsal attention network, and cingulo-opercular network (Kaiser et al. 2015). Intriguingly, the success of an increasingly common form of depression treatment, transcranial magnetic stimulation, is linked to the functional connectivity between the stimulation site in the left dorsolateral PFC and the subgenual cingulate (Fox et al. 2012). These studies emphasize the importance of understanding connectivity of frontal regions in at least some psychiatric conditions.

Challenges and New Frontiers for Large-Scale Networks of the Frontal Lobe

Network methods allow researchers to place frontal regions in the context of a distributed, stereotyped, and complex set of large-scale networks. This information provides a means to frame and constrain hypotheses about frontal function. However, a number of outstanding questions remain, posing challenges that must be met with further research.

Challenge 1: Linking Frontal Networks to Specific Executive Function Processes

Perhaps the largest outstanding challenge to the network perspective of the frontal lobe is an ongoing gap in understanding the "function" of each network. The presence of distinct, largely parallel networks suggests that there are distinct functions for each network that cause their segregated pattern of activity. Moreover, while regions within a given network are thought to share a common functional association, each region presumably makes unique contributions to the underlying processing, much like regions in the visual system each contribute to visual processing but have unique (and multifaceted) specialization.

However, the functional association of each "executive function" network and the processes that differentiate their subregions are still largely undetermined. Despite the advances reviewed above, unambiguous differentiation of the functions of the cingulo-opercular and frontoparietal networks (and the regions comprising these networks) has not yet been achieved to our knowledge.[3] Most task contrasts that tap executive function processes activate multiple frontal regions spanning both networks (Figure 11.3). As with task fMRI, task variables measured with electrophysiological recordings from nonhuman primates are also frequently represented across multiple PFC regions, showing low dissociation (see Rich and Averbeck, this volume). Thus, while clear distinctions are seen between the cingulo-opercular and frontoparietal network in functional connectivity, these distinctions have thus far been missing from task responses. It is unclear why sharp dissociations are seen in functional network measures, but not task responses.

Notably, many theories of executive processes do not seem to map those processes easily or clearly onto the major network in PFC nor their divisions. To take a few case examples: Friedman and Miyake (2017), on the basis of behavioral performance, have proposed that executive function includes both common processes (associated with a general executive function ability) and specialized properties associated with updating and shifting. Braver (2012) proposed that control functions can be separated into those associated with pro-active and re-active control. Yet others (Badre and Nee 2018; Koechlin et al. 2003) suggest that the lateral PFC encodes a progressive control hierarchy based on abstraction, that lateral PFC regions are associated with a progressive cascade of selection processes (Banich 2009), or that divisions of lateral PFC are associated with different forms of working memory processing (e.g., maintenance vs. manipulation) (D'Esposito et al. 1998a; Petrides 1994) or content (spatial vs. nonspatial) (Wilson et al. 1993). At present, these proposed divisions have not been cleanly mapped onto distinct networks identified with the network methods described above (e.g., Reineberg et al. 2018).

The lack of strong task dissociation for networks of the frontal lobe may reflect our limited understanding into the processes that differentiate these networks. Integration of information from computational modeling (e.g., see Koechlin and Wang, this volume) may help identify dimensions that are more likely to differ across these large-scale networks (see Shenhav et al., this volume). For example, recent work has used detailed tract-tracing results to develop sophisticated circuit-level models of executive function (for a review, see Wang 2022). These models have proven successful in reproducing a range of working memory and decision related responses in the PFC, suggesting that

3 Interestingly, separate domains of inquiry have reported functional dissociations along the medial wall of the PFC that may correspond to distinct functional networks (Shenhav et al. 2018; Venkatraman and Huettel 2012). For example, Ritz and Shenhav (2024) show two distinct areas of the dorsal medial PFC that encode distractor and target information, respectively. They also demonstrate that these areas correspond closely to the borders of "Salience" and "Control C" network representations from (Kong et al. 2019) (cingulo-opercular and salience in our terminology from Table 11.1). However, it is not clear that these tasks differentiate the cingulo-opercular from the frontoparietal network ("Control A" in (Kong et al. 2019); this network has a more dorsal aspect along the medial wall in group averages, but varies across individuals as shown in (Smith et al. 2021, Figure 3).

this may be a useful avenue for considering and uniting information regarding network structure and function.

Additional limitations may be driven by study approaches: fMRI task responses and functional network measures at rest are not frequently collected in the same participants and directly compared. More rapid progress is likely to be made with studies that combine resting-state fMRI with task data. Importantly, anatomical imprecision, as well as individual differences in network boundaries, can have a major impact on the ability to dissociate functions cleanly, since distinct networks lie anatomically juxtaposed with one another (Figure 11.6).

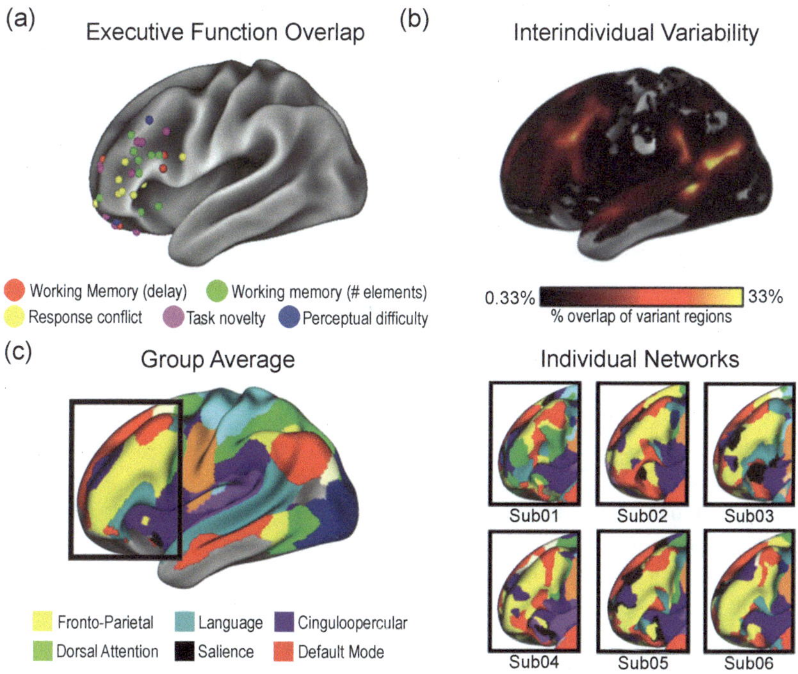

Figure 11.6 Despite many theories which distinguish types of cognitive control, it has been difficult to find functional dissociations in the PFC. (a) In a meta-analysis, Duncan and Owen (2000) showed that many different executive function tasks seem to have overlapping activation patterns in lateral PFC. (b) One potential confound is that the PFC is a particularly variable area of cortex across people, shown here in terms of locations with high variability between each individual and group average ("variants") (Seitzman et al. 2019). (c) Gordon et al. (2017) showed that each individual has a unique pattern of network organization that can be reliably mapped with sufficient data; here we highlight the variability in locations of the lateral PFC (in this case, based on 10 sessions of data from each participant). Panels (a) and (b) are reproduced with permission from (Smith et al. 2021).

Challenge 2: Understanding the Impact of Individual Differences of Brain Networks

Resting-state fMRI has often been collected in short, 5–10 minute scans from individual participants. With this amount of data, functional connectivity measures have low reliability (Gordon et al. 2017). Historically, this practice has led to a reliance on group approaches in functional connectivity studies, that average fMRI data across participants after anatomical normalization (Power et al. 2011; Yeo et al. 2011) and assume voxel-level correspondence across individuals. While these studies provide insights into typical patterns of network organization, they also obscure features that differ across individuals.

This is problematic, as the lateral frontal cortex is among the most variable in organization across individuals (see Figure 11.6b; Finn et al. 2015; Gordon et al. 2017; Kong et al. 2019; Seitzman et al. 2019). This individual variability suggests that group studies of the PFC are likely mixing together signals from different networks across individuals (Figure 11.6c). The mixing of signals across networks (and regions) in the PFC will limit our ability to know which aspects of function are dissociated or overlapping, and to use this knowledge to link the PFC to cognition, behavioral outcomes, and disease processes.

One way to address this issue is to shift toward individual-level mapping of brain networks. Precise individual-level mapping of networks depends on having sufficient high-quality resting-state fMRI data, that is properly de-noised from artifacts, to overcome inherent sampling variability and sources of systematic bias. This "precision" data can then be used to map individual brain networks with high reliability, yielding improved overlap with individual task activations and correspondence to anatomical features (Braga and Buckner 2017; Gordon et al. 2017).

Approaches using individual-level mapping have already provided improved understanding of networks in the frontal lobe (Braga and Buckner 2017; Braga et al. 2020; Gordon et al. 2017). For example, these approaches have identified subnetworks in the default (Braga and Buckner 2017) and cingulo-opercular networks (Gratton et al. 2022); subnetworks are likely to be present for other networks of the PFC as well. These initial studies suggest that these more fine-scale subnetworks link more clearly to function. For example, the two default subnetworks show a double dissociation for episodic projection and social cognition functions, respectively (DiNicola et al. 2020; see also Fedorenko et al. 2011; Michalka et al. 2015) for other specializations associated with frontal regions).

Challenge 3: Measuring Rapid Temporal Variation in Brain Networks

A final challenge in the study of large-scale networks is to improve our understanding of how network interactions (both within and between networks) change over time. In controlled behavior, different functions need to be united and updated flexibly to meet task goals, suggesting the need for modifications in how different regions interact with one another. It is natural to ask how network models support these rapid interactions, on the order of seconds, and whether rapid dynamics in functional networks may be revealing regarding these processes.

However, evidence suggests that functional networks measured with fMRI remain largely stable across fairly distinct tasks (e.g., Gratton et al. 2018a). Much of the variation in the magnitude of functional connectivity over shorter time windows (<1 min.) can be attributed to sampling variability or physiological artifacts from motion and respiration (Ladwig et al. 2022; Laumann et al. 2017; Liegeois et al. 2017).

Thus, the question remains of how large-scale networks support ongoing and flexible cognition. One possibility is that only relatively small differences in functional networks are needed for these processes. Indeed, small but significant variation in between-network interactions can be found across different task states, riding on top of a largely stable network backbone (e.g., Cohen et al. 2014; Gratton et al. 2016; Gratton et al. 2018a). These changes are consistent enough that they can be used to accurately decode task state from functional network patterns alone (e.g., Shirer et al. 2012). However, the precise links between these distributed changes and particular executive function processes remain unclear. One productive avenue of future work may be to unite network studies with computational models of brain state changes during executive function. Connectionist models may be well suited to this need, as they can help make inferences about the representational states of networks needed for the initiation of control as well as the need for optimization in the balance for cognitive flexibility and stability (Musslick and Cohen 2021).

A second possibility is that network function is changed at temporal or spatial scales that are not easily measured with standard fMRI techniques. Methods with higher spatial resolution (e.g., ultra-highfield 7T fMRI), spatiotemporal resolution (e.g., ECOG), or combinations across methods (e.g., fMRI + EEG) may be used to explore this possibility. Ultra-high field MRI, for example, has been used to show finer scale distinctions among subnetworks of the default network (Braga et al. 2019) and parietal memory network (Kwon et al. 2023). Layer-specific studies (Bandettini et al. 2021) may add further nuance to our understanding of the PFC (Finn et al. 2019) and its networks (Huber et al. 2021), by allowing investigators to separate feedforward and feedback connections found in distinct cortical layers (although challenges remain in the collection and analysis of whole-brain layer data (Bandettini et al. 2021; Huber et al. 2021). Networks have also been examined across multiple methods, finding similar parallels between fMRI findings and invasive electrophysiological recordings (Kucyi et al. 2018), stimulation (Fox et al. 2020), and noninvasive methods such as EEG/MEG (see review by Sadaghiani and Wirsich 2020). Interestingly, at least some of this evidence suggests that large-scale networks are largely stable across a wide range of frequency bands in ECOG recordings (e.g., Mostame and Sadaghiani 2021).

A final possibility is that flexible cognition is not mediated by changes in long-range cortical interactions, but is associated more with local interactions, perhaps aided by neuromodulatory signals. These possibilities are not mutually exclusive and must jointly be explored to improve our understanding of how flexible cognition can arise in the face of (what appears to be) a largely stable large-scale network architecture.

Conclusion

Network approaches add novel insights into frontal lobe organization and its contribution to executive function. These approaches have shown that the frontal lobe includes many distinct networks, connected with other cortical and subcortical regions. These networks show correlated, integrative activity, but are largely parallel to one another. These observations suggest that executive function is supported by multiple distinct networks, embedded within a complex architecture of whole-brain interactions, that have consequences for how damage and disease spread throughout the system. However, challenges remain in the study of large-scale networks of the frontal lobe. There is a need to improve our understanding of how network models relate to other cognitive and functional models of the frontal lobe and executive function, a need to address the substantial individual variability in large-scale network organization in the PFC, and a need to improve our understanding of temporal scales of variation in networks. Future advances are likely to come from studies with an increased focus on integrating different measures and obtaining reliable individual-level representations of networks in the frontal lobe.

Acknowledgments

Funding supporting this work was provided by NIH through R01MH118370 (CG), R01NS124738 (CG), T32NS047987 (DP), and T32AG020506 (ZL) as well as NSF CAREER 2305698 (CG).

Appendix 11.1: Measuring Large-Scale Brain Networks with Resting-State fMRI

The focus of this chapter is on networks measured with functional connectivity MRI, often during what is termed a "resting state." In resting-state fMRI, participants are asked to lie quietly at "rest" inside an MRI, typically with only a cross to fixate on, while letting their mind wander.[4] Experimenters then measure spontaneous BOLD activity across the brain and search for patterns of covariation between regions, termed "functional connectivity." Different measures can be used to quantify these statistical dependencies in brain activity (e.g., correlation, coherence, ICA, lagged covariance) (Sporns 2016). An advantage of resting-state fMRI is that it is relatively easy for experimenters to collect and participants to complete; this has led to it being widely adopted in consortia imaging projects and clinical populations.

While it appears quite unconstrained, resting-state fMRI can produce robust maps of large-scale brain networks, both at the level of groups (Power et al. 2011; Yeo et al. 2011) and individuals (Braga and Buckner 2017; Gordon et al. 2017), given sufficient data and appropriate de-noising methods. Notably, resting-state fMRI can simultaneously map networks both for sensorimotor regions as well as association regions like the frontal lobe (Power et al. 2011; Yeo et al. 2011). These spontaneous "resting-state" maps mimic activation patterns seen in a range of tasks (Smith et al. 2009) and can be used to predict individual activation patterns in many task contrasts (Gordon et al. 2017; Tavor et al. 2016). Indeed, recent work has highlighted that large-scale networks only differ subtly across task states (Gratton et al. 2018a), are stable across sessions of a participant (Gratton et al. 2018a) and are even present across states of consciousness (sleep, anesthesia), albeit with some alterations (Heine et al. 2012; Palanca et al. 2015; Sämann et al. 2011). Recent trends have led to improvements in the spatial and temporal resolution of resting-state fMRI (e.g., via multiband data acquisition), signal quality of noncortical regions (e.g., via multi-echo sequences), and the size and extent of samples (e.g., via large N consortia datasets such as the HCP, ABCD, and UK Biobank, as well as extended acquisition "precision" fMRI approaches of single individuals).

4 Participants likely engage in a range of internally-driven cognition during rest. However, it is unclear to what extent this internally-driven cognition alters functional connectivity measurements (Fox and Raichle 2007). Functional connectivity patterns (including each of the large-scale networks discussed in detail in this chapter) remain very similar during resting-state and a wide variety of explicit cognitive tasks (Gratton et al. 2018a). Thus, differences in internally-driven cognition during resting-state likely only have a subtle influence on functional connectivity.

12

Integrative Psychological, Computational, and Mechanistic Approaches to Frontal Lobe Function

Amitai Shenhav, Marie T. Banich, Christian Beste, Timothy J. Buschman, Naomi P. Friedman, Caterina Gratton, Etienne Koechlin, Nicolas Schuck, Xiao-Jing Wang, and John O'Doherty

Abstract

Since the earliest accounts of the prefrontal cortex (PFC), its core functions have remained elusive and hotly debated. Here, an attempt is made to bring order to these varied accounts and to account for the heterogeneous observations that have been made across methodologies and species. After cataloging the myriad functions that have been attributed to PFC and the approaches that have been taken to taxonomize these functions, a new framework is proposed for conceptualizing PFC function. This framework is based on a set of four canonical computations that is argued to collectively provide a more formal, coherent, and comprehensive account of existing findings regarding PFC function. These canonical computations include goal-directed integration, active maintenance, selection of task-relevant information, and monitoring. Discussion includes how previous PFC findings can be understood through one or more of these functions, and ways in which these computations may collectively form a motif that repeats throughout regions of PFC over different forms of inputs and outputs. Finally, critical directions for future research to validate or falsify this account of PFC functions are highlighted, including the leveraging of new and emerging directions for experimentation and analysis.

Group photos (top left to bottom right) Amitai Shenhav, John O'Doherty, Marie Banich, Naomi Friedman, Timothy Buschman, Nicolas Schuck, Etienne Koechlin, Caterina Gratton, Christian Beste, Xiao-Jing Wang, Marie Banich, Amitai Shenhav, John O'Doherty, Timothy Buschman, Christian Beste, Etienne Koechlin, Caterina Gratton, John O'Doherty, Xiao-Jing Wang, Nicolas Schuck, and Naomi Friedman

What Adaptive Functions Does Prefrontal Cortex Serve?

A Starting Point to Encapsulate PFC Function

Any integrative account of a given brain structure is destined to be incomplete and in need of revision, particularly when that brain structure subsumes the entirety of the prefrontal cortex (PFC). Even if the puzzle is not likely to be fully solved, one can at least strive to start to integrate as many of the critical pieces as possible. The challenge is knowing whether one is starting in the right place to solve the puzzle with the pieces that one has in hand, or whether one needs to start from scratch.

In seeking to compile and bring order to the functional mechanisms underpinned by the PFC, we will, therefore, start by asking: What is the broad range of psychological functions and phenomena in which this structure has been implicated? We then proceed to consider different approaches to taxonomizing and/or decomposing this array of functions, and what these taxonomies collectively reveal about possible canonical computations that unify or at least reduce the dimensionality of PFC functions. Finally, we discuss how the study of PFC function and its underlying computations can be improved by extending traditional methods and leveraging emerging experimental, analytic, and modeling approaches.

There are several sources of data that researchers have taken into account when attributing functions to the PFC (Table 12.1), including

- Cognitive impairments observed in individuals with PFC damage (e.g., lesions), inactivation (e.g., cooling or other noninvasive brain stimulation methods), and/or deterioration (e.g., frontotemporal dementia) of the PFC,
- PFC functions that are altered over the course of evolutionary development (across species) and/or ontological development (particularly over early development) along with development and maturation of these structures, and
- PFC functions whose engagement covaries with increased neural activity within and/or across prefrontal regions (as measured, e.g., via electrophysiology or neuroimaging).

Functions Commonly Ascribed to the PFC

Active Maintenance

Working memory is the ability to actively maintain a limited set of information in the absence of direct sensory input for short periods of time (e.g., 3–10 seconds). It is critical for complex cognition, allowing one to break free from the immediate world (i.e., simple stimulus-response associations) and to keep critical information at the ready. Working memory has been considered

Table 12.1 Semi-exhaustive list of functions commonly attributed to PFC.

Cluster	Sample Functions
Active maintenance	• Maintaining goals, values, task-relevant cognitive and emotional information • Buffering goals from interference
Selection	• Selecting/determining goals based on internal and external context • Distinguishing relevant vs. irrelevant information in the environment • Selecting relevant information from memory • Selecting specific information for prioritization • Emotion regulation and reframing
Versatility	• Suppressing prepotent responses (e.g., habits) • Shifting between goals or tasks • Arbitrating between hypotheses and strategies • Flexibility to novel, unfamiliar, or changed environments
Monitoring	• The environment for task-relevant information • Whether the correct action has been selected • Whether one's action led to the desired goal • Whether goals and actions align with values
High-level combinatorial processing	• Abstraction, generalization • Identifying novel or atypical strategies/solutions • Coordinating goals, learning, and memory • Constructing value • Processing for multiple tasks • Language • Reasoning
Simulation	• Envisioning novel solutions or courses of action • Simulating forward or backward in time • Hypothesis testing • Metacognitive processing • Social inference (e.g., theory of mind)

one of the canonical functions of the PFC. Decades of research, starting with groundbreaking recordings from Fuster and Goldman-Rakic, have promulgated the idea that the contents of working memory are actively maintained or referenced in the pattern of neural activity within the PFC, most notably in the face of distraction. Working memory is both capacity and time limited, enabling the maintenance of about 4–7 items for time periods up to about 10–15 seconds. Despite these limitations, it is highly flexible with regard to content. One can hold any type of information (e.g., verbal, spatial, emotional) in working memory, and neural correlates are likewise flexible in what they can represent. Neurons (or neural populations) in PFC have been found to actively represent sensory inputs (Fuster and Alexander 1971; Romo et al. 1999), motor actions (Mars and Grol 2007), the value or emotional significance of stimuli (Platt and Padoa-Schioppa 2008; Rolls et al. 2009; Salzman and Fusi 2010), actions (Barraclough et al. 2004; Shin et al. 2021), and task rules (Wallis et al. 2001; White and Wise 1999).

Selection

PFC has been implicated in selecting those representations and processes that are most relevant for the current task goals (e.g., Miller and Cohen 2001). For example, PFC may bias toward processing of specific relevant attributes of the external world (e.g., color, portions of space; Banich et al. 2000; Kastner and Ungerleider 2001) or types of information (e.g., linguistic; Snyder et al. 2014), memory (e.g., semantic; Wang et al. 2018), actions (e.g., action sequences; Zhang et al. 2021), emotion regulation (e.g., reappraisal; Braunstein et al. 2017), or abstract plans (e.g., steps required to traverse a subway system; Balaguer et al. 2016), all of which are selected in reference to current task goals.

The putative role of PFC in selection has also been exemplified in impairments observed during the selection of options in decision-making tasks. For instance, classic lesion studies in humans implicated ventral PFC (including the orbitofrontal cortex, OFC) in the selection of stimuli associated with varying reward values, especially following changes or reversals in reward associations (Murray et al., this volume; Bechara et al. 1997; Fellows and Farah 2003; Hornak et al. 2004; Noonan et al. 2010). There is accumulating evidence to suggest that ventral prefrontal regions, especially the OFC, may be especially important for selecting between stimuli based on the prospective rewards associated with them, whereas more dorsal parts of the PFC, including dorsal anterior cingulate cortex (ACC) and pre-supplementary motor area, may play more of a crucial role in making decisions over actions (Aquino et al. 2023; Camille et al. 2011b; Rudebeck et al. 2008b). PFC also appears to play a role in selecting between more abstract policies (e.g., different strategies, expert systems), which we discuss further below.

Versatility of Responding and Thought

Here we consider two aspects of the versatility of responding and thought: overcoming habitual patterns of responding and being able to switch flexibly between responses or thoughts. In terms of the former, let us consider Teuber's description of behaviors associated with frontal lobe damage, which he characterized as "bewildering" in variety (Teuber 1972:637) yet sharing elements of "compulsiveness" or "abnormally stimulus-bound behavior" (p. 640). That is, individuals with frontal lesions might be unable to avoid habitual responding in a given context in favor of less automatic responses which might be more appropriate in that context. Moreover, stimuli in the environment can trigger automatic responses; for instance, seeing a computer will engender starting to type on the screen (Lhermitte 1986; Lhermitte et al. 1986). This suggests that one ability enabled by the frontal lobes may be the ability to respond to stimuli in different ways beyond the stereotypical manner based on well-learned responses.

Another aspect of flexibility is the ability to change one's course of action or thought processes. Such a switch may be driven by external information that signals a change in certain processes is now possible or desirable, or with regard to external feedback about the utility of those processes under the current context or an internal evaluation of the efficacy of actions. Such abilities are compromised in individuals with damage to the frontal lobe (e.g., Adólfsdóttir et al. 2014; De Baene et al. 2019).

Monitoring

Critical to ensuring that one's actions and choices are efficacious in leading to a goal, one must evaluate or monitor outcomes or internal states, as in emotion regulation or memory retrieval. Monitoring refers to how an agent tracks its own behavior and/or the consequences of those behaviors in various situations (i.e., in the face of information obtained from the environment), which can impose varying demands on behavioral control (Botvinick et al. 2004; Holroyd and Coles 2002; Rushworth et al. 2004). Such monitoring processes depend on medial and superior PFC activity (Giller et al. 2020; Reinhart and Woodman 2014). Activity in these regions is increased in situations that are unexpected or deviate from one's goal (e.g., error commission). This suggests that increased monitoring during such situations is required to enable behavioral control. Such monitoring abilities are compromised after frontal lobe damage (e.g., Hochman et al. 2015). Importantly, the degree of monitoring has to be balanced to be able to cope with changes in situational requirements. This dynamic balancing in the degree of cognitive control monitoring has been termed "meta-control" (Eppinger et al. 2021; Hommel and Wiers 2017) and shown to be altered by disorders affecting frontal lobe functions, such as in obsessive-compulsive disorder and attention-deficit hyperactivity disorder (Colzato et al. 2022).

Higher-Level Combinatorial Processing

There is evidence that information maintained and selected by PFC can reflect a higher-level combination, or abstraction, of current sensory input or internal representations. Studies have, for instance, shown that categorization, which often requires a nonlinear combination of sensory variables, involves the PFC (e.g., Freedman et al. 2001; Seger and Miller 2010). Other studies have shown that populations of PFC neurons are engaged when animals switch between tasks that require the animal to focus on different aspects of the same stimulus

(Mante et al. 2013). These neurons show two important coding characteristics: (a) they exhibit mixed selectivity, meaning that a cell can be responsive to multiple cognitive features (e.g., Aoi et al. 2020); (b) they appear to be able to reduce information to underlying dimensions, such as being able to code information in discrete categories (e.g., Mack et al. 2020). The process of abstraction has also played a central role in research on value-based decision making (e.g., Cortese et al. 2021; De Martino and Cortese 2023), with orbitofrontal regions of PFC being implicated in representing "partially observable" information, such as context from past events, in the service of maximizing reward (Schuck et al. 2018; Wilson et al. 2014).

Simulation

Planning—a function known to be impaired after damage to PFC (Owen et al. 1990; Shallice and Burgess 1991)—relies on simulation. Simulation describes a process of bringing to mind (or "sampling from") potential future states of one's environment, including the potential positive or negative consequences of arriving in this state. The mental representation of these potential future states and outcomes is referred to as a world model. By mentally sampling a world model, one can identify valuable and efficient courses of action. As Sutton and Barto (1998) point out, this form of learning (model-based reinforcement learning, RL) can be equivalently viewed as moving forward into potential future states or as revising backward the courses of action which led to such states. It has thus been thought that PFC is critical for managing/controlling covert simulated behavior in the same way as overt behavior (Campbell et al. 2018). Additional evidence for this role, which we elaborate on later in our discussion of unifying features, comes from findings that regions throughout PFC track information related to the value of current and future states, as well as how these values are transformed to guide behavior.

Summary

We recognize that this listing of functions is likely not exhaustive. It also does not identify any new processes that have not been discussed previously in the literature. Nonetheless, it does identify core functions that involve the full extent of frontal regions.

Existing Approaches to Divide the Space of PFC Function

What Do We Want a Taxonomy of PFC Function to Accomplish?

The groupings offered above provide one form of functional taxonomy, but one whose boundaries are defined arbitrarily. To develop a better taxonomy, it is important to ask first what sorts of properties are needed to make such a taxonomy useful and effective. In other words, what are the criteria by which one might determine that they succeeded or failed in developing a good taxonomy of PFC function?

The first property that one might seek in a taxonomy of PFC function is its *descriptive* utility: How well does it capture variability in PFC function within an individual over time, and across individuals? To what extent does it capture deficits reported by PFC-damaged patients? How does it align with variability in prefrontal anatomy and physiology, including patterns of functional activation and connectivity? To what extent does it capture variability in PFC-related behavior, function, and structure over the course of development or in response to stressors?

The second property that one might seek is its *generative* utility. Can it be described in formal terms, and at a level of description that can be assessed across species and methods? Does it give rise to new assays (e.g., new tasks, metrics) that allow researchers to capture more precisely the sources of variability above? Does it identify ways of applying existing measures (e.g., behavior, physiology) and interventions (e.g., inactivation, pharmacology) to those assays to test new hypotheses? Does it point toward targeted treatments that alleviate deficits in patients with damage or dysfunction linked to PFC?

Forms of Taxonomy: Strengths and Limitations

Qualitative Description of Behavioral Impairment

Taxonomies drawn from observations of behavioral impairment after frontal lobe damage have a long and storied history, starting most famously with the case of Phineas Gage, a railroad construction foreman whose crew was excavating rock in 1848 to build a railroad line in Vermont. While using a tamping iron to pack an explosive into a borehole, a spark from the iron on the rock detonated the explosive, leading the rod to pierce the anterior portion of his left frontal lobe through the eye socket (Macmillan and Lena 2010). In the oft quoted description, changes in both social and cognitive characteristics were noted afterward. Socially he was no longer sensitive to others and could be profane, and while previously he had held the position of a construction foreman, he could no longer come up with a plan and systematically follow through on it. Other individuals who have suffered from frontal lobe damage in modern times have exhibited deficits on self-reports of their ability to deploy executive functions successfully in their daily lives (e.g., Løvstad et al. 2012).

Task Impairment

A more quantitative and systematic approach to understanding PFC-related impairments has focused on mapping out those regions where damage through lesions is commonly implicated in aberrant performance on well-characterized laboratory tasks (e.g., Godefroy et al. 2023; Meier et al. 2022). Such studies have a number of strengths and limitations. With regard to strengths, any taxonomy of frontal lobe function from patients is arguably most relevant for real-world behavior, as alterations to frontal lobe function are observed across a wide variety of neurological and psychiatric disorders. On the other hand, there is potential for reorganization of function between the time of damage and assessment. Moreover, lesions often span important morphological and functional boundaries in the brain, which can make determinations difficult and/or preclude studies from having large numbers of participants with damage to one particular brain region.

Factor Analysis of Performance across Tasks

As discussed by Duncan and Friedman (this volume), factor analysis has been used to evaluate whether performance on executive function and so-called "frontal lobe" tasks are influenced by a single or multiple underlying factors of ability. This question emerged from models of working memory, which suggested a central executive that controlled the contents of storage buffers (Baddeley 1986). In seeming contradiction to the notion of a unitary executive, executive function tasks showed low correlations. However, low correlations could arise even if there were a unitary central executive because executive tasks show low reliability and "task impurity." Because executive functions control other processes, executive tasks must include these non-executive functions, as differences in these can also influence performance (Miyake et al. 2000). Thus, Miyake et al. (2000) selected sets of tasks intended to tap three executive functions—response inhibition, working memory updating, and mental set shifting—that varied in these lower-level processes and used confirmatory factor analysis to extract latent variables. Latent variables are based only on shared variance across a set of tasks, so they can remove random measurement error as well as variance due to non-executive demands that differ across tasks (i.e., task impurity). They found that these latent variables showed moderate correlations, suggesting some shared variance, or "unity," but these correlations were significantly lower than 1, suggesting some distinct variance, or "diversity," even after accounting for task reliability issues. Thus, their conclusions, which were based on a sample of neurally intact college students, echoed conclusions of earlier studies that focused on frontal lobe damage (Duncan et al. 1997; Teuber 1972), which suggested "unity and diversity" of frontal lobe function.

Although this study might be described as creating a "taxonomy," it is important to note that Miyake et al. (2000) never intended this battery to capture "core" or "elemental" components of executive functions. They decided to focus on these three functions because they were among the most commonly examined executive functions at an intermediate level of analysis, but they explicitly noted that other executive functions likely existed and that functions could be conceptualized at different levels (e.g., planning might be composed of multiple sub processes). This study illustrates the principle that taxonomies can exist at multiple levels depending on the researchers' goals; this set of functions provided a tractable means with which to tackle the goal of evaluating whether commonly hypothesized executive functions could be considered unitary. That said, this has also proved useful in subsequent research to evaluate the relations of unity and diversity components to other constructs of interest, such as other cognitive processes, psychopathology, and neural areas (see Friedman and Miyake 2017).

A parallel approach is to utilize meta-analytic tools to find terms that are commonly associated with activation in prefrontal regions. For example, using a topic modeling approach, de la Vega et al. (2016, 2018) found that certain terms (e.g., inhibition, conflict, working memory, and decision making) are associated with studies that yield prefrontal activation. Terms could then be examined to determine with which regions of medial (de la Vega et al. 2016) and lateral (de la Vega et al. 2018) frontal cortex they are associated.

Theory-Driven Decomposition of Function

The set of functions attributed to PFC can be decomposed into interlocking functions that can be described along one axis by their control "effectors," that is, the distinct sets of controlled processes that are subsumed by each. For instance, different forms of control can be described as involving selective enhancement of particular processing streams (e.g., forms of selective attention; Desimone and Duncan 1995), directed search, and retrieval of information held in episodic or semantic memory (e.g., cued recall, prospection; Polyn et al. 2009; Schacter et al. 2008), transformation of information held in working memory (e.g., mental rotation, inference; Olivers et al. 2011; Shepard and Metzler 1971), and parameterizing one's decision process (e.g., response threshold; Bogacz et al. 2006; Leng et al. 2021; Wiecki and Frank 2013). Each of these define different forms or *types* of control that one can engage, many of which have been linked to regions of PFC (Duncan 2010; Miller and Cohen 2001; Shenhav et al. 2013, 2016).

However, the presence of these controllers alone is incomplete without an account of when, why, and *to what degree* (i.e., with what level of *intensity*) each of these are selectively engaged, disengaged, or modified (Hommel and Wiers 2017). Thus, an orthogonal level of functional description needs to provide at least a minimal account of the process by which each type of control is (a) selected (i.e., determining the appropriate amount/s and type/s of control to allocate), (b) executed (i.e., engaging the relevant control processes), and (c) monitored (i.e., identifying conditions under which control needs to be adjusted) (Botvinick and Cohen 2014).

Unlike the factor analytic approach described above, this form of functional taxonomy does not derive directly from quantitative task behavior, which is a limitation of the approach. It does, however, serve a similar purpose in providing a coherent lower-dimensional structure to the set of processes underpinning performance across those tasks. These taxonomies instead derive or take inspiration from a combination of psychological, neural, and/or computational evidence and are refined by the same (e.g., evidence of the neural and computational distinctiveness of different forms of control, dynamics of post-error performance adjustments). This approach serves as both a strength (drawing from convergent sources of evidence) and a limitation (affords a high level of subjectivity and flexibility in how to weigh the strength and plausibility of different sources of evidence).

Neurobiological Fractionation

A contrasting domain of approaches focuses on subdividing processes based on neurobiological criteria. These may be functional (e.g., fMRI or electrophysiological activations in particular tasks, correlations of functional signals across regions, changes in functional responses after damage to particular brain areas) or anatomical (e.g., macro-anatomic based on sulcal morphology or connectivity of major tracts, or micro-anatomic based on cytoarchitecture, receptor densities). Some common themes have emerged from this work, including the presence of specialized brain regions, and evidence that these brain regions join together to form large-scale brain networks (e.g., see chapters by Vertes et al., Gratton et al., and Murray et al. in this volume).

An advantage of these approaches is that they can provide a new way of conceptualizing

- Divisions in prefrontal function and constraints on theories of function (e.g., regarding the unity and diversity of functions or the types of processes that can be plausibly represented by neurobiology),
- How these divisions arise (e.g., via ties to evolution, development, and plasticity of neurobiology), and
- How different forms of brain damage can be biologically represented (e.g., via ties to particular regional functions, neurotransmitter modes of actions, models of interregional connectivity).

For example, as reviewed by Gratton et al. (this volume), resting-state functional connectivity has been shown to subdivide the cortex, including the PFC. In these descriptions, 10–17 networks are identified with fairly distinct spatial organization. The frontoparietal, cingulo-opercular/salience, default mode (A and B), dorsal attention, and ventral attention/language are the most studied "association" systems of PFC (see discussion in Gratton et al. on taxonomy and visualizations of these networks). The clear modularity exhibited by these networks (with high within-network connectivity and low between-network connectivity, replicable across groups, people, over time within a person) suggests that these may underpin fractionable functions within the frontal cortex. Indeed, these networks are associated with dissociations in task responses, anatomical features, electrophysiological response properties, neurodegenerative disease, and predictive of behavioral performance. These distinctions become even clearer in individual-level mapping that addresses issues of inconsistent spatial localization across people.

However, a limitation of these approaches is that they are largely descriptive and not closely tied to a mechanistic understanding of PFC function or cognition. While component processes identified with behavioral/cognitive measures show some overlap with neurobiological subdivisions (e.g., Duncan and Friedman, this volume), their alignment is not always clear; see discussion of cingulo-opercular and frontoparietal function in Gratton et al. (this volume). Thus, while neurobiological findings help to constrain theories, they may provide limited insights on their own regarding how functions are implemented in the PFC and give rise to differences in behavioral outcomes.

Computational Models: A Tool for Formalizing Taxonomies

One challenge for cognitive taxonomies is that verbal descriptions of functions are often vague, which can make them less useful for making predictions. Computational models can address this issue by recapitulating core aspects of behavior while providing a more formal, more reproducible, and less ambiguous description of the different functions, hence enabling quantitative predictions about behavioral and/or neural changes that result from arbitrary manipulations. Another challenge for most taxonomies is that a focus on the behavioral versus cognitive versus biological level can yield different results, while leaving unclear the translation between the different taxonomies. Computational descriptions could allow us to identify links between the different levels and help provide a mechanistic understanding that bridges the biological and cognitive levels, as illustrated by a recurrent circuit model of working memory and decision making (Wang 2002).

Computational models integrate multiple operations into a consistent functional system that can be used to investigate the empirical performance of individuals performing tasks described by that system. These simulations can then be used to test whether a model can reproduce subjects' behavior along with related neural activity, and to compare the degree to which distinct models can reproduce such empirical data so as to identify key computational operations within a consistent integrated system.

One of the advantages of a computational approach is that it can provide a common language that helps us bridge multiple levels of understanding and measurement. Computational models can, for instance, make predictions at the network level, about activation of a broad region, about patterns of neural activity within a region, and/or about distributions of receptors. In this way,

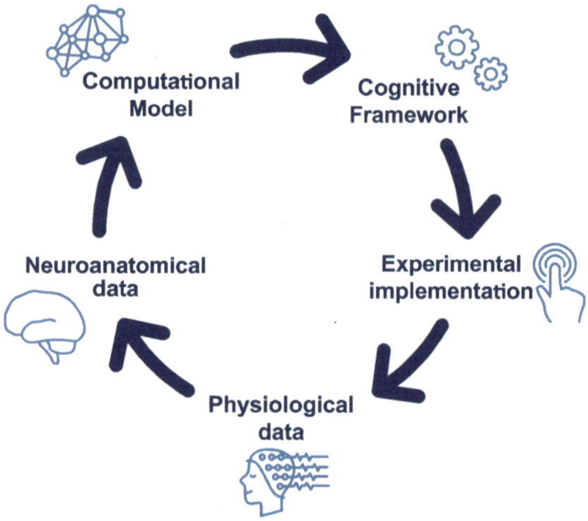

Figure 12.1 Interdependencies between theoretical and experimental approaches to investigating PFC function. Computational models help shape and formalize conceptual and theoretical frameworks for understanding cognition. Together, these serve to operationalize and form testable hypotheses, inspiring specific experiments for measuring relevant neural function and structure. Data collected from such experiments, in turn, serve to constrain preexisting models and/or adjudicate between multiple alternative models.

having a strong computational framework can allow researchers working with different methods and at different levels of description to communicate and inform one another.

Importantly, computational models work in tandem with and are directly informed by other levels of investigation (see Figure 12.1). A clear conceptual (e.g., cognitive) framework is necessary to allow the field to connect the insights gained from a computational model to the conceptual background that has been around in the field for a long time and has inspired well-validated experimental procedures. Physiological data (e.g., electrophysiological recordings during the experiment) and information about the neuroanatomy can then be further used to inform the computational approach taken.

How Can Psychological, Neurobiological, and Computational Approaches Constrain One Another?

Naturally, the process of identifying candidate functions, constructing computational models of those functions, and then mapping those functions onto biology is iterative and multidirectional. Identifying neurobiological mechanisms of prefrontal function will likely improve our understanding of what functions are important for cognition and how these are implemented in computational models of cognition. Conversely, identifying and specifying cognitive functions associated with executive control can motivate the design of new computational models (e.g., flexible working memory) which can, in turn, generate testable hypotheses for how these functions/computations are accomplished in the brain.

A good computational model of the frontal lobe must be able, for instance, to account for the presence of the known dissociable functional network architecture present in this region of the brain. Such a modular organization may emerge as a product of the modeling approach or may be necessary to implement to constrain the model. Different networks may act on different forms of information (indexed by their connectivity) but use similar canonical computations (discussed below). Alternatively, canonical computations may differ across networks (e.g., perhaps between the language and frontoparietal/multiple demand system).

Functional and neurobiological methods can also provide an understanding of the types of factors that a good model must be able to account for as well as estimates of their range/variability. For example, even in the normative population, substantial interindividual variability has been observed in executive function performance (Duncan and Friedman, this volume), in the spatial layout and extent of functional brain networks (Gratton et al., this volume), and in sulcal morphological characteristics (Murray et al., this volume).

What Are the Unifying Features of PFC Function?

What Are Essential or Canonical Computations Within PFC?

Next, we turn to understanding the canonical computations underlying the adaptive functions of the frontal lobe. Integrating theories drawn across the many taxonomies described above, we identified a set of four putative canonical computations performed by PFC (see Figure 12.2):

1. *Goal-directed integration* involves the ability to access, combine, and sequence information so it can be used effectively to create goals and subgoals, and is supported by the diverse anatomical connections of the frontal lobe, which allow it to integrate information across all cognitive domains.
2. *Maintenance of information* involves the ability to actively maintain representations over time, which supports the ability of the brain to sustain goals and direct cognition.
3. *Selection of task-relevant information* allows for the selection of information and representations, especially at a more abstract level, that are most relevant for current goals.
4. *Monitoring* enables the ability to compare expectations to outcomes, including the prediction of future outcomes, which enables the ability to monitor cognition and flexibly adapt to a changing world.

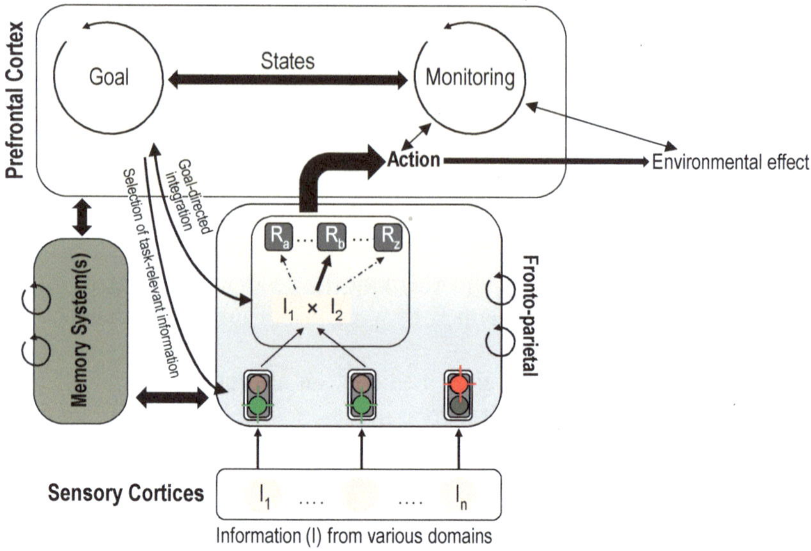

Figure 12.2 Illustration of canonical computations applied to an example of a cognitive task. Information about sensory features of current stimuli must be integrated to determine the appropriate response based on one's current goals and task set. This process requires actively maintaining representations of relevant stimulus features, actions, and/or goals in working memory. Goal-driven processes may act to bias processing of certain features or responses, particularly in cases where automatic processing of those features promotes responses inconsistent with one's current goal. Information indicating deviations from one's goal (e.g., errors, processing conflict) is monitored to modify ongoing and future control (e.g., biasing).

Before we briefly describe each of these canonical computations, it is important to note that these core computations were selected to be parsimonious: a simple set of functions that encompass the broad range of adaptive functions listed above. These functions can be applied broadly to give rise to cognition, across a variety of inputs, cognitive domains, and timescales. As we detail below, these computations do not act alone; complex cognition can only arise through the dynamic interaction and sequencing of these computations.

Goal-Directed Integration

A major innovation over the last few decades of research on PFC function was the proposal of a multiple demand (MD) system (Duncan 2010): a common set of brain regions in frontal and parietal cortices that are active across a variety of different cognitively demanding tasks. The MD system consists of distinct patches that can be found in both hemispheres and which span the lateral prefrontal regions, insular cortex, the dorsomedial frontal cortex, lateral and medial parietal cortices as well as temporal regions (Assem et al. 2022). As discussed by Duncan and Friedman (this volume), "with parts widely distributed through the cortex, strongly interconnected with one another, the core MD system is well placed to take in and integrate representations of many kinds and flexibly feed out the results for selective cognitive control," a process dubbed "attentional integration" (Duncan et al. 2020). Here, we expand on this conceptualization to identify specific forms of integration that occur within PFC.

Integration of sensory and motor representations. The PFC receives parallel streams of sensory and motor information. Superior parietal mechanisms contribute to the selection of motor responses (Bernier et al. 2012; Cisek and Kalaska 2002; Jaffard et al. 2008), possibly because the superior parietal cortex plays a central role in stimulus-response translation processes (Gottlieb 2007). There is, however, a well-known "binding problem" of how the sensory representations become connected to motor representations. To resolve this problem, the theory of event coding (TEC) (Hommel 2004; Hommel et al. 2001) draws on common coding principles to put forward the concept of an event file, which reflects the integrated representation of sensory and motor features that are themselves stored in distinguishable representations. According to TEC—and more recent derivatives thereof, which also consider functional neuroanatomical structures and neurophysiological mechanisms (Beste et al. 2023)—the coding and dynamic handling of event files involves structures in the parietal and PFC that strongly overlap with brain regions that constitute the MD system (Duncan 2010). Thus, commonalities between different instances of executive functions may become explainable through a smaller set of (computational) mechanistic principles relating to the integration of sensory and motor task sets.

Numerous lines of evidence suggest that the coding of integrated sensory and motor representations involves inferior and superior parietal areas, supplementary motor areas, the dorsolateral PFC, and the hippocampus (Chmielewski and Beste 2019; Dilcher et al. 2021; Kleimaker et al. 2020). Superior and posterior parietal areas integrate perception and action by binding sensory information into a common representation of the association between stimuli and responses (Gottlieb 2007). In a similar vein, regions of the temporoparietal junction contribute to this process by using environmental information to update these mental representations (Geng and Vossel 2013). So, through parietal mechanisms, the PFC is presented with different options for how to respond. The PFC then likely has to decide which of the different options to use and to connect with the appropriate motor program or task set that leads to observable behavior.

Integration of goals, values, schemas, memories, affect, and actions/policies. A primary challenge for the brain is to integrate the numerous aspects that make up cognitive functioning, such as goals, strategies, values, affective states, actions (and their affordances), sensory inputs and observations, and existing memories. We suggest that one canonical computation of the PFC lies in integrating these complex levels in a manner that serves to produce goal-oriented behavior or thought. This process entails integrating higher-level variables, such as one's goals and current affective state, to produce a course of action that best achieves these goals, which in turn can lead to changes in internal states (e.g., selective memory retrieval) and execution of particular action plans. This integration function is closely related to the ability to form complex combinations, as discussed above, and in particular to the idea that complex decision-making tasks require abstractions that can be thought of as a cognitive map or task set.

Task sets describe the relevant sensory information, representations, and actions needed to meet a specific goal under specific conditions. By analogy with the hippocampus, which has been shown to integrate multiple cortical representations into episodes (Eichenbaum 2017), a task set can be viewed as a large-scale neural frame integrating multiple representations distributed over cortical regions (e.g., stimulus-action mappings, action-outcome predictive models) that can be evoked collectively to form a consistent system that guides behavior. These task sets can, in turn, enable the PFC to regulate adaptive behavior. This notion of task sets or rules in PFC also relates to theories of RL discussed earlier, wherein it is proposed the PFC encodes a rich set of world models (e.g., of how objects and agents in our environment might interact). These world models can be flexibly applied to new situations via a probabilistic inference process about their relevance (Tomov et al. 2023; Tsividis et al. 2021).

The task sets that result from this integration process are closely linked to value signals and outcomes of RL in the brain (e.g., Schuck et al. 2016; Wilson et al. 2014), which also have been widely observed in ventromedial PFC (Adelhöfer and Beste 2020; Beierholm et al. 2011; Hampton et al. 2006; Hardung et al. 2017; Lee et al. 2014; Narayanan et al. 2013), but arguably extend to striatal and other brain areas (Sharpe et al. 2020). Some work has suggested that the computational function unique to the PFC, in particular the OFC, is to provide representations that go beyond merely observable information by adding relevant information of the past (context) (Niv 2019; Schuck et al. 2018; Wikenheiser and Schoenbaum 2016). It should be noted, however, that the integration performed by PFC goes beyond these processes and includes, for instance, integration of information across different strategies (e.g., Donoso et al. 2014b) and expert systems (Charpentier et al. 2020; Lee et al. 2014; O'Doherty et al. 2021). Moreover, the temporal scale across which integration is performed can be much longer than a single task, allowing the emergence of meta-learning.

Robust/Active Maintenance of Information across Functions

Maintaining information is critical to a wide array of cognitive functions. Classic studies focused on the maintenance of sensory inputs or the preparation of motor actions (Funahashi et al. 1989, 1993a; Fuster and Alexander 1971). This maintenance allows cognition to break free from the immediate world, integrating information over time and responding at the appropriate time. Active maintenance of information, however, is also critical for more "cognitive" variables, such as maintaining information about the current situation, the current task, one's goals, and the value of different options.

To support the integration functions above, different types of information must be integrated over many different timescales; while a current thought is only briefly maintained, goals can extend longer, from a few minutes of focusing on writing a manuscript to years of dutifully saving for retirement. These different timescales of integration are reflected in the variety of intrinsic timescales of individual neurons. The variety of timescales found in the frontal lobe may reflect the diversity of functionality; neurons with shorter time constants respond to stimulus inputs while neurons with longer time constants maintain that information in working memory (Wasmuht et al. 2018).

It is important to note that the maintenance of information is not passive. Rather, it is focused on task-relevant information. Part of the reason for this feature is that working memory has a severely limited capacity: we are able to hold only a few items (i.e., 4–7) "in mind" at once. Therefore, it is important for selection mechanisms to determine what information is allowed to enter working memory, often referred to as "gating" (O'Reilly and Frank 2006; Yang et al. 2016), as well as mechanisms to select individual memories to drive behavior, akin to attention to external stimuli (Gazzaley and Nobre 2012; Panichello and Buschman 2021). Of note, it has been shown that during such "gating" processes, similar brain regions and neurophysiological processes are in charge that are also relevant for the integration of sensory and motor representations, but via different pathways of information processing that terminates in frontopolar regions (Yu et al. 2022). Beyond overcoming limitations in capacity, focusing the contents of working memory on task-relevant information can also ensure that only goal-relevant information is represented, integrated, and acted upon, and that extraneous information does not intrude or interfere. This function requires a further type of canonical computation: selection.

Selection and/or Biasing/Regulation of Task-Relevant Information

The world is incredibly rich. At each moment in time, we are inundated with a flood of sensory information from the outside world: potential memories we could recall, thoughts we could manipulate, actions we could take. Filtering this flood is critical to cognition. It allows us to focus our behavior on those stimuli/memories/actions that are contextually relevant. Filtering also helps to focus learning on those representations that are believed to be important, helping to resolve which features of the environment are most predictive of potential outcomes (referred to as the credit assignment problem). These considerations suggest that a "selection" mechanism is a canonical computational function of PFC.

Selection Over Representations of the Outside World. Attention is perhaps the best studied form of selection. Decades of research suggests PFC plays a central role in internally directed attention. Neurons in PFC represent where attention is allocated in space and to what features (Buschman and Kastner 2015; Miller and Cohen 2001). Activity in these prefrontal regions is observed prior to activity in other brain regions, suggesting PFC plays a leading role in directing attention (Buschman and Miller 2007). Stimulating within PFC induces attention-like effects in visual cortex (Moore and Armstrong 2003), and inhibiting/lesioning causes deficits in tasks requiring attention (Bichot et al. 2019).

Attention acts to filter cognition by biasing representations in other brain regions. For example, directing attention to a spatial location increases the activity of visual cortex neurons with receptive fields at the attended location (Reynolds et al. 2000). This increase in activity acts through lateral inhibition to suppress other competing representations (Desimone and Duncan 1995; Reynolds et al. 1999; Reynolds and Heeger 2009). In this way, attention can selectively focus sensory processing on a subset of neural representations. Several alternative mechanisms have been proposed to achieve the same effect: synchronizing the activity of neurons can increase their impact on downstream neurons (Fries et al. 2001), decreasing noise correlations can improve the signal-to-noise ratio (Cohen and Maunsell 2009), and changing the geometry of neural representations may allow certain information to flow between brain regions (Panichello and Buschman 2021). In the end, top-down guided selection acts likely through a confluence of mechanisms to filter information in other brain regions.

Selection over internal representations. Selection is not limited to attention to sensory inputs. It can also act in other domains. For example, frontal cortex plays an important role in controlling recall from episodic memory. As reviewed by Eichenbaum (2017), animals and humans with prefrontal damage have trouble selectively recalling information from episodic memory because of intrusion of competing memories. This suggests that although PFC does not provide direct mono-synaptic inputs into the hippocampus, it plays an important role in selective recall from episodic memory. This also refers to the selection of integrated sensory-motor representation, which are also thought to be stored in episodic traces (Hommel 2009).

Selection can also filter representations within frontal cortex. As noted above, selection is critically important for protecting the limited capacity of working memory. A "gating" mechanism is thought to control what information enters working memory (O'Reilly and Frank 2006; Yang et al. 2016). Then, when multiple items are held in working memory, an "internal attention" mechanism acts to select one item and use it to guide behavior. Functional imaging has shown PFC regions that direct attention to internal representations (in working memory) also direct attention to external, sensory representations (Gazzaley and Nobre 2012). Consistent with these findings, recent electrophysiological recordings in monkeys show the same neural representation encodes both selection from working memory and sensory inputs (Panichello and Buschman 2021). Similar to selective attention, selecting an item from working memory biases the neural representation to improve the encoding of the selected item. These findings suggest that control representations in PFC may be domain-general, allowing the brain to select task-relevant information regardless of the source of information.

Selection of higher-order cognitive variables, including goals and information processing parameters (meta-control). Finally, selection may also act on higher-order cognitive representations that can influence neural processes themselves. For example, research has shown that people will adapt the parameters of learning and decision making depending on the current context (e.g., changing the decision threshold or affecting the time constant of integration) (Cavanagh et al. 2011; Dayan 2012; Leng et al. 2021; McGuire et al. 2014). These forms of "meta-control" may occur through the biasing of competition between potential strategies (O'Doherty et al. 2021) or by direct selection/adjustment of parameters governing the relevant learning and decision processes. Electrophysiological recordings suggest that this form of control adjustment may happen by selection acting on different cortical regions, for instance, amplification of neural representation in order to filter representations appears to occur in sensory cortex, while adjustment to decision criteria have been localized to the frontal cortex and/or basal ganglia (Beste et al. 2018; Cavanagh et al. 2011; Forstmann et al. 2008; Frank et al. 2015; Luo and Maunsell 2015, 2018).

The role of inhibition in selection. Inhibition is inherent in the concept of selection. Selecting one item is, by necessity, to the detriment of other representations. Projections from the frontal lobe are largely excitatory (although see interhemispheric inhibition in mice; Cho et al. 2023). This suggests inhibition occurs through local mechanisms in the circuit that is receiving the selection signals. One such mechanism would be local lateral inhibition (e.g., through parvalbumin-positive inhibitory interneurons; Cardin et al. 2009). In this way, selection can act positively to strengthen selected representation which would, in turn, act through lateral inhibition to suppress other representations. In the field of attention this mechanism is often referred to as the "biased competition model" (Reynolds and Heeger 2009), although it can be generalized to other domains (Carandini and Heeger 2012). It has also been argued that "inhibition" may be the byproduct of the top-down biasing done by the PFC, such as by maintaining a task set or goal, because such biasing does so to the detriment of other representations (Munakata et al. 2011).

Alternatively, selection may act through direct feedforward inhibition that specifically suppresses a particular representation or region. Such mechanisms may be important for inhibiting responses, thoughts, or memory recall (Depue et al. 2016; Hulbert et al. 2016).

With either mechanism, varying the strength of the inhibition could modulate the strength of selection. Moderate selection could allow multiple representations to co-exist, but with a bias toward the selected representation(s). In contrast, strong inhibition could lead to winner-take-all dynamics that select a single representation, which may be important when only one response can be emitted (Wilson et al. 2012).

Monitoring

Timescales. Monitoring is a key dimension of control. Monitoring processes evaluate the relevance and reliability of behavioral policies and cognitive strategies guiding behavior to identify the need to inhibit, enhance, or revise them to make behavior more adaptive and efficient. Monitoring processes are likely distributed over the PFC and have been proposed to operate on three main temporal dimensions: (a) *retrospectively* from actual action outcomes to reactively adjust control processes guiding ongoing behavior (e.g., within vmPFC or dACC), (b) *prospectively* from contextual cues to proactively adjust control processes before acting (e.g., within lateral PFC), and (c) *counterfactually*, regarding alternative behavioral policies/strategies that are not guiding ongoing behavior but might advantageously replace the current behavioral policy/strategy guiding ongoing behavior (e.g., within frontopolar PFC, Koechlin and Wang, this volume).

Sources. Dorsomedial PFC, including the dorsal ACC and the pre-supplementary motor cortex has long been found to encode error or conflict signals during performance of complex tasks. These signals were first observed in EEG studies in which the so-called error-related negativity has been found, localized to dorsomedial PFC, which has been argued to be related to an internal detection that an error has occurred (Fu et al. 2023; Gehring et al. 1993; Hauser et al. 2014). Similar error signals have also been found to occur at the time of feedback. One possible source of these error signals is the reward prediction error (Holroyd and Coles 2002; Schultz et al. 1997), which detects discrepancies between expected and actual outcomes, possibly reflecting the effect of dopaminergic innervation into medial frontal cortex. These kinds of error signals have also been found to be present in both pre-SMA and anterior cingulate neurons in both monkey and human studies, as well as in BOLD fMRI in humans (Debener et al. 2005; Phillips and Everling 2014; Shen et al. 2014; Wang et al. 2005).

From an electrophysiological perspective, all these signals share a reliance on theta oscillations in the medial frontal cortex, which, due to biophysical principles, are optimally suited to integrate information being processed in distant brain regions (Buzsáki and Draguhn 2004; Cavanagh and Frank 2014). These theta-related processes are believed to reflect a "surprise signal," indicating a need to adapt one's actions (Cavanagh and Frank 2014), for instance, when the executed action mismatched the correct one. These kinds of signals are thought to be important for providing a metric of how well one is performing on a task, whether it is in terms of successfully getting rewards or implementing intended actions. For this evaluation to occur, it is relevant to rely on a comparison process, according to which information about the expected effects or the action plan need to be retrieved—a process likely guided by theta as well as gamma band information (Beste et al. 2023). It is possible that these processes reflect integrated representations of stimulus and action features (Beste et al. 2023), and that these integrated representations reflect content-specific beta band activity, which changes from active to latent to reactivated states as needed (Spitzer and Haegens 2017; Wendiggensen et al. 2022). The interplay of theta and beta related activity is likely under the control of alpha band activity to flexibly balance between top-down and bottom-information (Beste et al. 2023; Wendiggensen et al. 2023). It is the interplay of these oscillatory activity patterns that is likely central for above-discussed canonical computations, referring to perceptual and motor task sets (Beste et al. 2023), and which may also give rise to dynamics and functions reflected within PFC and the broader MD system (Duncan 2010).

These monitoring signals are also likely important for facilitating changes in strategy. Reliability is another form of signal that is important for monitoring and evaluation, which goes beyond the punctate-based error signals based on single events. Reliability concerns how well a particular strategy is doing in terms of making predictions and can be considered to be related to the (inverse of) variance or degree of uncertainty in the predictions associated with a particular strategy (Daw et al. 2005; Lee et al. 2014; O'Doherty et al. 2021). One way to compute reliability is by integrating over prediction errors; for instance, if many reward prediction errors have occurred recently, then reliability of reward predictions can be said to be low, whereas if only few small errors have occurred, we can say that reward prediction reliability should be high. Reliability signals for different strategies (such as for model-free vs. model-based RL strategies or even between different ways of learning through observation) have been found to correlate with BOLD responses in ventrolateral PFC and frontopolar cortex in humans (Charpentier et al. 2020; Lee et al. 2014; O'Doherty et al. 2021), whereas reliability signals related to different possible model-based strategies (i.e., within the model-based system) have been found in ventromedial PFC and frontopolar cortex (see Koechlin and Wang, this volume). Thus, PFC appears to monitor performance at different levels of abstraction, from punctate error signals to strategy reliability signals.

Targets of adjustment. Another way in which forms of monitoring dissociate relates to the type of control they are supporting. Research has shown that a hierarchical gradient of control emerges in the lateral PFC, with more caudal areas of lateral PFC representing information lower in this hierarchy and more rostral regions representing information higher up in the hierarchy (Badre and D'Esposito 2009; Badre and Nee 2018; Koechlin et al. 2003). It was subsequently proposed that parallel regions along the medial wall may engage in forms of monitoring that subserve control at similarly increasing levels of response complexity (Taren et al. 2011; Venkatraman and Huettel 2012). For instance, caudal regions of dorsomedial PFC (potentially corresponding to the cingulo-opercular network) have been shown to track the amount of conflict between competing responses (e.g., should I respond left or right), whereas more rostral regions of dorsomedial PFC (potentially corresponding to the frontoparietal network) have been shown to track the amount of conflict between potential strategies or other higher-order goals (e.g., should I maintain my current strategy or switch) (Ritz and Shenhav 2024; Shenhav et al. 2018; Venkatraman et al. 2009a).

How Do These Canonical Computations Align with Behavior and Neurobiology?

Alignment with Behavior

Any one task likely involves all of the canonical computations outlined above: integration, maintenance, selection, and monitoring/evaluation. However, each task may place a different distribution of demands on these computations. As a result, the relative contribution of PFC to each of the relevant computations might vary across tasks. There can, for example, be tasks in which the monitoring/evaluation aspect takes more prefrontal computational resources than the other canonical computations or where this is the case for integration, maintenance, or selection. For instance, a typical response interference-based cognitive control task (e.g., Stroop, flanker, go/no-go) may place limited demands on integration of task-relevant information (e.g., linking stimulus features with appropriate responses) and/or maintenance (e.g., of relevant task rule), but greater demands on monitoring (e.g., for errors or processing conflict) and/or selection (e.g., biasing of task-relevant feature processing). Conversely, for a typical decision-making task (e.g., choosing between foods, goods, or gambles), the demands on goal-directed integration may be more substantial, requiring comparison across values of relevant features of the options and potential courses of action (Frömer and Shenhav 2022).

This involvement of multiple canonical computations with differences in their relative weighting might lead to observations of unity and diversity of functions (e.g., in individual differences in performances across tasks).

Alignment with Neurobiology

Differences in function versus differences in representation. It is possible that the various functions above are subsumed by distinct regions of PFC. Alternatively, it is possible that there are canonical computations that are repeated across subregions within PFC but with differing inputs and outputs. For example, there may be a cortical or subcortical circuit motif that actively maintains a representation. As noted above, this mechanism is broadly useful for sustaining stimulus, motor, or task representations. Therefore, the same circuit motif operating on differing inputs could serve different functions. This might explain observations of functional differences between regions (see Murray et al., this volume). For instance, spatial information is represented more strongly in lateral PFC than OFC, which may reflect anatomical differences in connectivity with parietal inputs to lateral PFC and insular, temporal, and amygdalar inputs to OFC (see Rich and Averbeck, this volume). Similarly, gradients in abstraction along the rostral-caudal axis may reflect positioning along the cortical hierarchy (Badre and D'Esposito 2007; Badre, this volume). Computational modeling has shown that repeating circuit motifs in a hierarchical structure, such that the output of one circuit feeds into the next, can describe the increase in time constants observed along the cortical hierarchy (Murray et al. 2014; Koechlin and Wang, this volume).

One advantage of this theory is that it is easier to conceptualize how the functional diversity within PFC could evolve or develop. Rather than needing mechanisms to generate unique circuits for different functions, the same circuit motif could be "copy-pasted" but still support different cognitive functions.

Which anatomical distinctions are less well-aligned with these computations? There is currently some debate as to whether specific regions of PFC are *not* specialized for the domain-general processes described above, but rather for more domain-specific processing, more specifically language. For 150 years, portions of the left inferior frontal cortex have been associated with language output. While some theories posit that the left inferior frontal gyrus is important for domain-general processing of relational and sequencing information (Fitch and Martins 2014; Pallier et al. 2011), others have argued that the left inferior frontal gyrus is organized such that these domain-general regions are interdigitated with more language-specific regions (Fedorenko and Blank 2020).

Another outstanding question is how the above taxonomy of frontal lobe function explains the functions of those portions of the frontal lobe that are associated with the default mode network. While the functions of some of these default mode regions are accounted for by the functions described above (e.g., value calculation by portions of ventromedial prefrontal function), exactly what the function of, for example, lateral DMN regions (e.g., area 8) is, and how they may or may not fit into the above taxonomy, remains unclear.

Computational Modeling in Interplay with Experimentation

Computational Building Blocks and Cross-Level Understanding

Computational modeling has always been an integral part of research on PFC function (Cohen et al. 1996) and has traditionally often distinguished between so-called algorithmic and implementational levels of modeling (cf. Marr 1982). We propose that the time is ripe to eschew this distinction and to conceptualize instead PFC-related models in terms of computational building blocks, their biological mechanisms and computational principles as laid out in the previous section. For some of these core processes, such as internal maintenance of working memory or time integration in decision making, it is possible to achieve cross-level understanding from cell types to recurrent neural population dynamics to behavioral performance (Arnsten et al. 2010; Goldman-Rakic 1995). For other, more complex cognitive functions, the underlying biological mechanisms remain poorly understood. Nevertheless, at a minimum, modeling serves as a tool, in close reciprocal interaction with experiments, to bridge phenomenological description at one level and explanation at another level.

Internal Maintenance and Manipulation of Information

Neural circuit models based on neurobiology have been developed for working memory and decision making (Wang 2002), suggesting a "cognitive-type" local circuit model of the PFC (Wang 2013). A neural network model can be designed by intuition or shaped by training using machine-learning algorithms. In the latter case, how the function is realized is not defined *a priori*; it emerges as a result of training connection weights, for instance, using a backpropagation algorithm. Building such a model for working memory-dependent tasks revealed that self-sustained persistent activity is necessary when information must not only be maintained but also manipulated to perform a task (Masse et al. 2019).

Such a model was designed to enable mechanistic understanding across multiple levels, with collective neural population dynamics described as attractor states providing an account of function/behavior, on the one hand, and enabling investigation of underlying cellular and molecular mechanisms on the other hand. In particular, a gating mechanism for filtering out distractors was proposed in terms of a microcircuit motif composed of three types of inhibitory neurons (Wang et al. 2004b). The dependence on NMDA receptors for recurrent excitation (Wang 1999) provided one clue as to why NMDA receptor signaling pathology might cause cognitive deficits in schizophrenia, one of the findings that prompted the emergence of computational psychiatry (Redish and Gordon 2017; Stephan and Mathys 2014; Wang and Krystal 2014).

Extending recurrent neural network models to rule-based tasks, such as the Wisconsin Card Sorting Test, led to the theoretical proposal of mixed selectivity of neuronal function (Rigotti et al. 2010). This was supported by experimental data (Rigotti et al. 2013) and suggests a computational advantage of complex neural firing patterns commonly observed in the PFC (Fusi et al. 2016).

Task Set Representation

The novel approach of training recurrent neural networks (Yang and Wang 2021) has also been used to realize a single network capable of performing many rule-based cognitive tasks (Bouchacourt et al. 2020; Yang et al. 2019). This approach makes it possible to investigate how task sets are represented, the (di)similarity between neural representations of different tasks, and suggests clues as to how the PFC may represent various task sets (Sakai 2008).

Monitoring and Evaluation of Performance and Outcomes

Monitoring and evaluating one's behavior in the service of task performance or learning is a fundamental aspect of PFC function. In many of the tasks discussed above, monitoring and evaluation reflects a continuous learning process that shapes future behavior based on previous outcomes. RL models have been the primary framework to computationally understand this monitoring and learning processes. At the heart of RL models is a process that monitors how achieved outcomes compare to expected outcomes and updates future expectations accordingly. RL models have been widely studied and validated as a model of the brain and behavior. Importantly, they can go beyond a simple outcome monitoring process in multiple ways, for instance, by incorporating cognitive maps that provide the model with planning abilities or by including state-inference or state-learning processes that can map observations onto abstract representations or learn the abstractions suitable for reward maximization, as is the case in deep Q network.

Within research on cognitive control, monitoring has been instantiated as a comparator that accesses information from a neural network-like architecture (e.g., levels of coactivation across response units), and it uses the result of this comparison process to modify ongoing processes across the network (Botvinick et al. 2001; Botvinick and Cohen 2014; Holroyd and Coles 2002). Recent work has augmented these monitoring algorithms to weigh additional factors relevant to the organism, including expected reward rate within the current environment and resource limitations, such as effort costs (Musslick and Cohen 2021; Shenhav et al. 2013). Cutting across research on decision making and cognitive control, an emerging theme has been increased focus on how such monitoring processes can be leveraged toward arbitrating between high-level action plans—between, for example, different strategies (Donoso et al. 2014b), model-based versus model-free decision making (Daw et al. 2005; Lee et al. 2014), expert systems (O'Doherty et al. 2021), or gradual versus state-inference based learning (Zika et al. 2023).

Large-Scale Cortical Network Model

Graph theoretical approaches have been used to analyze and model data associated with large-scale networks of the brain. For example, graph models can be developed for large-scale brain network architecture based on either structural or functional connectome data, and these models can then be lesioned in silico to generate predictions regarding the consequences of different forms of brain damage (Alstott et al. 2009; Honey and Sporns 2008; Sporns 2016). These models can then be tested to see whether their predictions are consistent with findings from human lesion patients (Gratton et al. 2012). Recent investigations have focused on multilayer modeling to represent linked changes in brain networks over time (Betzel and Bassett 2017; Gerraty et al. 2018; Muldoon and Bassett 2016), using control system models to form predictions about how different functional states can arise from a static structural connectome (Gu et al. 2015), and using dynamic oscillator models to link transient "events" to the development of a modular network architecture (Pope et al. 2021).

Using connectomic data, dynamical models have been developed for the large-scale primate cortex, both for monkeys (Chaudhuri et al. 2015) and humans (Deco et al. 2014; Demirtas et al. 2019). Among findings from this new line of research are the concept of macroscopic gradients of biological properties (Wang 2020) and a hierarchy of time constants along the cortical hierarchy (Chaudhuri et al. 2015; Murray et al. 2014), offering a mechanistic explanation for the PFC's capability of time integration in contrast to early sensory areas, which lack such a temporal mechanism. This model can be used to computationally explore how the PFC works together with the rest of the cortex, such as in working memory (Froudist-Walsh et al. 2021; Mejias and Wang 2022; Wang 2022).

Integrating across Modeling Approaches

Mutually Constraining Models across Levels of Detail

One way in which these various modeling approaches can be better integrated is by extracting information from mechanistic models and linking it to network models. This approach might be fruitful in the domain of individual differences. One could use computational models of executive processes to estimate individual-level parameters (e.g., learning rates), after which one could examine whether such parameters are associated with characteristics of brain networks. For example, one might hypothesize that individuals with faster learning rates show greater integration of information in the frontoparietal control network from various sources, as would be reflected in a higher value for the graph theoretic measure of participation coefficient.

Conversely, one might examine neural network models for properties expected based on graph models of brain function, such as the presence of large-scale modules and connector hubs (e.g., Gratton et al., this volume). These correspondences could be used as a criteria for model selection or incorporated more explicitly into model creation.

Incorporating Observations about Finer-Grained Structure

Functional brain organization differs systematically among individuals on a number of dimensions, including brain network topography, topology, areal size, and even morphological characteristics such as tertiary sulci (Gordon and Nelson 2021; Voorhies et al. 2021). Many of these differences have been linked to differences in brain function, such as task activations (DiNicola et al. 2020; Gordon et al. 2017; Seitzman et al. 2019; Tavor et al. 2016) and are predictive, in the sense of cross-validation, of individual differences in behavioral performance (e.g., Finn et al. 2015; Kong et al. 2019). It is unclear, however, why differences in the size, shape, or location of brain regions should necessarily be linked to performance. What processes benefit from access to additional neurons or particular neural circuits? Linking these observations of individual differences in structure and morphology to neural network models, such as local circuit models (Wang 2022), may provide additional deeper insights into the links between brain network organization and behavioral outcomes.

How Can These Models Be Used to Understand Unity and Diversity?

Confirming Mapping between Task Measures and Function

One benefit of models that formalize a given set of functions is that they allow you to simulate behavior on a given task and ask to what extent different parameters map onto different sources of variability in task performance. They also allow you to invert this process and ask to what extent a given measure of task performance selectively taps into a function of interest. For instance, Musslick et al. (2019) examined to what extent various common cognitive control task measures reflected individual differences in control capacity (i.e., how much control a hypothetical person might be able to maximally apply within a task), something that clinicians and developmental researchers often seek to index. These authors simulated a variety of task performance metrics for a given agent, including differences in performance between trials that (a) are incongruent versus congruent (congruency effect), (b) follow an incongruent versus congruent trial (conflict adaptation), and (c) follow a change versus repetition in task rule (switch costs). By simulating task performance across an array of artificial agents varying in control capacity as well as other model parameters (e.g., learning rate, task automaticity), they showed that the congruency effect, commonly used to tap into individual differences in capacity, is more likely to reveal individual differences in automaticity than capacity. At the same time, these theoretical analyses also revealed task measures that may provide a more sensitive measure of capacity (like conflict adaptation effects) and revealed more generally the extent to which these different parameters are likely to be confused with one another when using a given task measure. This approach can be extended to any of the modeling approaches described above, to aid in selection and development of tasks targeting different computations of interest.

Understanding Frontal Lobe Function through the Lens of Artificial Intelligence

As artificially intelligent agents evolve in the direction of generalized intelligence, they will likely have to overcome many of the same computational problems faced by the biological brain. The expansion of the frontal lobe over evolution has allowed for the expansion of cognition (Weiner et al., this volume). Therefore, it seems reasonable to expect that aspects of the evolution of cognition in artificial agents will involve the expansion of the same computational mechanisms that are served by the frontal lobe. Indeed, this is already reflected in many of the advances in artificial intelligence (AI) over the past several decades. Early neural network models were built using simple individual neurons with strict feedforward connectivity. While these networks were sufficiently flexible to capture complex cognitive processes, they were notoriously difficult to train to perform complex tasks. As techniques evolved, the introduction of recurrence allowed these networks to capture temporal dynamics and, importantly, begin to maintain memories of recent inputs. The next critical insight came from the introduction of selection-like mechanisms, whether it is gating of inputs into recurrent networks, such as long short-term memory (Hochreiter and Schmidhuber 1997) or using attention-like filters to selectively propagate task-relevant information, such as transformers (Vaswani et al. 2017). Around the same time, deep reward-learning networks were being trained to perform increasingly complex and diverse arrays of tasks (e.g., Mnih et al. 2015). It is notable that the evolution of intelligence observed in these network models reflects the iterative addition of each of the canonical computations described above. Parallel feedforward models are able to integrate information effectively; recurrent networks are able to maintain information actively; transformers and long short-term memory rely on selection of feedforward or recurrently maintained representations; and deep RL relies on monitoring to learn and update representations.

This evolution suggests that understanding the mechanisms supporting intelligence in artificial agents may provide a new angle to understanding human intelligence and the role of the frontal lobe, with the hope that some of these mechanisms will be similar to the ones observed in the brain. This development could be useful to gain mechanistic insight at a few levels.

First, computational models may provide insight into the mechanisms and functionality of the frontal lobe. By training computational models on increasingly complex, more "real-world" tasks, we can use the analytical approaches described above to decompose them into underlying computational motifs. Early attempts are already providing new insight into complex cognition: fully recurrent neural networks that are trained on complex context-dependent decision-making tasks show low-dimensional dynamics that are compositionally combined to perform more complex tasks (Yang et al. 2019). One difficulty is that it is often hard to understand how these dynamics emerge from the underlying circuit. In other words, are we simply swapping one complex system for another, slightly less complex system? One potential way to overcome this dilemma is to constrain these models to be low-dimensional (e.g., low rank connectivity) yet still recapitulate the function of more complex models. This approach often leads to more interpretable circuit mechanisms and can reveal computational motifs that align well with previous hand-built models, such as using gain modulation to do context-dependent computations (Dubreuil et al. 2022). This approach perhaps gives us some hope that complex models trained to perform complex behaviors could help us understand how previously known circuit and computational motifs are engaged during real-world behaviors.

Second, understanding AI may provide insight into the canonical computations that are critical for cognition. In other words, studying artificial agents may reveal new canonical computations that we have yet to consider. To a certain extent, such insights have been observed in the application of transformers to large-language models. While theoretical modeling focused on the learning of grammatical structure to generatively produce language, large-language models have demonstrated the power of a simple learning rule, predictive learning, in being able to learn and generate language (Piantadosi 2023). One could imagine similarly surprising insights emerging from AI agents trained to perform complex, real-world behaviors.

Experimental Approaches: Limitations, Opportunities, and Future Directions

Room for Improvement in the Assessment of Function

Goal Selection

In the vast majority of research, participants are either given their goal for a given trial explicitly (e.g., name the color of this word) or are able to infer it from the reward structure of their environment (e.g., it is currently most rewarding to focus on the shape feature). In real life, individuals typically have to set their own goals, including what task to complete and how to complete it. In failing to capture this element of ecology, these studies also fail to capture processes that are commonly impaired in patients with prefrontal lesions; namely, how a person selects their current goal and the subgoals that will help them achieve it (see Table 12.2). Patients with PFC lesions demonstrate substantial task initiation costs, goal neglect, and forms of apathy and avolition that could at least partially reflect an inability to settle on and sufficiently activate an immediate goal. Pathology aside, understanding goal selection can provide better insights into individual variability over development and across individuals in adaptation to the level of "goal scaffolding" within a person's environment (e.g., the extent to which their caretakers provide clear structure for their future aims).

There have been a number of attempts to lend further experimental insight into the process of goal selection, including the classic Multiple Errands Test (Shallice and Burgess 1991). Briefly, patients were sent out on their own to complete a series of errands of varying complexity around an area of London, including purchasing specific items and finding out particular types

Table 12.2 Directions for improvement in existing experimental approaches.

Domain	Common approach	Example novel directions
Goal selection	Explicit and/or well-constrained task goals	Choice of which task to perform when
Planning complexity	Planning over limited number of steps	Larger space of options and potential subgoals
Response complexity	Limited number of discrete and irrevocable actions	Continuous action space, reversible
Value of information	Limited opportunities for and scope of new information	More information-rich tasks and exploratory opportunities
Changes over time	Measures averaged over the course of a single session	Analyze temporal dynamics within/across many sessions
Naturalistic measures	Tasks performed in the lab	Tasks and other measures (EMA, physio, mobile EEG) measured out "in the wild"

of information from those items, such as the exchange rate of the French franc on the previous day. The errands required following specific rules to achieve those goals, such as not entering a shop without buying something. Compared to matched controls, the frontal lesion patients broke the rules more often and performed the task less efficiently, including exhibiting problems with the selection and implementation of subgoals. Qualitatively, a patient's behavior was unlike any of the controls. For example, one patient picked the wrong newspaper, did not pay for the item, and ended up being chased by the shopkeeper. Another focused on buying soap that she preferred rather than adhering to the instructed goal of obtaining the cheapest available soap. In essence, these patients appeared to exhibit difficulty both in adequately selecting and carrying out specific goals, as well as in implementing behaviors that were most appropriate to achieve those goals (i.e., rule breaking).

More recent experiments have examined much simpler forms of goal selection within the laboratory, by allowing, for instance, participants to choose freely which of a limited set of tasks to perform, and for how long, based on factors like expected reward and difficulty (Arrington and Logan 2004; Gilzenrat et al. 2010; Orr and Banich 2014; Parro et al. 2018; Westbrook et al. 2013). Some of these tasks have provided evidence of prefrontal involvement in task choice (e.g., Orr and Banich 2014; Westbrook et al. 2019; Wisniewski et al. 2015). Other work has examined another key dimension of goal selection; not *which* task to perform but *when* to perform it. For instance, Le Bouc and Pessiglione (2022) had participants perform laboratory choice tasks that assessed the extent to which they preferred exerting effortful tasks later rather than sooner, and then showed that these task-based estimates predicted how long participants would wait before returning a set of forms they had been asked to complete and return any time within the next month. These experiments offer instructive examples of studying the various dimensions of goal selection within a controlled environment. Nonetheless, they fall substantially short of capturing the complexity of real-world goal selection, as exemplified in the errands task above.

Planning Complexity

In line with the above-discussed desire to understand goal selection and changes in PFC function across time, tasks that require multistep planning (as would be required e.g., during cooking or playing Atari games), might be a particularly useful tool for studying PFC function. Planning tasks often require internal simulation before a choice is made, thus tapping into one of the main adaptive functions of PFC described above. Planning tasks can also incorporate a reward-learning process, which then opens a window into the relative roles of forward and backward simulations for planning and learning processes, as well as the goal-oriented cognitive map over which planning occurs (Mattar and Daw 2018). It might be particularly instructive to investigate forms of repeated planning that can be optimized over time, providing greater insight into the process by which subgoals are learned. In addition to the insights these tasks provide into planning itself, tasks like these have the added benefit that they involve a comparatively high number of options or action sequences (e.g., Eldar et al. 2020; Huys et al. 2012; Kurth-Nelson et al. 2016), which is another desirable property discussed further below.

Response Complexity

In general, behavioral tasks are conceived to be as simple as possible to expedite training for participants and make the space of analyses/interpretations as narrow and tractable as possible for the experimenter. Tasks that are too simple, however, might not engage prefrontal mechanisms, thereby obscuring anatomical and functional segregations, making it difficult to discriminate between computational models of prefrontal function. For instance, tasks that force a choice between two responses face a challenge disentangling between selection of one response and inhibition of the alternate response. Prefrontal functions are critical to manage real-life environments that feature high-dimensional, uncertain, changing and open-ended situations as well as continuous and often reversible behaviors. Investigating prefrontal functions certainly requires a consideration of behavioral paradigms that capture these complexities as much as possible.

The Value of Information

Another higher-level process that is believed to be supported by the frontal lobe is exploration (Badre et al. 2012; Domenech et al. 2020; Monosov and Rushworth 2022), including tracking properties of the environment that give rise to the antecedent experience of curiosity. This process serves to identify a conceptual space of potentially useful information or behaviors that *might* be relevant in the current context or potentially useful in the future. Acquiring knowledge about the environment to learn proper internal world models is central to efficiently fulfilling the ever-changing needs of the organism (Koechlin and Wang, this volume). Information-seeking is thus believed to constitute a primary drive of behavior and is potentially separate from reward-seeking. How the PFC arbitrates between reward-seeking and information-seeking motives, and the extent to which information-seeking serves to maximize expected future outcomes and/or minimize aversive uncertainty, awaits further research, both computational and empirical (cf. Cockburn et al. 2022; van Lieshout et al. 2019, 2021a, b). Doing so will benefit from novel experimental designs that incorporate a wider range of potential future states and varying motives for seeking out or avoiding those states, under varying levels of known or unknown uncertainty.

Changes in PFC Involvement over Time

Research into the broad set of functions laid out above often examines behavior and neural activity averaged over the course of an experiment (e.g., trials within a session). In doing so, these studies miss changes that occur over periods of time within a session that could offer critical insights into the drivers and dynamics of PFC function. For instance, over the course of a single experiment, attention, effort, and control demands may vary (e.g., due to boredom, mind-wandering, fatigue, practice, and/or fluctuations in mood); learning is likely to occur (shaping changes in task-relevant representations); and participants may shift between strategies for performing the task. These factors all raise the potential for increased measurement noise. More importantly, they also represent missed opportunities for understanding these functions at a finer grain (e.g., mechanisms of plasticity, distractor interference, influences of motivation and affect on controlled processes).

These dynamic changes have raised particularly acute concerns about the extensive training that occurs prior to nonhuman animals performing such tasks. This limitation also introduces opportunities, both for beginning to examine performance over the course of this extended training regime (e.g., Masís et al. 2023) and for examining human parallels to such extensive levels of training (e.g., Balci et al. 2010; Blain et al. 2016). For example, a recent study by Miller et al. (2022) carried out extended testing in human participants over the course of three months on both a working memory and serial reaction time task. Working memory performance improved throughout this time window, and significant evolution was seen in delay period activity patterns in the frontal lobe. More generally, though, these opportunities should be more regularly exploited over longer timescales, within both humans and other animal models, by studying how cognitive functions, neural anatomy, and physiology vary over the course of multiple experimental sessions, days, weeks, or months apart (e.g., Allen et al. 2022; Naselaris et al. 2021; Poldrack et al. 2015).

Measures of Naturalistic Behavior

As researchers, we are interested in understanding and predicting behavior outside the lab. Doing so inevitably will involve considering the greater diversity of environmental contexts that people experience. The real world is distracting, noisy, and variable in terms of resources. More naturalistic assessments may be helpful for understanding how differences in these and many other factors may affect PFC functioning. For example, having participants complete tasks in their homes or on their phones may provide a better understanding of real-world performance. Ecological momentary assessments (EMA), which prompt participants to answer questions about their experiences at that particular moment (e.g., their current goal, emotional state, or context), may provide insight into everyday behavior and variability (e.g., Hofmann et al. 2012b). One might also obtain measures relevant to function from passively collected data, such as global positioning satellite locations or accelerometry measures from wearable devices (e.g., Heller et al. 2020), which would help reduce participant burden and remove biases that may arise with self-report measures.

Advances have been made in the tools researchers have at their disposal to measure PFC activity out in the world, such as functional near infrared spectroscopy (fNIRS) or mobile EEG. With these measures, neural mechanisms, which until now have exclusively been investigated in laboratory environments, become translated to natural situations with the advantage that concepts about the neural implementation of PFC functions can be put to the test in natural environments. Such endeavors will significantly broaden the validity of the already established concepts about prefrontal neural processes, ultimately leading to a more holistic understanding of PFC function (see Table 12.2).

Leveraging Recent Advances in Data Analysis

Deep Convolutional Neural Networks/Deep Q Networks

One promising approach for probing the nature of the representations found in the PFC involves the use of network models imported verbatim or with small modifications from the AI literature. These models can either be pre-trained to perform specific tasks, such as object recognition (Kriegeskorte 2015), or trained from scratch to perform specific tasks, such as learning to play particular Atari games (Mnih et al. 2015). Although these models are very different from the architecture of the brain, both in terms of their physical structure and the rules used for modifying plasticity within them, they have been successful in revealing patterns of activity in their layers that seem to correspond broadly to patterns of activity in the brain (at the level of single neurons or populations) and fMRI activity, when applied to activity measured while animals or humans are performing the same tasks on which the network itself has been trained (Cross et al. 2021; Iigaya et al. 2023; Kriegeskorte 2015; Yamins et al. 2014). Though these approaches have been mostly used to date to illuminate representations in the ventral and dorsal visual stream as opposed to the PFC, it is likely that models incorporating more complexity, such as recurrency and/or multinetwork structure, may prove useful in explaining patterns of activity in the PFC as well (Perich and Rajan 2020). One important way to leverage these models is to explore whether some variants on their architecture can better account for neuronal activity than others. Furthermore, inducing lesions in those models and seeing to what extent particular components of the model are critical for behavior might also serve as a basis for refining hypotheses about causality regarding particular prefrontal areas, which could then be tested in future causal perturbation experiments, such as with inactivations, optogenetic or chemogenetic manipulations in animals, and/or transcranial magnetic stimulation.

Another approach would be to use deep convolutional neural networks—which have been optimized for the structure of neurophysiological (e.g., EEG) data (Lawhern et al. 2018)—to analyze such data in a way that makes it possible to delineate potential novel features that had been potentially overlooked by theory-driven approaches (Vahid et al. 2020). Through the use of explainable AI methods, the novel features identified could then be integrated into existing conceptual frameworks on the cognitive processes being examined in the study at hand. Moreover, other methods, such as generative adversarial networks (Goodfellow et al. 2014), may provide valuable insights into the neurophysiological principles underlying cognitive functions supported by the PFC. For example, these networks have been used to show that neurophysiological principles of two opposing instances of cognitive control processes or antagonistic behaviors can be transferred to each other (Vahid et al. 2022). Since such deep learning procedures are able to capture nonlinear interdependencies, these approaches may be well suited to examine the interrelation of neural principles that are associated with the above-mentioned canonical computations.

Combining across Data from Multiple Tasks

As outlined above, it is likely the interconnection between different canonical computations and the relative weighting of the computations are important to understand in PFC function. It is, therefore, important to abstract from the level of specific tasks and analyze neural data in a more overarching, task-invariant way. This approach has particularly been lacking within analyses of neural time series data. Principal component analysis (PCA) and independent component analysis (ICA) have been used to extract neurophysiological components but are optimized for two-dimensional data (e.g., covering spatial and temporal information of time series data). However, data from typical experiments with concomitant data recordings (e.g., EEG) can yield more dimensions: time, space, frequency, trial, condition, participant, and group (Cong et al. 2015). These dimensions can mathematically be described as tensors. Applying methods optimized for two-dimensional data (i.e., PCA and ICA) in the face of such data is only possible by reducing data dimensionality (e.g., by concatenating or stacking the data). This, however, leads to an inevitable loss of information (Cong et al. 2015).

Tensor decomposition techniques can capture additional dimensions of information contained in neural time series data (Cong et al. 2015). Through these techniques, factors such as "tasks" can directly be modeled in the data analysis, allowing one to look at possible distinct and common neural profiles across tasks. This approach may provide a necessary step toward a thorough examination of neural principles across tasks and probable distinct or common profiles of canonical computations mediated by the PFC. Crucially, this method also overcomes another important shortcoming of most strategies used in the analysis of neural time series data: the reliance on averaged parameters of neural activity, which tacitly assumes that neural processes do not change across time spent engaging in various aspects of canonical computations mediated by the PFC. Through tensor decomposition methods, it is possible to better model moment-to-moment variations and changes over time in the neural activity profile without losing the information of other relevant dimensions in neural time series data.

Data-Driven Identification of Functional Primitives/Common Motifs across Tasks

Classically, our understanding of the neural mechanisms underlying cognition has been "top-down". We use theoretical concepts to generate hypotheses which, in turn, drive experimental design and data analysis. Recent work has begun to take a more data-driven approach to identify processes of interest. For example, combining cutting-edge factorization and dimensionality-reduction techniques has allowed us to begin to decompose naturalistic behavior into a sequence of action primitives, referred to as "behavioral motifs." The transition between behavioral motifs has been related to striatal activity, providing a novel perspective for understanding the function of striatum (Markowitz et al. 2018, 2023; Wiltschko et al. 2015). Similarly, the spatiotemporal pattern of cortex-wide neural activity can be decomposed into a set of ~15 dynamic "neural motifs" (MacDowell and Buschman 2020). These motifs repeat over time, across tasks, and between individuals, suggesting they provide a canonical basis set of underlying patterns of neuronal firing (i.e., primitives) that aid in understanding the dynamics of neural activity across the cortex.

Can one take a similar approach to prefrontal function? To do so, it would be desirable to define PFC-dependent functional primitives objectively, in a data-driven way. Quantitative cognitive ontology is becoming possible with the help of large-scale population experiments and machine-learning aided data analysis (Eisenberg et al. 2019). While data-driven approaches will likely provide support for the theoretically motivated hypothesized cognitive functions, the hope is that they may also identify novel mechanisms that have not previously been considered.

Dynamical Systems, Subspaces, and Neural Geometry

A relatively new approach to describing neural representations is at the population level, providing new insight into the dynamics and representations of the frontal lobe. Large-scale recording approaches have allowed researchers to track the activity of an ever-increasing number of neurons. One can visualize patterns of activity across the entire population of neurons as a point in an N-dimensional space (where N is the number of recorded neurons and thus, very high). Recent work has begun to understand how the geometry of these representations may enable generalization and compositionality (Bernardi et al. 2020; Fu et al. 2022; Panichello and Buschman 2021; Weber et al. 2023).

Dimensionality-reduction techniques and classifiers can be used to identify low-dimensional subspaces within the high-dimensional neural space that encodes task-relevant variables. Understanding how these subspaces relate to cognition is a rapidly emerging field. For example, computational modeling suggests "learning-to-learn" is facilitated by creating a subspace within PFC that is shared across a series of tasks (Goudar et al. 2023).

The dynamics of neural activity can be quantified by considering the trajectory of neural activity in state space (Shenoy et al. 2013). This approach has provided insight into how neural representations evolve over time. For example, sensory representations have been found to rotate over time, eventually forming a short-term memory representation of the stimulus input (Libby and Buschman 2021). This rotation allows both sensory and memory information to be represented in independent subspaces. Importantly, because these subspaces are orthogonal to one another, sensory and memory representations do not interfere with one another. In the context of sequence learning, the coexistence of sensory and memory representations may be important for associative learning. More broadly, such rotations may be important for reducing interference in (a) working memory (Panichello and Buschman 2021), (b) between representations of targets of attention (requiring selective enhancement) and distractors (requiring selective suppression) (Ritz and Shenhav 2024), and (c) between tasks (Weber et al. 2023).

Finally, subspaces may allow for the routing of information between brain regions. Simultaneous recordings within visual cortex identified a subspace within V1 that "communicated" with V2: changing the neural activity within this subspace influenced downstream activity, while changing the neural activity outside of this subspace did not (Seitzman et al. 2019). Larger-scale recordings have shown that these communication subspaces are not one-to-one, but rather extend to broader networks of regions (MacDowell et al. 2023). This arrangement then may provide an ideal mechanism for cognitive control. Changing how information is represented within a given brain region could change how that information is propagated to other regions. For example, representing information in "private" dimensions (that are not communicated) could keep information local, while transforming that representation into a "shared" subspace could broadcast that information to other brain region(s).

Conclusions and Open Questions

Historically, research into the function of PFC has been driven by numerous conflicting theoretical accounts and insufficient and/or inconsistent evidence to constrain or adjudicate among them. We have sought to cut through these conflicts by offering an integrative perspective on the common computations that underpin previous findings within PFC, including both underlying patterns of neural activity and manifestations of damage to its regions. The four canonical computations that we have highlighted—integration, maintenance, selection, and monitoring/evaluation—offer a parsimonious account of the interlocking computations that are both necessary for goal-directed behavior and potentially sufficient for explaining the array of observations just noted.

The account we have provided is, of course, incomplete and in many ways demands further iterative refinement and revision. As one example, in accounting for the broad set of functions that we outlined at the start of this chapter (e.g., planning, flexibility, and active maintenance), we offered a set of common underlying computations that largely accorded with (and deliberately integrated over) ones that have been previously proposed across different literatures. Future work should seek to identify new computations to augment or even replace those we offered.

There are also a number of questions that we were unable to address but which will be critical for providing a comprehensive account of PFC function. How do PFC computations vary in the timing of their engagement within a given task trial (manifesting, e.g., as difference in proactive versus reactive control)? How are symbolic representations (e.g., language) incorporated into core computations (integration, selection, maintenance, and monitoring). and how does this emerge over evolutionary development (e.g., with human participants able to learn and adapt flexibly based on verbal instructions alone)? To what extent do these computations give rise to key elements of social cognition (e.g., mentalizing, perspective-taking), and in what ways are they supported by other core functions within or outside of PFC? In what ways is engagement of PFC functions experienced by the organism as effortful, and to what extent do these functions each depend on motivational input to carry out versus continuing automatically in the absence of motivation (Shenhav et al. 2017; Westbrook and Braver 2015)? Finally, how are PFC functions facilitated by and/or interfered with as mood and affective states vary (Kenwood et al. 2022; Pizzagalli and Roberts 2022)? Answers may provide important clues into the role of PFC dysfunction versus neuromodulation in generating versus alleviating symptoms of certain psychiatric disorders, such as major depression (see Rowe et al., this volume).

13

Pharmacological Modulation of Prefrontal Cortex in Affective Disorders

Emerging Therapies

Angela C. Roberts and Conor Liston

Abstract

There is an ever-expanding range of pharmacological treatments for psychiatric disorders but our understanding of their efficacy at the level of disorders, symptoms, and especially at the level of individuals is extremely limited. Neuroimaging studies reveal dysregulation in the higher-order cognitive and emotional control networks of the prefrontal and anterior cingulate cortices in patients suffering from affective disorders such as depression and anxiety. Moreover, successful treatment by antidepressants or anxiolytics is often associated with an amelioration of the dysregulation in these control networks. Treatment resistance is a common occurrence across patients, and without a detailed understanding of the neurobiological actions of efficacious pharmacotherapies, we are still far from being able to tailor the specific classes of pharmacotherapies to any one individual. Important insights into how the different prefrontal control networks may be differentially affected by different classes of antidepressants can be revealed by considering the marked heterogeneity in the neurochemical modulation of prefrontal and anterior cingulate cortices. For example, the distribution of receptors, transporters, and neuronal subtypes that are the targets of current antidepressants, including the monoamine, glutamate, GABA and opiate systems, differentially target those prefrontal and anterior cingulate regions involved in reward, affective, salience, executive, and default mode networks. However, while large-scale patient neuroimaging studies have implicated changes in activity within specific regions of prefrontal and anterior cingulate cortex (and associated networks) as mediators, predictors, and/or moderators of antidepressant efficacy, insight into the differential actions of the different classes of antidepressants has not been forthcoming. Experimental studies in animals, on the other hand, are beginning to provide important insights into cellular and molecular plasticity mechanisms within prefrontal cortex that may underlie antidepressant efficacy. Still, a major unanswered question is why there is such marked variation in efficacy between individual patients. Future work needs to directly compare the neuroimaging profiles of different classes of antidepressants in patients and take into account efficacy at the level of specific symptoms as well as treatment history. In addition, a greater focus on the comparison of the actions of different classes of antidepressants is needed in animals alongside a comparison of their actions within distinct regions of prefrontal and anterior cingulate cortex. Only then can we begin to identify the factors that may determine the treatment strategy for any given individual.

Introduction

Only 30–40% of individuals diagnosed with an affective disorder, such as anxiety or depression, show remission following first-line treatments, whether they be pharmacological or behavioral-cognitive-focused therapies. Moreover, even when treatment is successful, the underlying mechanism is poorly understood. As a consequence, it is currently not possible to tailor treatment strategies to individuals. Evidence from functional and structural neuroimaging, as well as postmortem studies of affected individuals highlights the marked alterations in the prefrontal cortex (PFC) that accompany these disorders. Perhaps not surprisingly, many of these alterations are reversed following successful treatment, but the neurobiological, neurochemical, and cognitive mechanisms by which this remission is achieved, and whether the effects of the treatment are due to direct or indirect targeting of prefrontal functioning, is still to be determined.

In this review, we consider the evidence that relates pharmacological treatment strategies with the modulation of prefrontal function, particularly in the context of depression. There have been a number of experimental approaches that have implicated the PFC. The most common in humans has been to image brain structure and function of affected individuals, either at the level of individual brain regions or at the level of connectivity patterns and circuit analysis. In some studies, imaging is performed before and after treatment, and posttreatment changes that are related to treatment efficacy are used to provide insight into the efficacious actions or mediators of the drug. Other studies focus on pretreatment and determine whether differences in brain structure and function between individuals can predict subsequent treatment efficacy. Limitations of all these studies include the issue of clinical heterogeneity, which can be offset somewhat by emerging approaches for stratifying patients according to clinical symptom profiles, behavior, or biological measures, although this is rarely achieved. Few studies compare against a placebo control group, and so for the majority of findings, it is not possible to identify drug-selective biomarkers of treatment response (moderators) separate from those of placebo. Even fewer studies have directly compared different antidepressants, such as selective serotonin inhibitors (SSRIs) versus noradrenergic inhibitors, important to tailor treatment strategies effectively. A less common approach is to study the action of pharmacological treatments in healthy controls. This, however, has the major limitation that baseline brain function almost certainly differs from that of affected individuals, thereby influencing the actions of any given pharmacological therapy and limiting the translatability of any results to the clinical condition.

In animals, the effects of antidepressants have either been studied in normal healthy controls, with the same caveats as raised above in humans, or alternatively investigated in chronic stress models that recapitulate some—but not all—behavioral features of clinical affective disorders. The latter include prolonged experience with social stressors (e.g., chronic social defeat, social isolation) or physical stressors (e.g., chronic restraint, chronic unpredictable mild stress) or prolonged treatment with the stress hormone, corticosterone. The efficacy of pharmacological agents to relieve these behavioral changes is then established. An additional dimension in animal studies is that the pharmacological agent can be given not only systemically, as in the clinic, but also centrally, targeting specific brain regions to provide insight into their target of action. Furthermore, rapidly developing approaches for recording and manipulating the activity of large populations of neurons in specific circuits and cell types (e.g., two-photon imaging, photometry, optogenetics) are enabling investigators to establish causal mechanisms linking the molecular effects of a given drug with circuit function and behavior. In all these studies, whether clinical or experimental studies in animals, particular insight is gained when treatments are directly compared with one another, including different pharmacological therapies (e.g., SSRIs versus dopamine transporter inhibitors) or different therapeutic approaches, such as pharmacological versus cognitive behavior therapy (CBT).

We begin with a brief summary of the most consistent alterations in PFC structure and function in depression as revealed by neuroimaging. We then consider the neurochemical signatures of prefrontal brain regions before reviewing our current understanding of their sensitivity to the actions of a range of classes of antidepressants. We focus, in particular, on those agents that target the monoamines as well as the more recently discovered rapid-acting antidepressants (RAADs).

Prefrontal Dysregulation in Depression

Neuroimaging tools have become a mainstay of studies aimed at identifying and characterizing pathological correlates of depression and other affective disorders. While a comprehensive review is outside the scope of this chapter, here we highlight major findings from structural and functional MRI studies of depression, which may be useful for contextualizing the findings reviewed in the following sections on pharmacological effects on PFC function. We focus on the most consistently replicated findings in large-scale studies. Three themes emerge from this literature. First, meta-analyses of structural MRI studies, such as those conducted by the Enhancing NeuroImaging Genetics through Meta-Analysis (ENIGMA) consortium, confirm that cortical thickness is consistently reduced in multiple areas of the PFC (Schmaal et al. 2017, 2020), including the medial orbitofrontal cortex, anterior cingulate cortex (ACC), and areas of the lateral PFC as well as areas outside the PFC, including the anterior insula, posterior cingulate cortex, and hippocampus. Reductions in cortical volume may be related to changes in the density of neurons or glia or stress effects on dendritic arborization, among other mechanisms (Davidson et al. 2002; Krystal and State 2014). Of note, these effects are modest (Cohen's d=0.14–0.17) and highly variable, but also highly reliable and statistically significant in this meta-analysis which involved over 1,700 patients with unipolar depression and over 7,000 healthy controls. Furthermore, these effects are not specific to unipolar major depression: a meta-analysis of voxel-based morphometry studies spanning six diagnostic groups (schizophrenia, bipolar disorder, major depressive disorder, obsessive-compulsive disorder, substance use disorders, and anxiety disorders) identified three areas with gray-matter volume reductions in all six groups: the dorsal ACC and the bilateral insular cortex (Goodkind et al. 2015).

Second, resting-state fMRI have identified a variety of alterations in functional connectivity in depression-related brain networks (Greicius et al. 2007; Sheline et al. 2010; Yan et al. 2019), some of which may be present only in subgroups of patients with this highly heterogeneous diagnosis (Drysdale et al. 2017; Price et al. 2017). One of the most consistent findings involves bidirectional alterations in functional connectivity seeded from dorsomedial prefrontal areas of the default mode network that are modulated by sex and may relate to a propensity for excessive rumination in some patients (Hamilton et al. 2011; Kaiser et al. 2015; Talishinsky et al. 2022; Yan et al. 2019). Another is reductions in functional connectivity between the salience network (with prefrontal nodes in the lateral PFC and ACC) and midline areas of the default mode network and frontoparietal control networks (Kaiser et al. 2015), which may relate to deficits in emotion regulation (Wager et al. 2008).

Third, the task-based fMRI literature is complicated to interpret owing to methodological differences across studies and smaller sample sizes. Overall, they implicate subcallosal anterior cingulate cortex (scACC) hyperactivity in both normal sadness and depression (Mayberg et al. 1999, 2005), excessive coupling between a hyperactive scACC and default mode network areas in rumination (Grimm et al. 2009; Hamilton et al. 2011), and hypoactivity in ACC, OFC, and striatum in anhedonia (Pizzagalli 2014).

Neurochemical Targets for Antidepressants

SSRIs (e.g., escitalopram, fluoxetine, sertraline, paroxetine) are the first-line treatment for anxiety and major depression in adults, but their efficacy in inducing remission is dependent on their chronic treatment in the order of 4–6 weeks. While these drugs target the serotonin transporter (5-HTT), they often have other actions. For example, sertraline is a weak dopamine transporter inhibitor, paroxetine is a weak noradrenergic inhibitor, and fluoxetine targets ion channels and has effects via the SNARE (SNAp REceptors) protein complex. If SSRIs are ineffective, then alternatives include combined serotonin and noradrenergic transporter inhibitors (SNRIs), noradrenergic transporter inhibitors, and combined noradrenergic and dopaminergic transporter inhibitors. In addition, there are mixed drugs, such as vortioxetine, an inhibitor of the 5-HTT but also a receptor antagonist at 5-HT3, 5-HT7, and 5-HT1D receptors and an agonist at 5-HT1A and 5-HT1B receptors. Also, trazadone, which besides inhibiting the 5-HTT is an antagonist at 5-HT2A, 2B, 2C, 2D receptors and at histamine and alpha-1 receptors. On the other hand, the relatively recently identified RAADs include ketamine, a dissociative anesthetic that acts as an antagonist at NMDA receptors but also interacts with binding sites for opioid, monoaminergic, cholinergic, nicotinic, and muscarinic receptors (Mion and Villevieille 2013) and psychedelics such as psilocybin which bind in particular to 5-HT-1A, 2A and 2C receptors as well as mTOR (mammalian target of rapamycin) and TrkB (Dodd et al. 2022).

Neurochemical Signatures of Prefrontal and Anterior Cingulate Cortices

Neurochemical parcellation studies of PFC in humans reveal the marked heterogeneity in the modulation of its regions, which can provide important insight into the likely target of different antidepressants (Figure 13.1a). In terms of overall cortical organization, there is an increase in the diversity of neurotransmitter receptor densities from sensory to association areas including PFC. Along the same sensory to association axis there is also an increase in the ratio of excitation to inhibitory receptor density and a gradient change in ionotropic and metabotropic receptors with ionotropic decreasing and metabotropic increasing (Goulas et al. 2021). This characteristic patterning, which was originally based on autoradiographic analysis, has since been corroborated using positron emission tomography (PET) (Hansen et al. 2022). Moreover, receptor pattern similarities between regions are not only greater between pairs of regions that are anatomically connected but also greater between regions within the same, compared to different intrinsic networks (Hansen et al. 2022). Of specific relevance to our discussion below on prefrontal targets for antidepressants, the scACC (particularly caudal regions including area 25) is a hotspot not only for 5-HTT (James et al. 2019; Palomero-Gallagher et al. 2009b) but also 5-HT1A receptors, although the latter are also dense across much of the rest of medial PFC, extending onto the dorsolateral surface. Since area 25 also sends dense projections into the dorsal raphe nucleus (Freedman

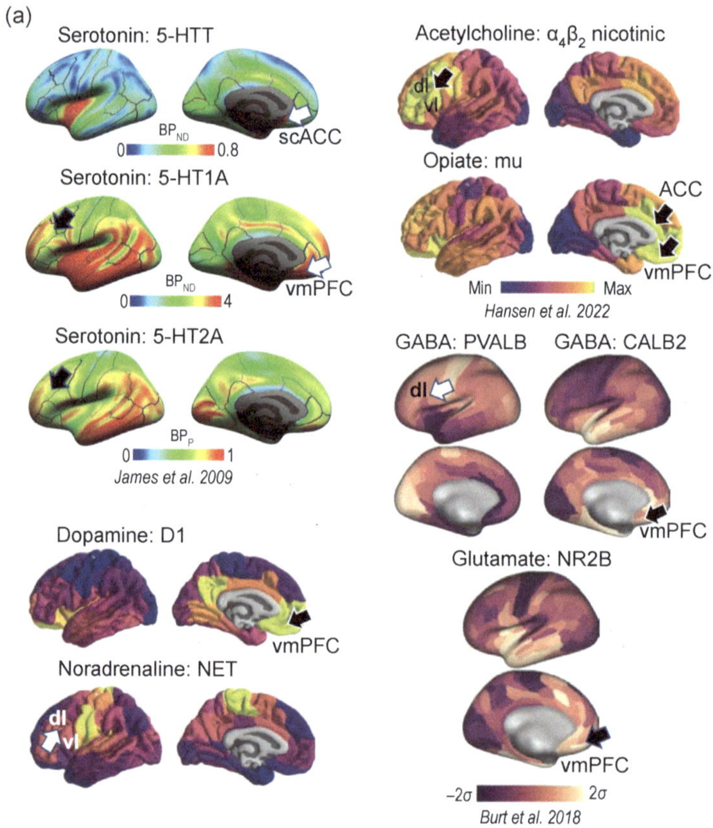

Figure 13.1 Targets of antidepressant actions in the PFC. (a) Distribution of receptors, transporters, and neuronal subtypes across the human brain based on PET (Hansen et al. 2022; James et al. 2019) and postmortem gene expression (Burt et al. 2018). All pictures reproduced with permission under a Creative Commons Attribution 4.0 International License. Arrows (white or black) highlight high densities of different receptors, transporters, and neuronal subtypes in anterior cingulate and prefrontal brain regions (and associated networks) of relevance to the efficacious actions of fast- and slow-acting antidepressants. Note, in particular, high densities of 5-HTT and 5-HT1A receptors in scACC, 5-HT2A receptors in dlPFC and vlPFC, D1 receptors in vmPFC, noradrenaline transporter (NET) in dlPFC and vlPFC, α4β2 acetylcholine nicotinic receptors in dl and vlPFC, mu opiate receptors in vmPFC and ACC, GABAergic pavalbumin (PVALB) and calbindin (CALB2) neurons in vmPFC and NR2B glutamate metabotropic receptors in vmPFC. ***σ*** – units plotted as standard deviation from the mean.

et al. 2000), which in turn sends serotonergic projections to much of the cortex, it likely has a marked impact on cortical serotonergic transmission more generally (Palomero-Gallagher et al. 2009b). Thus, SSRIs are likely to impact a dysfunctioning subcallosal network while newly discovered psychedelics, the targets of which include 5-HT1A receptors, may have effects that extend into higher-order cognitive networks as well.

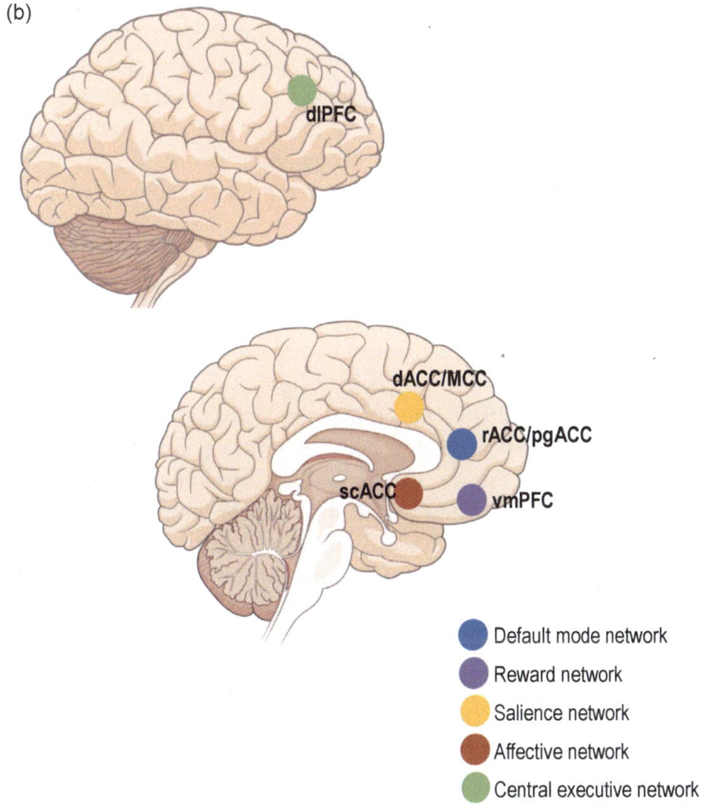

Figure 13.1 *(continued)* (b) Regions across PFC and ACC and their associated resting-state networks implicated in the mediation, prediction, and moderation of antidepressant efficacy.

Dopamine D1 and D2 receptors, alongside the dopamine transporter, also show relatively higher densities in medial PFC, including subcallosal cortex, compared to lateral regions while noradrenergic transporter (NET) and the $\alpha_4\beta_2$ nicotinic receptor show greater densities across dorsolateral and ventrolateral prefrontal regions. Thus, it might be predicted that antidepressants targeting, for example, NET are more likely to influence higher-order cognitive systems within the PFC. On the other hand, while area 25 has a marked abundance of glutamate receptors, including AMPA and NMDA, it has less abundance of metabotropic receptors compared to lateral PFC regions and also less $GABA_A$ and $GABA_B$ receptors compared to both medial and lateral PFC (Palomero-Gallagher et al. 2009b). Indeed, it has been suggested that the antidepressant actions of the RAAD, ketamine, acting as an NMDA antagonist may act early to inhibit an overactive scACC and on a slower timescale to enhance plasticity mechanisms in dorsolateral PFC in part via metabotropic receptor pathways (Arnsten et al. 2023; Opler et al. 2016). Ventromedial prefrontal regions, including scACC, also contain high levels of mu opiate receptors with less but still significant expression in OFC and lateral PFC (Hansen et al. 2022). Opiate pathways, as we shall see later, are linked to some of the actions of ketamine. Thus, overall, the distinct neurochemical signatures across regions of PFC and ACC provide a possible substrate for the differential efficacy of distinct classes of antidepressants (see Figure 13.1a).

It should be noted that the above descriptions are largely based on neurochemical receptor distributions in humans. How comparable they are across species, including monkeys and rodents, remains to be fully determined, although in comparison to rodents, macaque and humans have been shown to share a very similar profile of receptors at the regional and laminar level (Zilles and Palomero-Gallagher 2017). Specific examples of differences in the chemical microstructure of cortical circuits between primate and nonprimate species that would impact the efficacy of potential drug treatments include cortical cholinergic suppression (Disney and Robert 2019), of relevance to schizophrenia and Alzheimer disease, and mGluR3 localization in working memory circuit motifs in dorsolateral PFC (Datta and Arnsten 2018), of relevance to a range of psychiatric and neurodevelopmental disorders. For further details, see Izquierdo (this volume).

Monoamine-Targeting Anxiolytics and Antidepressants

Serotonin and Noradrenaline Targeting Therapeutics in Humans

Subcallosal Cingulate Cortex

An early study in which unmedicated patients suffering from major depression disorder (MDD) were imaged before and after treatment using PET (Drevets et al. 2002), a reduction in activity in rostral subcallosal cingulate cortex (scACC) at the level of the genu was revealed following chronic paroxetine (>4 weeks). When compared with CBT in unmedicated MDD patients, remission following 6 weeks treatment with paroxetine (5-HTT inhibitor and weaker NET inhibitor too) was associated with a similar reduction in activity in vlPFC (area 47) to CBT but to have opposite effects in dlPFC (area 9), which in the case of paroxetine were increases. Unique to paroxetine treatment, however, were decreases in the scACC (Goldapple et al. 2004), although the region was more caudal to that shown by Drevets et al. (2002). It was suggested that paroxetine reduced circadian and vegetative systems alongside increasing attentional-cognitive systems. Another study compared CBT with venlafaxine, a mixed serotonin and noradrenergic reuptake inhibitor (Kennedy et al. 2007). There were common decreases in activity in the right and left OFC and left dmPFC while venlafaxine-unique decreases were again seen in the caudal scACC, consistent with that seen for paroxetine. In contrast, unique increases in activity were seen in more rostral regions of scACC (area 32) associated with CBT. Comparison of brain activity pretreatment, on the other hand, has shown that increased activity within scACC-perigenual ACC border predicts nonresponse to both CBT and citalopram (Konarski et al. 2009; McGrath et al. 2014). It should be noted that scACC is quite an extensive region along the rostro-caudal axis composed of area 25, 24, and 32 (Öngür and Price 2000; Petrides et al. 2012); thus, alterations in activity within these scACC regions that accompany the treatment response or are predictive of treatment nonresponse may well be functionally distinct. Indeed, given the repeated reports of relationships between scACC, MDD, and its treatment, a more recent study focused on rostral scACC resting-state connectivity, this time comparing 12 weeks randomized treatment with escitalopram (SSRI) or duloxetine (SNRI) with CBT in treatment naïve patients (Dunlop et al. 2017). Again, the focus was on PFC activity pretreatment and the three regions that showed significant connectivity with rostral scACC; namely, vlPFC, vmPFC, and dorsal midbrain. Negative summed functional connectivity scores across all three regions were associated with remitters to medication and treatment failure to CBT, while positive summed scores were associated with remitters to CBT and treatment failure to medication.

In all the studies considered so far, there was no accompanying placebo group. Nevertheless, regions were identified that selectively predicted antidepressant outcomes compared to CBT or vice versa. Thus, placebo effects were unlikely to underlie these differential effects if it is assumed that placebo contributes relatively equally to both treatment strategies. In addition, Dunlop et al. (2017) also highlighted the importance of taking into account the current state of the patient at the time of treatment as a patient's brain state may be very different depending on whether they are treatment naïve or treatment experienced/resistant.

In summary, there is evidence for reductions in activity within caudal scACC (area 25) to accompany the treatment response to SSRI/SNRIs. It should be noted, however, that the associated increases in this region with MDD tend to be located more rostrally at the level of the genu. In addition, positive and negative connectivity, respectively, within this more rostral scACC region with other brain regions differentially predicts a treatment response to SSRI/SNRI and CBT.

Dorsal Anterior Cingulate Cortex

The structure and function of scACC are not, however, the only predictors of treatment outcome with SSRIs and SNRIs. The dorsal anterior cingulate cortex (dACC) has also been implicated. In this case, a task-based rather than resting-state fMRI study showed that increased positive connectivity from dACC to the amygdala (as opposed to negative connectivity), when viewing fearful versus happy facial expressions, was associated with the nonresponse to escitalopram, six weeks later (Vai et al. 2016). Reduced positive connectivity from the amygdala to the ACC and to the vlPFC was also reported in the same study. In contrast, changes in activity within dACC after just 1 week of SSRI treatment is predictive of a 6-week therapeutic response. Specifically, the hyperactivity of dACC to fearful facial expressions compared to happy facial expressions seen in depressed patients was reduced one week after treatment with the SSRI, escitalopram (Godlewska et al. 2016). More recently, a slightly more rostral region of ACC, around the level of the genu, showing increased activity to masked sad versus happy facial expressions at baseline, predicted later treatment response to escitalopram (Godlewska et al. 2018). A leave-one-out analysis suggested that activity in ACC was able to predict response status at the level of individuals. Davidson et al. (2003) also implicated dACC in the treatment response to venlafaxine, an SNRI, although here it had appeared that lower activation in this region to negative stimuli, pretreatment, was a predictor of success. Thus far, none of these studies were placebo controlled so the effects were only predictive of treatment responsiveness in general. Where dACC activity did differentiate, in this case, sertraline from placebo, it was under conditions of emotion conflict: the greater downregulation in activity a patient displayed in dACC (along with anterior insula and frontal pole), the better the outcome on sertraline (Fonzo et al. 2019).

Pregenual Anterior Cingulate Cortex

Even more rostrally in ACC, pregenual ACC is also implicated in SSRI treatment prediction (including escitalopram, sertraline, and SNRI, venlafaxine), with no significant differences between treatments. Specifically, intact functional connectivity between rostral anterior and posterior cingulate cortex (a major component of the default mode network) predicted an effective treatment response (Goldstein-Piekarski et al. 2018). This was shown to be independent of any other treatment response predictors such as comorbid anxiety, early life trauma, cognitive impairment, and body mass index. Indeed, structural changes in rostral ACC have been repeatedly identified as predictors of treatment response with SSRIs, including fluoxetine (Chen et al. 2007) and escitalopram (Gunning et al. 2009). Where such a relationship was not found with sertraline, increases in volume within the first week of treatment were significantly correlated with improvement at eight weeks (Bartlett et al. 2018). Functionally, rACC theta has been correlated with antidepressant response in two large trials using either rsMRI with the SSRI, sertraline (Pizzagalli et al. 2018) or EEG with three different medications: escitalopram, sertraline, or venlafaxine, a SNRI (Arns et al. 2015). However, converse results have been reported and the effects are not restricted to SSRIs but also placebo effects (Pizzagalli et al. 2018; Sikora et al. 2016). Thus, its utility for informing treatment selection appears limited. Moreover, greater consideration should be given as to whether the patients are relatively treatment resistant or not.

Dorsolateral Prefrontal Cortex

Outside of the ACC, dlPFC activity has been reported to be predictive of remission following SSRI and SNRI treatment. For example, greater activation within dlPFC (but not exclusive to this region) was reported in MDD patients compared to controls when performing correct rejections in a go/no-go task involving inhibition, activity that predicted posttreatment improvement in depressive symptoms with escitalopram (Langenecker et al. 2007). In addition, medication-free outpatients with MDD, who displayed remission in the iSPOT-D cohort, showed dlPFC activation during inhibitory no-go responses in a go/no-go task, similar to that seen in controls, whereas non-remitters showed hypoactivity (Gyurak et al. 2016). Of note, inferior parietal activation differentiated SSRI versus SNRI remitters: following SSRI treatment, remitters showed normal activation whereas non-remitters showed hypoactivation. The opposite was true for SNRI remission. This suggests that remission following SSRI and SNRI treatment is dependent on intact dlPFC functional activity. Consistent with this, greater dlPFC functional activation during working memory performance at pretreatment in a subset of patients in the iSPOT-D cohort predicted the extent of the antidepressant response (sertraline, escitalopram and venlafaxine) but only in patients without childhood maltreatment (Miller et al. 2015). Conversely, a volumetric study identified a cluster in the caudal sector of the left middle frontal gyrus that below a certain volume predicted a subset of non-remitters to sertraline, venlafaxine, or escitalopram (Korgaonkar et al. 2015). If reduced volume is taken to reflect reduced functioning then this result is still consistent with the hypothesis that intact functioning of dlPFC is necessary for successful treatment. The predictive value of dlPFC, however, has recently been brought into question in a large placebo-controlled trial. In the EMBARC study with over 100 patients in each group, improvements in the depression score for patients treated with sertraline occurred regardless of the connectivity values in a dlPFC resting-state network (derived from a focused dlPFC seed region), although high connectivity values did predict improvements following placebo and low connectivity differentiated sertraline from placebo (Chin Fatt et al. 2021). These results could be interpreted to suggest that positive treatment outcome for sertraline at high dlPFC connectivity reflected a placebo response, whereas the true impact of sertraline was only seen in those patients with low dlPFC connectivity. It should be noted, however, that the model chosen to describe the dlPFC relationship with placebo and sertraline was also dependent on activity being low within rostral scACC and high in nucleus accumbens and amygdala. Nevertheless, the overall result appears contrary to those studies described above, showing that greater dlPFC activity was predictive of an antidepressant treatment response. Still, the majority of these other studies measured task-based functional activity in dlPFC rather than resting state, which may have contributed to the contrasting effects. Importantly, those other studies were without placebo controls and so placebo effects may underlie the positive treatment outcomes. Indeed, additional support for dlPFC activity predicting the placebo response comes from another measure of brain activity within the same EMBARC patient population; namely, arterial spin labeling, rather than BOLD, which revealed that increased dlPFC perfusion only predicted a placebo and not a sertraline treatment response (Cooper et al. 2019). Thus, intact dlPFC activity is a likely prerequisite for placebo-induced improvements and is hypothesized to reflect active cognitive appraisal mechanisms contributing to the impact that expectation of mood enhancement can have on mood state (Zilcha-Mano et al. 2019).

Dopamine-Targeting Therapeutics

Although most monoaminergic antidepressants and anxiolytics target serotonin or noradrenaline signaling, at least two important drugs target dopamine as well. First, as noted above, bupropion is a noradrenaline-dopamine reuptake inhibitor and is among the most commonly prescribed drugs that target dopaminergic signaling in depression. Its antidepressant effects, however, are thought to be driven primarily by effects on noradrenergic signaling, due in part to the fact that its effects on dopamine reuptake are modest compared to its effects on noradrenaline. Very few studies to date have examined bupropion effects on PFC function in depression. In one such study, involving ten patients with unipolar depression scanned before and after an 8-week course of treatment, bupropion was found to reduce fMRI responses to negative emotional visual stimuli in the right OFC, left dmPFC, right vmPFC, right ACC, and right inferior frontal cortex (Robertson et al. 2007). Second, pramipexole is a relatively selective D2 receptor agonist, which is not indicated as a monotherapy for depression or anxiety but is frequently used as an augmentation strategy, especially for patients with pronounced anhedonia. Again, very few studies have examined pramipexole effects on PFC in depression, but those that have indicate that pramipexole may modulate prefrontal activity in the context of reward processing. For example, Whitton et al. (2020) found that in patients with depression, reward learning was slowed, with modestly blunted reward prediction error signals and modestly increased amphetamine-induced dopamine release as indexed by PET. Pramipexole improved depressive symptoms, including hedonic function, but had no direct effect on reward learning in the lab. Baseline reward learning, D2 receptor availability, and amphetamine-induced dopamine release did, however, predict greater improvements. As noted above, in both of these studies, there was no placebo control arm, so it is unclear whether changes in activity were related to bupropion or pramipexole treatment versus nonspecific improvements in mood.

Summary

Many regions across the prefrontal, orbitofrontal, and anterior cingulate regions have been implicated in monoamine-targeting antidepressant treatment responses in patients with MDD. In many cases, whether the brain changes that accompany or predict successful treatment are due to the antidepressant itself cannot be determined since a placebo control group has been lacking. Where placebo controls have been studied, it is evident that there is considerable overlap in the prefrontal circuitry predicting a placebo response and that predicting an antidepressant response. In some cases, the same brain region is implicated in both, differing only in the direction of the relationship. For example, while high levels of dlPFC activity predict a placebo response, low levels are more likely to predict a response to SSRIs compared to placebo (Chin Fatt et al. 2021), especially when levels in rostral subcallosal cingulate are also low. Activity in ACC is also variably associated with antidepressant response. Activity in pregenual regions is associated with placebo and so does not appear selective for antidepressants (Pizzagalli et al. 2018; Sikora et al. 2016), while at least one study shows differential task-based activity in dACC related to sertraline and not placebo (Fonzo et al. 2019). Finally, right inferior orbital frontal gyrus is selective for sertraline over placebo (Cooper et al. 2019).

Even less well established are differences between the varied monoamine-targeting antidepressants within PFC and ACC. This is somewhat surprising since the pattern of innervation of the monoamines differs markedly across the distinct regions of PFC (see above). One study compared sertraline, buproprion, and placebo but the only selective predictors for buproprion (a noradrenergic and dopaminergic uptake inhibitor) that were located in the PFC were higher anticipatory activity in the superior frontal gyrus and higher reward expectancy activation in the orbitofrontal cortex, both of which predicted less improvement with buproprion (Nguyen et al. 2022). Moreover, the caveat here was that patients who were moved on to buproprion failed to show a response to sertraline, so not only were numbers considerably lower but the cohorts distinct and thus comparison made difficult. When comparing SSRIs and SNRIs, little in the way of differences has been noted although opposing alterations in inferior parietal cortex did differentiate remitters from non-remitters between the two (Gyurak et al. 2016).

Cellular Mechanisms in Animals

There have been far fewer studies in experimental animals aimed at determining the prefrontal locus of action of monoaminergic antidepressants, and those that have are evenly spread across healthy controls and chronic stress models. Perhaps even more surprisingly, there have been very few studies that have compared different types of monoamine-targeting antidepressants, with the vast majority focusing on the relatively selective SSRI, fluoxetine. In most cases, fluoxetine is given systemically to match treatment regimes in the clinic, and medial regions of the PFC (mPFC) have been the primary focus. The rodent PFC is much less complex than in humans and other primate species, but anterior cingulate, prelimbic, and infralimbic cortex are thought to share some cytoarchitectural, functional, and anatomical features with the primate anterior cingulate, dorsomedial, and ventromedial PFC, respectively. However, very often the precise region within mPFC is not detailed, and rarely are different regions compared. Moreover, the OFC has been largely ignored, despite changes occurring within this region, both in patients with depression and in stress-induced models of depression in rodents. What is clear from these studies, though, is that fluoxetine has a marked impact on a range of measures of physiological function within mPFC. In intact animals, prolonged daily treatment with fluoxetine for anything between 2–4 weeks has been reported to alter the excitatory-inhibitory balance in the prelimbic cortex, with an increase in pyramidal cell firing and reduction in interneuron firing (Yin et al. 2021). In particular, chronic fluoxetine has been shown to reduce selectively parvalbumin but not other GABAergic interneurons within mPFC (Ohira et al. 2013; for opposite effects on mPFC parvalbumin neurons *in vitro*, see Zhong and Yan 2011). The accompanying reduction of perineuronal nets, a marker of neuronal maturation suggests one aspect of antidepressant action may be to reinstate a juvenile state of plasticity. A de-maturation of astrocytes has also been reported alongside dynamic changes in 5-HT1A receptors and upregulation of brain-derived neurotrophic factor (BDNF), which is also argued to be consistent with long-term neurotrophic effects (Song et al. 2021). Comparison of citalopram, a relatively selective SSRI, with the mixed antidepressant, trazadone (which is not only an SSRI but also a serotonin 2A/B receptor antagonist with effects on histamine and alpha-1 adrenergic receptors), found comparable effects on clock genes in mPFC but differentiable effects on BDNF and TrkB receptors. Only trazodone increased these in the mPFC while citalopram's effects were unique in the nucleus accumbens and amygdala, respectively (Carboni et al. 2022). When task-based firing patterns of mPFC have been investigated, chronic treatment with fluoxetine has been associated with overall reductions in firing related to the reward-predicting stimulus, likely related to a less redundant encoding capacity and a less robust encoding of information (Pereyra et al. 2021). The hypothesis that this may reflect increased flexibility, however, remains to be determined.

Of more relevance to our understanding of the actions of chronic treatment for ameliorating anxiety and depression are their effects on stress-induced models of anxiety and depression-like symptomatology in animals. In the majority of examples, regardless of the nature of the stressor (physical, social, or physiological), anxiety- or depression-like behavioral changes induced by the stressor are ameliorated or prevented by chronic treatment with the SSRI, fluoxetine. Such treatment can also ameliorate the accompanying changes in mPFC function brought about by the stressor, such as the downregulation of cytosolic proteins and upregulation of nonsynaptic mitochondria (Filipović et al. 2022) and reductions in BDNF protein levels (Misztak et al. 2021), the latter consistent with the effects of chronic fluoxetine in normal controls (Song et al. 2021). Moreover, such treatment also reverses the reduced gap junction function specifically within prelimbic cortex, reported to occur after chronic unpredictable mild stress (Xia et al. 2023) as well as the gliogenesis that occurs after chronic social defeat stress (Czéh et al. 2007). Where a mixed 5-HT drug has been used, vortioxetine, (targeting 5-HT receptors and the 5-HT transporter), this has been shown to reverse the inhibitory effects of chronic mild stress and chronic social defeat on mTORC1 signaling, important for protein synthesis and plasticity (Li et al. 2023). Chronic fluoxetine also reverses the desensitization of α2-adrenoceptors within mPFC following chronic slow release corticosterone (Horrillo et al. 2019) and inhibits microglial activation, regulates the Notch signaling pathway, and inhibits the inflammatory response within mPFC in a liposaccharide model of depression in Parkinson disease in rats (Zhang et al. 2022). In contrast, in a PTSD model involving severe acute footshock, the efficacy of chronic fluoxetine to reverse the subsequent increase in immobility in the forced-swimming test was only associated with its ability to also reverse the accompanying increases in the expression of the immediate early gene, cfos, in the amygdala, but not the prelimbic cortex or anterior cingulate (Cg1) (Yu et al. 2020).

A limitation of the above studies, which will be seen to be a reoccurring limitation throughout this review, is the lack of repeatability and comparability. The vast range of cellular mechanisms that have been studied across intact and stress-induced animal models makes it difficult to provide a comprehensive synthesis. However, effects on the variety of plasticity mechanisms available to the central nervous system is a common theme which likely underlies the changes in functional connectivity following successful treatment in patients.

The Rapidly Acting Antidepressant, Ketamine

The majority of individuals with depression will not show a full response to their first monoamine-targeting antidepressant trial (Rush et al. 2006). These limitations led investigators to pursue other antidepressant mechanisms that might yield more rapid responses, even in treatment-resistant individuals. Motivated by evidence that glutamatergic signaling in the PFC and other stress-sensitive brain regions may be altered in depression (Auer et al. 2000; Duman et al. 2019; Sanacora et al. 2004), these efforts led to clinical trials of ketamine, an NMDA receptor antagonist and dissociative anesthetic. In one early trial, seven patients with severe depression received an intravenous infusion of a subanesthetic dose of ketamine or saline on two separate days, and investigators observed potent antidepressant effects just six hours after treatment that persisted for at least three days (Berman et al. 2000). Larger-scale clinical trials with more robust placebo controls followed, confirming rapid and potent antidepressant effects (Cohen's d> 1.4) in both unipolar and bipolar depression that persisted in some individuals for up to a week (Diazgranados et al. 2010; Murrough et al. 2013; Zarate et al. 2006, 2012). This led in 2019 to FDA approval of esketamine, an intranasal formulation of ketamine's (S) enantiomer, for treatment-resistant depression.

Here, we consider insights gained into the molecular and circuit-level mechanisms of ketamine's actions within the PFC from studies in animal models before reviewing insights gained from neuroimaging studies.

Molecular Mechanisms

In preclinical rodent models, early studies in this field showed that ketamine's antidepressant properties are most likely mediated in part by effects on neuronal function and synapse formation in the mPFC. As has been the case with investigations into the monoamine-targeting antidepressants, mPFC has been the primary focus for many ketamine studies in mice and rats with no studies to date having examined these mechanisms in OFC. Studies have shown that ketamine causes a rapid increase in the expression of glutamatergic AMPA receptors, PSD95, and other synaptic markers in the prelimbic area of PFC in rats that correlated with changes in depression-related behavior (Li et al. 2010a). The same study in rats showed that these effects are mediated by NMDA receptor antagonism and are blocked by a prefrontal cortical infusion of rapamycin, implicating downstream effects on the mTOR signaling pathway. However, a subsequent study showed that unexpectedly, when rapamycin was systemically infused alongside ketamine in patients with depression, the antidepressant effects were not attenuated (Abdallah et al. 2020b). This may be related to confounding effects of a systemic infusion on inflammation, which may not occur with direct infusion into the PFC.

Ketamine's antidepressant effects are also driven by neurotrophic signaling. A parallel series of studies showed that ketamine's antagonism of NMDA receptor signaling enhances activity-dependent release of BDNF by de-suppressing its translation within neurons (Autry et al. 2011). Ketamine's effects on depression-related behavior, in turn, require BDNF and its receptor, TrkB (Autry et al. 2011; Lin et al. 2021). Enhanced activity-dependent release of BDNF may be especially important for sustaining ketamine's effects over time, through downstream effects on methyl-CpG-binding protein 2 (MeCP2) phosphorylation, which is required for maintaining ketamine's effects on behavior and long-term synaptic potentiation (Kim et al. 2021). In addition, at least one report indicates that ketamine's effects on BDNF signaling may be driven not only by NMDA receptor antagonism but also by direct binding to its receptor TrkB (Casarotto et al. 2021), an effect potentiated by astrocyte-derived cholesterol. Ketamine's interactions with the TrkB receptor facilitated BDNF signaling in active synapses and increased the expression of TrkB on dendritic spines. Conversely, mutating a specific ketamine-binding motif in the TrkB receptor blocked the effects of ketamine on depression-related behavior, synapse function, and plasticity. Interestingly, fluoxetine and a variety of monoamine-targeting antidepressants were also found to bind directly to TrkB, but different compounds accumulated at different rates in mPFC tissue (specific subregions were not studied here), suggesting one potential mechanism by which ketamine may elicit rapid antidepressant effects while fluoxetine and other SSRIs operate on slower time scales.

Importantly, ketamine is not a selective NMDA receptor antagonist; its effects on depressive symptoms and PFC function may also be driven by its other pharmacological properties. Recent studies have shown that hydroxynorketamine, an active metabolite of ketamine, may act to promote synapse formation and antidepressant effects through direct effects on AMPA receptors (Zanos et al. 2016), although other studies point to a role for NMDAR inhibition by hydroxynorketamine (Suzuki et al. 2017). Furthermore, ketamine is also a mu opioid receptor (MOR) agonist and its effects may be mediated in part by opioid receptor signaling systems (discussed further below). Together, these studies indicate that ketamine acts to relieve depressive symptoms rapidly via multiple molecular mechanisms, including NMDA receptor antagonism, activity-dependent BDNF release, and other neurotrophic signaling. Downstream effects on MeCP2 phorphorylation, in turn, play a critical role in synaptic potentiation and sustaining the antidepressant behavioral effects over time.

Circuit-Level Mechanisms

The studies reviewed above underscore a molecular mechanism involving NMDA receptor antagonism and activity-dependent BDNF signaling that culminates in prefrontal cortical synapse formation, implying a causal role for synaptogenesis in mediating its antidepressant effects. Until recently, it was challenging to test this hypothesis directly, but new approaches for *in vivo* imaging and optogenetics have made such studies possible (Figure 13.2). For example, two-photon laser-scanning microscopy combined with chronically implanted cranial windows or microprisms, which provide optical access to the PFC (Andermann et al. 2013; Low et al. 2014), have enabled researchers to characterize the time course of synaptogenesis after ketamine treatment precisely. One such study showed that ketamine has rapid effects on the formation of dendritic spines, microscopic protrusions from neuronal dendrites that usually contain functional synapses, and that these effects on prefrontal spinogenesis were rapid and persisted for at least two weeks (Phoumthipphavong et al. 2016), when the vast majority of dendritic spines will contain functional synapses (Holtmaat and Svoboda 2009; Knott et al. 2006). This study focused on the dorsal medial frontal cortex (also known as M2), which approximates the primate premotor cortex, indicating that ketamine's effects on synaptogenesis may be more generalized across cortical areas than previously appreciated.

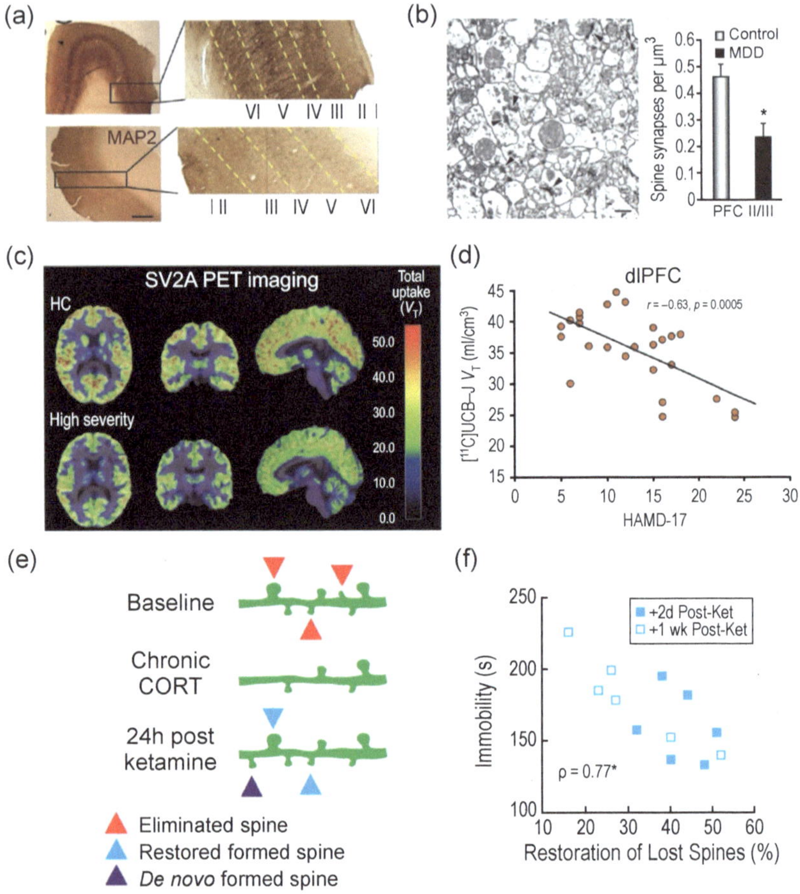

Figure 13.2 Ketamine rescues dendritic spine loss in PFC. (a) Immunohisto-chemistry images from postmortem human brain tissue showing reduced expression of the synaptic protein MAP2 in the dorsolateral PFC from a patient with depression (lower image) compared to an individual without depression (upper image). (b) Electron micrograph of synapses (marked by arrowheads) in layer II/III of the dlPFC in a depressed patient (left). Synapse density was reduced in MDD compared to control individuals (right). (c) SV2A PET imaging reveals reduced synapse density in patients with severe depression compared to healthy controls. (d) Synapse density (as indexed by [^{11}C] UCB-J V_T) correlated with depression severity (as indexed by HAMD-17). (e) Schematic showing how chronic corticosterone exposure eliminates postsynaptic dendritic spines in mice (red arrowheads), and ketamine restores spines to their original position (blue arrowheads). (f) Restoration of lost spines by ketamine was correlated with maintenance of ketamine's antidepressant-like effects on immobility in the tail suspension test. Panels (a) and (b) were adapted with permission from (Kang et al. 2012), panels (c) and (d) with permission from Holmes et al. (2019), and panels (e) and (f) with permission from Moda-Sava et al. (2019).

Of note, in almost all experiments from the studies reviewed above, ketamine was administered to "unstressed" mice or rats (i.e., in the absence of a chronic stress treatment), so it is unclear to what extent ketamine might engage prefrontal targets differently in a chronic stress state. Furthermore, it is unclear how ketamine effects on synapses and circuit function relate to those induced by stress. One recent study addressed these questions, showing that ketamine acts in a targeted way to reverse some effects of chronic stress and that prefrontal spinogenesis is required for supporting ketamine's antidepressant effects (Moda-Sava et al. 2019). Two-photon imaging showed that chronic stress causes spatially clustered, dendritic branch-specific synapse loss in the mPFC, and that ketamine acts in a targeted way to restore lost spines. These effects were observed in all three mPFC subregions, including ACC, prelimbic cortex, and infralimbic cortex. They were associated with parallel effects on functional connectivity and neuronal activity in multicellular ensembles that were disrupted in a neuroendocrine model of chronic stress, restored by ketamine, and required for driving motivated escape behavior. Unexpectedly, ketamine's effects on circuit function and behavior were evident just three hours after treatment and preceded its effects on spine formation, which did not emerge until 12 hours after treatment; this indicates that new spines were not required for initiating ketamine's antidepressant effects. However, using a newly developed optogenetic tool to selectively delete newly formed synapses, Hayashi-Takagi et al. (2015) showed that prefrontal synaptogenesis was required for sustaining ketamine's effects on prefrontal circuit function and behavior over time. Of note, these effects were specific: deleting newly formed synapses did not interfere with ketamine's effects on sucrose preference behavior, indicating a specific role for prefrontal synaptogenesis in sustaining effects on some depression-related behaviors (e.g., motivated escape behavior) but not others.

If prefrontal synaptogenesis is required only for sustaining ketamine's antidepressant effects, what then are the circuit-level mechanisms that initiate those effects? This is an outstanding question for the field, but converging data from several studies indicate that GABAergic interneurons may be involved. In one study, cell-specific deletion of GluN2B, an NMDA receptor subunit in somatostatin (SST)- or parvalbumin (PV)-expressing interneurons, but not glutamatergic pyramidal cells, in the prelimbic and infralimbic regions of mPFC was sufficient to block ketamine's effects on depression-related behavior (Gerhard et al. 2020). Another study went on to show that antidepressant-dose ketamine suppresses the activity of SST interneurons in the anterior cingulate and dorsal PFC, reducing dendritic inhibition and enhancing calcium signals in prefrontal pyramidal neurons (Ali et al. 2020). Together, these studies suggest that ketamine may initiate its antidepressant effects by silencing SST interneurons, disinhibiting prefrontal pyramidal neurons (Ali et al. 2020; Gerhard et al. 2020), and restoring multicellular ensemble events, which may in turn drive the formation of new synapses that sustain these effects over time (Moda-Sava et al. 2019). Future studies will be required to test this model.

An alternative approach, recently adopted in studies in nonhuman primates, has studied the efficacy of antidepressants on their ability to ameliorate specific symptoms induced by select targeted interventions highly associated with depressive illness. Specifically, systemic ketamine given 24 hours earlier ameliorated the anticipatory anhedonia (blunted appetitive arousal) but not the heightened anxiety induced by overactivation of caudal scACC (Alexander et al. 2019, 2020) in marmosets. By applying chemogenetics to obtain pathway specificity, it was shown that the anticipatory anhedonic effects could be localized to overactivation of the subcallosal-accumbens pathway and not the subcallosal-amygdala pathway, and that ketamine could ameliorate the anhedonia through its actions at the level of the nucleus accumbens (Wood et al. 2023). In the next section, we discuss how these findings are consistent with a recent imaging study in humans in which ketamine differentially blocked scACC hyperactivity to positive, but not negative, processing in depressed patients (Morris et al. 2020). Studies such as these open up the possibility of differentiating the actions of distinct classes of antidepressants on symptoms induced by specific network dysfunction.

Human Neuroimaging Correlates of Ketamine's Antidepressant Effects

Despite the relatively underdeveloped PFC of rats and mice upon which most experimental studies have been performed, converging data from human neuroimaging studies indicate that similar mechanisms may be operative in patients with depression. Magnetic resonance spectroscopy (MRS) is a tool that provides for the direct, noninvasive measurement of specific neurotransmitters in the living human brain. MRS studies have shown that depression is associated with a reduction in glutamate and glutamine availability in the dmPFC, ACC, and vPFC (Auer et al. 2000; Hasler et al. 2007; Moriguchi et al. 2019; Rosenberg et al. 2005). Ketamine acts to reverse these deficits, causing a rapid increase in glutamate availability in the PFC (Abdallah et al. 2018; Milak et al. 2016, 2020) and ACC (Rowland et al. 2005). While most MRS studies to date have not been able to resolve region-specific effects of ketamine on glutamate signaling in specific subregions of the PFC, future studies employing larger magnetic field gradients may be able to resolve such effects. This could be useful for characterizing associations between glutamate signaling and changes in specific PFC-dependent behavioral domains.

More recently, the development of new ligands for PET have enabled new approaches to studying synapse function directly and noninvasively in the human brain. PET studies of radioisotope binding to synaptic vesicle glycoprotein 2A (SV2A) have shown that depression is associated with reduced synapse density in the anterior cingulate and dlPFC (Holmes et al. 2019). A similar approach showed that ketamine reduces metabotropic glutamate receptor availability (mGluR5), which may be a compensatory response to a surge in glutamate release (Esterlis et al. 2018). These effects were most pronounced in the anterior cingulate, medial PFC, OFC, and striatum, among other areas. Unexpectedly, an SV2A-PET study of patients before and 24 hours after ketamine did not observe any significant effects on synapse density at the group level (Holmes et al. 2022). However, a post hoc exploratory analysis found that patients with lower prefrontal synapse density prior to treatment did show a significant increase in synapse density 24 hours after ketamine, consistent with effects reviewed above in rodent models, indicating that ketamine may act in a targeted way to restore synapses lost during chronic stress (Duman et al. 2019; Moda-Sava et al. 2019; Phoumthipphavong et al. 2016).

A host of effects on PFC function after ketamine treatment have been identified through fMRI. In one study, for example, anterior cingulate activity in response to fearful faces was reduced in depressed patients compared to healthy controls, and the magnitude of this effect correlated with increased likelihood of later responding to ketamine (Salvadore et al. 2009). In accord with its effects on synapse formation in rodents, ketamine appears to modulate functional connectivity in the human brain as well, as indexed by changes in the degree to which low-frequency fluctuations in the fMRI BOLD signal are correlated between brain regions. Previous work showed that an area of the dmPFC ("the dorsal nexus"), which is functionally coupled with three depression-related brain networks (the default mode network, the frontoparietal cognitive control, and the rostral affective network) exhibits increased functional connectivity in depression (Sheline et al. 2010), and subsequent work showed that ketamine rescues those effects, reducing dorsomedial prefrontal functional connectivity (Scheidegger et al. 2012). Indeed, ketamine has been shown to impact within and between connectivity of the default mode, affective, reward, central executive, and salience networks as well as for activity within these networks to act as biomarkers of treatment response (reviewed in Demchenko et al. 2022). "Global brain connectivity"—a distinct measure indexed by correlating the BOLD signal in a given region with every other area of gray matter and averaging across areas—has recently been employed to study treatment predictors in depression. Originally it was used to identify reductions in the PFC and increases in posterior midline structures, including the posterior cingulate cortex and precuneus in depression (Abdallah et al. 2017). It has subsequently identified a unique brain connectome fingerprint that predates and predicts the response to the slow-acting antidepressant, sertraline, and preliminary evidence suggests it also predicts response to ketamine (Nemati et al. 2020). Elaboration of this approach has since identified a ketamine-induced connectivity fingerprint from control subjects that at one week posttreatment predicts the success of sertraline at eight weeks (Abdallah et al. 2020a), highlighting the overlap of action of slow- and fast-acting antidepressants at the level of prefrontal connectivity.

In summary, the studies reviewed above indicate that ketamine's effects on molecular signaling, synapse formation, and circuit formation in rodent models are probably associated with pronounced effects on prefrontal network function and functional connectivity in the human brain. Most notable is the considerable overlap in the prefrontal networks affected by both ketamine- and monoamine-targeting antidepressants.

Mu Opioid Receptor Signaling As a Therapeutic Target

The studies reviewed above indicate that ketamine acts to restore lost synapses in the PFC by antagonizing NMDA receptor signaling and potentiating BDNF and TrkB signaling. Still, as noted above, ketamine has numerous other pharmacological properties that could also be involved. Among these is MOR agonism. In a recent study, Bonaventura et al. (2021) screened over 100 receptors and enzymes and found that ketamine had potent interactions of comparable magnitude with both MORs and NMDARs. To test whether MOR signaling might be required for mediating ketamine's antidepressant effects, Williams et al. (2018) co-treated depressed patients with intravenous infusions of ketamine and naltrexone, which antagonizes both mu and kappa opioid receptors, or with ketamine alone. They found that naltrexone blocked the antidepressant effects of ketamine without interfering with its dissociative properties (Williams et al. 2018), and it also disrupted ketamine's therapeutic effects on suicidal ideation (Williams et al. 2019). In a similar study of five patients with comorbid depression and alcohol use disorder, naltrexone did not interfere with the antidepressant effects of ketamine (Yoon et al. 2019), but it was unclear to what extent these benefits were attributable to ketamine versus naltrexone, which is an established treatment for substance use disorders. Thus, additional studies are required to resolve these discrepancies. Taken together, these results are consistent with the hypothesis that ketamine's antidepressant effects may involve MOR signaling, at least in patients without comorbid substance use disorders.

Preclinical studies lend further support to this hypothesis. In one study, Bonaventura et al. (2021) used esketamine (an S-ketamine enantiomer) to activate MOR signaling, and converging behavioral data showed that it was reinforcing in rats as measured by self-administration and conditioned place preference. PET studies in the same report showed that esketamine stimulated dopamine release in the mPFC, lending further support to an MOR-associated reinforcing mechanism. Likewise, Klein et al. (2020) showed that opioid antagonists blocked the effects of ketamine on depression-related behavior and hyperactivity in the lateral habenula in rats. Finally, Samuels et al. (2017) showed that tianeptine, an atypical antidepressant with an unknown mechanism of action, also requires MOR signaling for mediating its antidepressant behavioral effects. Interestingly, tianeptine-induced MOR signaling had opiate-like effects on reward processing and analgesia but did not lead to tolerance or withdrawal, indicating that distinct mechanisms—possibly involving distinct circuits or cell types—may be involved in mediating MOR-dependent antidepressant effects versus MOR-driven reinforcement and addiction potential.

Although these latter studies did not examine prefrontal function, they lend further support to the hypothesis that MOR signaling may be a viable target for developing new antidepressants and warrant further study.

Psychedelic Compounds

A growing body of work has begun to investigate the therapeutic potential of psilocybin and other psychedelic compounds, building on early work in the 1950s and 1960s (Vollenweider and Kometer 2010). Two randomized controlled trials published in 2016 triggered renewed interest in this topic, showing that psilocybin—the primary psychoactive compound in hallucinogenic Psilocybe mushrooms—had potent antidepressant and anxiolytic effects in patients with life-threatening cancer that emerged rapidly after a single dose and persisted for six months in many individuals (Griffiths et al. 2016; Ross et al. 2016). Subsequent small-scale open-label studies extended these antidepressant effects to individuals with severe treatment-resistant depression unrelated to a medical diagnosis (Carhart-Harris et al. 2016; Davis et al. 2021). In 2021, a larger study confirmed these observations in a randomized controlled trial, showing that psilocybin was statistically superior to escitalopram for achieving sustained remission (Carhart-Harris et al. 2021). Although the conclusions that can be drawn from these studies are associated with some important caveats—including small sample sizes (ranging from 12 to 59 patients) and technical difficulties in providing a convincing placebo control for a hallucinogenic drug—they are also an important step forward in efforts to develop other rapid-acting antidepressants in addition to ketamine.

Our understanding of the underlying mechanisms is still developing. Like ketamine, a single dose of psilocybin is sufficient to drive rapid and sustained increases in postsynaptic dendritic spine density, accelerated spine formation, and enhanced glutamatergic neurotransmission in a region of the dorsal frontal cortex in mice that is analogous to primate premotor cortex (Hesselgrave et al. 2021; Shao et al. 2021). These effects emerged within one day of treatment, correlated with antidepressant-like behavioral effects, and persisted in an attenuated form for at least one month. Interestingly, while the hallucinogenic and psychotomimetic effects of psilocybin in humans are widely understood to be driven by direct effects on serotonergic (5-HT2A) signaling (Kwan et al. 2022), the antidepressant effects may be driven by other mechanisms. For example, in mice, pretreatment with ketanserin, a potent 5-HT2A receptor antagonist, blocked the effects of psilocybin on head-twitch behavior (a commonly used screening assay for hallucinogenic potential) but did not interfere with effects on depression-related behavior or spine formation (Hesselgrave et al. 2021; Shao et al. 2021). Also, in accord with the hypothesis that the therapeutic and hallucinogenic properties of psychedelic compounds might be dissociable, other studies have identified structural analogs of psychedelic compounds that have no effect on head-twitch behavior in mice but retain their therapeutic effects on depression- and addiction-related behavior (Cameron et al. 2021; Dong et al. 2021).

Very few studies have systematically examined the network-level substrates of these effects in humans, but those that have suggest that psilocybin may alter functional connectivity in prefrontal cortical areas. In one such study, psilocybin or placebo was administered to 15 healthy volunteers and a significant decrease in functional connectivity between the dorsomedial PFC and posterior cingulate cortex was observed (Carhart-Harris et al. 2012). Subsequently, Carhart-Harris et al. (2017) showed that psilocybin treatment in 19 patients with treatment-resistant depression caused an increase in functional connectivity between ventromedial prefrontal and lateral parietal areas of the default mode network, as well as decreased cerebral blood flow in the amygdala and increased amygdala BOLD responses to emotional faces (Roseman et al. 2018). Finally, a third study in depressed patients showed that psilocybin treatment caused a rapid decrease in network modularity measures derived from functional connectivity data and involving multiple areas of the PFC—effects that may have been especially pronounced in 5-HT2A receptor-rich areas (Daws et al. 2022).

Current Limitations and Future Strategies

While considerable insights have been gained into the actions of antidepressants on PRC function at the molecular, cellular, network, and behavioral levels of analysis, we have not yet identified the critical factors that determine the differential responsivity of individual patients to antidepressants. Overwhelming evidence suggests that a wide variety of prefrontal regions and their associated circuits act as both mediators and predictors of antidepressant efficacy (Figure 13.1b) and that changes in plasticity and thus connectivity within and between functional circuits underlie symptom improvement. Although selected regions or circuits have been implicated at the level of individual studies, these differ across studies. One of the challenges in synthesizing findings across studies is that different approaches are used to acquire and analyze data. It is very rare to see one study attempt to prospectively replicate another—a major need for the field going forward. Furthermore, most studies tend to involve relatively small samples, on the order of tens of subjects, especially when an antidepressant treatment is involved; this may lead to false positives, inflated effect sizes, and varying results across studies (Elbau et al. 2023; Marek et al. 2022; Schmaal et al. 2020). Moreover, the primary outcome measure is nearly always a change in the global depression score with little focus on specific symptom recovery; the latter on occasion proving effective at identifying subtypes and parsing heterogeneity of depression (Drysdale et al. 2017; Goldstein-Piekarski et al. 2022; Spielberg et al. 2013, 2014; Williams 2016; Xia et al. 2018). In addition, the lack of placebo controls is very often a major caveat, alongside the relatively few studies that have directly compared antidepressant therapies. The ethical constraints on such studies is, of course, enormous because of the vulnerability of the patients under study, especially if they are treatment resistant. This makes direct comparison of rapidly acting antidepressants with the more traditional monoamine-targeting antidepressants, controlling for past treatment and overall depression severity, fraught with difficulties. This is where additional insights can be obtained from experimental studies in animals but surprisingly, direct comparisons of different antidepressants has so far been relatively rare. So, too, have comparisons across prefrontal brain regions, including the OFC, even though all these regions have been associated with stress-related changes; although not all in the same direction. For example, stress has been reported in some cases to potentiate synaptic plasticity and connectivity in the OFC compared to the reductions most often associated with medial PFC (reviewed in Pizzagalli and Roberts 2022). The extent to which antidepressant mechanisms are conserved across species is also unknown. Thus, future studies would benefit from a greater comparative approach, not only at the level of the different pharmacotherapies but also the distinct prefrontal/orbitofrontal regions and the distinct symptom-related behavioral functions.

14

Cognitive Interventions Targeting Executive Functions

How Do They Impact Prefrontal Circuits?

Susanne M. Jaeggi, Alexandru D. Iordan, and Juha Salmi

Abstract

Executive functions (EFs) are essential for everyday functioning. Implicated in many neurodevelopmental and psychiatric disorders, they are also highly susceptible to the effects of aging. There is a critical need to develop effective interventions to improve EFs. This chapter focuses on a particular type of intervention that directly targets EFs by repeatedly practicing on EF tasks using adaptive procedures. There is emerging evidence that such interventions are beneficial: not only do they improve skills related to the trained domain, but they also benefit other domains and symptoms as well as lead to changes in brain structure and function, especially in circuitry related to the prefrontal cortex. At the same time, little is known about the exact underlying mechanisms that drive behavioral and neural changes. Thus, a better understanding of individual differences and training-related factors that mediate and moderate training outcomes is needed to develop more effective interventions that take into account individuals' strengths and needs.

The Malleability of Executive Functions

Extensive research has demonstrated the malleability of executive functions (EFs) and related cognitive functions that rely on the integrity of the prefrontal circuits, demonstrating that these cognitive functions are susceptible to the effects of development and experience (Hsu et al. 2014; Mackey et al. 2013; Zelazo and Carlson 2020). Capitalizing on the plasticity of these circuits, there has been an increasing interest in interventions to remediate, improve, or maintain cognitive functions across the lifespan (Salmi et al. 2018; Tullo and Jaeggi 2022). Many cognitive interventions consist of repeated practice on a task or several tasks that target specific aspects of cognition, with the idea that this practice results in improvements not only in the targeted cognitive function but also translates to other domains related to the trained domain (Pahor et al. 2018). Although many types of cognitive interventions have been shown to impact prefrontal cortex (PFC) circuitry, including goal management therapy (e.g., Stamenova and Levine 2019) and other types of cognitive rehabilitation and remediation (e.g., Geraldo et al. 2023; Vita et al. 2021), we focus here on interventions which more narrowly and directly target EFs and related functions, such as working memory (WM). Typically, these types of interventions involve repeated practice on computerized EF tasks and are often referred to as "brain training" (e.g., Pahor et al. 2018). We note that our primary focus is on cognitive outcomes and their neural correlates, while acknowledging that outcomes which focus on social cognition, metacognitive processes, affect regulation, or self-control are equally important. The latter are, however, beyond the scope of this review; for further information see, for example, Course-Choi et al. (2017), du Toit et al. (2020), Philipp et al. (2019), Tang et al. (2022b), Vickery and Dorjee (2016), and Webb et al. (2012).

As illustrated elsewhere in this volume, EFs refer to a multidimensional construct that includes a set of cognitive mechanisms that control and regulate the contents of WM and action (cf. Murray and Constantinidis, this volume), as well as the ability to plan steps to a problem, ignore distracting information, monitor performance, override automatic responses, or control impulses and regulate emotions (Hsu and Jaeggi 2014). Overall, EFs facilitate purposeful and goal-directed behavior, which is especially critical in novel situations or tasks that have not been well learned (Norman and Shallice 1986). Not surprisingly, EFs are important for everyday life functions in that they predict school readiness, scholastic achievement, job productivity, and even physical health and quality of life (cf. Table 1 in Diamond 2013). EFs are also critically impaired in a range of clinical syndromes, such as depression, attention-deficit hyperactivity disorder (ADHD), addiction, and schizophrenia (Jones and Graff-Radford 2021). The development of EFs follows a distinct, inverted U-shaped trajectory across the human lifespan, typically yielding the best performance at young adulthood, followed by a gradual decline with aging (Hartshorne and Germine 2015). Although the structure of EFs and the extent to which the structure changes across the lifespan is still being debated (Karr et al. 2018), the most popular and well-established models explicate three primary subdomains that are intercorrelated—WM/updating, inhibition, and cognitive flexibility—each of which relies on distinct neural networks (Friedman and Miyake 2017; Friedman and Robbins 2022; cf. other chapters in this volume).

Given their relevance for cognitive and brain health across the lifespan, approaches to strengthen EF skills have appealed to many scientists and practitioners. It has been argued that strengthening specific EFs with targeted training might increase the efficacy of PFC circuitry functioning (Constantinidis and Klingberg 2016), and consequentially lead to performance benefits in domains that rely on the integrity of PFC functioning, especially if trained and nontrained tasks rely on overlapping neural circuitry (Bäckman et al. 2011; Salmi et al. 2018; Vartanian et al. 2022). Indeed, there is growing evidence that targeted, mostly computerized, EF training can improve performance in closely related domains ("near transfer"); this has been demonstrated in various clinical and nonclinical populations across the lifespan (Tullo and Jaeggi 2022). There are, however, persistent inconsistencies and controversies about whether and to what extent such targeted training reliably impacts cognitive functions beyond the trained domain or real-world behavior, such as success in school or ADHD symptoms ("far transfer"): several meta-analyses have demonstrated small effects (Au et al. 2015; Karbach and Verhaeghen 2014; Soveri et al. 2017) while others argue that such findings are essentially noise (Melby-Lervåg et al. 2016; Sala et al. 2019).

Our own view is more optimistic. We suggest that the inconsistency in results and controversy reflect not only the heterogeneity of EF training implementation, choice of outcome measures (Pergher et al. 2020b), and issues with measurement (Karr et al. 2018; Yangüez et al. 2023) but also variability across participants (Pahor et al. 2022). Importantly, those issues do not undermine the potential for EF training to improve cognitive and brain health. Instead, our groups argue that current research should focus on identifying and evaluating the underlying cognitive and neural mechanisms, as well as determine individual differences—the mediating and moderating factors that impact training efficacy (Jaeggi et al. 2011; Pahor et al. 2022)—using appropriate and evidence-based methodology (Green et al. 2019).

Critically, targeted EF training has provided new knowledge of how prefrontal circuits respond to experience and repeated practice using a causal approach and, as such, has contributed to a better understanding of brain plasticity and underlying mechanisms of learning across the lifespan. What makes targeted and computerized EF training particularly well suited to investigate brain plasticity in humans is that it relies on well-defined, widely used experimental tasks that are administered for both training and outcome measures, and which allow their implementation in a neuroimaging setting. This helps in interpreting the observed changes in neural functions and accounting for potential confounding factors (e.g., changes in sensorimotor processes or processing speed) that may also be affected by training (Salmi et al. 2018). Furthermore, most computerized EF interventions are relatively short, requiring ~20 sessions or less that are typically conducted over the course of a few days or weeks with minimal supervision. By contrast, more complex interventions (e.g., rehabilitation approaches in clinical settings, multimodal lifestyle, or educational interventions) typically take place over the course of months or even years, making them difficult to study using neuroimaging techniques.

Neural Correlates of Executive Function Training: Getting at Underlying Mechanisms

Several groups have advanced our understanding of underlying mechanisms of EF training using functional and structural neuroimaging to test whether and how training might impact the prefrontal cortices and/or broader brain circuits (e.g., Salmi et al. 2018; Vartanian et al. 2022). Others have used electrophysiological measures or neuromodulatory approaches to answer this question. Since other chapters in this volume address those issues, we focus here on functional/structural neuroimaging using magnetic resonance imaging (MRI).

Since the year 2000, dozens of functional brain imaging studies, most of them conducted using functional MRI (fMRI), have examined the neural underpinnings of EF training by comparing brain activation before and after the intervention, and a few studies have also focused on capturing trajectories during the intervention (e.g., Finc et al. 2020; Kühn et al. 2013). Over the past several years, converging evidence has started to emerge. For example, a meta-analysis conducted by Salmi et al. (2018), provided systematic evidence that EF training modulates activity in distributed brain areas, encompassing prefrontal, parietal, sensory, and subcortical areas such as striatal nuclei. By comparing these EF training effects to training studies that rely on sensorimotor or language-related tasks, they found that some of the changes in brain activity—especially those involving PFC circuits—are shared across different types of training regimens despite the fact that the comparison interventions were not specifically designed to target EFs. The role of the PFC, however, seems to be particularly critical in EF training. These findings are consistent with domain-general models of learning, where EFs are at the top of the hierarchy, influencing general attention control processes and ultimately lower-level representational systems (Chein and Schneider 2005; see also Shiffrin and Schneider 1977). In other words, EF training can impact large-scale cortical and subcortical neural networks that facilitate both domain-specific and domain-general cognitive processes, with prefrontal circuits driving most of EF-related learning and brain plasticity.

Differential Training-Related Changes in Activation Patterns: The Role of Brain Region and Time on Task

Turning to the specific changes in activation patterns as a function of EF training, an early meta-analysis (Chein and Schneider 2005) found that in most brain areas, task-related activation amplitudes decrease from pre- to post-test, although subsequent work also provided evidence for training-related activation increases (Buschkuehl et al. 2012), which seem to be associated with shorter interventions. More recently, relying on a larger number of papers and including a wider range of interventions, Salmi et al. (2018) demonstrated that specific brain areas seem to respond differently to the effects of training. In particular, converging evidence from fMRI and positron emission tomography studies suggests that two critical areas along the frontostriatal pathways show complementary responses to training. More specifically, prefrontal activations seem to decrease over the course of training, while the activations tend to increase in subcortical areas associated with skill learning (Bäckman and Nyberg 2013). These findings are consistent with a growing number of behavioral studies reporting that positive training outcomes are at least partially explained by acquisition of strategies that help to off-load the demand on EFs (Dunning and Holmes 2014; Forsberg et al. 2020; Laine et al. 2018). In a more recent meta-analysis focused specifically on WM training studies, however, the only consistent finding was training-related activation decreases, especially with longer interventions (Vartanian et al. 2022).

The lack of consistent subcortical effects, and thus any evidence for activation increases, might be partially related to the fact that several studies have not been optimally designed to examine activations in these restricted nuclei, which should ideally be segmented individually in each participant. Nonetheless, given the extensive evidence of the role of the striatum in learning and EFs (Packard and Knowlton 2002), as well as its powerful anatomical positioning within a hierarchical multilevel mosaic system and importance in facilitating the integration of information across cognitive, reward and motor functions (Haber 2016), it is critical to further elucidate how this system contributes to malleability of EFs.

Another PFC-related neural circuitry where EF training effects have been observed in several studies is the cerebro-cerebellar system (see Salmi et al. 2018). As in the case of striatum, training-related changes in the cerebellum have been associated with support processes that off-load the demand on EFs and facilitate the automatization of processes that the brain can learn to anticipate (e.g., timing, sensorimotor integration, regularities in the stimulus contents) (Boyden et al. 2004).

Of note, training-induced changes are not only evident in brain networks, as shown by functional changes discussed above, they have also been shown to impact underlying neurotransmitter systems (e.g., Bäckman et al. 2011; Dahlin et al. 2008). Of particular interest here is the dopaminergic system, which is known to be involved in EF performance as well as more broadly in learning and plasticity (Bäckman et al. 2006; Brehmer et al. 2009). Critically, using PET imaging, it has been demonstrated that EF training-related changes are mediated by dopaminergic modulation of the PFC, especially in older adults (Bäckman and Nyberg 2013; Dahlin et al. 2008; Klingberg 2010; Salmi et al. 2018).

Although there is evidence from several studies that EF training leads to activation decreases in prefrontal circuits with increased training time, several studies point to differential activation patterns depending on the brain region as well as time on task (i.e., intervention length; Kühn et al. 2013) or type of intervention (Belleville et al. 2014). The inconclusive evidence thus far might be related to the fact that the vast majority of studies have focused on regional effects, which may not fully capture the plasticity of neural systems that respond to EF training and that are involved involved in modulating the functioning of other brain networks (Braun et al. 2015; Finc et al. 2020). In particular, changes in structural and functional connectivity are commonly observed after targeted EF training (Colom et al. 2016; Iordan et al. 2021; Thompson et al. 2016) and, importantly, functional connectivity seems to be one of the markers for learning outcomes (Faraza et al. 2021; Kundu et al. 2013).

Training-Related Changes in Brain Activation: What do They Mean?

While there is evidence for training-related changes in amplitude depending on brain region, time on task, or intervention type, the meaning of such decreases and increases has yet to be established. For example, practice-related activation decreases may reflect gains in "efficiency" such that behavior becomes more automatic and well established, which reduces the cognitive load (Neubauer and Fink 2009). Alternatively, participants might figure out different, more appropriate strategies over the course of the training (Forsberg et al. 2020; Laine et al. 2018), which might be reflected in the recruitment of additional (or different) networks (Buschkuehl et al. 2012). In particular, activation increases might reflect the implementation of novel, more EF-demanding strategies (Salmi et al. 2018). The compensation-related utilization of neural circuits hypothesis (CRUNCH) has been used to explain training-related changes in brain activation (Lustig et al. 2009). It proposes a nonlinear (i.e., quadratic) relationship between WM load and brain activation, which is particularly relevant in aging (Reuter-Lorenz and Cappell 2008). Aging leads to decreased neural efficiency which older adults can partially counteract by over-recruiting or over-activating relevant brain regions, at least at lower levels of cognitive load (i.e., "compensatory over-activation"; Festini et al. 2018). With higher cognitive loads, however, a resource ceiling is reached to limit neural recruitment, which in turn, leads to a drop in performance (Cappell et al. 2010). Research has shown that healthy older adults reach their resource ceiling at lower loads, compared with young adults, which is illustrated by a demand-activation curve that is shifted leftward (Cappell et al. 2010). Importantly, CRUNCH makes clear predictions about how activation in regions critical to EF should change due to training (Lustig et al. 2009). Specifically, training should simultaneously

- reduce activation under low cognitive load, consistent with the idea of reduced need for compensatory over-activation with training, and
- increase activation under high cognitive load, consistent with the idea of enhanced dynamic range of activation (i.e., greater responsivity under high demand) with training (Kennedy et al. 2017).

In other words, as shown in Figure 14.1a, with EF training, CRUNCH predicts a rightward shift of the demand-activation curve, irrespective of age (Festini et al. 2018). In line with CRUNCH, Iordan et al. (2020) demonstrated that

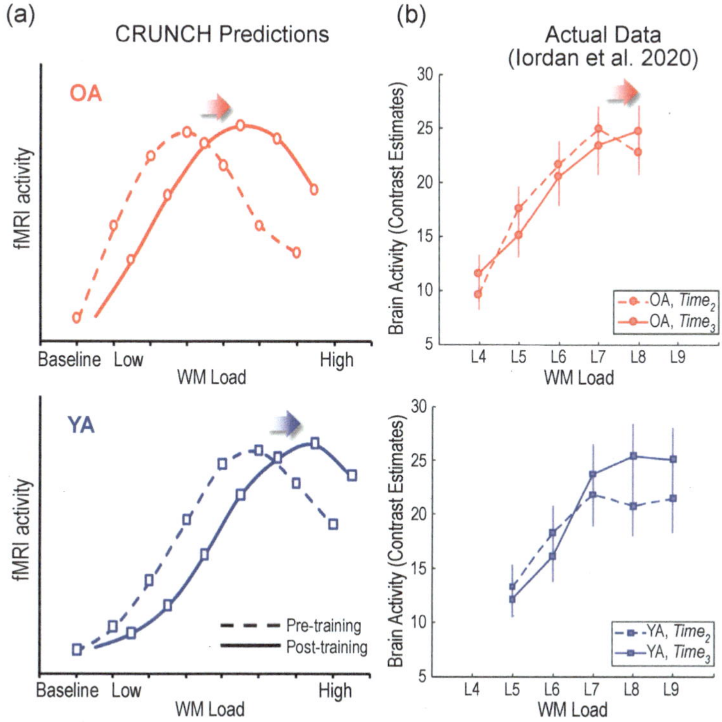

Figure 14.1 Hypothetical CRUNCH activation curves and supporting experimental data. (a) CRUNCH predicts a rightward shift of the neural recruitment curves with training, regardless of age. (b) Training effects in task-positive regions, associated with WM at baseline (Time 1), within each group. Both groups show greater activation at higher loads post (Time 3) relative to pre-training (Time 2). Panels reproduced from Iordan et al. (2020), with permission from Elsevier under a CC BY-NC-ND license.

training leads to activation increases in EF-related brain networks, specifically at higher memory loads, irrespective of age. Critically, training also shifts the demand-activation function rightward in older adults, consistent with a pattern of lower activation post- relative to pre-training for low cognitive loads, and greater activation post- relative to pre-training at higher cognitive loads (Iordan et al. 2020) (see Figure 14.1b). These results hold for both meta-analytically defined and age-group specific EF networks, comprising dorsolateral and ventrolateral PFC as well as lateral parietal cortices (Iordan et al. 2018, 2020, 2021).

Open Questions: Participant Motivation and its Role in Frontostriatal Plasticity

A critical aspect of neural mechanisms involved in EF training, which is still poorly understood, is the role of participant engagement and motivation.

According to behavioral studies, there is evidence that these factors do predict training gains (e.g., Carretti et al. 2011; Jaeggi et al. 2011, 2014). Therefore, it would be important to provide a better understanding of how engagement and motivation to practice influence not only behavior, but also the prefrontal circuitries, and how to disentangle these effects from core EF training effects. Given the important role of frontostriatal networks in motivation (see e.g., Wise 2004), this could be a key target system to investigate the coupling between learning and willingness to learn which might contribute to further enhance the interventions' efficacy. The link between motivation and other cognitive processes is, however, not only an empirical but also a conceptual challenge (cf. Braver et al. 2014).

Executive Function Training in Populations With EF Deficits

The malleability of brain functions as a result of EF training has been studied in various clinical populations where EF deficits are part of the core pathology. Such populations include ADHD (Lambez et al. 2020), schizophrenia (Reser et al. 2019), depression (Woolf et al. 2022), substance use disorders (Caetano et al. 2021), obsessive-compulsive disorder (cf. Duncan and Friedman this volume), and various neurodegenerative disorders, such as multiple sclerosis (MS), or age-related disorders such as Parkinson disease, Alzheimer disease, and related dementias (Lasaponara et al. 2021).

A recent transdiagnostic meta-analysis highlighted the critical role of the striatum, anterior insula, and the PFC, arguing that these are the core regions underlying EF deficits occurring in various syndromes (Yaple et al. 2021). The fundamental issues for brain imaging studies to address in clinical populations include whether and to what extent the dysfunctional neural processes can be influenced with behavioral interventions, and if so, whether there are specific malfunctioning neural circuitries that respond particularly well to training, and how such responses are manifested. In other words, the question is: Does PFC circuitry need to be intact to benefit from targeted EF training? On the behavioral level, even though there are mixed findings in the literature, there is emerging evidence that EF training provides greater benefits to phenotypes that express deficits in EFs, such as individuals with ADHD compared to those who do not (Karbach et al. 2017; Traut et al. 2021). This emphasizes both the need and potential for interventions that target and optimize PFC circuitry (Salmi et al. 2020).

In Neurodevelopmental Disorders: ADHD

Even though the literature on neural correlates of EF training in neurodevelopmental disorders is still scarce, some preliminary evidence on the effects of EF training have been reported, mostly on prefrontal, parietal, and temporal activity. For example, in two early studies that focused on ADHD, functional (Hoekzema et al. 2010) and structural (Hoekzema et al. 2011) changes were reported following 10-day cognitive training interventions which tapped multiple EF domains. However, like two other studies (de Oliveira Rosa et al. 2020; Stevens et al. 2016), sample size was small and other experimental issues limited the interpretation of findings (e.g., lack of control group with ADHD, add-on stimulant treatment). Using a slightly larger sample size, Salmi et al. (2020) examined changes in regional brain activity from pre- to post-test in a randomized controlled trial with dual n-back WM training. By including a group of neurotypical adults to the pretest session, they first extracted brain activity that was aberrant in ADHD adults and then demonstrated that some of these deviations in brain activity were restored during the training period. In this study, Salmi et al. also demonstrated that the neural modulations for trained and untrained (transfer) tasks were in the opposite direction: In trained tasks, they observed decreased activity, whereas in the untrained variant of the n-back task, activity increased. These findings could partially explain why reports of training-related activation increases versus decreases have not been systematic in the literature. As mentioned above, there has been a lot of variability in the training protocols and outcome measures across EF training studies, both in neurotypical populations as well as in clinical studies (Pergher et al. 2020b; Tullo and Jaeggi 2022); even in healthy participants, only very few brain imaging studies have included both trained and untrained variants of the EF task in the pre- to post-test battery. Despite these methodological limitations, the loci of activations have been consistent across these few ADHD studies, as training-related modulations have been systematically observed in overlapping parts of the prefrontal, parietal, and temporal cortices, which are among the areas that typically show aberrant brain activity in this clinical group (Cortese et al. 2012).

In Neuropsychiatric Disorders

The literature on training-related plasticity is more extensive in other neuropsychiatric disorders, in particular, schizophrenia. Here, even though a majority of studies seem to observe increased activations in left prefrontal regions, summarized by Mothersill and Donohoe (2019), there is also evidence for a more widely distributed pattern of activation across cerebro-cortical and subcortical areas after training. At the same time, the authors point out the extensive heterogeneity of these findings, which they attribute to the broad range of interventions implemented, making it difficult to extract a consistent and statistically significant pattern.

Focusing on psychiatric disorders more broadly, a recent meta-analysis of brain imaging studies of EF training, Li et al. (2022) reported consistent activation increases in the left inferior frontal gyrus and decreased activation in the precuneus and cuneus, when comparing pre- and post-test. These findings are supported by earlier meta-analytic findings reported by Salmi et al. (2018), who focused on EF training in nonclinical participants. In other words, the same core brain regions seem to respond to EF training in both clinical and nonclinical populations; however, the direction of the effects (i.e., decreases or increases in amplitude) seems to differ depending on the population. In general, and similar to pharmaceutical treatments (Kirkland and Holton 2019), EF training might restore the aberrant activity to a normal range (see Salmi et al. 2020).

In Neurodegenerative Disorders

Although fewer neuroimaging studies are available in neurogenerative disorders such as Parkinson disease or MS, the pattern is similar to neuropsychiatric disorders in that there seem to be (a) training-related increases in prefrontal, parietal, and cerebellar activity (cf. Prosperini and Di Filippo 2019), and (b) increased connectivity in the frontoparietal network and the default mode network, as captured by resting-state fMRI. With respect to MS, Prosperini and Di Filippo (2019) concluded that current evidence related to EF training-induced plasticity is fragmented, and that more evidence is needed on what would be the optimal brain imaging techniques in detecting the neural alterations relevant to MS, and how optimally to implement the training intervention (e.g., type, intensity, duration, combining behavioral, pharmacological treatment). As in other clinical conditions, one of the key avenues is to search for methods that will enable us to predict an individual patient's response to rehabilitation.

In Healthy Aging

Given that the typical course of aging is characterized by a decline in the functioning of core EFs, older adult populations have become a frequent target of EF training. Several meta-analyses have focused on the neural correlates of EF training in healthy aging (Duda and Sweet 2020; ten Brinke et al. 2017), but work has also synthesized the neural correlates of EF training in mild cognitive impairment and dementia (Beishon et al. 2020; van Balkom et al. 2020). Collectively, this work further highlights the heterogeneity in the type of EF training and outcome measures used, as well as the wide range of imaging methodology and analysis approaches being implemented. Still, in general, there seems to be evidence for training-related changes in regional activity and functional connectivity in the prefrontal and parietal areas overlapping with those reported in MS studies (Prosperini and Di Filippo 2019). Beyond the results showing increased functional connectivity, there are findings of decreased connectivity after training (Beishon et al. 2020) as well as reports that demonstrate more pronounced segregation of frontoparietal and default mode brain networks after training in younger but not in older adults (Iordan et al. 2021). Similar to the issue of training-induced activation increases versus decreases, our understanding of the changes in the strength of the functional connectivity and topological patterns in the large-scale neural networks is still limited (Baniqued et al. 2019).

Summary

The effects of EF training on brain activity and connectivity in populations with impaired EFs seem to be in line with compensatory mechanisms (Lövdén et al. 2012), in particular, the CRUNCH model (Reuter-Lorenz and Cappell 2008). In this context, neural compensation refers to alterations in neural functioning that offset the effects of age-related neural decline or pathology and facilitate elevated levels of cognitive and behavioral output. Specifically, older adults or otherwise compromised individuals frequently show greater and more widespread frontal lobe activity and less functional network segregation (Iordan et al. 2020, 2021). Under conditions of equivalent cognitive performance, the interpretation is that additional activation and network integration may serve a compensatory function (Reuter-Lorenz and Cappell 2008; Reuter-Lorenz and Iordan 2018). When neural plasticity is compromised (e.g., due to more advanced neural decline or pathology), CRUNCH predicts that individuals with EF needs, relative to healthy controls, would show lower, flattened demand-activation curves within frontal regions, responding only to small cognitive loads. When training succeeds, CRUNCH predicts a potential recovery of activation in frontoparietal regions, with a leftward shift of the demand-activation curve (see Figure 14.1). Furthermore, emerging evidence suggests that EF training can remediate aberrant neural activity in various clinical conditions. At the same time, given the considerable heterogeneity of EF-related disorders as well as the extensive interindividual differences across patients, the question of who is likely to benefit from training and which factors mediate positive training outcomes are of particular importance from a clinical point of view and has begun to receive more attention in recent years (Tullo and Jaeggi 2022).

Conclusions, Outstanding Questions, and Implications

Current research on EF training in various clinical and nonclinical populations across the lifespan suggests that behavioral and neural effects differ as a function of the tasks, intervention length, or populations, as well as other variables related to individual differences. In particular, there is variability in training benefits with respect to both training-specific gains as well as transfer, suggesting that there is no "one-size fits all" approach for EF training. The most salient questions that need to be addressed in current and future research concern how to determine which type of training works for whom, and why (Jaeggi et al. 2011; Pahor et al. 2022).

There are several avenues to address these issues; in particular larger sample sizes are needed to uncover and replicate the relevant individual-difference factors that mediate and moderate the training outcome (Ørskov et al. 2021; Pahor et al. 2022). Here, it could be beneficial for research groups to use common methods (e.g., intervention types, outcome measures) to facilitate the acquisition of larger and more diverse datasets and allow for generalization beyond individual experiments and labs (Pergher et al. 2020b). To test the impact on training outcomes, an alternative approach would be to pick participants selectively according to certain characteristics, such as populations with or without EF needs—e.g., young verus older adults, individuals with and without ADHD (Iordan et al. 2021; Salmi et al. 2020). The recent literature has increasingly focused on those issues, demonstrating the relevance of certain individual-difference factors, ranging from preexisting cognitive abilities to performance during training, personality characteristics or demographic variables, along with motivational factors (Katz et al. 2021; Ophey et al. 2020; Ørskov et al. 2021), biomarkers including brain modularity (Gallen and D'Esposito 2019), or genotype (e.g., Bellander et al. 2011; Feng et al. 2015; Hernes et al. 2021; Zhao et al. 2020).

For instance, brain network modularity has been proposed as a biomarker of intervention-related plasticity, with particular relevance for aging (Gallen and D'Esposito 2019). Specifically, whereas high pre-training modularity, particularly during resting state, may reflect a more "optimal" functional network organization that promotes cognitive improvements with training (e.g., Gallen et al. 2016; Iordan et al. 2018), older adults (as well as clinical populations) may be less able to increase network segregation with training, as an expression of overall diminishing neural plasticity (Park and Reuter-Lorenz 2009; Reuter-Lorenz and Park 2014). Another possibility is that modularity may be generally beneficial for cognitive functioning, and local declines in brain function due to aging or neurodegeneration may be compensated by a more integrated workspace.

At the behavioral level, we and others have demonstrated that baseline abilities are among the key predictors for training-specific benefits (Jaeggi et al. 2011, 2014). Interestingly, while some work has shown evidence for compensation effects (i.e., individuals with lowest initial performance gain most from training or, in other words, catch up to the others), other work has found evidence for magnification effects (i.e., the rich get richer phenomenon) (Jaeggi et al. 2011; Karbach et al. 2017; Ørskov et al. 2021). It is currently unclear whether those differences are attributable to specific populations (e.g., age or patient groups) or related to training paradigms and the outcome measures studied, or a combination thereof (Feng et al. 2023). Our research findings emphasize the importance of paying attention to participants' performance in the training task themselves, as well as whether and to what extent they improve in nontrained variants of the training tasks ("near transfer"). Specifically, in several studies, we have demonstrated that those who improve during training and/or improve in nearest transfer measures are also more likely to show far transfer effects (Jaeggi et al. 2011; Jaeggi et al. 2014; Pahor et al. 2022; Parong et al. 2022). The key issue here is to figure out how to engage participants optimally during the intervention so that they can reap the full benefits of the training. Here, we and others have emphasized the role of good game design, along with motivational features which take into account participants' interests and demographic backgrounds (Deveau et al. 2015; Pasqualotto et al. 2022). Overall, it seems critical to account for individual differences that might affect adherence and persistence with cognitive training interventions (Tullo and Jaeggi 2022; Tullo et al. 2023). We also need to get a better understanding of the cognitive and neural mechanisms underlying training success, as well as the extent to which they might change as a function of EF training (Gallen et al. 2016; Kühn et al. 2011; Pahor et al. 2022; Parong et al. 2022). Furthermore, it is important to recognize that there is considerable overlap between the PFC circuits engaged in task performance and those involved in motivation and effort (Braver et al. 2014; Haber 2016). This poses an interesting and unique challenge for EF interventions: the targeted skills are also (or closely related to) the abilities required to engage effectively with the intervention itself; this is especially salient in ADHD, where issues with motivation and persistence are part of the core symptoms (Arnsten and Rubia 2012; Shen et al. 2020; Sibley 2020). As such, the issue is whether the effectiveness of EF interventions might benefit from incorporating additional tasks or components that purposively engage motivation- and/or effort-related circuits. Indeed, some groups have started to implement such approaches, e.g., by adding metacognitive and/or motivational components with promising effects (e.g., Carretti et al. 2014; Jaeggi et al. 2023; Vranic et al. 2013).

Another approach to maximize training benefits by capitalizing on potential additive effects could include the combination of EF training with physical exercise (Daugherty et al. 2018; Karssemeijer et al. 2017), brain stimulation (Au et al. 2022), mindfulness meditation (Course-Choi et al. 2017), or by more broadly incorporating multidomain lifestyle factors, as demonstrated by the FINGER study (Rosenberg et al. 2018). Such approaches likely implicate brain regions beyond the PFC networks and thus might increase the likelihood for broader/generalized and possibly, more sustained effects. At the same time, it is important to keep in mind that such multimodal interventions are typically much more demanding in terms of time, logistics, and personnel as compared to unimodal interventions, and it is not always clear what components work best and in what combination. As such, getting a mechanistic understanding of the intervention efficacy is even more challenging.

In conclusion, emerging research points to the relevance of personalized training approaches that take into account participants' strengths and needs, which can be derived from their preexisting EF skills, as well as their demographics, personality, interests, and biomarkers (e.g., brain network modularity, genotype, dopaminergic functions). The cognitive training literature might benefit from taking inspiration from precision medicine (Lenze et al. 2021). Furthermore, getting a better understanding of the intervention-related factors and the ideal combination of intervention components is critical for the design of effective and sustainable interventions to benefit a broader range of populations.

15

Neurosurgery for Intractable Obsessive-Compulsive Disorder

A Window into Prefrontal Cortical Function in Humans

Steven A. Rasmussen

Abstract

Recent advances in functional and diffusion imaging, as well as neuromodulatory devices that can both stimulate and record, have opened up new avenues for advancing hypothesis-driven circuit-based treatments of neuropsychiatric disorders. Neuromodulatory treatments will also expand our understanding of prefrontal cortical function in humans. Across neuropsychiatric illnesses, obsessive-compulsive disorder (OCD) stands out as being a condition where we have an initial understanding of the neural circuitry associated with the illness. Converging evidence has implicated ventrolateral orbital, rostral anterior cingulate, and medio-orbitofrontal and medial frontopolar cortex in OCD psychopathology. Expanded interest of basic scientists in the clinical phenomena of OCD and interdisciplinary collaboration will be essential to further delineate the neurobiologic basis of the illness as well as the mechanism of action of circuit-based treatments.

Introduction

What evidence implicates prefrontal frontostriatal loops in the pathogenesis of obsessive-compulsive disorder (OCD)? How have invasive neurosurgical procedures for OCD contributed to our understanding of prefrontal cortical function in health and disease? In this chapter, I will examine two main bodies of evidence: (a) studies of OCD psychopathology and (b) findings from invasive neurosurgical studies for intractable OCD. Although neurosurgical interventions were designed to develop novel approaches to treatment for patients with intractable neuropsychiatric conditions, they also offer an unparalleled opportunity to expand our understanding of prefrontal cortical function in humans.

This chapter focuses on OCD and its symptoms, yet it is important to keep a transdiagnostic perspective in mind. Most OCD patients suffer from comorbid anxiety and depression (Rasmussen and Tsuang 1986). Most depressive episodes are preceded by stress, uncertainty, worry, and anxiety (Barlow and Campbell 2000). The mainstay pharmacologic treatment for all of these conditions are the selective serotonin reuptake inhibitors. A significant percentage of OCD and anxiety disorder patients suffer from behavioral inhibition and separation anxiety during childhood (Rosenbaum et al. 1988). MacLean (1958) pointed to the important role medial prefrontal cortex (PFC) plays in separation anxiety, and as highlighted below, cingulotomy and capsulotomy were used to treat intractable depression as well as intractable OCD. More recently, corticofugal fibers from medial PFC that traverse the anterior limb of the internal capsule have been targeted for deep brain stimulation (DBS) of depression as well as OCD. Haber et al. (2021) have pointed to the connectional similarities and differences in the ventral anterior limb of the internal capsule (vALIC), subthalamic nucleus (STN), subgenual and superolateral branch of the medial forebrain bundle (slMFB) targets for DBS in the treatment of depression and OCD. Future studies will help to clarify the neuroanatomical pathways that are shared as well as unique that correlate with the symptoms of depression, anxiety, and OCD.

Sixty years ago, MacLean put forward the hypothesis that from an evolutionary perspective, the pathogenesis of most neuropsychiatric illnesses were due to abnormalities in the medial PFC and its connections with subcortical regions (Maclean 1958). He posited that these recently developed neocortical regions had not been subject to the process of natural selection, compared to older more highly conserved circuitry. Since then, converging data from studies of brain imaging, cognitive-affective neuroscience, neuromodulation, and animal models suggests that OCD represents a neural network-based disorder (Milad and Rauch 2012; Yuste 2015) involving the dysregulation of cortico-striato-thalamo-cortical (CSTC) loops. The seminal contributions by Alexander et al. (1986) led OCD researchers to examine certain parallel segregated CSTC loops, subserving different motor or cognitive functions, as the neuroanatomical basis for obsessive-compulsive behavior (Baxter, Jr. et al. 1988; Breiter et al. 1996; Saxena and Rauch 2000). Revisions to this model have demonstrated a more complex picture of the organization of CSTC loops (Haber et al. 2020) and show more overlap and functional integration between loops than previously thought. Both positron emission tomography (Schwartz et al. 1996) and functional magnetic resonance imaging (fMRI) studies have shown increased activation in regions of the orbitofrontal cortex (OFC), anterior cingulate cortex (ACC), and portions of the basal ganglia in the symptomatic state compared to healthy controls (Figee et al. 2013). These areas of abnormal activation normalize following successful treatment with either pharmacotherapy or behavioral therapy. Successful treatment of OCD with DBS (Figee et al. 2013), surgical ablation (Zuo et al. 2013), and transcranial magnetic stimulation (Dunlop et al. 2016; Nauczyciel et al. 2014) has also been associated with reductions in brain activity in these regions compared to baseline. An important development in circuit-based theories of OCD has been a shift in focus from static regions of interest to investigation of functional networks underpinning different cognitive or behavioral functions related to the symptoms of OCD.

Frontostriatal Loops and the Psychopathology of OCD

The limitations of symptom- and diagnosis-based approaches in understanding the cause of anxiety and obsessive-compulsive spectrum disorders strongly suggest the importance of looking beyond symptoms and symptom dimensions toward underlying dimensional endophenotypes. We have identified two core constructs underlying the symptoms of anxiety and obsessive-compulsive spectrum disorders, harm avoidance and incompleteness, that have demonstrated clinical face validity but for which the underlying neural basis is poorly understood. In 1992, we proposed a conceptual model of these two core constructs of OCD that integrated symptom subtypes, temperament, and neurocircuitry to explain the marked comorbidity between OCD, OC spectrum disorders, and anxiety disorders (Rasmussen and Eisen 1992). We defined incompleteness as the need to have a thought or action perfect before moving onto the next motor or cognitive action. In contrast, we defined harm avoidance as the urge to perform an active avoidance behavior so as to prevent something bad from happening to self or others. These two constructs have demonstrated clinical face validity, and we postulated that harm avoidance was differentially associated with avoidance/punishment circuitry (Dalley et al. 2011) and incompleteness with action selection/reward circuitry. While active avoidance has been investigated in numerous studies of anxiety disorders (Kampman et al. 2014), there has been comparatively little recognition of incompleteness in psychopathology or its public health significance. Incompleteness, or the wish to finish something completely or perfectly before moving on to the next task, is familiar to us all. Clinical manifestations of incompleteness can range from procrastination and perfectionism with excessive attention to detail, to marked difficulty with planning and lost productivity, leading to the inability to sustain goal-directed behavior. Harm avoidance and incompleteness share disruptions in goal-directed action control, (Gillan et al. 2011), with harm avoidance minimizing uncertainty and exploratory behavior in favor of security (Hinds et al. 2010; Szechtman and Woody 2004) and incompleteness minimizing speed and productivity in favor of accuracy and precision. The neurocircuitry underlying goal-directed planning, action control, and emotion is complex and widely distributed across multiple large-scale networks (Cocchi et al. 2012; de Wit et al. 2009; Everitt and Robbins 2005; Seeley et al. 2007; Valentin et al. 2007). It remains unclear whether abnormalities in the top-down disinhibition of frontostriatal circuits or bottom-up limbic activation of frontostriatal circuits are abnormal in excessive harm avoidance and incompleteness, or if there is a dynamic interaction between the two.

Although frontostriatal circuitry and abnormalities in goal-directed behavior have long been implicated in the pathogenesis of OCD, the relationship of the symptoms and core features of the illness to three key large-scale prefrontal networks remains unclear: the salience network (Seeley 2019; Uddin 2016), the cognitive control network (Badre 2008; Koechlin et al. 2003), and the task execution network (Badre 2020; Badre and D'Esposito 2007; Dosenbach et al. 2006). The representation of value, hierarchical action selection, and action-outcome monitoring that form the basis of goal-directed behavior are made up of highly distributed interacting networks that rely on parallel processing. We define cognitive control as the ability to select mappings between states and actions based on internally maintained representations of context, goals, and anticipated outcomes (Shenhav et al. 2013). Control over action requires maintaining context at different levels of abstraction and over varying timescales (Holroyd and Yeung 2012). Rostral ACC is at the juncture between cortical structures that represent current and inferred states related to salience and valuation (insula and OFC), structures involved in communicating information about current and past environment and context (temporal lobe), as well as structures that are responsible for task execution (lateral PFC) (see Figure 15.1) (Gläscher et al. 2012; Holroyd and Verguts 2021; Monosov et al. 2020; Shackman et al. 2011; Shenhav et al. 2016).

Early theories of hierarchical sequential control of the goal-directed behavior involved in task execution focused on cortico-cortical interactions in which rostrolateral PFC affected submodular processing in premotor cortex (Badre and Nee 2018). More recent work has suggested that hierarchical cognitive control may emerge from the interaction of nested frontostriatal loops, where action selection at one cortico-striatal level is constrained or gated by inputs from more anterior levels (Badre and Frank 2012; Frank and Badre 2012). Neuroanatomical evidence has pointed to areas of convergence of these frontostriatal loops at the level of the striatum and thalamus (Haber and Calzavara 2009; Haber et al. 2006). The full range of cognitive control over action is likely to reflect a continuous integrative cascade of processing, from valuation to monitoring, to task specification, and finally to regulation of task execution (Chatham et al. 2014). One can easily see how impairment in these frontostriatal loops could cause problems with action initiation and termination (Heilbronner and Hayden 2016; Hinds et al. 2012; Woody et al. 2005), as well as excessive attention to subgoals and making implicit subgoals explicit versus overall goals leading to the cardinal symptoms associated with OCD. Several lines of evidence point to the idea that these behaviors may result from abnormalities in connectivity between the rostral-most portions of the frontal cortex

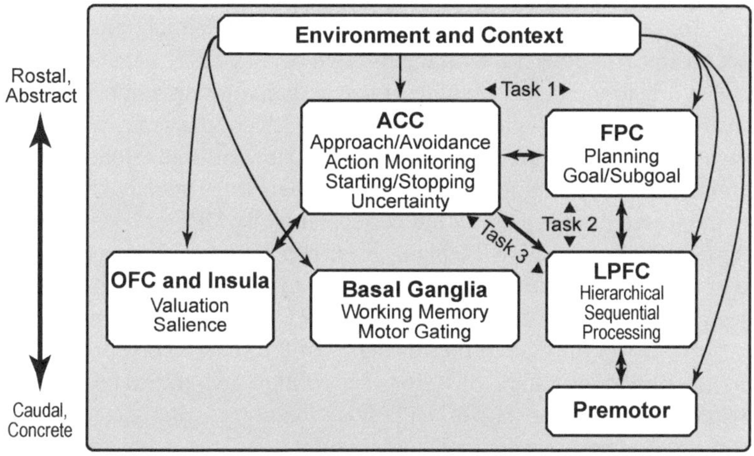

Figure 15.1 Conceptual diagram linking neurocognitive tasks involving goal-directed behavior to the neurocircuitry of incompleteness on three tasks: (1) the Archer task, (2) the sequential control task, and (3) MCIT task.

and its connection with the rostral cingulate and OFC (Braver and Bongiolatti 2002; Burgess et al. 2007; Koechlin et al. 1999; Mansouri et al. 2017; Ramnani and Owen 2004; Rouault et al. 2019). Neuronal ensembles in the frontal pole encode chosen goals during feedback, which suggests that it promotes learning about which kinds of goals and goal-generating processes produce particular costs and benefits. The frontopolar cortex (FPC) selectively mediates the human ability to hold goals in mind while exploring and processing secondary goals. Medial anterior PFC in association with ventral striatum is preferentially engaged when subjects execute tasks in sequences that were expected, whereas FPC was involved preferentially when subjects performed tasks in sequences that were contingent on unpredictable events (Corkin 1979). Taken together, these findings suggest that connections between these regions/networks may underlie deficits in goal-directed behavior associated with OCD.

How Has Invasive Neurosurgery for OCD Contributed to Our Understanding of the PFC?

The extensive literature on the effects of frontal lobotomy on behavior provide a unique source of clinical observations as do the rigorous studies that attempted to correlate neuropathologic examination of the placement and size of the lesions with therapeutic outcomes as well as the correlation of adverse behavioral effects with frontal lobe function. These studies can now be reinterpreted in light of recent findings about PFC function. Freeman et al. (1942) pointed to the role of the highly distributed nature of cortico-cortical connections in preserving behavior. He noted that following lobotomy, the frontal cortex is isolated from the thalamus and no longer receives thalamic input except by indirect means. Nonetheless, with the passage of time many patients who had undergone lobotomies were capable of social and working adjustments, thus demonstrating that even with such a profound lesion, the brain was capable of reorganization due to the highly distributed nature of frontal lobe function.

Reminiscent of the single neurotransmitter theories of schizophrenia in the 1980s, the search has been on for the perfect target for DBS or lesions in OCD and depression. Recognizing the highly distributed nature of, for example, salience or value, most investigators have begun to focus on the effect of stimulation or lesions on networks as opposed to individual white matter tracts or nuclei. Rylander (1948) noted the resilience of the frontal cortex to injury as well as the important effects of lesions on volition and prefrontal control of autonomic function. Patients who underwent lobotomy using local anesthesia revealed no changes after the incisions in the frontal lobe were completed on either side, or after incisions to both upper halves or both lower halves of the two frontal lobes. When a third quadrant was sectioned, there was a notable falloff in the length of the replies and in the display of emotion connected to them as well as no evidence of spontaneous speech. When the fourth quadrant was sectioned, the patient became unresponsive except to urgent questions: his face was expressionless and his orientation was lost, and any preexisting anxiety was no longer present, with corresponding effects on pulse rate and blood pressure.

In a long-term twenty-year follow-up of motor deficits in leucotomized schizophrenic patients, Benson et al. (1981) found no signs of praxis, no elementary motor dysfunction, and no frontal release signs. Patients with schizophrenia with the largest prefrontal damage by structural imaging learned and performed a three-step sequence task better than schizophrenic subjects with less or no bifrontal damage and as well as controls. Most of the subjects with sizeable bifrontal damage could complete go/no-go and alternation of response tests as well as normal controls. Although standardized tests have not uncovered deficits, there is something to be learned from the careful clinical observations made by investigators working with frontal lobotomy patients. This is particularly true in the effects on social awareness of the effects of their actions on others as well as in goal-directed planning. As Robinson (1946) noted:

> They have become not so much social as gregarious, not more interested in the thoughts of others merely less in their own. They have no hint of ulterior motives. Past and future seem telescoped into the present. It is the capacity for deliberateness that they have missing.

A recent paper pointed to the role of the frontal pole in episodic future thinking as well as monitoring action outcomes in the past. Many of our patients are stuck either anticipating the future or regretting the past. Freeman and

Watts (1950) pointed to the role of the frontal cortex in social interaction and anxiety, delayed discounting as well as its role in projecting the image of the self into the future:

> The frontal lobes are important for insight, for subtlety, for postponing pleasure and for projecting the individual self into the future. They are essential for the elaboration of a vivid picture of the future with all its deviations all its implications all its difficulties and dangers all its triumphs and disasters…the operated patient lives in a perpetual present, his interests in the outside world being much more vivid than his interests and reactions to them.

Rylander (1948) was among the first to report that a slight but fateful intellectual reduction that was difficult to demonstrate with ordinary intelligence tests, but that affected abstract reasoning and the ability to plan was present in long-term follow-up after surgery. Simple planning tasks like the Tower of London failed to show pre-post changes. More complex, real-life planning tasks have yet to be tested (Burgess 2000). Advances in ecological momentary assessment and digital phenotyping should make this possible. The success at completing habitual albeit complex goal-directed behaviors as opposed to the deficits in completing novel goal-directed behaviors suggests we need to look more closely at defining goal-directed versus habitual behavior than the simple dichotomy of model-based versus model-free behavior. It also argues for testing of the cognitive challenges that are found in daily life versus those in a laboratory setting.

Given the size and extent of the lesions of anterior cingulotomy, anterior capsulotomy, and limbic leucotomy, one cannot help but wonder why they do not result in more deficits in executive function and neuropsychological testing. Following cingulotomy, patients exhibited no change in error monitoring. These data are consistent with current neuropsychological studies that show either no deficits or improvements in tests of executive function following DBS or lesions.

Cingulotomy

In 1975, an independent group of psychologists at MIT, experts in neuropsychological function and brain trauma, were enlisted by the U.S. Congress to study cingulotomy patients as part of a white paper on psychosurgery (Valenstein et al. 1977). They followed 18 patients prospectively through a series of in-depth interviews of the patients and their families, as well as an intensive battery of 24 neuropsychological tests administered preoperatively, four months prior to surgery, as well as four months to ten years after surgery. Analysis of the life history data and interview material failed to disclose any major adverse effects from the intervention. There were no lasting effects of the cingulotomy per se on the 24 behavioral tasks. For other indicators of frontal lobe function, there was no change in verbal fluency, nonverbal fluency, Porteus maze, or delayed alteration tasks.

A series of follow-up papers reported on the long-term follow-up of 64 patients who had anterior cingulotomy for intractable OCD between 1989–2009, with a mean follow-up at 64 months (Baer et al. 1995; Dougherty et al. 2002; Jenike et al. 1991; Sheth et al. 2013). Thirty-six patients had a single pair of lesions and 28 had a triple pair of lesions located along the cingulate bundle, stretching from the genu of the corpus callosum posteriorly. No significant difference in outcome was observed between those who had single- and triple-paired lesions. Using the Yale-Brown Obsessive Compulsive Scale (Y-BOCS), 22 patients showed a greater than 35% drop in outcome, with an additional five patients showing a 25% drop. Jung (2006) reported on the one- and two-year follow-up of 17 patients who had anterior cingulotomy for OCD and found that eight patients had a greater than 35% drop in the Y-BOCS at follow-up. Lesions were placed slightly more anterior in the cingulate bundle than in the Massachusetts General Hospital cohort.

Cohen et al. (1994, 1999a, b, 2001) reported on 12 patients who had a single bilateral cingulate lesion, with follow-up at 3 and 12 months post-surgery. Immediately after cingulotomy, mutism, akinesis, blunted affect, lethargy, and apathy were common. These severe symptoms resolved quickly, however, and 3 months post-surgery most patients had returned to baseline with regard to language, visual, motor, memory and intellectual functioning. Despite this preservation of function, many families reported that subtle personality and functional changes remained, particularly continued behavioral passivity. Deficits of executive control and attention also persisted, with spontaneous response production most affected (i.e., spontaneous utterances, object construction, design fluency), a pattern of impairment frequently observed among patients with frontal lobe damage. Patients continued to show performance variability, slowed processing, and vulnerability to interference. Cingulotomy did not affect performance on tasks that placed primary demands on sensory selective attention (e.g., letter cancellation), attention span, and working memory (e.g., digit span). Learning and memory were also intact.

More recently, Banks et al. (2015) reported on 14 OCD patients who had cingulotomy as well as high-resolution structural and diffusion imaging scans. They identified a gray matter cluster just anterior to the lesion in the right anterior cingulate that correlated with poor response using voxel-based morphometry. Using diffusion connectivity measures, they also found increased right-sided connectivity between the lesion site and the caudate that predicted enhanced treatment response.

Intraoperative single or multiunit recordings as well as stimulation prior to making a lesion offers a unique opportunity to extend findings about electrophysiologic studies of the PFC to humans. While most of these studies have been conducted in the context of DBS trials, several electrophysiologic and behavioral studies have been published about cingulate function in humans in

OCD patients undergoing cingulotomy. The small number of subjects tested limits the conclusions that can be drawn from these results. They are generally in agreement with electrophysiologic studies done in nonhuman primates as well as imaging studies in humans. These studies point to the anterior cingulate's key role in action initiation and monitoring and their relationship to salient events in humans.

Gentil et al. (2009) tested preoperative stimulation at the cingulate and subcaudate target sites and found that stimulation was accompanied by increased autonomic arousal as measured by skin conductance but not heart rate acceleration. Srinivasin et al. (2013) studied the immediate effects of anterior cingulate ablation on action initiation in six OCD patients. Three patients had preoperative and immediate postoperative simple reaction time tests, whereas another three patients completed a pre- and postoperative reward-based decision task. The frequency of false starts following a visual cue increased in the simple reaction task.

Sheth et al. (2012) demonstrated that the modulation of current dorsal ACC activity by previous activity produces a behavioral adaptation that accelerates reactions to cues of similar difficulty to previous ones and retards reactions to cues of different difficulty. This conflict adaptation was abolished after surgically targeted ablation of the dorsal ACC. Sheth et al. concluded that the dorsal ACC provides an updated prediction of expected cognitive demand to optimize future behavioral responses. In situations with stable cognitive demands, this signal promotes efficiency by hastening responses; however, in situations with changing demands, it engenders accuracy by delaying responses.

Sklar et al. (2017) tested nine OCD patients undergoing cingulotomy, identifying a population of rostral ACC neurons that respond differentially or in a graded manner to cognitively demanding high- and low-conflict Stroop tasks, including those with emotional valence (Davis et al. 2005). Their data suggested that rostral ACC neurons may be acting as salience detectors when faced with conflictual or emotional stimuli, consistent with neuroimaging results of rostral ACC responses to abrupt novel, task-relevant, or painful stimuli.

Anterior Capsulotomy

Mindus and Myerson reported on the outcome of two capsulotomy cohorts: one with severe intractable anxiety, the other with intractable OCD. Patients were either lesioned with thermocapsulotomy or with a noninvasive radiosurgical instrument called the gamma knife. Twenty-four patients with intractable anxiety were followed at 3, 6, 9, and 12 months after the procedure as well as a long-term follow-up of a mean of 8 years. Nyman and Mindus (1995) administered an extensive neuropsychological battery to 17 of these patients. Tests showed either an improvement or a stable pattern following capsulotomy, with the only exception being the Wisconsin Card Sorting Test, which showed an increased number of perseverative errors in five of the 17 patients (Nyman et al. 2001). In a separate study, Mindus et al. (1999) gave the Karolinska Scale of Personality to 24 patients at baseline and one year following thermocapsulotomy. At the one year follow-up, significant decreases (toward normality) were found in eight of the scales. Impulsiveness hostility and aggressiveness were within the normal range.

Zhang et al. (2017) administered the Iowa Gambling Task to 24 OCD patients preoperatively and 3–5 months following bilateral anterior capsulotomy and observed no significant differences in decision making between the preoperative and 3–5 month follow-up groups. At the long-term follow-up, one to three years afterward, decision-making abilities of patients had improved on par with healthy controls. Rück et al. (2003) conducted an independent long-term follow-up (mean of 13.5 years) of 26 bilateral thermocapsulotomy patients with severe anxiety who had no OCD. In their study, seven of 17 patients were rated as having significant adverse effects: the major symptoms were apathy and dysexecutive behavior. Using a simple scale that measured executive function apathy and disinhibition, one of the patients was rated as severe in all three measures, two moderate in executive function and apathy, and one severe in executive function and apathy. These patients also made more perseverative errors on the Wisconsin Card Sorting Test. Though many patients benefited from the procedure, Rück et al. concluded that a minority were left with significant long-term adverse cognitive effects.

In a separate long-term follow-up (mean of 10.9 years) of the OCD cohort, Rück et al. (2008) studied 25 patients with intractable OCD who had had anterior capsulotomy using either thermocapsulotomy or the gamma knife. Twelve out of 25 patients had sustained a greater than 35% drop in the Y-BOCS. Significantly, none of the patients was working at the time of the follow-up. Two patients suffered from severe executive dysfunction, apathy, and disinhibition while six had at least moderate impairment in one of these domains.

In an effort to minimize adverse effects, we began doing ventral gamma capsulotomies located 8–10 mms anterior to the posterior border of the anterior commissure in the coronal plane, and which targeted fibers connecting the orbital and medial frontal cortices with the thalamus and brainstem but left the dorsolateral cortical fibers that ran in the dorsal portion of the capsule intact (Rasmussen et al..2018). At three year follow-up, we found that 31 of the 55 patients (56%) had an improvement in the primary efficacy measure, the Y-BOCS, that was greater than or equal to 35%. Standard neuropsychological testing found that patients' performance on each of these tests improved at follow-up. Four patients exhibited increased postoperative apathy that improved during the year following the procedure. In addition, three patients experienced the development of cysts around the target site at five years follow-up: two patients were asymptomatic, the third case was associated with radionecrosis. The majority of patients returned to work and/or school and at the 20-year follow-up were leading productive lives as physicians, judges, writers, engineers and other professions, all of which required intact executive function. Two additional recent reports of OCD patients with thermal capsulotomies from Eastern Europe have documented capsulotomies efficacy and safety with no impairment seen in frontal function (Csigó et al. 2010; Krámská et al. 2021).

Kim et al. (2018) reported on the use of high-intensity focused ultrasound to make ventral capsulotomy lesions in 11 OCD patients. At 12 months, six (54.5%) patients were responders and three (27.3%) patients were partial responders. At 24 months, six patients were responders, two (18.1%) were partial responders, and one had achieved full remission. The mean Memory Quotient score improved significantly across the24-month follow-up period: $F3, 6.5=236.3$, $p<0.001$. In addition, no significant changes were observed in K-WAIS, COWAT, Stroop, or Digit Span scores. Davidson et al. (2020a) created a single 7 mm lesion using high-intensity focused ultrasound to study the cognitive effects of a single lesion in the anterior capsule in ten patients with refractory OCD or depression. They followed patients at 6 and 12 months, utilizing tests of executive function, memory, and processing speed. Patients endorsed fewer symptoms of apathy at 6 and 12 months and fewer overall frontal symptoms at 12 months. Kim et al. also used high-intensity focused ultrasound to make lesions placed in the same location as the gamma knife lesions in the Rasmussen et al. study: seven (58%) of the 12 patients showed a greater than 35% drop in the Y-BOCS, no adverse cognitive effects were noted at 6- and 12-month follow-up with improvement in the Memory Quotient Scale and no change in frontal measures.

Following the gamma knife lesion studies that targeted the ventral half of the anterior limb of the internal capsule (vALIC), DBS that targeted the same white matter tract was found to be beneficial in three of four cases of intractable OCD (Nuttin et al. 1999). Since then, DBS of the ALIC (Abelson et al. 2005) or neighboring targets (i.e., the ventral striatum or nucleus accumbens, a subregion of the ventral striatum) have shown response rates in the range of 40–70% (Goodman and Alterman 2012; Goodman et al. 2010; Greenberg et al. 2010). In 2010, the FDA approved a Humanitarian Device Exemption for vALIC DBS in intractable OCD. Recently, progress has been made in trying to define more precisely the anatomy of exactly where these fibers run in the capsule in macaques. High-resolution diffusion tensor imaging was combined with anterograde and retrograde tracers in the same animal and then used to extrapolate to high-resolution diffusion tensor imaging in humans (Haber et al. 2020; Jbabdi et al. 2013).

The optimal "target" for the DBS electrode or lesion has been a matter of debate. Some studies have focused on deep gray matter structures (e.g., the ventral striatum, nucleus accumbens, or bed nucleus of the stria terminalis) as critical mediators of response (Luyten et al. 2016). Others have suggested that these nuclei are useful guideposts, but that the white matter fibers connecting PFC and thalamus, which course through the vALIC superjacent to these nuclei, are critical as they convey the influence of neuromodulation to the wider symptomatic network (Figee et al. 2013). The fact that DBS targeting similar white matter pathways in disparate brain regions (e.g., ventral capsule/ventral striata, subthalamic nucleus) achieves comparable results provides support for the white matter hypothesis. Li et al. (2020a) analyzed data from four cohorts of patients (N = 50) who underwent DBS targeting at either the ALIC, nucleus accumbens, or subthalamic nucleus and identified a specific white fiber tract that was associated with optimal clinical outcome. This bundle connects frontal regions directly to the subthalamic nucleus and may represent a unified connectomic target for successful clinical response to DBS in OCD. However, as noted by Robbins et al. (2019), while DBS in the vALIC led to improved mood, DBS in the subthalamic nucleus site significantly improved cognitive flexibility.

Converging evidence suggests the ventral internal capsule white matter tracts connecting the rostral cingulate and ventrolateral PFC to thalamus and brainstem are the optimal target for clinical efficacy across multiple DBS targets for OCD. Recently, Cui et al. (2023) examined which prefrontal regions and underlying cognitive processes might be implicated in the effects of capsulotomy by using both task fMRI and neuropsychological tests to assess OCD-relevant cognitive mechanisms known to map across prefrontal regions connected to the tracts targeted in capsulotomy. Post-capsulotomy OCD subjects showed improved OCD symptoms, disability and quality of life, and no differences in cognitive task performance on a battery of executive, inhibition, memory, and learning tasks. Task fMRI revealed post-capsulotomy decreases in the nucleus accumbens during negative anticipation, as well as in the left rostral cingulate and left inferior frontal cortex during negative feedback. In spite of these lesions, there were remarkably few changes in cognitive function, particularly given the overall therapeutic impact on OC symptoms. These data suggest that we may be looking in the wrong place for deficits. The prefrontal network that underlies the social brain, that involves the discrimination of social context, language, and action, is one place to focus. There have been almost no studies of the effect of capsulotomy or DBS on complex context-related social decision making, real-world planning, or probabilistic approach avoidance paradigms. Some of the astute clinical observations made of behavioral changes following prefrontal lobotomy may provide additional clues.

The symptoms of OCD involve a complex interaction at the interface between emotion, cognition, and action. Freeman commented on that intersection, OCD, and the frontal lobe. (McLardy 1950):

> We have compared emotion to the fixing agent that prevents a photographic image from fading back into obscurity. Remove the emotion and the image gradually fades. In the obsessive state, prefrontal lobotomy reduces or abolishes the feeling tone attached to the obsessional ideas. The ideas continue and the compulsions often last a long time but the anxiety or tension associated with them is no longer present. One patient said it is as though the painful idea which used to be in the center of the circle of my attention has receded to the periphery.

There is an interesting parallel between the subjective experiences of patients who underwent cingulotomy for pain and those who had a similar procedure for OCD. Both report that the awareness of pain or obsessional anxiety continued to be present but that it somehow did not bother them as much; it was easier to divert the obsessional thought or pain into the periphery of their attention. Interestingly, obsessional patients who respond to serotonin reuptake inhibitors also reported that they do not seem to feel as strong of an urge to complete the compulsion and that the obsessional cue does not carry the affective weight that it did prior to treatment. Similarly, patients treated with serotonin reuptake inhibitors or surgery often notice they are much less likely to cry or feel strong negative or positive emotions.

De Haan et al. (2015, 2017) have made a careful qualitative assessment of the long-term effects of vALIC DBS on the lived experience and personality of 18 patients with intractable OCD. Many of their observations are eerily reminiscent of earlier lesion studies: some patients reported less concern about the social consequences of self-motivated behaviors and even changes in interest in music and reading. For the most part, patients and their significant others describe these changes as beneficial and allowing them to grow into their "true selves." Continued qualitative observation of these patients, in combination with more defined task-based approaches to changes in frontal lobe function, are needed to understand how DBS and lesions effect both symptoms as well as an understanding of self.

NIH-funded studies are underway using next generation DBS devices that can record local field potentials as well as deliver neurostimulation (NCT03457675, NCT03244852). The feasibility of recording local field potentials in OCD patients chronically implanted with a DBS device that can both stimulate and sense was recently demonstrated (Sheth and Mayberg 2023). These types of studies may yield insights into the neural signatures of behavioral states associated with changes in OCD symptom severity. Implantation of stereotactic electrodes designed to find the network associated with compulsive urges or the anxiety accompanying obsessive thoughts are currently in progress. Such future studies will surely advance our understanding of frontal lobe function in humans as well as contribute to our growing understanding of the neural network underlying OCD, anxiety, and depression. The ability to record neural data from patients in their natural environment, time locked with behavior and physiology, offers a unique research opportunity to test hypotheses about the neurocircuitry of OCD, prefrontal brain networks, and the resulting remarkable resilience of the human brain to injury. This hodologic model of frontal function emphasizes the redundancy of cortical function and the importance of white matter cortical subcortical connections, and has been validated with electrical stimulation studies of patients undergoing frontal resections for low-grade gliomas (Duffau 2012).

In summary, advances in imaging, device engineering as well as increased understanding of the anatomy, electrophysiology, and behavior associated with

PFC and its connections are likely to lead to innovative approaches to the treatment of neuropsychiatric conditions like OCD, depression, and anxiety disorders. The relative homogeneity of OCD as well as an emerging consensus about the neural network underlying its symptoms make it a logical place to focus our translational research efforts. Emerging evidence of rostral to caudal continuums, from the abstract to concrete in lateral PFC, cingulo-opercular, and OFC, have implications for our understanding of the abnormalities of goal-directed behavior seen in OCD. The relationship of prospective expected value to overvalued ideation in obsessions, schizophrenia, and delusional disorders merits further investigation. Expanding our understanding of how the prospective expected value associated with future consequences relates to action-outcome monitoring and getting compulsively stuck on motor rituals will be key to developing a working model of the distributed neural network that underlies OCD. These findings should lead to novel hypothesis-driven approaches to treatment. Continued collaborative interaction between basic scientists interested in disease and clinicians interested in basic science is needed to advance the field in this most promising area for future investigation.

16

Translating Prefrontal Cortex Insights to the Clinic and Society

James B. Rowe, Dibyadeep Datta, Christian J. Fiebach, Susanne M. Jaeggi, Conor Liston, Beatriz Luna, Steven A. Rasmussen, Angela C. Roberts, Rajita Sinha, and Suzanne N. Haber

Abstract

The prefrontal cortex (PFC) is implicated in a wide range of neuropsychiatric disorders. Many of these become manifest in adolescence (e.g., anxiety, obsessive-compulsive disorders, addiction, attention-deficit hyperactivity disorders) while others arise from selective neurodegeneration of the frontal lobe in later life. A major challenge to research into the disorders associated with the PFC has been the lack of one-to-one mappings between clinical syndromes, their underlying pathophysiology, and root neurobiological causes. Here, we propose a multilevel framework in which syndromes can be linked to symptom profiles, symptoms to cognitive processes, and cognitive processes to pharmacological and computational processes embedded in PFC and its associated networks. This approach explains the frequency of multi-morbidity of neuropsychiatric disorders. The multilevel framework has enabled animal models of underlying biology and psychological processes to inform the understanding and treatment of clinical disorders without necessitating full recapitulation of the complexity of human neurological and psychiatric disorders. Discussion include the causes and treatment potential of the prefrontal cortical circuit disorders, based on convergent evidence across animal and human studies of the mechanisms of action of lesion, stimulation, pharmacological and cognitive behavioral therapies. Challenges are emphasized in the development, validation, and precision-medicine application of such treatments and

Group photos (top left to bottom right) James Rowe, Suzanne Haber, Dibyadeep Datta, Rajita Sinha, Angela Roberts, Christian Fiebach, Steven Rasmussen, Susanne Jaeggi, Conor Liston, Beatriz Luna, James Rowe, Dibyadeep Datta, Christian Fiebach, Angela Roberts, Beatriz Luna, Suzanne Haber, Conor Liston, Susanne Jaeggi, Rajita Sinha, Beatriz Luna, and Steven Rasmussen

consideration given to the prefrontal systems and prefrontal disorders in the context of global opportunities for education, health and social policy.

The Challenge of Disorders of the Prefrontal Cortex

The PFC is implicated in many neurological and psychiatric disorders, arising from developmental variants, neurodegeneration and focal injury. Despite their diversity of etiology, the clinical manifestations and therapeutic strategies can be understood in terms of systems cognitive neuroscience. In this chapter, we illustrate this approach, drawing on examples from obsessive-compulsive disorder (OCD), attention-deficit hyperactivity disorder (ADHD), addiction, anxiety and depression, schizophrenia, stroke, and dementia.

We propose a layered dimensional framework to study the disorders and guide treatment approaches, mapping between diagnostic groups, underlying symptoms, core cognitive processes and their neuronal mechanisms (Figure 16.1). This provides a parsimonious explanation of multi-morbidity and the effects of stress and development on mental health while opening transdiagnostic insights and treatment potential. We also propose that each level of analysis is associated with gradients across the PFC and its connections. The core cognitive processes and their neuronal mechanisms enable cross-species comparisons and bidirectional translation between animal models and clinical disorders. An additional challenge, however, concerns a principled method to improve the effectiveness of treatments, or combinations of treatments, tailored to individual differences in symptoms and causes. Looking beyond individual treatment, we consider in the final section the advances in prefrontal cortical science in relation to wider societal issues of equity, public engagement, and education.

This approach to the disorders of PFC is agnostic to common but arbitrary professional boundaries (e.g., neurology, psychiatry, psychology, education). We advocate for an interdisciplinary approach, in which mechanisms and treatments in the context of one condition can facilitate the understanding and treatment of another. The benefits of this approach may be apparent especially in mental health and developmental disorders where the genetic, molecular, and lesion bases for disease are less well characterized than in classical neurological disorders.

Mapping Syndromes and Symptoms to Processes and Etiology

Syndromes are defined by a composite of symptoms and signs, each of which are a function of changes in one or more component cognitive processes. These component cognitive processes are in turn the result of, or moderated by, a complex array of underlying neural, metabolic, pharmacological or genetic processes. Figure 16.1 summarizes this analytical framework, using distinct levels of analysis: syndrome, symptom, process, and mechanism.

Examples of the neuropsychiatric syndromes include ADHD, OCD, anxiety disorders, depression, and addiction. Their high rate of comorbidity is not the mere chance intersection of separate pathophysiologies. Rather, it emerges from a finite set of signs or *symptoms and signs* (e.g., anxiety or poor inhibitory and attentional control). Each of these symptoms and signs, in turn, can arise from relative impairments in a small set of fundamental cognitive *processes*, such as response inhibition, set shifting, action-reward association, and fear-conditioning. These cognitive processes are mediated by specific *mechanisms* which can be characterized in terms of neural circuits, neurotransmitters, and genetic variants.

In this multilevel framework, a one-to-one linear mapping from syndrome through to mechanisms is unusual; more commonly, there is divergence and convergence between each level. A structural change in the network mediating

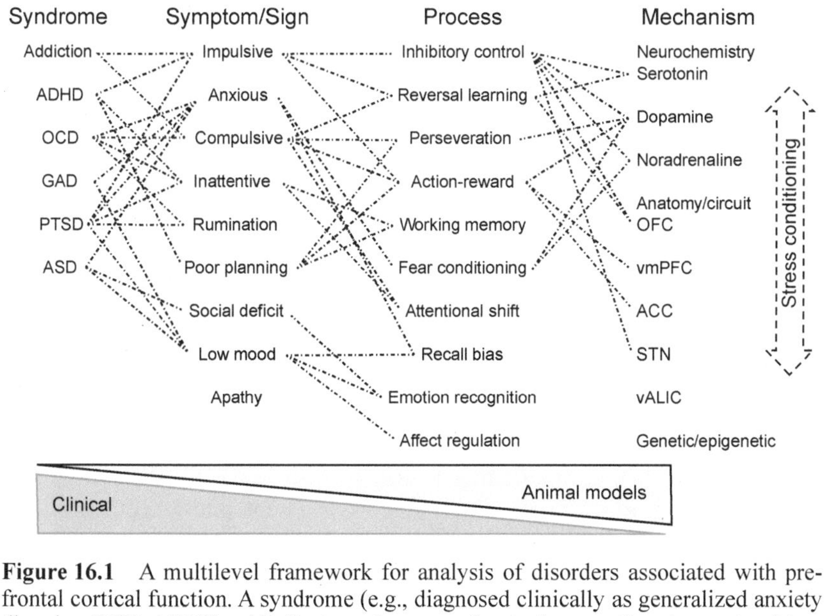

Figure 16.1 A multilevel framework for analysis of disorders associated with prefrontal cortical function. A syndrome (e.g., diagnosed clinically as generalized anxiety disorder, OCD, ADHD, addiction) can be mapped onto the constituent symptom/sign. Symptoms and signs are attributable to a finite set of underlying cognitive processes (e.g., inhibitory control, habit formation, attentional control, cognitive flexibility), which in turn are dependent on specific neurotransmitters and anatomical circuits. The exemplar symptoms, processes, and mechanisms, and their connections, are illustrative not exhaustive. The anatomical and neurochemical substrates are dynamic, with developmental trajectories through adolescence and vulnerability to conditioning effects of stressors, such that risk exposure creates a deferred as well as immediate risk of illness. Clinical studies and animal studies are differentially represented over these four levels, but not exclusively so.

a specific process, or a genetic variant affecting a given receptor type, will have its effect propagated up through the process level, so as to influence many symptoms and therefore contribute to many syndromes.

A corollary of this framework is that animal studies are more readily applicable to the levels of cognitive process and mechanism, whereas clinical studies are more readily applicable to syndromic and symptomatology descriptions. However, the formal linkage between levels increases the potential for translation: to pull preclinical insights forward to understand clinical disorders, and to select appropriate animal models, which we discuss further below.

The manifestations of adult neuropsychiatric disorders are influenced by multifactorial determinants, including processes during embryonic and postnatal development and environmental factors and stressors. These influences can be described by epidemiological associations at the upper levels (e.g., between a developmental exposure and prevalence of a given syndrome). However, to understand the mechanisms of developmental and environmental influences, it is necessary to examine their moderation of the lower levels - their influence on cognitive processes supported by specific circuits, cell types and receptors.

The emphasis on the process level of analysis, rather than by diagnosis or symptom, has some similarity to the RDoC initiative (Cuthbert 2014). Our proposal encompasses the RDoC concept of disease dimensions. One of the challenges, however, is to ensure that studies of human and animal PFC include data/assays on enough of the relevant processes in their task array to enable a systematic and comparative analysis.

Comorbidity

The neuropsychiatric syndromes associated with the PFC are highly heterogeneous. Accounting for this analytically is critical for understanding the role of PFC pathophysiology in modulating the underlying cognitive processes, behaviors, and symptoms. Comorbidities are the rule, not the exception, in population prevalence studies across the life span as well as disease-focused studies (Caspi et al. 2020; Kessler et al. 2003). Having multiple diagnostic labels does not imply the existence of separate diseases or distinct neuropathologies. Rather, multiple diagnoses can reflect different expressions of a single underlying disease entity within an individual (Crossley et al. 2014; Drysdale et al. 2017; Goodkind et al. 2015; Tokuda et al. 2021; Xia et al. 2018)

There are two main challenges to progress in understanding the mechanistic basis of multiple diagnoses. First, there is typically a gulf between studies with extremely large sample numbers but very limited phenotyping: genome-wide association studies often consist of $n > 10{,}000$ whereas studies with deep phenotyping consist of a much smaller number of cohorts, typically $n < 50$ for neuroimaging and bespoke psychophysical tasks. The former have the scale required to identify the cumulative effect of multiple weak risks, whether genetic polymorphisms or environmental exposures, but often lack the range of questions or tasks required for deep characterization of the underlying cognitive processes and neural mechanisms. The latter use in-depth tools sufficient for mechanistic detail but lack the scale to identify small effect sizes of risks factors and moderators that underlie individual differences. In principle, large-scale deep phenotyping is possible. Detailed assessment of prefrontal structure and function has been attempted with $500 < n < 5000$ in studies such as the ABCD, ALSPAC, CamCAN, and IMAGEN (Barnett et al. 2007; Shafto et al. 2014; Volkow et al. 2018; Whelan et al. 2012). Still, new studies are required with even larger samples and deeper phenotyping to enable (a) data-driven approaches to resolve heterogeneity together with (b) theoretically informed hypothesis testing.

Second, there is a paucity of longitudinal studies of prefrontal cortical function and disorders associated with pathophysiology of PFC. Longitudinal rather than cross-sectional studies are less vulnerable to cohort differences such as intergenerational differences in schooling, nutrition, or social media. Longitudinal studies are also more suitable for the analysis of causality (e.g., via mediation analysis). These are particularly important given the dynamic nature of cognitive and neural development through adolescence and incidence of diagnostic expression of neuropsychiatric disorders. The influence of sex and gender differences on brain, cognitive, and clinical development through adolescence highlight the challenges for cross-sectional data in understanding the origins of neuropsychiatric disorders.

Larger studies increase the power of data-driven methods to study comorbidity. For instance, despite the multiplicity of neuropsychiatric diagnoses, psychopathology may have a very low dimensionality in the population. This can be summarized as a single dominant "P-factor" or small set of dimensions revealed, for example, by principal components analysis or confirmatory factor analysis (Sprooten et al. 2022). A core deficit (or psychopathology spectrum) would explain the clustering of disorders, within individuals as well as families. Where larger studies have gathered genetic or neuroimaging data, the dimensions of diagnostic comorbidity map onto common neural and genetic dimensions. Similarly, low dimensionality of neuropsychiatric symptom manifestations and corollary prefrontal structural change is observed with frontotemporal lobar degeneration syndromes (Murley et al. 2020). This calls for a transdiagnostic approach, to which we now turn.

Comorbidity and Transdiagnostics

The orbitofrontal cortex (OFC), anterior cingulate cortex (ACC), ventrolateral prefrontal cortex (vlPFC), and insula are strongly connected (Öngür and Price 2000), as part of a circuit that mediates value-encoding and goal-directed behaviors (Haber and Behrens 2014; see also Figure 16.2). Their association with these cognitive processes suggest that any of diverse pathologies affecting

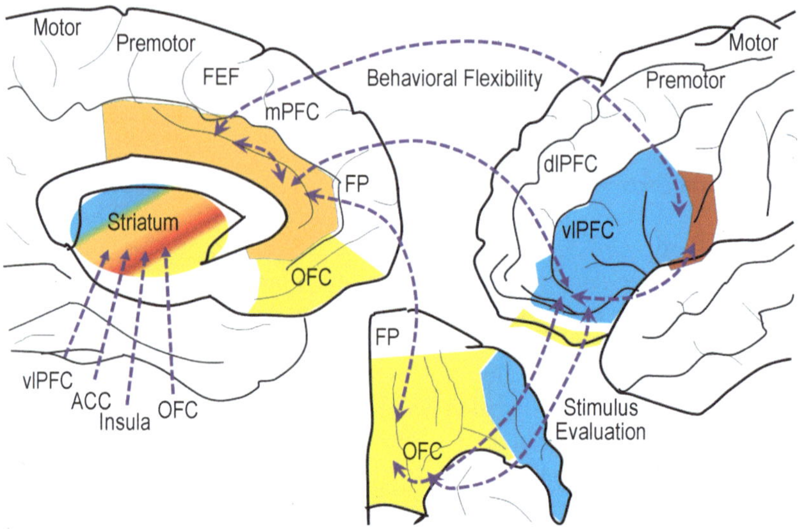

Figure 16.2 The network in which abnormalities are associated with several mental health disorders (Haber and Behrens 2014). OFC (yellow), ACC (orange), vlPFC (blue), and insula (brown) are strongly connected with each other directly and to the striatum. This network mediates value-encoding and goal-directed behaviors. Its association with these fundamental cognitive processes suggest that diverse pathologies impacting on these connections are likely to be associated with diagnoses with overlapping signs and symptoms.

these circuits are likely to be associated with overlapping signs and symptoms. A consequence of the disruption of such circuits is that insights into the mechanisms and cognitive processes associated with the circuit will be of relevance to multiple clinical disorders. This provides a strong motivation for the transdiagnostic approach to understand and treat disorders: clusters of diseases identified under "comorbidity" lend themselves to similar treatment with multiple benefits. The same drug (e.g., an SSRI), same target (e.g., noradrenergic alpha2 receptors), or same surgical site (e.g., capsulotomy) may have cognitive benefits for people with any of a wide set of diagnoses.

Therapeutically effective targeting does not necessarily require resolution of the "injury" or abnormality, merely the recovery of function of the system as a whole. Obsessions, for example, may have different neurocognitive antecedents in OCD and frontotemporal dementia, or depression may have different antecedents in stroke, adolescents, or aging populations (Costello et al. 2023). Nonetheless, there may be a common treatment for the symptom, despite variation in underlying processes or mechanisms, especially where the treatment targets convergent frontal cortico-subcortical circuits (Rasmussen, this volume; Greenberg et al. 2003).

Despite homologies in the anatomy and pharmacology of parallel frontal cortico-subcortical circuits, there is a rostro-caudal gradient in the local intra-cortical connections in PFC (see Murray et al., this volume). This means that information can transfer rapidly between the OFC, medial frontal, and lateral frontal areas of PFC and converge on polymodal areas of the PFC (Figure 16.2). The proximity and strength of connectivity among these regions means that the temporal separation of the signals is very short, approximately 20 msec. This short latency implies highly efficient parallel processing rather than sequential or independent functions. By these routes, information on object recognition can be associated with hippocampal, insular, and amygdala representations of current and past value experience. The expected and future value, encoded in ventromedial and orbitofrontal cortex, can be shared with dorsal ACC, whereby action selection and monitoring are influenced directly by emotion and expected action outcomes (Shenhav et al. 2016).

Stress and Trauma

The frontal lobes are critically involved in adaptive function, comprising the major foci for facing and adapting to challenging and novel environments. Focusing on a goal in the face of challenge and stress can draw on several strategies. To adapt to unstable environments and avoid dangers, one may use executive cognitive skills such as planning, problem solving, switching between subgoals and generating options, or redirecting attention. People may also take action to seek emotional support or reduce effort/costs by accepting things one cannot change. Each of these strategies has been associated with the PFC. Highly stressful situations and traumatic events may overwhelm this ability of the PFC and its networks to optimize goal-directed behaviors. Stress impairs dynamic flexibility and responsiveness, with a shift to habitual or sensorimotor responding (Roberts 2011). This may occur in acute events that are threatening, challenging, uncontrollable, and unpredictable and may include the maladaptive phenomenon of "shutting down." Similar failure of PFC adaptive mechanisms may occur in response to chronic adverse, uncontrollable, or volatile situations in which there are no clear options. High chronic stress goes beyond adaptive "healthy" stress responses with OFC and hippocampal changes that relate to physical and psychological health symptoms (Seo et al. 2014).

Stressful and traumatic events for humans are common including, for example, physical, sexual and emotional abuse and neglect in children (e.g., Centers for Disease Control and Prevention Behavioral Risk Factor Surveillance System[1]), domestic violence, assaults, loss of close relationships (by death or divorce), or loss of one's home due to war, migration, or climate change. In a general population survey of 24 countries, 70% of respondents reported having

---

[1] https://www.cdc.gov/brfss

experienced at least one traumatic event, and over 30% had experienced multiple events (Benjet et al. 2016).

To understand the impact of major stressors on the risk and expression of neuropsychiatric disorders, we need to consider their influence on the processes underlying symptoms and the neural circuits that mediate those processes. As illustrated in Figure 16.1, stressors may condition multiple PFC-mediated processes and therefore be indirectly manifest in the increased risk of multiple disorders.

Stressors in early development years, or in adulthood, change the structure and connectivity of PFC in terms of structural gray matter volume reductions and connectivity as well as functional brain responses to stress (Bartholomeusz et al. 2013; Chen et al. 2018; Goldfarb et al. 2020; Hanson et al. 2012, 2021). The effect of stressors is not uniform across regions: changes are especially common in OFC, ventromedial, rostral ACC, dlPFC, and their immediate connections to striatum and insula (Ansell et al. 2012). Animal studies of stress show consonant changes in homologous or analogous regions to the human studies (discussed further below). The global COVID pandemic provided a "natural experiment" to study the impact of compound stressors, and there is emerging evidence of post-pandemic increases in the rates of addictive behaviors (e.g., alcohol, cannabis, illicit drug use, gambling), anxiety, eating disorders, and other maladaptive behaviors. This may reflect the effects of stress on long-term function and plasticity of the PFC.

Different stressors may act divergently or convergently. Some of the clearest evidence comes from the effects of violence and trauma, with recent data on social deprivation (Dash et al. 2023; Pollak et al. 2010; Xiao et al. 2023). However, further characterization of other stressor effects is required. The greater the stress from an event or condition, in terms of uncontrollability, unpredictability, acuity/intensity, and chronicity (relentlessness), the greater the deleterious effect on the PFC. Moderate levels of stress can be advantageous for learning, memory encoding, and cortical plasticity. However, nonlinearity of dose-response relationships applies to the effect of stress as much as the effect of selective monoaminergic medications.

There are multiple mechanisms by which stress affects prefrontal processes, including changes in dopamine, noradrenaline, cannabinoids, and corticotrophin-releasing factor receptor modulators (Cools and Arnsten 2022; Datta and Arnsten 2019; Tomassini et al. 2022; Uliana et al. 2023). Physiological circadian oscillations in glucocorticoid signaling are critical for supporting developmental pruning and learning-induced plasticity (Liston et al. 2013; McGaugh 2004), whereas severe stressors and chronically elevated glucocorticoids in humans and animal models lead to excessive synapse pruning, dendritic atrophy, and associated cognitive deficits (Izquierdo et al. 2006; Liston and Gan 2011; Liston et al. 2011; Liston et al. 2009; McEwen et al. 2015). Macroscale human neuroimaging shows loss of prefrontal flexibility under high acute stress, affecting ventromedial, orbitofrontal and dorsolateral cortices (Sinha et al. 2016).

Other monoamine neurotransmitter systems may mitigate the effects of stress. For example, serotonin is an important regulator of cognitive flexibility and adaptive responses to negative feedback in human, nonhuman primate, and rodent models (den Ouden et al. 2013; Roberts 2011). Serotonin also interacts with the HPA axis to regulate sleep, appetite, social interactions, and mood, thus indirectly influencing the response to stressors. However, individual differences in serotonergic mitigation of stress involve a complex interaction of genetics, neurochemistry, and behavior.

The prefrontal cortical consequences of stressors are linked to diverse rather than selective cognitive processes: each of these processes may, in turn, lead to a common set of symptoms, such as anxiety. For example, stress-related effects on PFC alter working memory, motor control, and cognitive control. The acute induction of stress in otherwise healthy individuals has been used in addition to the post-stress evaluation of chronically stressed individuals and those with established psychiatric disorders (Luo et al. 2018; Seo et al. 2013). Stress-related symptoms and signs can be classified as cognitive (forgetting, working memory, attention, rumination, negative bias), behavioral (habitual, maladaptive behaviors, avoidant and repetitive behaviors), emotional and affective (anxiety, hyperarousal), and physical health (e.g., sleep, food intake, pain, gastrointestinal distress). The mechanisms by which these signs and symptoms emerge are beginning to be characterized. Such circuit-level changes underlying anxiety (sACC), pain (vmPFC, dACC, insula), gastrointestinal symptoms (ventromedial and orbitofrontal), and behavioral decisions (ventromedial and orbitofrontal) (Dundon et al. 2021; Hollunder et al. 2023; Wood and Nee 2023; Zeredo et al. 2019). In the future, more mechanistic studies of this nature would be of benefit. The link to physical symptoms may be mediated by cognitive maladaptive changes, especially of functions related to PFC (Atlas et al. 2014; Eijsbouts et al. 2021; Woo et al. 2017).

The effects of stress on PFC function may not be immediately apparent. Stress may provide an enduring "first hit" that alters the future susceptibility to a "second hit," whether that second occurrence is another stressor or a distinct neurobiological injury. In other words, stress affects long-term resilience of the cortex. Multiple hits by cumulative or sequential stress exposure has dose-dependent effects on gray matter volume. It changes functional responsivity of PFC to adaptive stress with progressive loss of resilience and increasing risk for stress-related illnesses. A multiple hit may also be seen in gene-by-trauma exposure effects, such as on the depression and anxiety risks in response to stress (Caspi et al. 2010, 2003). The stress-signaling pathways may themselves be moderated by genetic variants. Further research on repeat or combined stressors is required, especially in relation to periods of higher vulnerability during child and adolescent development.

Summary

Many neuropsychiatric and neurodegenerative disorders are primarily associated with deficits in the function of PFC and its subcortical pathways. There is, however, no one-to-one mapping between syndromes and specific symptoms, specific cognitive deficits, and specific root biological causes in terms of gene, receptor, or anatomy. Instead, there is extensive comorbidity and overlapping etiology. This can be understood in terms of a multilevel approach to disease, with convergence and divergence across a wide spectrum of syndromes, in terms of their underlying symptoms, processes, and etiology. This approach accommodates not only the complexity (and weakness) of clinical-pathological correlations, but also the diverse effects of development and stressors.

Gradients across Prefrontal Cortex in Health and Disease

Gradients of the PFC

The structure and functional organization of the PFC is not merely a juxtaposition of discrete entities. Instead, there is a set of intersecting spatially distributed gradients that can be characterized by their direction, content (Badre, this volume; Vertes et al., this volume), or the mechanisms underlying cognitive processes. The content of a gradient may be described in terms of the progression or hierarchy of cognitive processes based, for example, on their complexity, abstractness, or temporal scale. The gradient may also express differences in physiological properties of the neurons, cytoarchitectonic difference, or connectivity patterns, or the spatial patterns of gene transcriptomic variance and receptor density, as illustrated in Figure 16.3.

There is an advantage to analyzing gradients rather than discrete functions of structures, in part due to spatially smooth variance in the biological substrates of prefrontal function, rather than discontinuities. In addition, the effects of common developmental, neuropsychiatric, and degenerative disorders are typically spatially distributed rather than discrete (in contrast to stroke or surgical lesions). The historical emphasis on discrete regions made an important contribution to understanding cortical and subcortical inhomogeneity and maximized the insights from sparse data. It may be tempting to follow Plato, for whom "...our best theories will be those which carve nature at its joints." However, the brain and its disorders are complex. Reducing natural gradients to arbitrary categories is to disregard much of the variance in the biological information used to understand risk and expression of disease. As for other modeling methods, when trying to identify statistical dependencies among continuous variables, it is preferable to retain variance in the model rather than the error terms. Thus, it is important to consider gradients of PFC: how they relate to each other as well as to the dimensions of disease.

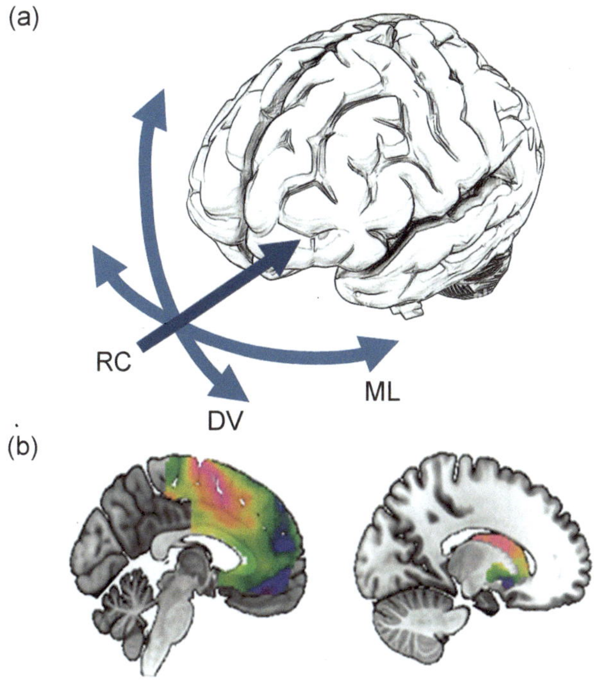

Figure 16.3 Prefrontal gradients are observed receptors, cytoarchitecture, connectional patterns, function, and transcriptomics. (a) Spatially distributed gradients in mechanisms and processes may have rostro-caudal, medial-lateral, and dorsal-ventral directions. (b) Functional gradients may result, for example in representation of switching (hot) and repetition (cool) of abstract and concrete rules in frontal cortex and striatum respectively (Kehagia et al. 2017). Figure 16.3 continues on pp. 330–331.

Morphology and Pharmacology

There are important differences in neuronal morphology, circuit architecture, and physiological properties across the cortical hierarchy (Gilman et al. 2017; Wang 2020). These include factors that promote the persistent neuronal firing that benefits some forms of higher cognitive functions, such as increasing local recurrent circuits with corresponding spine density and increasing numbers of regulatory interneurons (Elston 2000; Elston et al. 2006; Gonzalez-Burgos et al. 2019; Torres-Gomez et al. 2020). Whereas MRI macroscale imaging gradients are associated with transcriptomic variance, there are also transcriptomic gradients across cortical hierarchies of genes that have prima facie relevance to synaptic transmission and plasticity. There is increased reliance on magnified calcium signaling (e.g., calbindin, NMDA GluN2B) as one moves up the neurocognitive hierarchy (Burt et al. 2018).

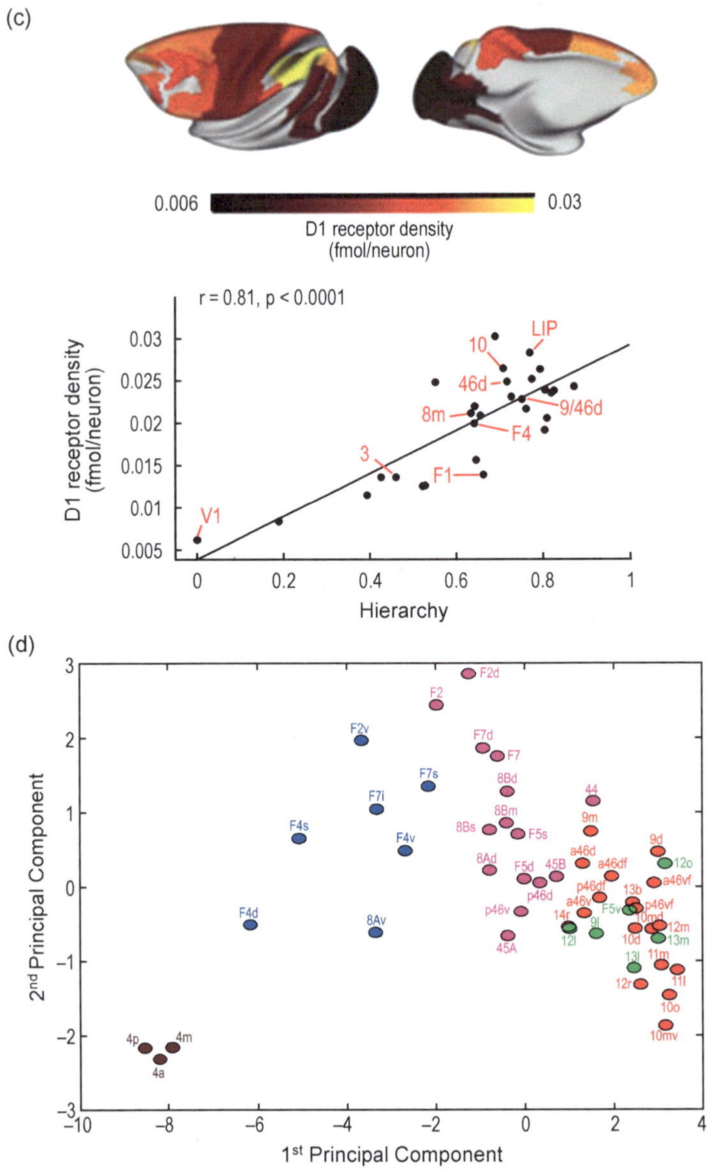

Figure 16.3 (continued) Prefrontal gradients are observed receptors, cytoarchitecture, connectional patterns, function, and transcriptomics. (c) Anatomical gradients of receptor density can be seen across the frontal lobe, illustrated with D1 receptors that control a working memory hierarchy (Froudist-Walsh et al. 2021). From multiple receptor densities (e.g., AMPA, kainate, NMDA, $GABA_A$, $GABA_B$, M1, M2, M3, α1, α2, 5-HT1A, 5-HT2, and D1), multidimensional scaling (d) and hierarchical clustering (e) of "receptor fingerprints" reveal a rostro-caudal gradient over prefrontal cortex (Rapan et al. 2021, 2023).

(e)

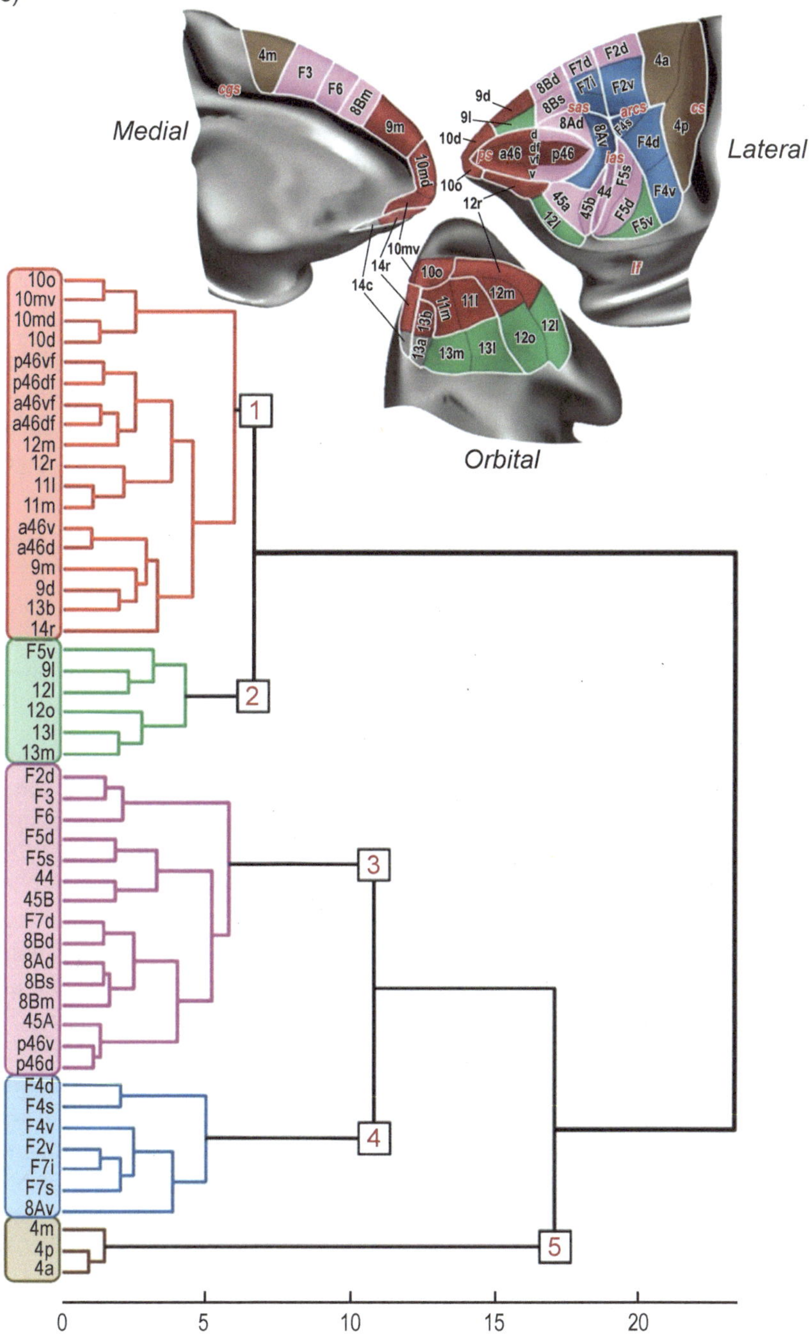
Medial
Lateral
Orbital
4m
F3
F6
8Bm
9m
10md
cgs
9d
9l
10d
10o
12r
10mv
14r
14c
ps
a46
p46
d
df
vf
v
8Bd
8Bs
8Ad
F7d
F7i
F2d
F2v
4a
4p
cs
sas
arcs
8Av
F4s
F4d
ias
F5s
44
45a
45b
F5d
F4v
F5v
12l
lf
10o
11l
11m
12m
13b
13a
13m
13l
12o
12l
10o
10mv
10md
10d
p46vf
p46df
a46vf
a46df
12m
12r
11l
11m
a46v
a46d
9m
9d
13b
14r
F5v
9l
12l
12o
13l
13m
F2d
F3
F6
F5d
F5s
44
45B
F7d
8Bd
8Ad
8Bs
8Bm
45A
p46v
p46d
F4d
F4s
F4v
F2v
F7i
F7s
8Av
4m
4p
4a
1
2
3
4
5
0
5
10
15
20

Similarly, the D1-receptor distribution shows gradients across PFC (Froudist-Walsh et al. 2021). These gradients encompass the multivariate fingerprint based on a large panel of receptors (Rapan et al. 2023), whereby there is a gradual progression of neurochemical functionality from central sulcus to frontopolar cortex (Figure 16.3). Such neurochemical gradients shape the anatomical mediation of psychopharmacological treatments for cognitive and psychiatric disorders and modulate the connectivity of regions.

Connectivity Gradients

PFC does not operate in isolation but acts via partially dissociable cortical-subcortical-thalamo-cortical loops for which the functional properties also form a gradient. These large-scale functional networks vary between individuals (Gratton et al., this volume), and the integration of network perspectives with the processes associated with symptoms can elucidate individual differences in vulnerability, resilience, or treatment opportunities. Connectivity gradients have been demonstrated at different levels of analyses:

1. Cortico-cortical connections based on cytoarchitectonic organization (Goulas et al. 2018),
2. Spatial gradients in which there is high connectivity between adjacent cortical areas that decreases with distance, and
3. Anatomic functional connectivity, which creates links, for example, limbic to cognitive to motor regions (Tang et al. 2019; Trambaiolli et al. 2022).

An example of the latter is the ACC, an area of particular interest for its association with depression, anxiety, and OCD. The ACC is anatomically heterogeneous and can be divided into subgenual (sACC), rostral (rACC), and dorsal (dACC) regions (Morecraft et al. 2012; Öngür and Price 2000). The sACC and vmPFC, which also includes ventral area 10 and 14m, are a central part of the motivation network. The vmPFC is strongly connected to OFC, amygdala, rACC, and the shell of the nucleus accumbens (Haber and Behrens 2014). It supports visceral and emotional functions in motivation (Alexander et al. 2019, 2020; Woods et al. 2023) and is critical for determining value (Camille et al. 2011a; Jocham et al. 2012; Kolling et al. 2016b). The sACC is tightly connected to the rACC, which in turn is connected with the dACC, dlPFC, and vlPFC (Tang et al. 2019). The rACC is associated with cognitive control and choice of action (Kolling et al. 2018). Caudally, the dACC is connected with the action network consisting of motor control areas, including frontal eye fields and premotor areas (Morecraft et al. 2012; Öngür and Price 2000). The dACC is associated with motor planning and action execution (Caruana et al. 2018; Picard and Strick 1996). Thus, through these anatomic connections, the ACC can use value-based information to help regulate flexibility, adaptation, and top-down control (Etkin et al. 2015; Kolling et al. 2016b; Shenhav et al. 2016).

Importantly, there are no clearly defined borders between these three anterior cingulate divisions based on their anatomical connections. Instead, there is a gradual transition in the information content in the projections, gradually changing from limbic to cognitive and finally motor systems (Tang et al. 2019). Cortico-striatal and cortico-thalamic connections follow a similar gradient. Thus, although frontostriatal projections are organized in a general functional topographic manner, forming a ventromedial/dorsolateral gradient, there is a great deal of overlap between projections from these different areas. For example, inputs from OFC, sACC, and rACC converge extensively in the medial striatum. rACC, dorsal ACC, and OFC fibers converge with those from the dlPFC and vlPFC in more central caudate and putamen regions, particularly at rostral levels. Hence, cortical connections from distant regions converge within the striatum (Averbeck and Costa 2017; Giarrocco and Averbeck 2023). These areas of convergence are likely important regions for integrating information across diverse functional domains.

The concept of functional networks predate modern-day technical developments and maps. In the 18th century, Franz Joseph Gall recognized the importance of white matter connectivity between brain regions that were assigned specific functions (Zola-Morgan 1995). In the 19th century, Carl Wernicke thought the connectivity between brain regions, rather than location, was central to function (Catani and Ffytche 2005). In the mid-20th century, Norman Geschwind supported the notion that higher cognitive functions depended on a combination of localized function and their connectivity, leading to the idea that the brain was comprised of complex anatomic networks supporting cognitive and emotional processes (Geschwind 1965). More recent advances in neuroimaging have been combined with graph theory approaches to define brain networks. Whole-brain functional magnetic resonance imaging (fMRI) networks have been subdivided into functionally specialized resting-state networks, many of which include the PFC such as the default mode network, frontoparietal control and attention networks. Within such networks, a subset of regions serve as "hubs" to bring information together, either within or between networks. The term "hub," first coined by Marsel Mesulam to describe transmodal cortical areas that serve as anatomic and computational epicenters for large-scale cognitive networks, is now used in human network analyses to describe specific regions that serve as information integration centers (for review, see Haber et al. 2022). Such hubs are dynamic over the life span, with prefrontal hubs stabilizing in adolescence in concert with maturation of many cognitive systems (Hwang et al. 2013; Marek et al. 2015; Satterthwaite et al. 2013). Although important for efficiency of integrative processing, hubs also create vulnerability for dysfunction (Bassett et al. 2018; Crossley et al. 2014).

Cognitive Gradients

PFC can be viewed as a gateway to therapeutic interventions. Behavioral, pharmacological, and target-specific invasive and noninvasive interventions, such as deep brain stimulation (DBS) and transcranial magnetic stimulation (TMS), need to be understood in terms of the mechanisms and circuits of underlying cognition. For example, executive functions can be summarized as belonging to three principal groups:

1. Working memory: the ability to maintain task-relevant information over brief periods of time and manipulate this information if necessary
2. Cognitive flexibility: the ability to switch flexibly between tasks and/or goals
3. Inhibition: the ability to resist interference and inhibit inappropriate actions and behaviors

Despite behavioral evidence for such a functional fractionation, functional neuroimaging in humans remains equivocal on the strength of a corresponding functional-neuroanatomical dissociation. For example, large-scale quantitative meta-analysis of 193 functional neuroimaging studies indicated largely overlapping brain systems for these three "core" executive functions, spanning wide areas of lateral and medial PFC and their subcortical connections (Niendam et al. 2012). This does not mean that the PFC is undifferentiated. Behavioral evidence suggests that inhibition may contribute to tasks primarily designed to probe working memory and cognitive flexibility, and such a common executive function might be captured by prefrontal multiple demand systems (see Duncan and Friedman, this volume). Gradient models offer a parsimonious account of PFC that accommodates both task commonalities (the apparent co-localization in multiple demands) and smooth functional variation along axes of anatomical organization, with heterogeneity associated with variation in cytoarchitecture, connectivity patterns, and neurochemistry.

The direction of a cognitive gradient may lie along dorso-ventral, rostro-caudal, or medio-lateral axes (see Figure 16.3). For example, on lateral PFC, there is a gradient of organization as one records progressively more rostrally, in terms of activity or connectivity (Badre, this volume; Badre and D'Esposito 2007). The cognitive processes associated with this rostro-caudal gradient in response and connectivity have been described in several terms. Hierarchical control models have been proposed for lateral PFC, according to different types of representations or control signals that vary in the degree of abstraction (Badre 2008). The rostro-caudal gradient may also reflect a functional hierarchy in the timescales across episodic, contextual, and event-based determinants of behavioral decisions. While posterior regions control behavior and actions driven primarily by direct motor affordances of a current stimulus, mid-rostral regions are associated with more abstract cognitive control (e.g., contextual control of stimulus-driven behaviors according to transient abstract task sets), and more rostral regions mediate controlled behavior depending on past experiences or future long-term goals (episodic control; Koechlin and Summerfield 2007). The highest level of behavioral control, often attributed to the frontopolar cortex, has been associated with the management or monitoring of multiple goals and subgoals in parallel (Mansouri et al. 2017).

The temporal scale of cognitive processes also maps onto a spatial gradient of PFC. This is seen in the temporal dynamics of intrinsic fluctuations in neuronal spiking in nonhuman primate and human cortex, whereby sensory cortical areas have shorter timescales and PFC association areas have longer timescales (Demirtas et al. 2019; Murray et al. 2014). Such a gradient in temporal dynamics influences the cognitive-physiological properties supported across the gradient. For example, primary visual cortex (V1) requires a short timescale to accurately decode the onset and offset of a visual stimulus, while sensory association cortices (e.g., MT/V5 or LIP) use longer timescales to integrate and analyze information to facilitate recognition, and dlPFC uses still longer timescales to maintain and manipulate information for many seconds without sensory stimulation (Funahashi et al. 1993b; Leavitt et al. 2017; Wang and Krystal 2014). Lateral and medial prefrontal rostro-caudal gradients also reflect the temporal span of task-relevant representations (e.g., immediate action, contextual task set, episodic influence and enduring normative social rules) and temporal extent of influence of motivational signals (immediate rewards, context-dependent motivational signals, longer-term episodic goals) (Kouneiher et al. 2009; Wood et al. 2023).

Control demands may vary along the ventral-to-dorsal axis. The classical proposal of the organization of working memory systems in lateral PFC is that ventral regions host the sustained maintenance of task-relevant information, whereas dorsal regions are engaged when cognitive load increases beyond capacity limits or when actions are required on working memory contents (manipulation, updating, selection; cf. D'Esposito et al. 1998b). A dorsal-to-ventral axis is observed along the medial prefrontal cortex, as tasks or their underlying representations vary in the degree of emotional control (vmPFC, sACC, and pregenual ACC) or cognitive control including the monitoring of conflict and uncertainty (dorsal ACC) (Bush et al. 2000; Sheth et al. 2012).

The lateral-to-medial axis has correlates in the processing of value signals, with differential responses to negative (punishment) versus positive (reward) value (Kringelbach and Rolls 2004), that may guide avoidance versus approach behaviors. A medial-to-lateral gradient has also been proposed for the degree to which lateral regions are oriented toward external states and goals while medial PFC is oriented to internal states (e.g., Denny et al. 2012). On this basis, frontopolar cortex might be involved in switching between such externally versus internally guided controlled behavior (e.g., the gateway hypothesis; Burgess et al. 2007).

The existence of orthogonal gradients creates a "matrix" of PFC functions with which to understand the nature of prefrontal deficits in neuropsychiatric disorders. A very large set of regions with specific properties can be efficiently created from a small set of macroscopic gradients: each conjunction of gradients defines areas with apparent "localization" of functions, leading to apparent localization of the correlations with symptom, such as contextual control signals in lateral prefrontal cortex (Barbalat et al. 2011). A hierarchical organization of cognitive control may result in asymmetric deficits, such that impairments in episodic control (e.g., due to traumatic experiences) may indirectly impact hierarchically "lower" stages of contextual or sensory control, even though these in themselves could be unaffected (e.g., at the level of brain-structural integrity or neurochemical modulation). Understanding cognitive contributions to psychiatric disorders at such a fine-grained level of resolution requires a systematic approach to experimental psychopathology research with new classes of experimental paradigms built on cognitive control theory. It has the potential to link cognitive phenotypes of a disorder to underlying mechanisms, not only in terms of local effects but in terms of the statistical dependency between cognitive, physiological, and pharmacological gradients.

Brain imaging by structural and functional MRI often contains graded information, with graded rather than discontinuous variation in activity or connectivity. Unfortunately, published brain imaging maps are typically thresholded, creating the impression of discrete functional areas. To get around this limitation, the raw data or unthresholded maps should be shared. An alternative approach is to use statistical tools that express gradients in structural and functional imaging data (Bethlehem et al. 2020). Such system-level gradients are not restricted to atrophy or fMRI connectivity but can be generated for microstructural differentiation so as to reveal the pattern of change in adolescence or aging (Bethlehem et al. 2022b). These gradient mapping methods are well suited to characterize multidimensional hierarchical functional systems. These gradients are not restricted to imaging modalities but can be directly linked to spatial variation in receptor density or gene expression, linking the macroscale imaging of disorders to genetic regulators of neurons, glia or endothelium (Altmann et al. 2020). Across multiple neuropsychiatric disorders, the spatial patterns of cortical anatomy changes in adolescence correlate with spatial expression of copy number variation genes in neurotypical adults. Such genetic gradients provide a mechanism to mediate the mapping of genetic risk onto regional brain changes in neurogenetic disorders (Seidlitz et al. 2020). They are likely to contribute to the strong polygenetic influence on developmental trajectories of brain structure and connectivity (Bethlehem et al. 2022a) and establish developmental gradients.

Developmental Gradients and Critical Periods

The dynamic nature of PFC during development confers a particular risk to disruption and, in turn, increased risk for psychopathology. The developmental timing of stress exposure is similarly important. These prefrontal cortical circuits are undergoing significant specialization during adolescence, including decreases in frontostriatal (Parr et al. 2021) and fronto-amygdala connectivity (Jalbrzikowski et al. 2017) and increases in fronto-hippocampal connectivity (Calabro et al. 2020). Given the sex differences in adolescence and brain development, the age of stress may lead to differential risks of psychopathology in later life. The dynamic nature might also confer resilience to recovery, following the termination of stressors (McEwen 2013).

Development can be seen as a process of accumulation through childhood and long into traditional definitions of adulthood. Cell division, migration, and axonal connections are well established by birth. The brain achieves 95% of adult size and weight by 7–11 years of age, and full adult weight by adolescence (Caviness et al. 1996; Giedd et al. 1996). However, developmental trajectories are not equivalent across the PFC, with peak cortical thickness achieved last in vmPFC and insula/vlPFC (Bethlehem et al. 2022a). During postnatal development synaptogenesis, synaptic pruning and myelination become the dominant means of plasticity (Huttenlocher 1990). Synaptic pruning in PFC begins in childhood and continues into the 30s (Petanjek et al. 2011, 2023). Functional connectivity decreases in frontostriatal and fronto-amygdala systems (Jalbrzikowski et al. 2017; Parr et al. 2021) reflective of dampening of activation from subcortical regions (Murty et al. 2018). Myelination begins during gestation and continues through adulthood. Myelination of sensorimotor tracts is in place by childhood but major tracts that provide connectivity for lateral PFC regions, such as the superior longitudinal fasciculus, mature throughout adolescence. Those providing connections to ventral PFC systems, including the cingulum and uncinate fasciculus as well as myelination of endpoints in the gray matter, continue to mature into adulthood (Lebel and Beaulieu 2011; Simmonds et al. 2014). Myelination is not confined to white matter tracts: using magnetization transfer ratio, layer 5 and 6 of human cortex reveals increases in intracortical myelination up to 24 years of age (Whitaker et al. 2016).

With neuronal maturation comes the development of cognitive abilities. For example, the trajectory of executive function from childhood through to adulthood mirrors anatomical maturation (Luna et al. 2015; Tervo-Clemmens et al. 2023). The efficiency of executive systems increases in parallel: activations of ACC and lateral PFC decreases from childhood to adolescence during inhibitory control and working memory tasks (Ordaz et al. 2013; Simmonds et al. 2017). By adolescence, essential neural systems are in place, with spatial gradients and specialization finessing performance toward the adult level of executive function. A corollary of this development of cognitive abilities associated with PFC is the development of the risks for major psychopathology (see Figure 16.4) (Gogtay et al. 2004; Han et al. 2021; Paus et al. 2008; Solmi et al. 2022; Uhlhaas et al. 2023). Understanding the neural mechanisms of maturation of prefrontal cortical systems may explain the emergence of mental

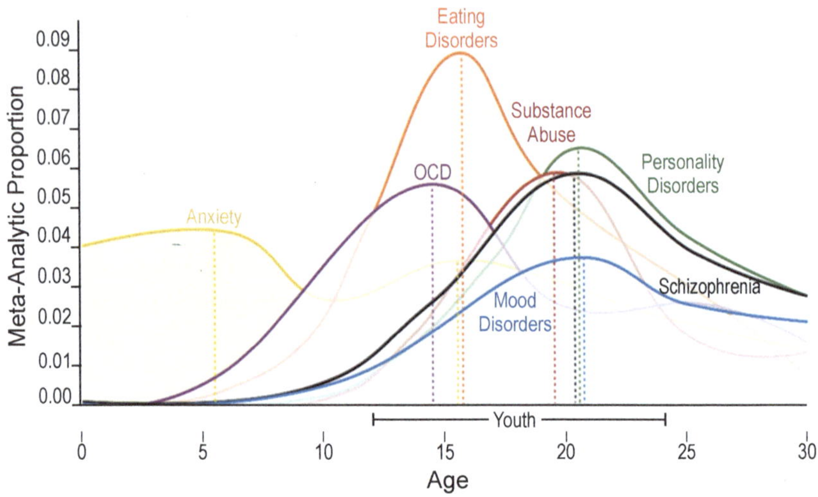

Figure 16.4 The risk of developing neuropsychiatric disorders varies with age and peaks during adolescence as the neural systems underlying the relevant cognitive processes themselves reach maturation (based on a meta-analysis by Solmi et al. 2022).

illness as an expression of a neurobiological predisposition or an impairment of normal developmental plasticity.

The phased maturation of cortical and subcortical circuits creates critical periods for the neural risks of mental health disorders. The expression of psychopathology emerges at different developmental periods: with ASD emerging in infancy, ADHD and initial OCD in early childhood, anxiety in mid-childhood, and psychosis, bipolar, and mood disorders in adolescence. This temporal sequence is influenced by the hierarchical maturation in terms of neurogenesis, synaptogenesis, synaptic pruning, and myelination, thus creating critical windows. The neurobiological basis of critical windows in development has been established most comprehensively for the visual system (Toyoizumi et al. 2013) but similar principles apply to prefrontal cortex. Critical period plasticity is underscored by increases in glutamatergic excitatory function, breaking its balance with inhibitory GABAergic function. This triggers change in inhibitory circuitry, such as parvalbumin neurons that dampen spontaneous excitatory neural activity returning excitatory-inhibitory balance (Dorrn et al. 2010; Hensch and Fagiolini 2005; Toyoizumi et al. 2013). Similar processes occur in animal and human postmortem studies. In adolescence, for instance, GABAergic parvalbumin cells increase (Caballero et al. 2014; Larsen and Luna 2018) in parallel with decreases in prefrontal glutamatergic signaling (Henson et al. 2008; Hoftman et al. 2018). *In vivo* high-field 7T MRI spectroscopic imaging has identified the progression of prefrontal glutamate-GABA balance into adulthood, supporting an adolescent critical period of plasticity (Perica et al. 2022).

The presence of plasticity through adolescence creates a particular susceptibility to environmental influence. For example, a genetic neurobiological predisposition for psychopathology may be more strongly expressed within a stressful environment so as to foster the phenotypic behaviors of diverse mental illnesses. In this way, the mechanisms underlying plasticity need not be impaired as such; they merely need to adjust to experience. Critical periods may vary in duration, with prolonged critical period plasticity or precocious termination, according to glutamatergic and GABAergic systems status and external factors such as stress. Chronic stressors during adolescence decrease excitatory activity and plasticity in frontal cortex in animal models (Novick et al. 2016; Urban and Valentino 2017; Yuen et al. 2012). Similarly, chronic stress in adolescence destabilizes and dampens inhibitory activity and peri-neuronal nets (Bicks et al. 2020; Tzanoulinou et al. 2016).

Stressors in the fetal period reduce critical GABAergic processes (Suwaluk and Chutabhakdikul 2022a, b) and protein phosphorylation affecting prefrontal cortical maturation. This is associated with anxiety and depression as well as risk for mental health disorders later in life. Stress during infancy and childhood also affects prefrontal circuits, such as fronto-amygdalar connectivity (Morin et al. 2020), and altered expression of immediate early genes and myelin-related genes (Blaze et al. 2013; Teissier et al. 2020). By adolescence, stress, especially social stress, affects cortico-limbic regions involved in emotion and stress regulation, including amygdala structure and social circuitry (Godfrey et al. 2023; King et al. 2023). Not all stressors are equivalent in their consequences: rodent and human studies show that short-term acute stress can have enhancing effects on cognition and excitation, whereas long-lasting chronic stress generally dampens excitatory and inhibitory processes undermining critical period plasticity and increasing the risk for psychopathology.

Gradients of Disease Expression and Treatment Outcome

The evidence of cognitive gradients comes from human functional neuroimaging and therapeutic lesion outcomes. The evidence of morphological, receptor, and transcriptomic gradients in nonhuman primates suggests the likely existence of analogous functional gradients. However, the type of gradient based on abstraction hierarchies has yet to be demonstrated in nonhuman primates. This partly reflects the challenge of training and performing multiple tasks in other species. So, while animal models have established the molecular, pharmacological, and microanatomical underpinnings of critical cognitive processes, we can also learn from the syndromic associations of regionally defined disorders and focal interventions.

How do gradients in cognitive hierarchies across the PFC link to neuropsychiatric syndromes? Consider the rostro-caudal gradients in lateral, medial, and cingulo-opercular networks described above. The temporal scaling property along this axis is ideally suited to support the gradient from simple

Pavlovian stimulus-bound value (caudal) to value associated with local context (mid) and to enduring representations of prospective expected value associated with episodic future thinking (rostral). Symptoms related to OCD and anxiety disorders may mirror this spatiotemporal gradient and distinguish, for example, those present at the stimulus level (e.g., a contaminated object), the local contextual level (e.g., holding a knife in the presence of a child associated with aggressive obsessions), or an extended abstract future-oriented consequence (e.g., my parents might go to hell if I don't complete this ritual). The neurobiological basis of symptoms can, in principle, be mapped onto the rostro-caudal gradient, either in the OFC representation of expected value or medial cingulate monitoring of action outcome.

The contiguity of such gradients through cortico-striato-thalamo-cortical circuits can inform the selection of sites for therapeutic surgery or focal stimulation by DBS or TMS. By this means, a very focal lesion may affect the function of a much larger swathe of PFC.

Neurodegenerative Gradients

Thus far we have focused on disorders that emerge during adolescence and young adult life, including the clustered psychopathologies of autism syndrome disorders, ADHD, anxiety, OCD, and addiction. However, phenomenologically analogous syndromes can arise from focal neurodegeneration. Developmental and degenerative disorders are not exact homologues, but they are mutually informative and have critical cognitive and behavioral similarities. This is most evident in the family of syndromes caused by frontotemporal lobar degeneration. This leads to a progressive rostro-caudal gradient of synaptic and neuronal loss, beginning in mid-to-later life. In the behavioral variant of frontotemporal dementia (FTD), for example, there is early synaptic and neuronal loss in insula, orbitofrontal, and ventromedial regions with later progression to ventrolateral and anterior cingulate cortex. Symptoms include repetitive and obsessional behaviors, poor executive function, impulsivity, risk-taking, and cognitive inflexibility. There are additional changes to affective cognition, with loss of social cognitive skills, poor empathy and a reduction of goal-directed behaviors (i.e., apathy). The autosomal dominant genetics, molecular pathology and prominent atrophy in these associated disorders has contributed to their classical designation as "neurological" rather than "psychiatric" disorders. This professional distinction can obscure the phenomenological similarity between behavioral variant FTD and developmental or young adult psychiatric and neuropsychiatric disorders. The genetic risks and structural change may be more subtle with the latter group, but despite the scarcity of autosomal dominant etiology of psychiatric disorders, the heritability of cortical, subcortical gray, and white matter volumes is very high (Bethlehem et al. 2022a).

Other focal and multifocal "neurological" disorders affecting the frontal lobe can lead to similar cognitive and behavioral change, whether from leukodystrophy, stroke, tumors and their excision, inflammatory lesions, or traumatic brain injury. Despite myriad etiologies, the lens of systems cognitive neuroscience can be used to understand the clinical presentations and guide therapy (Passamonti et al. 2018). Not all therapeutic approaches have been through disease-specific randomized controlled trials, but anecdotal reports, case series, and early phase trials support the translational relevance of the schema illustrated in Figure 16.1 to dementias (Holland et al. 2021; Murley and Rowe 2018).

The impulsivity and cognitive inflexibility arising from behavioral variant FTD has several contributory factors. The FTD-related atrophy of ventrolateral and orbitofrontal cortex is associated with impulsivity (Lansdall et al. 2017, 2018), while the loss of induced beta-power from lateral prefrontal cortical microcircuits correlates with everyday challenging behaviors (Hughes et al. 2018). There is also a severe loss of serotonergic innervation of the PFC (Murley and Rowe 2018), resonant with the serotonergic role in perseveration and impulsivity in marmoset and rodent models (Clarke et al. 2004, 2005, 2007; den Ouden et al. 2013). Although the atrophy cannot yet be rectified, serotonergic reuptake inhibition has been shown to partially restore neurophysiological functions of the PFC in FTD (Hughes et al. 2015). Serotonergic reuptake inhibition is widely used in the clinic for challenging behaviors, even in the absence of depression or anxiety. A related frontotemporal lobar degeneration syndrome of note is progressive supranuclear palsy (PSP). In addition to motor deficits, people with PSP are impaired in response inhibition (Zhang et al. 2016), cognitive flexibility (Robbins et al. 1994), social cognition (Ghosh et al. 2012), and goal-directed behavior (Murley et al. 2020). People with PSP have modest atrophy of medial PFC but severe atrophy of subcortical nuclei (locus coeruleus and subthalamic nuclei and pallidum) and severe synaptic loss across the PFC that correlates with clinical decline (Holland et al. 2023). PSP causes early and severe noradrenergic deficits arising from degeneration of the locus coeruleus, leading to impulsivity and apathy (Kaalund et al. 2020; Ye et al. 2023a), in part by loss of noradrenergic-dependent connectivity between prefrontal cortical regions and their subcortical pathways (Tomassini et al. 2022). Given the robust noradrenergic influence on inhibition and set shifting across species (Bari et al. 2011; Chamberlain et al. 2006; Rae et al. 2016; Robinson et al. 2008; Ye et al. 2023a), noradrenergic strategies are now in clinical trials for cognitive and behavioral consequences of neurodegeneration. The noradrenergic hypothesis provides an example of the value of cross-species and transdiagnostic approaches, based on systems cognitive neuroscience: bootstrapping noradrenergic therapies for attentional and cognitive control in ADHD (Elliott et al. 2020), addiction (NCT00218543), Alzheimer disease (David et al. 2022; Eudract 2016-002598-36), and parkinsonism (ISRCTN99462035). Future studies in neurological disorders can draw on new insights into the regulation of PFC and look to ameliorate symptoms through restorative pharmacology aimed at the range of processes outlined in Figure 16.1. The noradrenergic hypothesis also illustrates the direct line of sight from rodent and NHP models through psychopharmacological probe studies in humans with neuroimaging support, and then to clinical therapeutics.

Summary

Much of the complexity of PFC function can be explained in terms of the intersection of gradients. Individual gradients may reflect fundamental neural variance (e.g., receptor density, anatomical connection patterns, and myelination). They may also reflect information content of encoded information and the temporal scales to which they refer. The trajectory of development of these gradients gives rise to critical windows for the risk and manifestation of psychopathology. An important corollary of prefrontal gradients is their cross-species homologies that can inform therapeutic strategies and prediction of outcomes.

Animal Models Related to Human Disorders

Role

There are two broad aims for animal models of human disorders. First, they may seek to recapitulate the pathology (e.g., through genetic manipulation, cytotoxic lesions, pharmacology or environmental insults such as stress). Second, the animal model may seek *construct equivalence*, to study specific symptoms related to particular parts of the pathophysiology or psychopathology of the disease. Gaining such an understanding of the basic neurobiological mechanisms of specific processes, the dysregulation of which lies at the core of clinical symptoms, is of enormous value.

Animal models can be designed and used so as to aid the understanding of human disorders. However, responsibility lies in both directions. Those studying clinical phenotypes also need to ask the right questions and record the right variables with human volunteers so one can learn from the insights emerging from the animal literature. This is especially important for the neuropsychiatric disorders associated with the PFC, where cross-species homologies can be challenging.

There are clear examples of animal models that are helpful in understanding the prefrontal circuitry and its dysregulation associated with neuropsychiatric symptoms:

- The disruptions in goal-directed behavior in rats (Balleine 2019), marmosets (Duan et al. 2021) and macaques (Murray and Rudebeck 2018) that are also seen in people suffering from OCD.

- The effect of cingulate lesions in macaques on monitoring of social consequences, relevant to social phobia (Rudebeck et al. 2008a, b).
- The platform avoidance task of Quirk and colleagues related to OCD and other anxiety disorders that manifest active avoidance as a prominent symptom (Martinez-Rivera et al. 2023).
- Impaired inhibitory control in stop signal reaction time, related to impulsivity seen in ADHD (Eagle et al. 2008a).

Such experimental studies in animals are often better placed to determine whether alterations in activity associated with a particular disorder are compensatory or causal to the disorder and its symptoms: the hyperactivity of orbitofrontal cortex in OCD or hyperactivity of subcallosal cingulate cortex in depression. Overactivation of subcallosal cingulate cortex can induce behavioral changes in monkeys similar to symptoms of anxiety and anhedonia reported in depression (Alexander et al. 2019). Similarly, overactivation of OFC has been shown in rodents to cause compulsive-like grooming behavior of relevance to the compulsivity seen in OCD (Price et al. 2021). Evidence that hyperactivity in a disorder is compensatory may require an experimental second hit (e.g., lesion or inhibitory stimulation), which is usually clinically not advisable.

As stress is a known contributor to the onset of many clinical disorders, another approach in experimental studies in animals has been to study the impact of stress on prefrontal function. For example, diverse types of psychological, social, or physical stressors affect the prefrontal physiology underlying clinically relevant cognitive processes. These include plasticity mechanisms and related behaviors including cognitive flexibility, goal-directed behavior, working memory, and reactivity to negative and positive reinforcers (see Roberts and Liston, this volume). These stress manipulations can recapitulate some of the clusters of symptoms seen transdiagnostically. The psychological or physical nature of the stressor may differentially influence specific prefrontal circuits (Bondi et al. 2008; Danet et al. 2010). Moreover, when these stressors are induced during development, the pattern of behavioral changes seen can also vary depending upon the timing of the stressor. For example, in rats, maternal deprivation in infants produces a different phenotype to social deprivation in juveniles/adolescents, indicating distinct neurobiological substrates for stress-related disorders, depression, and ADHD (Matthews and Robbins 2003; Robbins et al. 1996). This highlights the contribution that animal studies can provide to our understanding of neurodevelopmental processes in general and effects of stress in particular.

We discussed the developmental trajectory of the human frontal lobe, with respect to myelination, synaptic pruning, and circuit connectivity. Analogous trajectories are seen in animals, particularly nonhuman primates (Sawiak et al. 2018; Scott et al. 2016). Even in marmosets, the neural substrate of individual differences in cognitive development can be seen in the trajectories of prefrontal gray matter volume (Sawiak et al. 2018).

There are, however, limits to cross-species comparisons. For example, the sex differences in brain development that are evident in humans have not been reliably replicated in marmosets. This may be true null result in a species or may reflect the obstacles to large studies of nonhuman primates: compare n > 130,000 humans scanned individuals collated by Bethlehem et al. (2022a) with nonhuman primates studies typically n < 10 and rarely 10 < n < 50.

Animal studies can also provide insight into the prefrontal mechanisms that confer vulnerability or resilience to brain disorders. For example, with respect to vulnerability, distinct behavioral traits, such as hyperactivity, poor flexibility, or impulsivity in rats, can lead to different aspects of drug-seeking, drug-taking, and drug-dependency behavior and related prefrontal disturbance, which is of relevance to our understanding of addiction (Belin et al. 2016). On the other hand, rats or mice which fail to display anxiety-like or depression-like symptoms following chronic social defeat stress have the potential to provide insight into mechanisms of resilience (Krishnan et al. 2007). Further insights into resilience can be gained by not excluding non-responders. Some studies may exclude animals that do not express psychopathological responses to stressors, such as social stress. Such natural variation in trait vulnerability offers an important opportunity to determine the mechanisms of vulnerability and resilience (Lorsch et al. 2021; Nasca et al. 2019).

Selection of Models and Tasks to Support Translation

Different species may be better suited to translate specific aspects of disorders associated with prefrontal function, their etiology, and treatment. There are critical decisions for the research team regarding the processes and regions of interest and the nature of any intervention. The complementarity of models rests in part on the intrinsic capacity of species to support a cognitive process in a recognizably homologous cortical area. For example, an animal study of hierarchical representations across the prefrontal gradient, akin to that described by Badre (this volume), requires a species with a highly differentiated dlPFC; in other words, a macaque and less so a marmoset, where the dlPFC is less differentiated, and not a rodent, where it appears nonexistent. By contrast, a study of auditory social interactions may be more appropriate with marmosets. This does not mean a lack of ambition for animal models. Even rats can be used, for example, to study confidence estimates, previously suggested to require "metacognition" and conscious awareness. The decision of species and task to study the relevant process are intimately connected. Complementarity also extends to the mode of intervention: skull morphology, brain size, or nucleus volume may critically determine the feasibility of focal surgery.

The availability of established models of behavior, disease, and risk is an important consideration. For example, the degree to which a physical or psychological stressor is recognized for a given species and the degree to which the animal behavior is interpretable for a given species varies. Even where models (e.g., for stress) exist and are transferable across species, the optimal readouts of the model may differ between species.

A further choice lies in the selection of a task to compare across species. Some tasks have been extensively studied and validated across species, such that the task can be run with formal equivalence in animal and human laboratories. Examples of this type of task are the stop signal task of inhibitory control (Eagle et al. 2008a) and intra/extra-dimensional shift tasks (Chamberlain et al. 2021). These can be operationalized with equivalence across species and have major homologies in terms of functional anatomy and psychopharmacology across mice, rat, marmoset, macaque, and human species. Care is still required to determine the possible differences in cognitive strategies by which an animal or human might approach the same task, because even within a species, there can be differences in the strategy used by an individual. Nonetheless, these tasks have shown how comparisons can be sustained, and they support translation of pharmacological interventions, such as the noradrenergic hypothesis discussed above.

Despite limitations of cross-species homology, animal models offer many advantages. These include experimental methods that are not practical or ethical with human participants, such as the ability to systematically manipulate genetic variants by breeding of traits or CRISPR technology as well as the control of neuronal function by optogenetics or pharmacology using DREADDs (designer receptors exclusively activated by designer drugs). A much wider range of pharmacological interventions is available for animal research, relevant to prefrontal function, such as selective D1 agonists that are not yet available for human use to study working memory systems. Animal models also enable a wider range of readouts than is available for clinical studies, both *in vivo* (e.g., physiological recording or calcium imaging) and postmortem (at any stage of development).

This experimental control over the baseline state of the PFC, before a stressor or drug, is a powerful tool to study and accommodate baseline dependency. For many stressors and pharmacological interventions, response depends markedly on the baseline state of the organisms. For example, the effect of dopaminergic manipulations of impulsivity, risk-taking, and working memory depends on the individuals' baseline performance and baseline dopaminergic function. This contributes to nonlinear dose-response curves and heterogeneous responses to standardized interventions. It may fully obscure the group-wise effect of intervention, unless one controls for individual differences. Such baseline differences are quantifiable in humans but are less easy to control experimentally.

Selection of Clinical Evidence

Animal studies demand critical decisions regarding the selection of model, task, and intervention for them to be relevant to human prefrontal function and its disorders. Likewise, critical decisions are also required of human normative and clinical studies; however, this challenge arguably receives less consideration. Are human studies recording the information required to make use of the data emerging from animal models? We need to rethink not only the approach to animal models of biological processes and behaviors that are relevant in PFC-related neuropsychiatric disorders. It is equally important to optimize clinical trials and human neuroscience studies to ensure that they are recording the information and data types required for integration with insights from animal models. Three principles should guide this work in the future.

First, it will be critical to refine clinical ratings scales to maximize data quality. Clinician-rated scales have some advantages over self-report assessments of psychiatric symptoms, but they typically depend on rigorous training to ensure robust and reproducible results. Conversely, the validity of patient-rated scales may not hold in the context of some PFC-related disorders (Williams et al. 2023). The field would benefit from a greater understanding of the factors that influence data quality, validity, and reliability.

Second, it is important to optimize clinical scales and trial designs to ensure they are quantifying the right variables, especially those that can also be studied in animal models. For example, there has been significant progress in recent years toward understanding the prefrontal circuit mechanisms that regulate reward-seeking, motivation, incentive salience, and effort valuation. These constructs are, however, rarely assessed in detail in large-scale clinical studies. Similarly, it would be valuable to quantify symptoms in multiple domains in a standardized way across different clinical disorders rather than diagnosis-specific rating scales; for example, to assess OCD symptoms, compulsive behaviors, and cognitive flexibility in studies focused on depression, and vice versa.

Third, studies should not rely unduly on subjective clinical scales but also include objective behavioral assessments. These can complement clinical symptom measures. The advantage of the objective behavioral assessments is that they can be designed to capture similar functions across species. This will greatly strengthen the translational bridges across species and models and accelerate the development of clinical therapeutics informed by preclinical model systems with a wider range of methods than can be applied in human studies. One needs to remain mindful of the fact that a human might solve the same problem differently than a mouse or marmoset.

Limitations

Animal models of clinical disorders do not need to be exact homologies to be useful. The closer the approximation to critical clinical phenomenology, the easier it may be to see a pathway for translation from laboratory model to clinical therapeutics. This, however, is not essential, provided that researchers avoid naive interpretations of tasks and look behind the superficial interpretations of clinical phenomena. For example, challenging behaviors from prefrontal cortical degeneration in FTD may be called impulsivity or disinhibition, when in fact they arise from a loss of contextual knowledge to indicate social norms (O'Callaghan et al. 2016; Restrepo-Martinez et al. 2023). In other words, a semantic deficit may be misinterpreted as impulsivity. Similarly, apathy as an observed deficiency of goal-directed behavior may be misinterpreted as depression, even in the absence of a mood disorder. Training and cross-disciplinary collaboration mitigates this risk.

Are any cognitive processes and domains off-limits in animal research? At first glance, it may seem that some human cognitive functions cannot be studied in rodents or even primate models. However, through the adoption of construct equivalences and new model-based approaches, few cognitive domains are out of bounds.

Language may at first seem exclusively human, yet critical aspects of language are amenable. For example, marmosets can be used to study the vocal sensorimotor integration in real time (Pomberger et al. 2020; Takahashi et al. 2015). Also, in nonfluent aphasia, the excessive precision of speech priors in ventral PFC undermining comprehension is part of a wider deficit in predictive coding, which in turn is amenable to preclinical models (Cope et al. 2017; Kocagoncu et al. 2021). While social behaviors may not be manifested in the same way in humans and macaques, there are close similarities in the underlying constructs to enable detailed assessment of PFC regions in social cuing, inference, and behavior. The representations and functional anatomy of face identify, face emotion, eye gaze, rewards associated with social partners, and social decisions establish strong equivalent constructs across species. Moreover, the cooperative breeding style and allomaternal care of marmosets mirrors that of humans, as distinct from other primate species (e.g., chimpanzees and macaques), and is an excellent model for studying sociocognitive brain development (Hrdy and Burkart 2022). To understand the representation of events that have not happened is challenging. However, this challenge is not limited by species. Prefrontal representation of counterfactual events and their value can be studied in macaques as well as humans (Fouragnan et al. 2019).

Summary

There is a balance to be struck between the simplicity of a model whose components are readily understood and the complexity of a model that may afford greater ecological relevance. Progress in translational neuroscience is facilitated by the use of complementary models, and tasks, referring to a common set of underlying constructs. We have illustrated how trans-species constructs at the level of processes and mechanisms can be used to understand the symptoms and syndromes associated with human prefrontal function. Whether this approach is robust enough to understand the mechanisms of psychotherapy, via animal models, remains to be seen but should not be ruled out.

Improved Targeting of Treatments, with Combinations and Prediction

The treatment of disorders associated with PFC and its associated circuits (e.g., frontostriatal and fronto-amygdala) might seek to reverse the deficit directly, for example, by replacing deficient neurotransmitters and improving symptom severity in an individual, thereby improving their quality of life. This chain of therapeutic effects cannot be assumed, even where the intervention engages the intended target. Moreover, reversal of the pathophysiological deficit itself may not be required. Instead, an effective treatment may engage other areas of the cortex or frontostriatal circuits, so as to compensate for the deficit rather than reverse it. Many individuals with neuropsychiatric syndromes struggle with prefrontal-related cognitive tasks (e.g., executive functions), which may underlie and/or exacerbate other problems (e.g., emotion regulation) and functional disadvantages (e.g., scholastic achievement). There is a pressing need for interventions that address and remediate cognitive processes and psychiatric illness. Both curative and symptom-mitigation treatments aspire to improve quality of life for the affected individual.

Treatments can be considered as focal or diffuse in their mode of application. Focal treatments in clinical use include neurosurgery, TMS, focused ultrasound stimulation, and DBS. Their benefit may nonetheless be mediated by diffuse systems, in the case of wide projections from the site of intervention. The effects of diffuse treatments, including pharmacology and cognitive behavioral therapies (CBTs), may nonetheless be exerted by their action on a focal system or circuit (see Roberts and Liston, this volume).

Psychological and Behavioral Therapies

As discussed by Jaeggi et al. (this volume), CBTs are representative of a wider body of evidence-based psychological interventions for psychiatric disorders and behavioral health symptoms. Here we set them in context of PFC circuits and other interventional approaches. Note that some are inherently diffuse in their cognitive processes and in the presumed functional anatomical associations (e.g., mindfulness) while others are cognitively and by extension anatomically constrained (e.g., cognitive training, exposure therapy, or goal management training). Classical CBT methods lie midway in this spectrum.

CBT methods share a structured, time-limited, problem-focused, and goal-oriented form of psychotherapy, through partnering with the client for symptom reduction. This includes a detailed assessment of the key symptoms, their antecedents and consequences of the symptoms or problems, and the contexts in which they occur. CBT has a strong evidence base in depression (Hofmann et al. 2012a). A common clinical model is to have an 8–12 weeks course that focuses on patient-specific symptoms. Specific interventions are adjusted to the problem, but they use a common underlying methodology: monitoring, tracking, antecedents, behavior, and consequences. Specific sessions and interventions may be implicitly or explicitly focused on processes associated with PFC such as self-monitoring; chain analysis of thoughts; feelings and actions in context; goal setting, planning and problem solving; and developing new strategies with greater cognitive control. In other words, CBT is goal directed, seeking adaptive strategies, reminiscent of the functions of the PFC itself. CBT may also include relaxation training, participating in pleasant activities, exposure to contexts and situations causing distress, the toleration of distress, and other exercises. The goal is to target maladaptive cognitive and behavioral processes and achieve a better understanding of one's symptoms and their drivers, together with training to reduce symptoms via adaptive cognitive, emotional, and behavioral responses. CBT protocols and their variations have been adapted for specific psychiatric disorders (e.g., major depression, anxiety disorders, OCD, addictions) and management of health symptoms such as insomnia, chronic pain, stress and anxiety management, binge eating, and weight gain may occur in isolation or co-occur with neurologic and other medical illnesses. There is extensive support of efficacy of CBT approaches in the treatment of these conditions, with response rates in the range 30–60%, depending on the illness, condition, and severity.

Prefrontal involvement in the working of CBT interventions has been shown via functional neuroimaging and cognitive testing. For example, neuroimaging studies have shown improvement and normalization of amygdala-prefrontal activation and connectivity during exposure to sad versus neutral faces, when comparing pre- to posttreatment in major depressive disorder. Such task-specific improvements are seen after treatment of posttraumatic stress disorder (Malejko et al. 2017), OCD (Cyr et al. 2020), social anxiety (Whitfield-Gabrieli et al. 2016; Young et al. 2019), and addiction disorders (Yip et al. 2019). Collectively, these studies provide evidence that CBT improves prefrontal neural circuit function along with symptoms.

Other psychotherapeutic approaches have been developed and tested with similar positive efficacy to CBT in the treatment of neuropsychiatric disorders and health symptoms. Examples include mindfulness training based on mindfulness-based stress reduction, acceptance and commitment therapy (Hayes 2019), prolonged exposure (Foa and McLean 2016), and cognitive processing therapy. These approaches maintain the principle of focusing on the present symptoms and context and typically use sensory, emotion, interoceptive, and behavioral stimulation with the reexperiencing of subjective states so as to promote adaptive functioning. From a neural circuit perspective, they may be seen as bottom-up approaches configured to revisit the symptoms and context in different ways to promote new, more adaptive learning and functioning. When combined with MRI, they suggest that ACC changes in response to fear images but to date, the neural evidence is less developed than for standard CBT.

There is increased activation of ventromedial and anterior pregenual cortex in OCD and depression, which is diminished following successful pharmacologic or behavioral treatment. These findings, as well as previous stereotactic neurosurgical interventions, support the use of ventral anterior limb of the internal capsule (vALIC) and subgenual targets to treat refractory OCD and depression, respectively. In spite of the relatively large size of the cingulotomy and ventral capsulotomy lesions as well as wide electrical fields affected by DBS in the capsule and subgenual regions, few neuropsychological deficits have been reported. This reflects the highly distributed nature of PFC and its functional resilience to focal injury. In OCD and depression, lesions and DBS target ventromedial OFC and ACC hyperactivity and the longitudinal white matter pathways that connect these top-down cortical control regions with thalamic, subthalamic, and brainstem structures as well as the reciprocal connections to PFC. Ongoing studies are in progress to identify the fiber tracks that are most predictive of a positive treatment outcome. These refinements in individual lesion targeting are facilitated by improvements in the resolution of diffusion imaging and the ability to image patients safely with implanted DBS devices. There appear to be few major adverse neuropsychological effects on prefrontal function from modern lesion or DBS procedures. Careful assessment is needed, however, of real-life tasks, particularly in the social and planning realms, as deficits in these areas may be overlooked by traditional methods of assessment.

Cognitive interventions can also be focused (for detailed discussion, see Jaeggi et al., this volume). Typically, they are designed to target a specific process (e.g., working memory, inhibitory control) with the idea that training such tasks or processes strengthens the underlying circuitry or systems. In ADHD, where targeted (computerized/app-based) executive function training is often implemented (mostly to supplement pharmacological treatments), training-specific executive function tasks aim to improve not only those trained cognitive domains, but ultimately to have a broader impact on domains that rely on the integrity of those cognitive functions (e.g., ADHD symptoms, well-being, self-efficacy, scholastic achievement), thus benefitting the quality of everyday life.

Despite growing popularity, not all individuals benefit from these approaches and often, the benefits are more proximal (restricted to the trained domain). The heterogeneity of outcomes likely reflects individual differences in cognitive strengths and needs, the heterogeneity of symptoms (Nigg et al. 2020) as well as the heterogeneity of approaches (Pergher et al. 2020a, b). As such, we need to increase understanding of the underlying mechanisms of an intervention (i.e., mechanisms of action) and individual differences in patients/participants to stratify treatment and improve efficacy (personalized medicine). A growing literature focused on improving understanding of individual differences, mediators, and moderators can inform efforts to determine training efficacy at the cognitive level, thus illustrating how baseline cognitive ability as well as training engagement and improvement are powerful predictors for training benefits and treatment response (Karbach et al. 2015, 2017). Other work has focused on biomarkers, such as brain modularity (Gallen and D'Esposito 2019), which might reflect the brain's "readiness to learn," as an example of the potential for the development of personalized approaches.

A key issue is the motivational readiness to engage in treatment, which itself is a function associated with PFC. Here, combined interventions that include a focus/supplement on motivation and participant buy-in could be particularly powerful (e.g., Jaeggi et al. 2023), as could those that include pharmacological components to get participants to a level where they are ready to engage (e.g., with exposure therapy, CBT) and work synergistically. Such combined approaches may results in broad impacts (due to multiple targets) and more sustained effects, since individuals have the chance to capitalize on what is learned and continue to "practice" in various environments/circumstances, which would promote long-term learning or the process of "learning to learn" (Beck 2011).

Focal Lesions and Stimulation

The therapeutic response to focal lesions may not be immediate: for OCD, it can take 6–12 months to fully respond to DBS or lesions of the vALIC (Rasmussen et al. 2018). Qualitatively, individuals experience a gradual lessening of the anxiety associated with obsessions and the corresponding urge to complete compulsions (Barrios-Anderson et al. 2022). This is accompanied by a recognition that the extensive effort needed to undertake a compulsion may not be worth it. This sets in motion a process that enables individuals to approach stimuli and contexts, which they previously avoided at all costs, and to engage in exposure-based treatments (see Rasmussen, this volume). This learning process, however, takes time.

There is converging evidence that the addition of exposure-based CBT to pharmacologic or neuromodulatory interventions in OCD and other anxiety/depressive disorders leads to the improved outcomes (Franklin et al. 2011; Strawn et al. 2022). As for capsulotomy, the benefit of combination may take several months to emerge and be influenced by baseline clinical severity. One reason for the therapeutic delay is that these interventions lead to a greater willingness to approach feared stimuli or contexts; still, they cannot replace the action-outcome effect of being exposed to the feared consequence followed by not experiencing the feared consequence. In other words, pharmacologic and neuromodulatory interventions may enable learning to take place, and it is the effect of learning that reduces symptoms. Again, this learning process takes time.

A focal treatment alternative to neurosurgical lesions is high-intensity ultrasound, which has been FDA approved for the treatment of essential tremor (Martinez-Fernandez and Pineda-Pardo 2020). It has also been tested in the vALIC as the target for OCD with promising preliminary results on clinical OCD benefit without major cognitive side effects (Davidson et al. 2020a). Focality of the target has been limited by attenuation and dispersion of the beam through the skull, making the total energy delivered to the target and therefore the size of the lesion variable (Davidson et al. 2020b). Technical limitations in targeting and regulatory concerns, however, present significant challenges for blinded treatment trials for neuropsychiatric conditions.

TMS modulates neurons in a relatively focal, superficial area of cortex by delivering potent, high-frequency magnetic field pulses that elicit electric field fluctuations and depolarize neurons at the target site. TMS is already used to treat a variety of neuropsychiatric conditions, such as depression, OCD, addiction, and chronic pain (Zhao et al. 2023). Understanding of its therapeutic mechanisms has evolved rapidly over the past two decades, particularly from work in depression, and provides insights in three main areas. The first concerns the success of cross-species modeling. Early TMS treatment protocols emerged from insights derived from patients with left dorsolateral prefrontal strokes and analogous studies in marmosets. For example, left dlPFC inactivation increases anxiety, which may be due to interhemispheric imbalance, that can be mitigated by TMS (Lefaucheur et al. 2014).

Second, TMS has confirmed the hypothesis that depression is a network disorder. Functional connectivity between subgenual cingulate and a dlPFC target site modulates the TMS response and connectivity, such that therapeutic effects are driven in part by effects on downstream targets. Although connectivity in a single circuit account for only a small percentage of variance, the combination of prefrontal circuits mediates additive benefits (Elbau et al. 2023). These effects of TMS accord with lesion mapping studies to suggest that network-level functional connectivity patterns are important to predict depression after stroke, as well as to identify TMS targets (Hollunder et al. 2022; Siddiqi and Fox 2023). It is not fully understood how TMS engages downstream areas that are remote from the local prefrontal target, and animal models together with concurrent TMS/fMRI/EEG studies are required to selectively manipulate neuronal responses and determine causal mechanisms.

Third, TMS studies highlight the individual variation in response to treatment, which may be explained, and predicted, in terms of network connectivity. For example, functional mapping has revealed robust and reproducible individual differences in the topology of functional networks in the human PFC (Fox et al. 2012; Gratton et al. 2014; Siddiqi et al. 2020). Personalized approaches can be attempted that allow investigators to determine the optimal TMS target site and coil orientation to engage selectively a specific network while avoiding others (Lynch et al. 2022).

Further studies are needed to develop accelerated protocols and enhance responses with optimal dosage. This is likely to require large clinical studies using systematic approaches to target stimuli and readouts so as to optimize and individualize treatments over a high-dimensional parameter space. The degree to which TMS "rescues" or "compensates" for biological and behavioral deficits remains unresolved, both at the physiological and process level. The functional connectivity features that predict treatment response may not be abnormal; rather, variance of intact functional connectivity of the TMS site may determine the capacity to engage downstream targets and to manipulate systems that mediate one cognitive domain (e.g., primary sensory motor representations of pain) in order to improve another (e.g., depression).

Pharmacological Approaches

Drug interventions in humans are macroscopically diffuse, even though they are pharmacologically specific and thereby microscopically restricted to specific cell types and, in some cases, highly restricted receptor distributions. Experimental studies in animals shed light on the underlying mechanisms of drug treatment, focused on the basic molecular, cellular, network, and behavioral analysis of chemical systems in the PFC on which pharmaceutical treatments, such as guanfacine and ketamine, act (Robbins and Arnsten 2009). This not only indicates the likely targets of current treatments but also potential novel targets for treatment. In addition, fundamental neuroanatomical studies have provided insight into the neural pathways likely to be affected by DBS or ablative lesions, used for treating disorders such as depression or OCD (Rasmussen and Eisen 1997; Rasmussen and Goodman 2022). Moreover, DBS or tract lesions in animals can provide further insight into the underlying functional networks that are engaged. Other animal studies have used stressors (in development or adulthood) to elicit symptom-relevant behaviors (e.g., anhedonia or anxiety) and reveal the physiological and behavioral mediators of pharmacological treatments, such as serotonin reuptake inhibition or ketamine (Roberts and Liston, this volume). A recurrent theme of these animal-pharmacology studies is the prefrontal plasticity that follows treatment.

Evolutionary Expansion of mGluR3-NAAG-GCPII Signaling

Based on decades of research in rhesus macaque dlPFC, the prevailing notion is that intracellular calcium–cAMP–PKA–K^+ mechanisms must be tightly regulated to maintain network connectivity and cognitive function (Arnsten 2009; Arnsten et al. 2021, 2022). Their biochemical feedforward nature can otherwise rapidly generate elevated levels of cytosolic calcium and cAMP, with deleterious effects. Specifically, the receptors that inhibit cAMP production via $G_{i/o}$ signaling (e.g., mGluR3 and noradrenergic α2A-AR) are localized on dendritic spines in layer III of dlPFC, and both enhance delay cell firing and working memory performance via inhibition of cAMP–PKA–K^+ channel signaling. Genetic predispositions in *GRM3*, which encodes metabotropic glutamate receptor type 3, are associated with elevated risk of schizophrenia based on genome-wide association studies. The mGluR3 receptors are selectively activated by NAAG, which is a highly prevalent neurotransmitter co-released with glutamate. NAAG is catabolized by glutamate carboxypeptidase II (GCPII). The mGluR3s are also localized on astrocytes, where they augment glutamate uptake through excitatory amino acid transporters (Neale et al. 2011). Based on experiments in rodents, mGluR3s reside on presynaptic terminals and reduce glutamate release, playing key a role in neuronal microcircuits. They have traditionally been seen as providing negative feedback on glutamate signaling and protective against excitotoxicity (Cao et al. 2016). Recent studies, however, support the hypothesis that their action in pyramidal neurons has changed and expanded with cortical evolution across phylogeny. In rhesus monkey dlPFC, for example, mGluR3 and GCPII have an evolutionarily novel role in higher cortical circuits: strengthening the connectivity of layer III dlPFC circuits that mediates working memory (Jin et al. 2018; Yang et al. 2022). This may partially explain their genetic predilections to human cognition and cognitive disorders.

Ultrastructural studies using immunoelectron microscopy of the rhesus monkey layer III dlPFC show that mGluR3s are concentrated postsynaptically on dendritic spines, which is strikingly different from their classic location on presynaptic terminals in rodent circuits. The mGluR3s are also localized on astrocytes in primate dlPFC, but the presynaptic receptors on glutamate axon terminals are exclusively mGluR2 rather than mGluR3 (Jin et al. 2017, 2018). Relevant to therapeutics is the finding that NAAG-mGluR3 signaling in primate dlPFC can enhance neuronal firing related to working memory by attenuating cAMP–PKA–K^+ channel signaling (Arnsten 2015; Arnsten et al. 2021; Birnbaum et al. 2004; Gamo et al. 2015). Therefore, NAAG–mGluR3 signaling strengthens the connectivity of higher cortical glutamatergic circuits and increases dlPFC neuronal firing in primates, opposite to the decrease in glutamate release typically associated with mGluR3 presynaptic actions in rodents. These mechanisms influence prefrontal cortical function and provide a further mechanism for the effect of stress on cognition via exacerbated catecholamine release (Jin et al. 2018; Yang et al. 2022).

Noradrenergic Therapeutic Strategies for PFC-Associated Cognitive Impairment

Studies across many species highlight the critical role for noradrenergic neurotransmission in prefrontal function, and have, for example, resulted in

- the selective norepinephrine (NE) α2A-adrenoceptor (α2A-AR) agonist, guanfacine (Intuniv™),

- selective noradrenergic reuptake inhibitor, atomoxetine (Strattera™), and
- the nonselective modulator of noradrenaline and dopamine, methylphenidate (Ritalin™).

These drugs provide a clear example of translational success (Holland et al. 2021; Robbins and Arnsten 2009): based on clinical trials that followed preclinical studies of noradrenergic attentional and inhibitory control in animal models and preclinical human studies, they are approved in many countries to treat ADHD. They are also widely used off-label to treat additional mental disorders that involve impaired functioning of PFC, including stress-related disorders such as substance abuse (Levin et al. 2009), schizotypal cognitive deficits, and traumatic brain injury (NCT00702364; Ripley et al. 2014). Clinical trials in neurodegenerative disorders such as Alzheimer disease and PSP are underway (e.g., NCT03116126, ISRCTN99462035). At the level of neuronal microcircuits, pioneering work has revealed that guanfacine acts within the PFC via postsynaptic α2A-AR on dendritic spines to inhibit cAMP–PKA–K^+ channel signaling, thus consolidating network connectivity, improving prefrontal cortical neuronal firing, and enhancing prefrontal cognitive functions (Hains et al. 2015). Although guanfacine's beneficial effects on attentional and inhibitory control are present in rodents, they are especially evident in primates where the PFC greatly differentiates and elaborates during evolution. Therefore, NE α2A-AR-mediated actions by guanfacine or atomoxetine can fine-tune top-down control by prefrontal networks, which may explain their therapeutic efficacy in a variety of mental disorders (Arnsten 2020; Hains et al. 2015). It is interesting to note that the use of the drugs in this context is to improve symptoms and function, not to resolve the root mechanisms underlying risk or vulnerability to illness. The normalization of function does not necessitate normalization of the underlying neurobiology. This distinction is relevant to drug and nondrug interventions, whether the intention may be curative (e.g., phobias) or ameliorative (e.g., OCD severity).

An important caveat for pharmacological strategies to target prefrontal networks is drug dosage. For example, both NE α1-AR and DA D1R have a nonlinear inverted-U dose-response effect on dlPFC persistent firing and working memory function. Mediated by activation of calcium–cAMP signaling in dendritic spines, moderate levels are essential; excessive levels significantly reduce firing and cognition by opening nearby K^+ channels (Datta and Arnsten 2019; Datta et al. 2019; Jin et al. 2018; Wang et al. 2019). Optimal levels of stimulation may strengthen persistent firing by magnifying calcium near the postsynaptic density and/or by phosphorylation of NMDA receptors to amplify their synaptic actions (Li et al. 2010b; Skeberdis et al. 2006). Paradoxically, higher levels of stimulation as a result of uncontrollable stress or medication reduces neuronal firing and impairs working memory by opening HCN and KCNQ channels (Birnbaum et al. 2004). Excessive levels of catecholamines strengthen more primitive circuits, such as the amygdala (Ferry et al. 1999), switching control of behavior to more unconscious habitual and instinctive responses. With chronic stress exposure, sustained weakening of network connections by calcium–cAMP–PKA–K^+ signaling leads to removal of spines and dendrites (Hains et al. 2009; Moda-Sava et al. 2019; Radley et al. 2006), an observation seen in humans (Ansell et al. 2012). Clinical applications of diffuse drug treatments are made more complex by these nonlinear dose dependencies, thus requiring stratified or even individualized dosing decisions for comparable effects. Higher doses may not only fail to confer added benefit, they may be counterproductive. The nonlinear dose-response relationships and baseline dependency of effects may explain a proportion of apparent non-responders.

Summary and Future Considerations

Combination treatments are often used in practice, by either combining a drug with a behavioral therapy or through the use of two or more drugs. A systematic approach to combinatorial therapies is required in preclinical and clinical studies, but it has proven challenging to implement in practice. From a theoretical perspective, drug combinations might be rational: one drug may open a patient's receptiveness to another treatment or amplify efficacy. However, clinical polypharmacy is often not a combinatorial science. It is highly complex in view of the multiplicity of neurotransmission and deficits in the PFC.

Looking ahead, we see four areas for research focus in therapeutics. First, rigorous placebo-controlled studies are essential, whether in clinical trials of humans or animal studies. This requires animal models of the pathophysiological processes of the disorder as well as the candidate intervention.

Second, the systematicity of pharmacological interventions, and their combinations, needs to be linked to systematic phenotyping of patients with heterogeneous syndromes. Only this type of systematic, inclusive approach to disorders will resolve the dimensional complexity of neuropsychiatric illness. Within such systematic phenotyping, sex differences should be a factor of special interest, not merely a confound.

Third, there is a pressing need for targeting or precision medicine, based on models that predict response to a given treatment. These models might include genetic, phenotypic, neurochemistry, activity, or connectivity imaging data, or even the response to a test dose. Computational psychiatry approaches (see Koechlin and Wang, this volume) are attractive foundations for such predictive models, although simpler modality-specific data may be sufficient to predict, for example, remission from depression in response to diverse treatment approaches, according to PET or MRI measurements of overactivity in area 25 (McGrath et al. 2014).

Finally, to improve the understanding of underlying biological mechanisms disease, heuristically predictive models should be compared with biophysical or neurocognitive informed models. This necessitates a cross-disciplinary approach to research, unfettered by historical professional boundaries or historical boundaries between funding bodies and healthcare services. Precision medicine in the future should aspire to be informed by mechanisms of disease, adapted to developmental stages, and attentive to individual differences, including their "windows of opportunity" for maximal therapeutic efficacy.

Focal lesions and pharmacological treatments provide complementary and additive clinical benefit for a range of neuropsychiatric syndromes. By targeting specific neurochemical mechanisms and prefrontal networks, they can influence the core cognitive processes underlying multiple symptoms. As shown in Figure 16.1, this leads to potential clinical benefits in multiple diagnostic groups, while remaining subject to individual differences in severity, demographics, and comorbidity. Looking ahead, a systematic approach is required to guide therapeutic combinations and participant phenotypic variation, and enable accurate prediction models as a foundation for precision medicine.

Prefrontal Cortex and Society

How can insights about PFC function and its contribution to mental health be harnessed for the benefit of our global society? Mental health, climate change, conflict, and communication: these are all areas of intersection between the neuroscience of PFC and society.

Executive functions of the PFC may provide a highly effective, sensitive singular marker of brain health—a sort of "canary in the coal mine" for societal brain health. Basic markers of executive function could identify people at risk of diverse mental health disorders, akin to the century-old height and weight growth charts for children, or blood-pressure and cholesterol surveillance in mid-life. Growth charts and developmental milestones are sensitive to myriad risks, diseases, nutrition, and stress and provide early warnings for investigation and treatment. Many psychiatric conditions involve PFC dysfunction, with deficits acting as a powerful early warning system (e.g., Diamond 2013). Since major psychopathology emerges during adolescence (Paus et al. 2008), monitoring PFC development may provide a strong risk marker for atypical development and pathways toward psychopathology. To determine the integrity of PFC function at a large (societal) scale, executive functions that require just a few minutes to complete (e.g., on mobile devices) could provide an initial screening, for example, of motor skills, vision and hearing, and social skills, which are tested at regular intervals in children. Similar approaches, for example by the Brain Health Project at UT Dallas,[2] are being evaluated at scale in adults. If screenings indicate impaired maturation against a "cognitive growth chart" of normative development, additional assessments may be warranted. Further psychological measures and interviews may then be targeted

2 https://centerforbrainhealth.org/project

to identify problems and, if appropriate, lead to interventions to improve resilience and decrease the risk for adverse developmental trajectories. Increasing resilience as well as an individual's chance for successful school outcomes would be a major step forward in tackling the disastrous effects of socioeconomic inequality on individual developmental opportunities.

Such neurocognitive screening needs to be accompanied by appropriate and personalized interventions that are accessible to individuals and communities (e.g., leveraging school and family support). To realize this at scale, a new range of education technologies may be required (e.g., the "EF+Math" program[3]) as will novel ways to engage and prioritize traditionally underserved populations. Education technology offers a powerful means to improve the accessibility of assessments and interventions, yet have historically been preferentially accessible to high socioeconomic groups that are disproportionately white and geographically uneven. Examples for interventions that might be implemented at scale include web-based CBT and app-based computerized intervention "games" that can be played on low-cost devices (Iyadurai et al. 2018). Preliminary evidence indicates that remote interventions and assessments can work as well as in-person interventions, and they have the potential to be more cost-effective and accessible.

The challenge is to make them also equitable and purposefully designed by being sensitive to and taking into account of the relevant cultural background of the target population. This benefits from a co-design approach, as implemented in the "EF+Math" program mentioned above: a focus on strengths rather than deficits, while capitalizing on patient resources to maximize participant buy-in and agency (Fluckiger et al. 2023). Lessons need to be learned from historical misuse and divisiveness related to IQ testing, systemic disadvantages, and loss of trust arising from a failure of cultural embedding of assessments. Better cultural embedding is one means of linking neuroscience advances to global challenges.

Climate change and massive population displacements from war and famine represent major global challenges. They are a cause of chronic stress for many individuals, with enduring consequences for neuropsychiatric health. They also represent a collective failure of control, restraint, forward planning, and value-based decision making (cf. functions of our prefrontal cortex). For decades, we have failed to adjust our decisions and actions in in the service of global goals, despite the existence of abundant knowledge about the potential risks of global warming. Immediate adjustments ranging from individual actions to political regulatory measures may seem obvious, yet the majority of the world's population has problems overcoming long-established patterns of behavior. Lack of behavioral regulation continues to happen across levels: from individual consumer behaviors to large-scale commercial organizations to governmental policy. Beyond an analogy with the functions of PFC:

[3] https://aerdf.org/programs/ef-math

- How can understanding the brain basis of decision making contribute to resolving obstacles to behavioral change?
- To what extent is it valid to map decision making across levels (i.e., from individual consumer decisions to organizations and policy), with geographically and temporally distant consequences?
- Can we improve current behaviors based on an understanding of long-term goals of future generations, making them more motivationally relevant for current decisions and actions?

Such broad-scale questions go beyond the traditional scope of neuroscience, but cognitive neuroscience can contribute to an interdisciplinary research agenda that promotes adaptive and anticipatory behavior at large scales, in response to global challenges (Aron et al. 2020; Castiglione et al. 2022).

To study and engage population-based approaches to health and mental health, the language of neuroscience may need to change. Words matter as we consider communicating about the role of frontal cortex and prefrontal processes in neuropsychiatric and other brain disorders. Mental health disorders are already associated with stigma (Rose et al. 2007). Do we help or hinder patients when the terminology of our research framework is based on phrases that have strong negative connotations, such as "cognitive control," "suppression," or "management"? Such phrases can alienate the public and get in the way of a research-based approach to illness and health conditions by reducing engagement in prevention and treatment strategies (Bailey 1999; Burns and Rapee 2006; Volger et al. 2012; Young et al. 2008). It should be possible to use terminology acceptable to individuals from diverse backgrounds, races, ethnicity, and cultures. Identifying people as patients, defined by their illness, is often perceived as pejorative, stigmatizing, or less desirable. It may push individuals away from engaging with information about the illness, their associated underlying mechanisms and participation in treatment and prevention efforts (Volkow et al. 2021). For example, there is broad-based consensus across diseases and medical conditions for the use of *first person* language when describing individuals with an illness: *persons with depression* or *individuals with obesity* are preferred and not "depressed patients" or "obese people" across clinical, scientific, or public health contexts (Volkow et al. 2021). Furthermore, words such as "mental" or "mental health" may convey emotionality and mental weakness; "suppression," "control," and "management" may convey messages of colonial or social dominance. Alternative term such as "self-regulation," "stress," or "resilience" are regarded more favorably and may denote higher acceptability in conveying concepts PFC function to a wider audience.

Together, these issues of education, resilience, global policy and inclusive language are important to advance global health and economic well-being. They speak to social determinants of brain health, and therefore public policies to improve brain health. They speak to ways to reduce illnesses associated with prefrontal function, reducing trauma, stress, and developmental risk. They also speak to active steps that can be taken through education and health services to prevent, detect, and treat disorders associated with prefrontal cortical function.

Conclusion

In this chapter, we began with consideration of the complexity of the PFC, and the many levels at which its function and disorders can be analyzed. We proposed a way through this complexity that involves (a) multiple explanatory levels with divergence and convergence between syndrome, symptom, process, and biological mechanisms, and (b) spatial and temporal gradients across the PFC. Together, the levels and gradients provide an explanatory framework that links animal and human studies in such a way as to inform therapeutic strategies.

The recent evolutionary expansion of the PFC in humans and nonhuman primates has been subject to natural selection for a relatively short time period, from an evolutionary perspective. This expansion of neocortical regions and their subcortical connections has clearly led to selective advantages but may also have created vulnerability to mental health disorders. Converging evidence implicates PFC circuitry and its connections in many neuropsychiatric conditions. Basic cognitive functions such as working memory, decision making, selective attention, and executive control depend on the same prefrontal regions and associated circuits that are abnormal in psychiatric and neurological disorders.

Theoretical, laboratory, and clinical neuroscientists can work together to understand prefrontal function and its deficits. New models of computational psychiatry (Wang and Krystal 2014) as well as advances in experimental tools and big data will further help establish a solid biological foundation for the diagnosis and treatment of diseases of PFC. To realize the full potential of this endeavor requires highly cross-disciplinary collaborative and translational research, with improved career pathways, regulatory recognition, and training. With this success, insights about prefrontal cortical function can be harnessed for the benefit of our global society, with equity of access to evidence-based health and education.

17

Pathways Forward Toward an Understanding of Frontal Lobe Function

Marie T. Banich, Suzanne N. Haber, and Trevor W. Robbins

In this concluding chapter we examine some of the cross-cutting themes that emerged from the Forum. Here, we consider issues that transcend the individual working groups, which we believe are ripe for further discussion and investigation, notably translation from animal to human models, the role of connectivity in frontal function, and unique aspects of human cognition that are supported by the frontal lobes.

Animal Models: Utility, Limitations, and Future Potential

Comparisons across species were a recurrent theme at the Forum, figuring to variable extents in each of the four working groups. There are essentially two major reasons for this. First is the intriguing issue of how prefrontal cortex and its associated functions evolved, which is inevitably bound up with consideration of what were the main drivers of human evolution. Second, is the more pragmatic issue of how studies on infra-human animals can inform the understanding and possible treatment of clinical neurological and psychiatric disorders associated with the frontal lobe, through "animal models." Although human brain imaging methods are constantly being refined, in terms of resolution, modality, and analytic sensitivity, to provide sophisticated regional and functional network maps of the human cortex, they still cannot provide detailed information at the cellular, molecular, and circuit levels. This information can only be provided by animal models, which are essential for understanding the underlying the causal mechanisms that lead to effective treatments of human disorders.

Comparing possible behavioral functions across species can also be problematic given the complexity of human cognition. However, it is a useful exercise to identify test procedures, such as the stop signal reaction time task, that procedurally appear to utilize comparable requirements across species, for example, in terms of contingencies and inferred requirements for perception, memory, and behavioral control. Thus, classical tests of prefrontal function in humans were reported at the Forum to have several parallels in nonhuman primates (NHPs) and rodents. However, operational parallels in performance may not necessarily be matched by the quite the same psychological processes in humans and other animals, given the additional capabilities of humans in language, insight, and rapid learning. Nevertheless, it was deemed reasonable to assume that comparable performance in test paradigms in experimental animals may at least identify some of the "cognitive building blocks" of more advanced functions in humans. This notion of "building blocks" is supported by the evident hierarchical nature and rostral-caudal gradients of organization of the prefrontal cortex across species, an important theme of the Forum.

This comparison can, of course, be strengthened by finding factors that appear to affect behavioral performance in the same manner across species, including importantly neural mediation. If the same brain regions cross-species can be shown to be necessary for mediating behavioral performance, that then heightens the likelihood of effective translation of findings trans-species. This translational approach runs into special difficulties, however, in the case of the prefrontal cortex, in terms of the homology of its component regions. Homology (i.e., shared origins of structure among species) was a major theme for the first discussion group (Weiner et al., Chapter 4). This group considered several criteria for establishing the principle of homology, which entails conservation of structures during evolution, and agreed that the two most important were (a) the detailed histological (cytoarchitectonic) composition of different (pre)frontal regions and (b) their neural connectivity, not only within themselves but also with other brain regions. The latter highlighted a major recurrent theme; namely, networks within prefrontal regions as well as between prefrontal regions and other areas of the brain are likely very important in influencing and determining prefrontal function.

Overall, the classical position was generally supported: only some parts of the rodent prefrontal cortex—mainly posterior orbitofrontal cortex, prelimbic and infralimbic (medial prefrontal cortex) and some parts of dorsal anterior cingulate—have obvious counterparts in the primate brain, whereas counterparts to the dorsolateral (dlPFC) and ventrolateral prefrontal cortex and frontopolar cortex are not so evident. Vertes et al. (Chapter 3) provide an intriguing alternative proposition, positing "two prefrontal streams": one based on orbitofrontal cortex and the other based on anatomical observations made on the tree shrew prefrontal cortex. This may provide some supporting evidence of dlPFC precursors existing in rodent medial prefrontal cortex, but this position is clearly controversial and it would need considerably more evidence to overturn the classical view that rodents do not have a dlPFC. Overall, infra-human primates, such as the macaque rhesus monkey, undoubtedly provide the best structural model of the human brain and may even indicate precursors of specialized human regions, such as Broca's area, whereas New World Monkeys, such as the marmoset, may provide an economically effective, compromise option for translational research involving NHPs (see Rowe et al., Chapter 16).

Neuroimaging and neural network analysis in humans can be paralleled quite readily in NHPs. Murray and Constantinidis (Chapter 6) and Rich and Averbeck (Chapter 5) illustrate the potential of these animal models in providing a more refined analysis of prefrontal organization and function at cellular levels through multiunit recordings. Ideally, these types of recordings should now be performed concurrently in several prefrontal regions to elucidate the interactions among different (pre)frontal regions as well as the connectivity of these regions to noncortical areas (e.g., as in the frontostriatal pathways). In a complementary manner, investigations using sensitive anatomical tracing methods that help to define prefrontal connectivity relevant for human prefrontal networks (see Weiner et al., Chapter 4) will continue to be essential. However, NHPs were considered not to be entirely optimal as pragmatic animal models of human mental health disorders because NHPs cannot be used in the large numbers required to match those employed in large-scale human clinical or neuroimaging studies nor for drug discovery (see Roberts and Liston, Chapter 13), especially as such animal models frequently require an intervention such as a stressor or genetic manipulation which are ethically, as well as technically, difficult to employ in NHPs.

The utility of rodent models is that they allow for more invasive perturbations, such as genetic manipulations (e.g., genetic knock-out preparations), lesions, or stress as well as richer and more sophisticated procedures for tracking the activity of functional neural circuitry, including optogenetics, chemogenetics, or fiber photometry to measure calcium or neurotransmitter fluxes (see Izquierdo, Chapter 2). Although promising examples of the use of NHPs in some these techniques were described by Weiner et al. (Chapter 4) and Rowe et al. (Chapter 16), the general observation was that some studies are pursued more effectively and economically in rodents, especially mice.

Limitations

This partial dependence on rodents for understanding aspects of prefrontal organization and function leads to two major problems. First, an obvious one, is that only some "primate" prefrontal regions may be homologous in the rodent brain; hence rodents cannot effectively model the roles of such regions in primates. Second, there are instances where homology did appear to apply, but the underlying functions studied in both rodents and primates did not align, either because they were distinct or did not appear to operate in the same manner (for an example, see Weiner et al., Chapter 4). Of course, it is conceivable that some (pre)frontal structures that appear to be homologous based on cellular organization and connectivity may nonetheless have evolved to perform different functions, but this possibility is not helpful for the triangulation approach to translation. It is then often necessary to focus on behavioral similarities across species and the common effects of other variables such as drug treatments (for examples, see Rowe et al., Chapter 16), genetic expression and environmental challenges such as stress (including during early life and forms of social deprivation) to achieve translational validity.

Future Potential: Linkage to Clinical Issues

Several interfaces have to be negotiated when translating findings from rodents to NHPs to humans for clinical use. Moreover, this translation should be bidirectional. Clinical observations and issues should not only be inspirational, they should also influence the precise questions that are posed and used to design preclinical studies. For example, detailed understanding of the neurobehavioral basis of distinctive symptoms should be pursued using current theoretical psychological or cognitive conceptions as this will also help to test the utility of those theories. We suggest that this goes beyond the Research Domain Criteria (RDoC) approach (Kozak and Cuthbert 2016) which identifies a matrix of relevant constructs (e.g., inhibition/suppression) applied to specific psychiatric disorders to identify commonalities. This modified RDoC approach is potentially useful when considering transdiagnostic and "dimensional" approaches to neuropsychiatric nosology; for example, the existence of impulsive-compulsive symptoms in many diagnoses (from ADHD and OCD to neurodegenerative disorders such as progressive supranuclear palsy), in which dysfunction of frontal regions or frontal connectivity is implicated. Commonality of symptoms across disorders may indicate some commonalities in approaches to their treatment, based on the fact that they may have overlapping impairments in neural circuitry (see Rowe et al., Chapter 16).

The availability of some modern techniques, such optogenetic or chemogenetic stimulation, may enable, at the level of cellular resolution, the simulation of particular deficits in neural network function with much greater specificity than hitherto possible (see Izquierdo, Chapter 2). Such manipulations may be important, as a gross malfunction of a neural network could result from a variety of different deficits at the level of its contributory nodes or from impairments in distinct molecular or cellular components. Hence, superficially similar behavioral phenotypes may result from deficits in different mechanisms. In addition, the moot question of how many treatments—including pharmacological (Roberts and Liston, Chapter 11), neuromodulatory (e.g., deep brain stimulation, and rTMS (Rasmussen, Chapter 15), and psychological interventions (Jaeggi et al., Chapter 14)—actually work in neural network terms is currently not well understood. Better understanding of their mode of operation and underlying mechanisms may help lead to more refined versions of those treatments. One futuristic projection from the first working group (Weiner et al., Chapter 4) was that combined pharmacological/surgical procedures involving chemogenetic interventions may hold promise in the treatment of human psychiatric and neurological disorders, given improved knowledge and precision of their neural correlates. Reaching this goal in the future will depend upon a phylogenetic, anatomical, and functional "vertical integration" of work on prefrontal cortex extending across species.

How Does Connectivity of Frontal Regions Enable Its Functions?

Throughout the Forum, we confronted this question in different ways and consider it to be a major gap that is ripe for future investigation. As alluded to above, connectivity can be examined at multiple levels: between cells, between subdivisions of frontal cortex, between frontal regions and other cortical areas, as well as between frontal regions and subcortical/noncortical areas. Some of these aspects of connectivity have been addressed more than others, but it remains obscure as to how they enable the function and computations performed by frontal cortex (for an overview, see Shenhav et al., Chapter 12).

Although there has been extensive research and knowledge gleaned regarding the organization and functioning of cells in sensory and motor regions, such heuristics may be unlikely to provide a suitable framework to understand the cellular organization and functioning of prefrontal cortex. For example, although the types of information being represented by cells in sensory cortex have been well delineated (e.g., contrast between light and dark in primary visual cortex, motion in area MT), prefrontal cortex is fundamentally different as prefrontal cells appear to have a multiplex coding of information, integrating multiple dimensions. Moreover, the same cell can flexibly change its coding scheme depending on task demands as well as code for abstract categories that are not constrained by physical characteristics (e.g., visual features). Hence, searching for "the element" of information that is encoded by prefrontal cells may be a futile or frustrating endeavor, yet this may be the exact reason why this region of cortex is so adept at modifying and modulating the functioning of other portions of the brain.

At the cellular level, at least part of (pre)frontal function, as is also true for other regions of the brain, is governed by its intrinsic connectivity patterns. It may be the connectivity pattern of the cells themselves that are important. For example, the ability to maintain representations over time, which are critical for working memory and maintaining goals, may depend on the ability of prefrontal cells to sustain activity via recurrent activity (see Koechlin and Wang, Chapter 10). On a different level, the ability of prefrontal cells to code abstract information may depend on the pattern of inputs from sensory, more posterior aspects of cortex, and subcortical regions. Once again, like other regions of cortex, the function of prefrontal cells is likely to vary by the input they each receive, the way that input is coded/weighted, and the context in which such information occurs (see Murray et al., Chapter 8). However, unlike portions of the visual system, we currently lack a detailed understanding of the "wiring diagram" of frontal cortex. Hence a better understanding of the motifs of input to prefrontal cortex will be desirable, as well as a greater understanding of the degree to which such motifs may vary depending on the source of inputs—other prefrontal regions, more distant regions of cortex, or subcortical/noncortical regions—and in what contexts.

One notable aspect of (pre)frontal cells is that the same cells appear to be able to code different types of information flexibly at different times, often referred to as mixed selectivity. This characteristic may also distinguish them, in part, from other brain regions that act as convergence zones. For example, the hippocampus has some cells that code an animal's location in space (place cells), and others that code the time at which events occurs (time cells). These populations of cells appear to be somewhat distinct as cells in the CA3 subfield of the hippocampus appear to encode information about space only, whereas those in CA2 encode information about time only (Eichenbaum 2017). (It should be noted that, nonetheless, at least some mixing of sensitivity to both of these dimensions does occur in the CA1 subfield).

The question that then arises is what schema or mechanism can allow the same cell to represent different information at different times. One topic highly relevant to this question is a consideration of how oscillatory firing may enable aspects of (pre)frontal function. At the cellular level, it is possible that firing of frontal neurons may be able to be disentangled into different frequency bands (e.g., via a Fourier transform), enabling separate simultaneous channels of communication to distinct target regions. For example, it has been suggested that high theta oscillations and gamma oscillations may play an important role in prefrontal functions including cognitive control (Cavanagh and Frank 2014), insight (Bieth et al. 2024), and problem solving (Bieth et al. 2024; Lin et al. 2023). This issue is a topic ripe for future investigation (for further information, see Weiner et al., Chapter 4).

With regards to connectivity patterns between prefrontal regions and the information flow between them, less is known. A number of models argue that connectivity is somewhat hierarchical in nature, with information computed at one level passed onto the next which overcomes the computational limitations at that prior level (see Koechlin and Wang, Chapter 10) or assumes that particular regions of prefrontal cortex (i.e., dlPFC) sit at the top of the hierarchy based on asymmetries of input and output connections (see Badre, Chapter 7). Aspects of such models are appealing as they mimic characteristics of the organization of posterior brain regions in which regions further along a hierarchy of a circuit aid in building up a representation (e.g., spots of light are detected in early visual areas, which are linked together to allow the detection of edges, which are linked together to detect lines of different orientations). In the case of prefrontal hierarchies, it is generally the case that the representations become progressively more abstract along a caudal-rostral lateral prefrontal axis; hence presumably elements of the earlier representations are lost as the abstract representation develops; in this sense, the hierarchies for visual and prefrontal representation are parallel. However, exactly how the connectivity pattern of frontal cortex enables its specific functions is not well understood.

With regards to cortical-noncortical interactions, there were three main systems discussed at the Forum on which research has focused: the thalamus, the basal ganglia, and the amygdala. Understanding these interactions in more detail is likely to aid in understanding the influence of frontal regions on function. An interesting issue is whether the different regions of prefrontal cortex interact with these subcortical structures in an essentially similar manner. Although not much discussed at the Forum, another region of importance in this regard in the hippocampus. We consider each in turn.

While it has been known for quite some time that the role of the thalamus is to filter, sort and relay information reaching the cortex, more recent research has suggested that it may play a role in aspects of higher-level cognition through its connectivity with prefrontal cortex (Hwang et al. 2020). And through reciprocal loops of course, the cortex may in turn influence what information it "receives" from the thalamus.

Connectivity between the basal ganglia and the cortex has been argued to act a "gate" that is kept closed when the need to maintain information in working memory is high, but then opened when information in working memory needs to be updated (Hazy et al. 2007) with control over the gate learned through reinforcement learning driven by dopaminergic inputs. These basal ganglia inputs to cortex appear to have a specific topology, which could enable the gating of different types of information in parallel but interacting striatal-frontal loops (Rusu and Pennartz 2020) and has implications for diseases such as Parkinson disease (Wapstra et al. 2022) as well as for normal development (Parr et al. 2022).

Another aspect of prefrontal connectivity concerns connectivity with regions involved in emotional processing, most notably the amygdala. Prefrontal control over the amygdala can occur through several cortico-amygdala connections, which is of obvious importance for mental health and psychiatric disorders. For example, a dlPFC–vmPFC–amygdala circuit is thought to be involved in the ability to reframe information or experiences of emotional significance to enable more adaptive behavior (Denny et al. 2023). These connections are, of course, bidirectional; there are times when salience of information needed for survival detected by the amygdala necessitates more automatic action than a more thoughtful and planned response that would involve prefrontal control. Exactly which parameters govern such "interrupts" remain unknown.

An aspect of connectivity that was not discussed much at the Forum involves connectivity between frontal regions and the hippocampus. This is clearly of importance for the types of functions that frontal cortex supports, both in terms of higher-order learning and adaptive behavior. Once again, it is important to consider both directions of connectivity; that is, from the prefrontal cortex to the hippocampus as well as from the hippocampus to prefrontal cortex. With regards to the former, there is evidence that connectivity from frontal regions to the hippocampus are involved in aiding to highlight specific features of an event at encoding (Kim 2011), in strategic retrieval of memories (Blumenfeld and Ranganath 2019), and even control including the inhibition of memory retrieval itself (dlPFC) (Anderson et al. 2016). Moreover, connectivity between inferior medial prefrontal regions and the hippocampus may allow for a contextual rubric that integrates information across diverse episodes that are then used for inference and higher-order thinking (Morton and Preston 2021). In the opposite direction, connectivity from the hippocampus provides a rich source of information that can be used for prefrontal function. For example, work suggests that connectivity from the hippocampus to prefrontal cortex aids in mental simulation of future events (Campbell et al. 2018), and with integration with information from vmPFC allows for the affective valuation of future possibilities (Benoit et al. 2019).

Finally, different portions of frontal cortex are linked to different sets of cortical areas. In the human brain, this is most notable with regards to different (pre)frontal regions belonging to distinct intrinsic connectivity networks (see Gratton et al., Chapter 11, and Duncan and Friedman, Chapter 9). These different networks have been linked to somewhat different functions (e.g., cognitive control by the frontoparietal network; salience detection and evaluation by the ventral attention/cingulo-opercular networks; and “internal cognition” by the default mode network). However, these broad functions appear rather vague and in need of further definition. It is also unclear how precisely they interact to mediate executive function. Nonetheless, multiple networks of this sort allows different regions of frontal cortex to participate in distinct networks that code different aspects of information, thus providing parallel representations of information used in service of higher-order thought (DeRosa et al. 2024). One issue for further investigation is how “hubs” (see Rowe, Chapter 16) within the prefrontal cortex may participate in several independent neural networks.

Consistent with the patterns observed for prefrontal connectivity, it has been suggested that such connectivity is important for top-down inhibitory control, which is often considered a core aspect of prefrontal function (Friedman and Miyake 2017). It has been suggested that prefrontal regions maintain a task set and use that information to modulate processing in distant brain regions that are relevant for the current goal. Maintenance of such task sets is of particular importance in “inhibitory” tasks in which the maintained task set must override more prepotent or automatic responses (Munakata et al. 2011). Empirical evidence suggests a specific role of different portions of the lateral prefrontal cortex of the right hemisphere in exerting such “inhibitory” control. For example, it was observed within the same individuals that functional connectivity between the dlPFC and the hippocampus is associated with an individual’s ability to inhibit memory retrieval, functional connectivity between this region and the amygdala is associated with the ability to suppress emotional responsivity and functional connectivity between this region and the inferior frontal gyrus and subthalamic nucleus predicts the ability to inhibit motor responses (Depue et al. 2016).

Social Processing, Higher-Order Cognitive Skills, and Development

On display at the Forum were diverse and complicated aspects of cognitive and emotional processing supported by (pre)frontal mechanisms. Many other topics, however, did not receive much attention but make the frontal lobes an intriguing and important brain region to study and understand.

One such issue is how frontal neural mechanisms allow for a more developed understanding of the self, one's relationship to others (including Theory of Mind), and the ability to use emotions for a higher purpose than self-survival. These abilities include self-evaluation and extend as empathy, moral reasoning, and judgment. Evidence exists to suggest that these abilities rely at least in part on medial prefrontal regions that form part of the default mode network (Andrews-Hanna et al. 2010), but also most likely require an exquisite coordination of information from across the brain to be integrated in frontal regions. As such, a better understanding of the underlying motifs of frontal organization and function, as well as the essential computations performed by frontal regions, are likely to be necessary to understand how the frontal lobe contribute to those skills and abilities that make us uniquely human.

While some of these higher-order abilities can be examined in animal models (e.g., associative inference), many aspects of human cognition will not benefit from cross-species comparisons (Levy 2023). In general, these abilities derive from the role that the prefrontal cortex plays in temporal processing and integration (Fuster 2001). In humans, the more extended capacity to consider and integrate information over longer time spans allows for abilities such as higher-order planning, analogical reasoning, and likely even historical perspective.

While certain aspects of prefrontal function can be explained by reinforcement learning algorithms (see Shenhav et al., Chapter 12), in which learning is based on whether an action leads to a reward or not, some do not rely on having a prior experience. These aspects of human frontal function allow not only for simulations based on past experience but for the novel, innovative, and creative aspects of thought. They cannot be guided by past experience of reward (or lack thereof), but only by one's imagination, insight, or conceptual vision.

Finally, another issue of great importance is the development of prefrontal function and the factors that influence its development (Rowe et al., Chapter 16). Prefrontal cortex undergoes protracted development through the late teens into the early 20s and supports many of the abilities generally considered to make one an "adult." At a cellular level, these processes includes cell division, migration axonal connections, synaptogenesis, pruning, and myelination. At a larger scale, these processes include changes in network coherence and connectivity. Understanding how various insults, such as stress, impacts these processes is critical for evaluating functional deficits and developing strategies for early intervention. For example, stress during early development can cause structural and connectivity changes. While touched on briefly by Rowe et al. (Chapter 16), this topic needs further discussion. Developmental issues were not emphasized, in part, because they are currently the focus of much work internationally, including the Adolescent Brain Cognitive Development (ABCD) Study in the United States (https://abcdstudy.org) and the IMAGEN project in Europe (Schumann et al. 2010). Nonetheless, such work is likely to be a rich source of information to address many of the issues raised above, from how cortical and subcortical interaction lead to increase cognitive control, to the understanding of how we as individuals can infer other's likely internal thoughts.

Summarizing Statement

The Forum provided excellent and much needed discussion and analysis of where we stand today in our scientific understanding of this critically important brain region for humans as well as other animals. Moreover, the concluding chapters of each section describing the group discussions provide a roadmap for the conceptual and practical issues that will need to be addressed to further enhance our understanding of the frontal lobes. A general conclusion across many of the group discussions at the Forum was the realization that, while the frontal lobes are a very active area of research, much of this work occurs within "silos," and there is not as much communication amongst researchers as might be desirable. We may be divided by species, prefrontal region, specific function or technique. Animal researchers may focus on a particular species and not be integrating knowledge with findings found in other species; individuals may focus on their particular frontal region of interest and not consider how that region interfaces with other regions of the frontal cortex or indeed the rest of the brain; researchers may focus on a particular function (e.g., language) of a region of prefrontal cortex (e.g., left ventrolateral prefrontal cortex), and neglect whether that region might be involved in other aspects of cognitive or emotional processing. Finally, it is evident that the methods we employ for analyzing the frontal cortex vary enormously in terms of their spatial and temporal resolution, but that more effective cross-disciplinary integration of the data they acquire is required, for example to understand how neural networks measured using fMRI in humans relate to anatomical and electrophysiological findings at the cellular level. The need to engage with, and resolve, the complexities of the techniques themselves also lead to methodological "silos," that are obstacles to broader discussion.

A take home message for us as convenors of the Forum, therefore, was that more cross-talk would go a long way to speeding up our understanding of the frontal lobes, and we were so very grateful that the Forum provided an opportunity to take a meaningful step in this direction.

Bibliography

Note: Numbers in square brackets denote the chapter in which an entry is cited.

Abbie, A. A. 1940. Cortical Lamination in the Monotremata. *J. Comp. Neurol.* **72**:429–467. [5]

———. 1942. Cortical Lamination in a Polyprotodont Marsupial, Perameles Nasuta. *J. Comp. Neurol.* **76**:509–536. [5]

Abdallah, C. G., K.-H. Ahn, L. A. Averill, et al. 2020a. A Robust and Reproducible Connectome Fingerprint of Ketamine Is Highly Associated with the Connectomic Signature of Antidepressants. *Neuropsychopharmacol.* **46**:478–485. [13]

Abdallah, C. G., L. A. Averill, K.S A. Collins, et al. 2017. Ketamine Treatment and Global Brain Connectivity in Major Depression. *Neuropsychopharmacol.* **42**:1210–1219. [13]

Abdallah, C. G., L. A. Averill, R. Gueorguieva, et al. 2020b. Modulation of the Antidepressant Effects of Ketamine by the mTORC1 Inhibitor Rapamycin. *Neuropsychopharmacol.* **45**:990–997. [13]

Abdallah, C. G., H. M. De Feyter, L. A. Averill, et al. 2018. The Effects of Ketamine on Prefrontal Glutamate Neurotransmission in Healthy and Depressed Subjects. *Neuropsychopharmacol.* **43**:2154–2160. [13]

Abdallah, M., G. Zanitti, V. Iovene, and D. Wassermann. 2022. Functional Gradients in the Human Lateral Prefrontal Cortex Revealed by a Comprehensive Coordinate-Based Meta-Analysis. *eLife* **11**:e76926. [4, 7]

Abe, H., and D. Lee. 2011. Distributed Coding of Actual and Hypothetical Outcomes in the Orbital and Dorsolateral Prefrontal Cortex. *Neuron* **70**:731–741. [5]

Abelson, J. L., G. C. Curtis, O. Sagher, et al. 2005. Deep Brain Stimulation for Refractory Obsessive-Compulsive Disorder. *Biol. Psych.* **57**:510–516. [15]

Adelhöfer, N., and C. Beste. 2020. Pre-Trial Theta Band Activity in the Ventromedial Prefrontal Cortex Correlates with Inhibition-Related Theta Band Activity in the Right Inferior Frontal Cortex. *Neuroimage* **219**:117052. [12]

Adólfsdóttir, S., J. Haász, E. Wehling, et al. 2014. Salient Measures of Inhibition and Switching Are Associated with Frontal Lobe Gray Matter Volume in Healthy Middle-Aged and Older Adults. *Neuropsychol.* **28**:859–869. [12]

Aggleton, J. P., N. F. Wright, D. L. Rosene, and R. C. Saunders. 2015. Complementary Patterns of Direct Amygdala and Hippocampal Projections to the Macaque Prefrontal Cortex. *Cereb. Cortex* **25**:4351–4373. [8]

Aguirre, C. G., J. H. Woo, J. L. Romero-Sosa, et al. 2023. Dissociable Contributions of Basolateral Amygdala and Ventrolateral Orbitofrontal Cortex to Flexible Learning under Uncertainty. *bioRxiv* **04**:2023.2004.2003.535471. [4]

Ahmari, S. E., T. Spellman, N. L. Douglass, et al. 2013. Repeated Cortico-Striatal Stimulation Generates Persistent OCD-Like Behavior. *Science* **340**:1234–1239. [4]

Akaishi, R., N. Kolling, J. W. Brown, and M. Rushworth. 2016. Neural Mechanisms of Credit Assignment in a Multicue Environment. *J. Neurosci.* **36**:1096–1112. [4]

Akam, T., I. Rodrigues-Vaz, I. Marcelo, et al. 2021. The Anterior Cingulate Cortex Predicts Future States to Mediate Model-Based Action Selection. *Neuron* **109**:149–163. [4]

Alexander, G. E., M. R. DeLong, and P. L. Strick. 1986. Parallel Organization of Functionally Segregated Circuits Linking Basal Ganglia and Cortex. *Annu. Rev. Neurosci.* **9**:357–381. [8, 15]

Alexander, L., P. L. R. Gaskin, S. J. Sawiak, et al. 2019. Fractionating Blunted Reward Processing Characteristic of Anhedonia by over-Activating Primate Subgenual Anterior Cingulate Cortex. *Neuron* **101**:307–320. [4, 8, 13, 16]

Alexander, L., C. M. Wood, P. L. R. Gaskin, et al. 2020. Over-Activation of Primate Subgenual Cingulate Cortex Enhances the Cardiovascular, Behavioral and Neural Responses to Threat. *Nat. Commun.* **11**:5386. [4, 8, 13, 16]

Alexander, L., C. M. Wood, and A. C. Roberts. 2023. The Ventromedial Prefrontal Cortex and Emotion Regulation: Lost in Translation? *J. Physiol.* **601**:37–50. [3]

Alexander, W. H., and J. W. Brown. 2011. Medial Prefrontal Cortex as an Action-Outcome Predictor. *Nat. Neurosci.* **14**:1338–1344. [1]

Ali, F., D. M. Gerhard, K. Sweasy, et al. 2020. Ketamine Disinhibits Dendrites and Enhances Calcium Signals in Prefrontal Dendritic Spines. *Nat. Commun.* **11**:72. [13]

Allen, E. J., G. St-Yves, Y. Wu, et al. 2022. A Massive 7T fMRI Dataset to Bridge Cognitive Neuroscience and Artificial Intelligence. *Nat. Neurosci.* **25**:116–126. [12]

Allen Institute for Brain Science. 2011. Allen Reference Atlas: Mouse Brain. atlas.brain-map.org. (accessed Jan. 19, 2024). [2]

Alsiö, J., O. Lehmann, C. McKenzie, et al. 2020. Serotonergic Innervations of the Orbitofrontal and Medial-Prefrontal Cortices Are Differentially Involved in Visual Discrimination and Reversal Learning in Rats. *Cerebral Cortex* **31**:1090–1105. [4]

Alstott, J., M. Breakspear, P. Hagmann, L. Cammoun, and O. Sporns. 2009. Modeling the Impact of Lesions in the Human Brain. *PLoS Comput. Biol.* **5**:e1000408. [12]

Altmann, A., D. M. Cash, M. Bocchetta, et al. 2020. Analysis of Brain Atrophy and Local Gene Expression in Genetic Frontotemporal Dementia. *Brain Commun.* **2**:fcaa122. [16]

Amarante, L. M., M. S. Caetano, and M. Laubach. 2017. Medial Frontal Theta Is Entrained to Rewarded Actions. *J. Neurosci.* **37**:10757–10769. [4]

Amiez, C., R. Neveu, D. Warrot, et al. 2013. The Location of Feedback-Related Activity in the Midcingulate Cortex Is Predicted by Local Morphology. *J. Neurosci.* **33**:2217. [4]

Amiez, C., and M. Petrides. 2014. Neuroimaging Evidence of the Anatomo-Functional Organization of the Human Cingulate Motor Areas. *Cereb. Cortex* **24**:563–578. [4]

Amiez, C., J. Sallet, W. D. Hopkins, et al. 2019. Sulcal Organization in the Medial Frontal Cortex Provides Insights into Primate Brain Evolution. *Nat. Commun.* **10**:3437. [4]

Amiez, C., J. Sallet, J. Novek, et al. 2021. Chimpanzee Histology and Functional Brain Imaging Show That the Paracingulate Sulcus Is Not Human-Specific. *Commun. Biol.* **4**:1–12. [4]

Amiez, C., C. R. E Wilson, and E. Procyk. 2018. Variations of Cingulate Sulcal Organization and Link with Cognitive Performance. *Sci. Rep.* 8:13988. [4]

Amit, D. J., and N. Brunel. 1997. Model of Global Spontaneous Activity and Local Structured Activity during Delay Periods in the Cerebral Cortex. *Cereb. Cortex* **7**:237–252. [10]

Ammons, C. J., M. E. Winslett, J. Bice, et al. 2021. The Mid-Fusiform Sulcus in Autism Spectrum Disorder: Establishing a Novel Anatomical Landmark Related to Face Processing. *Autism Res.* **14**:53–64. [4]

Amso, D., C. Salhi, and D. Badre. 2019. The Relationship between Cognitive Enrichment and Cognitive Control: A Systematic Investigation of Environmental Influences on Development through Socioeconomic Status. *Dev Psychobiol* **61**:159–178. [7]

Amunts, K., M. Lenzen, A. D. Friederici, et al. 2010. Broca's Region: Novel Organizational Principles and Multiple Receptor Mapping. *PLoS Biol.* **8**:e1000489. [4]

Amunts, K., A. Schleicher, U. Bürgel, et al. 1999. Broca's Region Revisited: Cytoarchitecture and Intersubject Variability. *Journal of Comparative Neurology* **412**:319–341. [4]

Amunts, K., A. Schleicher, and K. Zilles. 2004. Outstanding Language Competence and Cytoarchitecture in Broca's Speech Region. *Brain and Language* **89**:346–353. [4]

Anastasiades, P. G., and A. G. Carter. 2021. Circuit Organization of the Rodent Medial Prefrontal Cortex. *Trends Neurosci.* **44**:550–563. [2]

Andermann, M. L., N. B. Gilfoy, G. J. Goldey, et al. 2013. Chronic Cellular Imaging of Entire Cortical Columns in Awake Mice Using Microprisms. *Neuron* **80**:900–913. [13]

Anderson, M. C., J. G. Bunce, and H. Barbas. 2016. Prefrontal–Hippocampal Pathways Underlying Inhibitory Control over Memory. *Neurobiol. Learn. Mem.* **134**:145–161. [17]

Andrews-Hanna, J. R., J. S. Reidler, J. Sepulcre, R. Poulin, and R. L. Buckner. 2010. Functional-Anatomic Fractionation of the Brain's Default Network. *Neuron* **65**:550–562. [11, 17]

Ansell, E. B., K. Rando, K. Tuit, J. Guarnaccia, and R. Sinha. 2012. Cumulative Adversity and Smaller Gray Matter Volume in Medial Prefrontal, Anterior Cingulate, and Insula Regions. *Biol. Psych.* **72**:57–64. [16]

Aoi, M. C., V. Mante, and J. W. Pillow. 2020. Prefrontal Cortex Exhibits Multidimensional Dynamic Encoding during Decision-Making. *Nat. Neurosci.* **23**:1410–1420. [12]

Aponik-Gremillion, L., Y. Y. Chen, E. Bartoli, et al. 2022. Distinct Population and Singleneuron Selectivity for Executive and Episodic Processing in Human Dorsal Posterior Cingulate. *eLife* **11**:e80722. [4]

Apps, M. A. J., M. F. S. Rushworth, and S. W. C. Chang. 2016. The Anterior Cingulate Gyrus and Social Cognition: Tracking the Motivation of Others. *Neuron* **90**:692–707. [4]

Apšvalka, D., C. S. Ferreira, T. W. Schmitz, J. B. Rowe, and M. C. Anderson. 2022. Dynamic Targeting Enables Domain-General Inhibitory Control over Action and Thought by the Prefrontal Cortex. *Nat. Commun.* **13**:274. [9]

Aquino, T. G., J. Cockburn, A. N. Mamelak, U. Rutishauser, and J. P. O'Doherty. 2023. Neurons in Human Pre-Supplementary Motor Area Encode Key Computations for Value-Based Choice. *Nat. Hum. Behav.* **7**:970–985. [12]

Armbruster, B. N., X. Li, M. H. Pausch, S. Herlitze, and B. L. Roth. 2007. Evolving the Lock to Fit the Key to Create a Family of G Protein-Coupled Receptors Potently Activated by an Inert Ligand. *PNAS* **104**:5163–5168. [2, 4]

Armstrong, E., A. Schleicher, H. Omran, M. Curtis, and K. Zilles. 1995. The Ontogeny of Human Gyrification. *Cereb. Cortex* **5**:56–63. [4]

Arns, M., A. Etkin, U. Hegerl, et al. 2015. Frontal and Rostral Anterior Cingulate (Racc) Theta EEG in Depression: Implications for Treatment Outcome? *Eur. Neuropsychopharmacol.* **25**:1190–1200. [13]

Arnsten, A. F. T. 2009. Stress Signalling Pathways That Impair Prefrontal Cortex Structure and Function. *Nat. Rev. Neurosci.* **10**:410–422. [16]

———. 2015. Stress Weakens Prefrontal Networks: Molecular Insults to Higher Cognition. *Nat. Neurosci.* **18**:1376–1385. [16]

Arnsten, A. F. T. 2020. Guanfacine's Mechanism of Action in Treating Prefrontal Cortical Disorders: Successful Translation across Species. *Neurobiol. Learn. Mem.* **176**:107327. [16]

Arnsten, A. F. T., D. Datta, and M. Wang. 2021. The Genie in the Bottle-Magnified Calcium Signaling in Dorsolateral Prefrontal Cortex. *Mol. Psych.* **26**:3684–3700. [16]

Arnsten, A. F. T., M. K. P. Joyce, and A. C. Roberts. 2023. The Aversive Lens: Stress Effects on the Prefrontal-Cingulate Cortical Pathways That Regulate Emotion. *Neurosci. Biobehav. Rev.* **145**:105000. [13]

Arnsten, A. F. T., C. D. Paspalas, N. J. Gamo, Y. Yang, and M. Wang. 2010. Dynamic Network Connectivity: A New Form of Neuroplasticity. *Trends Cogn. Sci.* **14**:365–375. [12]

Arnsten, A. F. T., and K. Rubia. 2012. Neurobiological Circuits Regulating Attention, Cognitive Control, Motivation, and Emotion: Disruptions in Neurodevelopmental Psychiatric Disorders. *J. Am. Acad. Child Adolesc. Psych.* **51**:356–367. [14]

Arnsten, A. F. T., E. Woo, S. Yang, M. Wang, and D. Datta. 2022. Unusual Molecular Regulation of Dorsolateral Prefrontal Cortex Layer III Synapses Increases Vulnerability to Genetic and Environmental Insults in Schizophrenia. *Biol. Psych.* **92**:480–490. [16]

Aron, A. R., R. B. Ivry, K. J. Jeffery, et al. 2020. How Can Neuroscientists Respond to the Climate Emergency? *Neuron* **106**:17–20. [16]

Aron, A. R., and R. A. Poldrack. 2006. Cortical and Subcortical Contributions to Stop Signal Response Inhibition: Role of the Subthalamic Nucleus. *J. Neurosci.* **26**:2424–2433. [9]

Aron, A. R., T. W. Robbins, and R. A. Poldrack. 2014. Inhibition and the Right Inferior Frontal Cortex: One Decade on. *Trends Cogn. Sci.* **18**:177–185. [4]

Arrington, C. M., and G. D. Logan. 2004. The Cost of a Voluntary Task Switch. *Psychol. Sci.* **15**:610–615. [12]

Assem, M., M. F. Glasser, D. C. Van Essen, and J. Duncan. 2020. A Domain-General Cognitive Core Defined in Multimodally Parcellated Human Cortex. *Cereb. Cortex* **30**:4361–4380. [7, 9]

Assem, M., S. Shashidhara, M. F. Glasser, and J. Duncan. 2022. Precise Topology of Adjacent Domain-General and Sensory-Biased Regions in the Human Brain. *Cereb. Cortex* **32**:2521–2537. [9, 12]

———. 2024. Basis of Executive Functions in Fine-Grained Architecture of Cortical and Subcortical Human Brain Networks. *Cereb. Cortex* **34**:bhad537. [9]

Atlas, L. Y., M. A. Lindquist, N. Bolger, and T. D. Wager. 2014. Brain Mediators of the Effects of Noxious Heat on Pain. *Pain* **155**:1632–1648. [16]

Au, J., E. Sheehan, N. Tsai, et al. 2015. Improving Fluid Intelligence with Training on Working Memory: A Meta-Analysis. *Psychon. Bull. Rev.* **22**:366–377. [14]

Au, J., R. N. Smith-Peirce, E. Carbone, et al. 2022. Effects of Multisession Prefrontal Transcranial Direct Current Stimulation on Long-Term Memory and Working Memory in Older Adults. *J. Cogn. Neurosci.* **34**:1015–1037. [14]

Auer, D. P., B. Pütz, E. Kraft, et al. 2000. Reduced Glutamate in the Anterior Cingulate Cortex in Depression: an *in Vivo* Proton Magnetic Resonance Spectroscopy Study. *Biol. Psychiatry* **47**:305–313. [13]

Autry, A. E., M. Adachi, E. Nosyreva, et al. 2011. NMDA Receptor Blockade at Rest Triggers Rapid Behavioural Antidepressant Responses. *Nature* **475**:91–95. [13]

Averbeck, B. B. 2022. Pruning Recurrent Neural Networks Replicates Adolescent Changes in Working Memory and Reinforcement Learning. *PNAS* **119**:e2121331119. [5]

Averbeck, B. B., and V. D. Costa. 2017. Motivational Neural Circuits Underlying Reinforcement Learning. *Nat. Neurosci.* **20**:505–512. [16]

Averbeck, B. B., J. Lehman, M. Jacobson, and S. N. Haber. 2014. Estimates of Projection Overlap and Zones of Convergence within Frontal-Striatal Circuits. *J. Neurosci.* **34**:9497–9505. [5, 11]

Averbeck, B. B., and E. A. Murray. 2020. Hypothalamic Interactions with Large-Scale Neural Circuits Underlying Reinforcement Learning and Motivated Behavior. *Trends Neurosci.* **43**:681–694. [5, 8]

Averbeck, B. B., J. W. Sohn, and D. Lee. 2006. Activity in Prefrontal Cortex during Dynamic Selection of Action Sequences. *Nat. Neurosci.* **9**:276–282. [8]

Azevedo, F. A. C., L. R. B. Carvalho, L. T. Grinberg, et al. 2009. Equal Numbers of Neuronal and Nonneuronal Cells Make the Human Brain an Isometrically Scaled-up Primate Brain. *J. Comp. Neurol.* **513**:532–541. [4]

Azuar, C., P. Reyes, A. Slachevsky, et al. 2014. Testing the Model of Caudo-Rostral Organization of Cognitive Control in the Human with Frontal Lesions. *Neuroimage* **84**:1053–1060. [7, 8]

Bäckman, L., and L. Nyberg. 2013. Dopamine and Training-Related Working-Memory Improvement. *Neurosci. Biobehav. Rev.* **37**:2209–2219. [14]

Bäckman, L., L. Nyberg, U. Lindenberger, S. C. Li, and L. Farde. 2006. The Correlative Triad among Aging, Dopamine, and Cognition: Current Status and Future Prospects. *Neurosci. Biobehav. Rev.* **30**:791–807. [14]

Bäckman, L., L. Nyberg, A. Soveri, et al. 2011. Effects of Working-Memory Training on Striatal Dopamine Release. *Science* **333**:718. [14]

Baddeley, A. 2012. Working Memory: Theories, Models, and Controversies. *Annu. Rev. Psychol.* **63**:1–29. [6]

Baddeley, A. D. 1986. Working Memory. Oxford: Clarendon Press. [12]

Badre, D. 2008. Cognitive Control, Hierarchy, and the Rostro-Caudal Organization of the Frontal Lobes. *Trends Cogn. Sci.* **12**:193–200. [4, 7, 15, 16]

———. 2020. On Task: How Your Brain Gets Things Done. Princeton: Princeton Univ. Press. [7, 15]

Badre, D., A. Bhandari, H. Keglovits, and A. Kikumoto. 2021. The Dimensionality of Neural Representations for Control. *Curr. Opin. Behav. Sci.* **38**:20–28. [7]

Badre, D., and M. D'Esposito. 2007. Functional Magnetic Resonance Imaging Evidence for a Hierarchical Organization of the Prefrontal Cortex. *J. Cogn. Neurosci.* **19**:2082–2099. [7, 12, 15, 16]

———. 2009. Is the Rostro-Caudal Axis of the Frontal Lobe Hierarchical? *Nat. Rev. Neurosci.* **10**:659–669. [4, 6–8, 12]

Badre, D., B. B. Doll, N. M. Long, and M. J. Frank. 2012. Rostrolateral Prefrontal Cortex and Individual Differences in Uncertainty-Driven Exploration. *Neuron* **73**:595–607. [7, 8, 12]

Badre, D., and M. J. Frank. 2012. Mechanisms of Hierarchical Reinforcement Learning in Cortico-Striatal Circuits 2: Evidence from fMRI. *Cereb. Cortex* **22**:527–536. [7, 15]

Badre, D., M. J. Frank, and C. I. Moore. 2015. Interactionist Neuroscience. *Neuron* **88**:855–860. [8]

Badre, D., J. Hoffman, J. W. Cooney, and M. D'Esposito. 2009. Hierarchical Cognitive Control Deficits Following Damage to the Human Frontal Lobe. *Nat. Neurosci.* **12**:515–522. [7, 8]

Badre, D., A. S. Kayser, and M. D'Esposito. 2010. Frontal Cortex and the Discovery of Abstract Action Rules. *Neuron* **66**:315–326. [7, 10]

Badre, D., and D. E. Nee. 2018. Frontal Cortex and the Hierarchical Control of Behavior. *Trends Cogn. Sci.* **22**:170–188. [6, 7, 11, 12, 15]

Baer, L., S. L. Rauch, H. T. Ballantine, Jr., et al. 1995. Cingulotomy for Intractable Obsessive-Compulsive Disorder. Prospective Long-Term Follow-up of 18 Patients. *Arch. Gen. Psych.* **52**:384–392. [15]

Bailey, S. 1999. Young People, Mental Illness and Stigmatisation. *Psychiatr. Bull.* **23**:107–110. [16]

Baird, B., J. Smallwood, K. J. Gorgolewski, and D. S. Margulies. 2013. Medial and Lateral Networks in Anterior Prefrontal Cortex Support Metacognitive Ability for Memory and Perception. *J. Neurosci.* **33**:16657–16665. [6]

Bakhurin, K. I., V. Goudar, J. L. Shobe, et al. 2017. Differential Encoding of Time by Prefrontal and Striatal Network Dynamics. *J. Neurosci.* **37**:854–870. [4]

Bala, P. C., B. R. Eisenreich, S. B. M. Yoo, et al. 2020. Automated Markerless Pose Estimation in Freely Moving Macaques with Openmonkeystudio. *Nat. Commun.* **11**:4560. [2]

Balaguer, J., H. Spiers, D. Hassabis, and C. Summerfield. 2016. Neural Mechanisms of Hierarchical Planning in a Virtual Subway Network. *Neuron* **90**:893–903. [12]

Balci, F., P. Simen, R. Niyogi, et al. 2010. Acquisition of Decision Making Criteria: Reward Rate Ultimately Beats Accuracy. *Atten. Percept. Psychophys.* **73**:640–657. [12]

Baldassano, C., J. Chen, A. Zadbood, et al. 2017. Discovering Event Structure in Continuous Narrative Perception and Memory. *Neuron* **95**:709–721. [4]

Balleine, B. W. 2019. The Meaning of Behavior: Discriminating Reflex and Volition in the Brain. *Neuron* **104**:47–62. [4, 16]

Balleine, B. W., R. W. Morris, and B. K. Leung. 2015. Thalamocortical Integration of Instrumental Learning and Performance and Their Disintegration in Addiction. *Brain Res.* **1628**:104–116. [4]

Balleine, B. W., and J. P. O'Doherty. 2009. Human and Rodent Homologies in Action Control: Corticostriatal Determinants of Goal-Directed and Habitual Action. *Neuropsychopharmacol.* **35**:48–69. [4]

———. 2010. Human and Rodent Homologies in Action Control: Corticostriatal Determinants of Goal-Directed and Habitual Action. *Neuropsychopharmacol.* **35**:48–69. [2, 4]

Balsters, J. H., V. Zerbi, J. Sallet, N. Wenderoth, and R. B. Mars. 2020. Primate Homologs of Mouse Cortico-Striatal Circuits. *eLife* **9**:e53680. [8]

Bandettini, P. A., L. Huber, and E. S. Finn. 2021. Challenges and Opportunities of Mesoscopic Brain Mapping with fMRI. *Curr. Opin. Behav. Sci.* **40**:189–200. [11]

Banich, M. T. 2009. Executive Function: The Search for an Integrated Account. *Curr. Dir. Psychol. Sci.* **18**:89–94. [1, 11]

———. 2019. The Stroop Effect Occurs at Multiple Points Along a Cascade of Control: Evidence from Cognitive Neuroscience Approaches. *Front. Psychol.* **10**:2164. [11]

Banich, M. T., M. P. Milham, R. Atchley, et al. 2000. fMRI Studies of Stroop Tasks Reveal Unique Roles of Anterior and Posterior Brain Systems in Attentional Selection. *J. Cogn. Neurosci.* **12**:988–1000. [12]

Baniqued, P. L., C. L. Gallen, M. B. Kranz, A. F. Kramer, and M. D'Esposito. 2019. Brain Network Modularity Predicts Cognitive Training-Related Gains in Young Adults. *Neuropsychologia* **131**:205–215. [14]

Banks, G. P., C. B. Mikell, B. E. Youngerman, et al. 2015. Neuroanatomical Characteristics Associated with Response to Dorsal Anterior Cingulotomy for Obsessive-Compulsive Disorder. *JAMA Psych.* **72**:127–135. [15]

Barbalat, G., V. Chambon, P. J. Domenech, et al. 2011. Impaired Hierarchical Control within the Lateral Prefrontal Cortex in Schizophrenia. *Biol. Psych.* **70**:73–80. [16]

Barbas, H. 2000. Connections Underlying the Synthesis of Cognition, Memory, and Emotion in Primate Prefrontal Cortices. *Brain Res. Bull.* **52**:319–330. [8]

Barbas, H., T. H. Henion, and C. R. Dermon. 1991. Diverse Thalamic Projections to the Prefrontal Cortex in the Rhesus Monkey. *J. Comp. Neurol.* **313**:65–94. [8]

Barbas, H., S. Saha, N. Rempel-Clower, and T. Ghashghaei. 2003. Serial Pathways from Primate Prefrontal Cortex to Autonomic Areas May Influence Emotional Expression. *BMC Neurosci.* **4**:25. [3]

Barbey, A. K., R. Colom, J. Solomon, et al. 2012. An Integrative Architecture for General Intelligence and Executive Function Revealed by Lesion Mapping. *Brain* **135**:1154–1164. [9]

Bari, A., A. C. Mar, D. E. Theobald, et al. 2011. Prefrontal and Monoaminergic Contributions to Stop-Signal Task Performance in Rats. *J. Neurosci.* **31**:9254–9263. [4, 16]

Barlow, D. H., and L. A. Campbell. 2000. Mixed Anxiety-Depression and Its Implications for Models of Mood and Anxiety Disorders. *Compr Psychiatry* **41**:55–60. [15]

Barlow, R. L., J. Alsiö, B. Jupp, et al. 2015. Markers of Serotonergic Function in the Orbitofrontal Cortex and Dorsal Raphé Nucleus Predict Individual Variation in Spatial-Discrimination Serial Reversal Learning. *Neuropsychopharmacol.* **40**:1619–1630. [4]

Barnett, J. H., J. Heron, S. M. Ring, et al. 2007. Gender-Specific Effects of the Catechol-O-Methyltransferase Val108/158Met Polymorphism on Cognitive Function in Children. *Am. J. Psych.* **164**:142–149. [16]

Barraclough, D. J., M. L. Conroy, and D. Lee. 2004. Prefrontal Cortex and Decision Making in a Mixed-Strategy Game. *Nat. Neurosci.* **7**:404–410. [12]

Barreiros, I. V., H. Ishii, M. E. Walton, and M. C. Panayi. 2021a. Defining an Orbitofrontal Compass: Functional and Anatomical Heterogeneity across Anterior-Posterior and Medial-Lateral Axes. *Behav. Neurosci.* **135**:165–173. [2, 4]

Barreiros, I. V., M. C. Panayi, and M. E. Walton. 2021b. Organization of Afferents Along the Anterior-Posterior and Medial-Lateral Axes of the Rat Orbitofrontal Cortex. *Neurosci.* **460**:53–68. [2, 4]

Barrios-Anderson, A., N. C. R. McLaughlin, M. T. Patrick, et al. 2022. The Patient Lived-Experience of Ventral Capsulotomy for Obsessive-Compulsive Disorder: An Interpretive Phenomenological Analysis of Neuroablative Psychiatric Neurosurgery. *Front. Integr. Neurosci.* **16**:802617. [16]

Bartholomeusz, C. F., S. L. Whittle, A. Montague, et al. 2013. Sulcogyral Patterns and Morphological Abnormalities of the Orbitofrontal Cortex in Psychosis. *Prog. Neuropsychopharmacol. Biol. Psych.* **44**:168–177. [16]

Bartlett, E. A., C. DeLorenzo, P. Sharma, et al. 2018. Pretreatment and Early-Treatment Cortical Thickness Is Associated with SSRI Treatment Response in Major Depressive Disorder. *Neuropsychopharmacol.* **43**:2221–2230. [13]

Bartra, O., J. T. McGuire, and J. W. Kable. 2013. The Valuation System: A Coordinate-Based Meta-Analysis of BOLD fMRI Experiments Examining Neural Correlates of Subjective Value. *Neuroimage* **76**:412–427. [4]

Basile, B. M., J. L. Schafroth, C. L. Karaskiewicz, S. W. C. Chang, and E. A. Murray. 2020. The Anterior Cingulate Cortex Is Necessary for Forming Prosocial Preferences from Vicarious Reinforcement in Monkeys. *PLoS Biol.* **18**:e3000677. [4]

Bassett, D. S., C. H. Xia, and T. D. Satterthwaite. 2018. Understanding the Emergence of Neuropsychiatric Disorders with Network Neuroscience. *Biol. Psych. Cogn. Neurosci. Neuroimag.* **3**:742–753. [16]

Baum, G. L., D. R. Roalf, P. A. Cook, et al. 2018. The Impact of in-Scanner Head Motion on Structural Connectivity Derived from Diffusion MRI. *Neuroimage* **173**:275–286. [11]

Baxter Jr., L. R., J. M. Schwartz, J. C. Mazziotta, et al. 1988. Cerebral Glucose Metabolic Rates in Nondepressed Patients with Obsessive-Compulsive Disorder. *Am. J. Psych.* **145**:1560–1563. [15]

Baxter, M. G., D. Gaffan, D. A. Kyriazis, and A. S. Mitchell. 2009. Ventrolateral Prefrontal Cortex Is Required for Performance of a Strategy Implementation Task but Not Reinforcer Devaluation Effects in Rhesus Monkeys. *Eur. J. Neurosci.* **29**:2049–2059. [8]

Bechara, A., H. Damasio, D. Tranel, and A. R. Damasio. 1997. Deciding Advantageously before Knowing the Advantageous Strategy. *Science* **275**:1293–1295. [12]

Beck, J. A. 2011. Cognitive Behavior Therapy: Basics and Beyond. New York, NY: Guilford Press. [16]

Behrens, T. E., M. W. Woolrich, M. E. Walton, and M. F. Rushworth. 2007. Learning the Value of Information in an Uncertain World. *Nat. Neurosci.* **10**:1214–1221. [10]

Beierholm, U. R., C. Anen, S. Quartz, and P. Bossaerts. 2011. Separate Encoding of Model-Based and Model-Free Valuations in the Human Brain. *Neuroimage* **58**:955–962. [12]

Beishon, L., K. Intharakham, D. Swienton, et al. 2020. Neuroimaging Outcomes in Studies of Cognitive Training in Mild Cognitive Impairment and Early Alzheimer's Disease: A Systematic Review. *Curr. Alzheimer Res.* **17**:472–486. [14]

Belin, D., A. Belin-Rauscent, B. J. Everitt, and J. W. Dalley. 2016. In Search of Predictive Endophenotypes in Addiction: Insights from Preclinical Research. *Genes Brain Behav.* **15**:74–88. [16]

Bellander, M., Y. Brehmer, H. Westerberg, et al. 2011. Preliminary Evidence That Allelic Variation in the LMX1A gene Influences Training-Related Working Memory Improvement. *Neuropsychologia* **49**:1938–1942. [14]

Belleville, S., S. Mellah, C. de Boysson, J.-F. Demonet, and B. Bier. 2014. The Pattern and Loci of Training-Induced Brain Changes in Healthy Older Adults Are Predicted by the Nature of the Intervention. *PloS One* **9**:e102710. [14]

Benjet, C., E. Bromet, E. G. Karam, et al. 2016. The Epidemiology of Traumatic Event Exposure Worldwide: Results from the World Mental Health Survey Consortium. *Psychol. Med.* **46**:327–343. [16]

Benoit, R. G., P. C. Paulus, and D. L. Schacter. 2019. Forming Attitudes via Neural Activity Supporting Affective Episodic Simulations. *Nat. Commun.* **10**:2215. [17]

Benson, D. F., D. T. Stuss, M. A. Naeser, et al. 1981. The Long-Term Effects of Prefrontal Leukotomy. *Arch. Neurol.* **38**:165–169. [15]

Berendse, H. W., Y. Galis-de Graaf, and H. J. Groenewegen. 1992. Topographical Organization and Relationship with Ventral Striatal Compartments of Prefrontal Corticostriatal Projections in the Rat. *J. Comp. Neurol.* **316**:314–347. [3]

Berger, M., N. S. Agha, and A. Gail. 2020. Wireless Recording from Unrestrained Monkeys Reveals Motor Goal Encoding Beyond Immediate Reach in Frontoparietal Cortex. *eLife* **9**:e51322. [2]

Berman, R. M., A. Cappiello, A. Anand, et al. 2000. Antidepressant Effects of Ketamine in Depressed Patients. *Biol. Psychiatry* **47**:351–354. [13]

Bernardi, S., M. K. Benna, M. Rigotti, et al. 2020. The Geometry of Abstraction in the Hippocampus and Prefrontal Cortex. *Cell* **183**:954–967. [5, 6, 12]

Bernier, P.-M., M. Cieslak, and S. T. Grafton. 2012. Effector Selection Precedes Reach Planning in the Dorsal Parietofrontal Cortex. *J. Neurophysiol.* **108**:57–68. [12]

Bertolero, M. A., B. T. T. Yeo, D. S. Bassett, and M. D'Esposito. 2018. A Mechanistic Model of Connector Hubs, Modularity and Cognition. *Nat. Hum. Behav.* **2**:765–777. [11]

Bertolero, M. A., B. T. T. Yeo, and M. D'Esposito. 2015. The Modular and Integrative Functional Architecture of the Human Brain. *PNAS* **112**:E6798–6807. [11]

Beste, C., C. K. E. Moll, M. Pötter-Nerger, and A. Münchau. 2018. Striatal Microstructure and Its Relevance for Cognitive Control. *Trends Cogn. Sci.* **22**:747–751. [12]

Beste, C., A. Münchau, and C. Frings. 2023. Towards a Systematization of Brain Oscillatory Activity in Actions. *Commun. Biol.* **6**:137. [12]

Bethlehem, R. A. I., C. Paquola, J. Seidlitz, et al. 2020. Dispersion of Functional Gradients across the Adult Lifespan. *Neuroimage* **222**:117299. [16]

Bethlehem, R. A. I., J. Seidlitz, S. R. White, et al. 2022a. Brain Charts for the Human Lifespan. *Nature* **604**:525–533. [16]

———. 2022b. Publisher Correction: Brain Charts for the Human Lifespan. *Nature* **610**:E6. [16]

Betzel, R. F., and D. S. Bassett. 2017. Multi-Scale Brain Networks. *Neuroimage* **160**:73–83. [12]

Bhandari, A., and D. Badre. 2018. Learning and Transfer of Working Memory Gating Policies. *Cognition* **172**:89–100. [7]

Bichot, N. P., R. Xu, A. Ghadooshahy, M. L. Williams, and R. Desimone. 2019. The Role of Prefrontal Cortex in the Control of Feature Attention in Area V4. *Nat. Commun.* **10**:5727. [12]

Bicks, L. K., K. Yamamuro, M. E. Flanigan, et al. 2020. Prefrontal Parvalbumin Interneurons Require Juvenile Social Experience to Establish Adult Social Behavior. *Nat. Commun.* **11**:1003. [16]

Bieth, T., M. Ovando-Tellez, A. Lopez-Persem, et al. 2024. Time Course of EEG Power during Creative Problem-Solving with Insight or Remote Thinking. *Hum. Brain Mapp.* **45**:e26547. [17]

Birnbaum, S. G., P. X. Yuan, M. Wang, et al. 2004. Protein Kinase C Overactivity Impairs Prefrontal Cortical Regulation of Working Memory. *Science* **306**:882–884. [16]

Birrell, J. M., and V. J. Brown. 2000. Medial Frontal Cortex Mediates Perceptual Attentional Set Shifting in the Rat. *J. Neurosci.* **20**:4320–4324. [4]

Bissonette, G. B., G. J. Martins, T. M. Franz, et al. 2008. Double Dissociation of the Effects of Medial and Orbital Prefrontal Cortical Lesions on Attentional and Affective Shifts in Mice. *J. Neurosci.* **28**:11124–11130. [4]

Blackman, R. K., D. A. Crowe, A. L. DeNicola, et al. 2016. Monkey Prefrontal Neurons Reflect Logical Operations for Cognitive Control in a Variant of the AX Continuous Performance Task (AX-CPT). *J. Neurosci.* **36**:4067–4079. [5]

Blain, B., G. Hollard, and M. Pessiglione. 2016. Neural Mechanisms Underlying the Impact of Daylong Cognitive Work on Economic Decisions. *PNAS* **113**:6967–6972. [12]

Blanke, O., S. Morand, G. Thut, et al. 1999. Visual Activity in the Human Frontal Eye Field. *Neuroreport* **10**:925–930. [8]

Blaze, J., L. Scheuing, and T. L. Roth. 2013. Differential Methylation of Genes in the Medial Prefrontal Cortex of Developing and Adult Rats Following Exposure to Maltreatment or Nurturing Care during Infancy. *Dev. Neurosci.* **35**:306–316. [16]

Bliss-Moreau, E., V. D. Costa, and M. G. Baxter. 2022. A Pragmatic Reevaluation of the Efficacy of Nonhuman Primate Optogenetics for Psychiatry. *Oxf. Open Neurosci.* **1**:kvac006. [2]

Bludau, S., S. B. Eickhoff, H. Mohlberg, et al. 2014. Cytoarchitecture, Probability Maps and Functions of the Human Frontal Pole. *Neuroimage* **93**:260–275. [4]

Blumenfeld, R. S., and C. Ranganath. 2019. Chapter 12: The Lateral Prefrontal Cortex and Human Long-Term Memory. In: Handbook of Clinical Neurology, ed. M. D'Esposito and J. H. Grafman, pp. 221–235, vol. 163. Amsterdam: Elsevier. [17]

Blumenfeld, R. S., T. G. Lee, and M. D'Esposito. 2014. The Effects of Lateral Prefrontal Transcranial Magnetic Stimulation on Item Memory Encoding. *Neuropsychologia* **53**:197–202.

Bogacz, R., E. Brown, J. Moehlis, P. Holmes, and J. D. Cohen. 2006. The Physics of Optimal Decision Making: A Formal Analysis of Models of Performance in Two-Alternative Forced-Choice Tasks. *Psychol. Rev.* **113**:700–765. [12]

Bolkan, S. S., J. M. Stujenske, S. Parnaudeau, et al. 2017. Thalamic Projections Sustain Prefrontal Activity during Working Memory Maintenance. *Nat. Neurosci.* **20**:987–996. [3, 4]

Bonaventura, J., S. Lam, M. Carlton, et al. 2021. Pharmacological and Behavioral Divergence of Ketamine Enantiomers: Implications for Abuse Liability. *Mol. Psychiatry* **26**:6704–6722. [13]

Bondi, C. O., G. Rodriguez, G. G. Gould, A. Frazer, and D. A. Morilak. 2008. Chronic Unpredictable Stress Induces a Cognitive Deficit and Anxiety-Like Behavior in Rats That Is Prevented by Chronic Antidepressant Drug Treatment. *Neuropsychopharmacol.* **33**:320–331. [16]

Boorman, E. D., T. E. Behrens, M. W. Woolrich, and M. F. Rushworth. 2009. How Green Is the Grass on the Other Side? Frontopolar Cortex and the Evidence in Favor of Alternative Courses of Action. *Neuron* **62**:733–743. [7]

Borst, G., A. Cachia, J. Vidal, et al. 2014. Folding of the Anterior Cingulate Cortex Partially Explains Inhibitory Control during Childhood: A Longitudinal Study. *Dev. Cogn. Neurosci.* **9**:126–135. [4]

Botvinick, M. M. 2008. Hierarchical Models of Behavior and Prefrontal Function. *Trends Cogn. Sci.* **12**:201–208. [7]

Botvinick, M. M., T. S. Braver, D. M. Barch, C. S. Carter, and J. D. Cohen. 2001. Conflict Monitoring and Cognitive Control. *Psychol. Rev.* **108**:624. [11, 12]

Botvinick, M. M., and J. D. Cohen. 2014. The Computational and Neural Basis of Cognitive Control: Charted Territory and New Frontiers. *Cogn. Sci.* **38**:1249–1285. [12]

Botvinick, M. M., J. D. Cohen, and C. S. Carter. 2004. Conflict Monitoring and Anterior Cingulate Cortex: an Update. *Trends Cogn. Sci.* **8**:539–546. [1, 12]

Bouchacourt, F., S. Palminteri, E. Koechlin, and S. Ostojic. 2020. Temporal Chunking as a Mechanism for Unsupervised Learning of Task-Sets. *eLife* **9**:e50469. [12]

Boulougouris, V., J. W. Dalley, and T. W. Robbins. 2007. Effects of Orbitofrontal, Infralimbic and Prelimbic Cortical Lesions on Serial Spatial Reversal Learning in the Rat. *Behav. Brain Res.* **179**:219–228. [4]

Boyden, E. S., A. Katoh, and J. L. Raymond. 2004. Cerebellum-Dependent Llearning: The Role of Multiple Plasticity Mechanisms. *Annu. Rev. Neurosci.* **27**:581–609. [14]

Bradfield, L. A., A. Dezfouli, M. van Holstein, B. Chieng, and B. W. Balleine. 2015. Medial Orbitofrontal Cortex Mediates Outcome Retrieval in Partially Observable Task Situations. *Neuron* **88**:1268–1280. [4, 7]

Braga, R. M., and R. L. Buckner. 2017. Parallel Interdigitated Distributed Networks within the Individual Estimated by Intrinsic Functional Connectivity. *Neuron* **95**:457–471. [11]

Braga, R. M., L. M. DiNicola, H. C. Becker, and R. L. Buckner. 2020. Situating the Left-Lateralized Language Network in the Broader Organization of Multiple Specialized Large-Scale Distributed Networks. *J. Neurophysiol.* **124**:1415–1448. [11]

Braga, R. M., K. R. A. Van Dijk, J. R. Polimeni, M. C. Eldaief, and R. L. Buckner. 2019. Parallel Distributed Networks Resolved at High Resolution Reveal Close Juxtaposition of Distinct Regions. *J. Neurophysiol.* **121**:1513–1534. [11]

Braun, U., A. Schäfer, H. Walter, et al. 2015. Dynamic Reconfiguration of Frontal Brain Networks during Executive Cognition in Humans. *PNAS* **112**:11678–11683. [11, 14]

Braunstein, L. M., J. J. Gross, and K. N. Ochsner. 2017. Explicit and Implicit Emotion Regulation: A Multi-Level Framework. *Soc. Cogn. Affect. Neurosci.* **12**:1545–1557. [12]

Braver, T. S. 2012. The Variable Nature of Cognitive Control: A Dual Mechanisms Framework. *Trends Cogn. Sci.* **16**:106–113. [11]

Braver, T. S., and S. R. Bongiolatti. 2002. The Role of Frontopolar Cortex in Subgoal Processing during Working Memory. *Neuroimage* **15**:523–536. [15]

Braver, T. S., M. K. Krug, K. S. Chiew, et al. 2014. Mechanisms of Motivation–Cognition Interaction: Challenges and Opportunities. *Cogn. Affect. Behav. Neurosci.* **14**:443–472. [14]

Bray, S., A. Rangel, S. Shimojo, B. W. Balleine, and J. P. O'Doherty. 2008. The Neural Mechanisms Underlying the Influence of Pavlovian Cues on Human Decision Making. *J. Neurosci.* **28**:5861–5866. [4]

Brehmer, Y., H. Westerberg, M. Bellander, et al. 2009. Working Memory Plasticity Modulated by Dopamine Transporter Genotype. *Neurosci. Lett.* **467**:117–120. [14]

Breiter, H. C., S. L. Rauch, K. K. Kwong, et al. 1996. Functional Magnetic Resonance Imaging of Symptom Provocation in Obsessive-Compulsive Disorder. *Arch. Gen. Psych.* **53**:595–606. [15]

Brincat, S. L., J. A. Donoghue, M. K. Mahnke, et al. 2021. Interhemispheric Transfer of Working Memories. *Neuron* **109**:1055–1066. [8]

Brito, G. N., G. J. Thomas, B. J. Davis, and S. I. Gingold. 1982. Prelimbic Cortex, Mediodorsal Thalamus, Septum, and Delayed Alternation in Rats. *Exp. Brain Res.* **46**:52–58. [4]

Brodmann, K. 1909. Vergleichende Lokalisationslehre der Großhirnrinde in Ihren Prinzipien Dargestellt auf Grund des Zellbaues. Leipzig: Johann Ambrosius Barth Verlag. [3–5]

Bromberg-Martin, E. S., and O. Hikosaka. 2009. Midbrain Dopamine Neurons Signal Preference for Advance Information About Upcoming Rewards. *Neuron* **63**:119–126. [2]

Bruce, C. J., and M. E. Goldberg. 1985. Primate Frontal Eye Fields. I. Single Neurons Discharging before Saccades. *J. Neurophysiol.* **53**:603–635. [4, 8]

Bruce, C. J., M. E. Goldberg, M. C. Bushnell, and G. B. Stanton. 1985. Primate Frontal Eye Fields. II. Physiological and Anatomical Correlates of Electrically Evoked Eye Movements. *J. Neurophysiol.* **54**:714–734. [8]

Brunel, N., and X. J. Wang. 2001. Effects of Neuromodulation in a Cortical Network Model of Object Working Memory Dominated by Recurrent Inhibition. *J Comput Neurosci* **11**:63–85. [10]

Brydges, C. R., C. L. Reid, A. M. Fox, and M. Anderson. 2012. A Unitary Executive Function Predicts Intelligence in Children. *Intelligence* **40**:458–469. [9]

Buckner, R. L., and L. M. DiNicola. 2019. The Brain's Default Network: Updated Anatomy, Physiology and Evolving Insights. *Nat. Rev. Neurosci.* **20**:593–608. [4]

Bugeon, S., J. Duffield, M. Dipoppa, et al. 2022. A Transcriptomic Axis Predicts State Modulation of Cortical Interneurons. *Nature* **607**:330–338. [2]

Burgess, P. W. 2000. Strategy Application Disorder: The Role of the Frontal Lobes in Human Multitasking. *Psychol. Res.* **63**:279–288. [15]

Burgess, P. W., I. Dumontheil, and S. J. Gilbert. 2007. The Gateway Hypothesis of Rostral Prefrontal Cortex (Area 10) Function. *Trends Cogn. Sci.* **11**:290–298. [15, 16]

Burguière, E., P. Monteiro, G. Feng, and A. M. Graybiel. 2013. Optogenetic Stimulation of Lateral Orbitofronto-Striatal Pathway Suppresses Compulsive Behaviors. *Science* **340**:1243–1246. [4]

Burman, K. J., and M. G. P. Rosa. 2009. Architectural Subdivisions of Medial and Orbital Frontal Cortices in the Marmoset Monkey (Callithrix jacchus). *Journal of Comparative Neurology* **514**:11–29. [4]

Burns, J. R., and R. M. Rapee. 2006. Adolescent Mental Health Literacy: Young People's Knowledge of Depression and Help Seeking. *J. Adolesc.* **29**:225–239. [16]

Burt, J. B., M. Demirtaş, W. J. Eckner, et al. 2018. Hierarchy of Transcriptomic Specialization across Human Cortex Captured by Structural Neuroimaging Topography. *Nat. Neurosci.* **21**:1251–1259. [4, 6, 13, 16]

Buschkuehl, M., S. M. Jaeggi, and J. Jonides. 2012. Neuronal Effects Following Working Memory Training. *Dev. Cogn. Neurosci.* **2**:S167–179. [14]

Buschman, T., and E. K. Miller. 2007. Top-Down Versus Bottom-Up Control of Attention in the Prefrontal and Posterior Parietal Cortices. *Science* **315**:1860–1862. [12]

Buschman, T., M. Siegel, J. E. Roy, and E. K. Miller. 2011. Neural Substrates of Cognitive Capacity Limitations. *PNAS* **108**:11252–11255. [8]

Buschman, T. J., and S. Kastner. 2015. From Behavior to Neural Dynamics: An Integrated Theory of Attention. *Neuron* **88**:127–144. [12]

Bush, G., P. Luu, and M. I. Posner. 2000. Cognitive and Emotional Influences in Anterior Cingulate Cortex. *Trends Cogn. Sci.* **4**:215–222. [16]

Bussey, T. J., S. P. Wise, and E. A. Murray. 2002. Interaction of Ventral and Orbital Prefrontal Cortex with Inferotemporal Cortex in Conditional Visuomotor Learning. *Behav. Neurosci.* **116**:703–715. [8]

Butter, C. M. 1969. Perseveration in Extinction and in Discrimination Reversal Tasks Following Selective Frontal Ablations in *Macaca mulatta*. *Physiol. Behav.* **4**:163–171. [4]

Buzsáki, G., and A. Draguhn. 2004. Neuronal Oscillations in Cortical Networks. *Science* **304**:1926–1929. [12]

Caballero, A., D. R. Thomases, E. Flores-Barrera, D. K. Cass, and K. Y. Tseng. 2014. Emergence of GABAergic-Dependent Regulation of Input-Specific Plasticity in the Adult Rat Prefrontal Cortex during Adolescence. *Psychopharmacol.* **231**:1789–1796. [16]

Cachia, A., G. Borst, C. Tissier, et al. 2016. Longitudinal Stability of the Folding Pattern of the Anterior Cingulate Cortex during Development. *Dev. Cogn. Neurosci.* **19**:122–127. [4]

Caetano, T., M. S. Pinho, E. Ramadas, et al. 2021. Cognitive Training Effectiveness on Memory, Executive Functioning, and Processing Speed in Individuals with Substance Use Disorders: A Systematic Review. *Front. Psychol.* **12**: [14]

Cai, W., S. Ryali, T. Chen, C.-S. R. Li, and V. Menon. 2014. Dissociable Roles of Right Inferior Frontal Cortex and Anterior Insula in Inhibitory Control: Evidence from Intrinsic and Task-Related Functional Parcellation, Connectivity, and Response Profile Analyses across Multiple Datasets. *J. Neurosci.* **34**:14652–14667. [4]

Cai, X., and C. Padoa-Schioppa. 2012. Neuronal Encoding of Subjective Value in Dorsal and Ventral Anterior Cingulate Cortex. *J. Neurosci.* **32**:3791–3808. [8]

———. 2014. Contributions of Orbitofrontal and Lateral Prefrontal Cortices to Economic Choice and the Good-to-Action Transformation. *Neuron* **81**:1140–1151. [5, 8]

Calabro, F. J., V. P. Murty, M. Jalbrzikowski, B. Tervo-Clemmens, and B. Luna. 2020. Development of Hippocampal-Prefrontal Cortex Interactions through Adolescence. *Cereb. Cortex* **30**:1548–1558. [16]

Cameron, L. P., R. J. Tombari, J. Lu, et al. 2021. A Non-Hallucinogenic Psychedelic Analogue with Therapeutic Potential. *Nature* **589**:474–479. [13]

Camille, N., C. A. Griffiths, K. Vo, L. K. Fellows, and J. W. Kable. 2011a. Ventromedial Frontal Lobe Damage Disrupts Value Maximization in Humans. *J. Neurosci.* **31**:7527–7532. [16]

Camille, N., A. Tsuchida, and L. K. Fellows. 2011b. Double Dissociation of Stimulus-Value and Action-Value Learning in Humans with Orbitofrontal or Anterior Cingulate Cortex Damage. *J. Neurosci.* **31**:15048–15052. [2, 8, 12]

Campbell, K. L., K. P. Madore, R. G. Benoit, P. P. Thakral, and D. L. Schacter. 2018. Increased Hippocampus to Ventromedial Prefrontal Connectivity during the Construction of Episodic Future Events. *Hippocampus* **28**:76–80. [12, 17]

Cao, Y., Y. Gao, S. Xu, et al. 2016. Glutamate Carboxypeptidase II Gene Knockout Attenuates Oxidative Stress and Cortical Apoptosis after Traumatic Brain Injury. *BMC Neurosci.* **17**:15. [16]

Cao, Y., and K. Tsetsos. 2022. Clarifying the Role of an Unavailable Distractor in Human Multiattribute Choice. *eLife* **11**:e83316. [10]

Cappell, K. A., L. Gmeindl, and P. A. Reuter-Lorenz. 2010. Age Differences in Prefontal Recruitment during Verbal Working Memory Maintenance Depend on Memory Load. *Cortex* **46**:462–473. [14]

Carandini, M., and D. J. Heeger. 2012. Normalization as a Canonical Neural Computation. *Nat. Rev. Neurosci.* **13**:51–62. [12]

Carboni, L., L. Rullo, F. F. Caputi, et al. 2022. Chronic Trazodone and Citalopram Treatments Increase Trophic Factor and Circadian Rhythm Gene Expression in Rat Brain Regions Relevant for Antidepressant Efficacy. *Int. J. Mol. Sci.* **23**: [13]

Cardenas, V. A., T. C. Durazzo, S. Gazdzinski, et al. 2011. Brain Morphology at Entry into Treatment for Alcohol Dependence Is Related to Relapse Propensity. *Biol. Psych.* **70**:561–567. [4]

Cardin, J. A., M. Carlén, K. Meletis, et al. 2009. Driving Fast-Spiking Cells Induces Gamma Rhythm and Controls Sensory Responses. *Nature* **459**:663–667. [12]

Carhart-Harris, R., B. Giribaldi, R. Watts, et al. 2021. Trial of Psilocybin versus Escitalopram for Depression. *N. Engl. J. Med.* **384**:1402–1411. [13]

Carhart-Harris, R. L., M. Bolstridge, J. Rucker, et al. 2016. Psilocybin with Psychological Support for Treatment-Resistant Depression: an Open-Label Feasibility Study. *Lancet Psych.* **3**:619–627. [13]

Carhart-Harris, R. L., D. Erritzoe, T. Williams, et al. 2012. Neural Correlates of the Psychedelic State as Determined by fMRI Studies with Psilocybin. *PNAS* **109**:2138–2143. [13]

Carhart-Harris, R. L., L. Roseman, M. Bolstridge, et al. 2017. Psilocybin for Treatment-Resistant Depression: fMRI-Measured Brain Mechanisms. *Sci. Rep.* **7**:1–11. [13]

Carlen, M. 2017. What Constitutes the Prefrontal Cortex? *Science* **358**:478–482. [2]

Carlson, M., and K. A. Fitzpatrick. 1982. Organization of the Hand Area in the Primary Somatic Sensory cortex (SmI) of the Prosimian Primate, Nycticebus Coucang. *J. Comp. Neurol.* **204**:280–295. [3]

Carmichael, S. T., and J. L. Price. 1994. Architectonic Subdivision of the Orbital and Medial Prefrontal Cortex in the Macaque Monkey. *J. Comp. Neurol.* **346**:366–402. [4, 5]

———. 1996. Connectional Networks within the Orbital and Medial Prefrontal Cortex of Macaque Monkeys. *J. Comp. Neurol.* **371**:179–207. [8]

Carpenter, P. A., M. A. Just, and P. Shell. 1990. What One Intelligence Test Measures: A Theoretical Account of the Processing in the Raven Progressive Matrices Test. *Psychol. Rev.* **97**:404–431. [9]

Carretti, B., E. S. Borella, M. Zavagnin, and R. De Beni. 2011. Impact of Metacognition and Motivation on the Efficacy of Strategic Memory Training in Older Adults: Analysis of Specific, Transfer and Maintenance Effects. *Arch. Gerontol. Geriatr.* **52**:e192–e197. [14]

Carretti, B., N. Caldarola, C. Tencati, and C. Cornoldi. 2014. Improving Reading Comprehension in Reading and Listening Settings: The Effect of Two Training Programmes Focusing on Metacognition and Working Memory. *Br. J. Educ. Psychol.* **84**:194–210. [14]

Cartoni, E., B. Balleine, and G. Baldassarre. 2016. Appetitive Pavlovian-Instrumental Transfer: A Review. *Neurosci. Biobehav. Rev.* **71**:829–848. [4]

Caruana, F., M. Gerbella, P. Avanzini, et al. 2018. Motor and Emotional Behaviours Elicited by Electrical Stimulation of the Human Cingulate Cortex. *Brain* **141**:3035–3051. [16]

Casarotto, P. C., M. Girych, S. M. Fred, et al. 2021. Antidepressant Drugs Act by Directly Binding to Trkb Neurotrophin Receptors. *Cell* **184**:1299–1313.e1219. [13]

Caspi, A., A. R. Hariri, A. Holmes, R. Uher, and T. E. Moffitt. 2010. Genetic Sensitivity to the Environment: The Case of the Serotonin Transporter Gene and Its Implications for Studying Complex Diseases and Traits. *Am. J. Psych.* **167**:509–527. [16]

Caspi, A., R. M. Houts, A. Ambler, et al. 2020. Longitudinal Assessment of Mental Health Disorders and Comorbidities across 4 Decades Among Participants in the Dunedin Birth Cohort Study. *JAMA Netw. Open* **3**:e203221. [16]

Caspi, A., K. Sugden, T. E. Moffitt, et al. 2003. Influence of Life Stress on Depression: Moderation by a Polymorphism in the 5-HTT Gene. *Science* **301**:386–389. [16]

Castiglione, A., C. Brick, S. Holden, E. Miles-Urdan, and A. R. Aron. 2022. Discovering the Psychological Building Blocks Underlying Climate Action-a Longitudinal Study of Real-World Activism. *R. Soc. Open Sci.* **9**:210006. [16]

Catani, M., and D. Ffytche. 2005. The Rises and Falls of Disconnection Syndromes. . *Brain* **128**:2224–2239. [16]

Catania, K. C., C. E. Collins, and J. H. Kaas. 2000. Organization of Sensory Cortex in the East African Hedgehog (*Atelerix Albiventris*). *J. Comp. Neurol.* **421**:256–274. [3]

Cavada, C., and P. S. Goldman-Rakic. 1989. Posterior Parietal Cortex in Rhesus Monkey: II. Evidence for Segregated Corticocortical Networks Linking Sensory and Limbic Areas with the Frontal Lobe. *J. Comp. Neurol.* **287**:422–445. [8]

Cavanagh, J. F., and M. J. Frank. 2014. Frontal Theta as a Mechanism for Cognitive Control. *Trends Cogn. Sci.* **18**:414–421. [12, 17]

Cavanagh, J. F., T. V. Wiecki, M. X. Cohen, et al. 2011. Subthalamic Nucleus Stimulation Reverses Mediofrontal Influence over Decision Threshold. *Nat. Neurosci.* **14**:1462–1467. [12]

Caviness, V. S., Jr., D. N. Kennedy, C. Richelme, J. Rademacher, and P. A. Filipek. 1996. The Human Brain Age 7–11 Years: A Volumetric Analysis Based on Magnetic Resonance Images. *Cereb. Cortex* **6**:726–736. [16]

Chafee, M. V., and P. S. Goldman-Rakic. 1998. Matching Patterns of Activity in Primate Prefrontal Area 8a and Parietal Area 7ip Neurons during a Spatial Working Memory Task. *J. Neurophysiol.* **79**:2919–2940. [5]

Chakraborty, S., N. Kolling, M. E. Walton, and A. S. Mitchell. 2016. Critical Role for the Mediodorsal Thalamus in Permitting Rapid Reward-Guided Updating in Stochastic Reward Environments. *eLife* **5**:e13588. [8]

Chamberlain, S. R., A. Hampshire, U. Müller, et al. 2009. Atomoxetine Modulates Right Inferior Frontal Activation During Inhibitory Control: A Pharmacological Functional Magnetic Resonance Imaging Study. *Biol. Psych.* **65**:550–555. [4]

Chamberlain, S. R., U. Muller, A. D. Blackwell, et al. 2006. Neurochemical Modulation of Response Inhibition and Probabilistic Learning in Humans. *Science* **311**:861–863. [16]

Chamberlain, S. R., J. E. Solly, R. W. Hook, M. M. Vaghi, and T. W. Robbins. 2021. Cognitive Inflexibility in OCD and Related Disorders. *Curr. Top. Behav. Neurosci.* **49**:125–145. [16]

Chan, M. Y., D. C. Park, N. K. Savalia, S. E. Petersen, and G. S. Wig. 2014. Decreased Segregation of Brain Systems across the Healthy Adult Lifespan. *PNAS* **111**:E4997–E5006. [11]

Chan, S. C., Y. Niv, and K. A. Norman. 2016. A Probability Distribution over Latent Causes, in the Orbitofrontal Cortex. *J. Neurosci.* **36**:7817–7828. [10]

Charnov, E. L. 1976. Optimal Foraging: The Marginal Value Theorem. *Theor. Popul. Biol.* **9**:129–136. [2]

Charpentier, C. J., K. Iigaya, and J. P. O'Doherty. 2020. A Neuro-Computational Account of Arbitration between Choice Imitation and Goal Emulation during Human Observational Learning. *Neuron* **106**:687–699. [12]

Chatham, C. H., and D. Badre. 2012. Parts to Principles: Anatomical Origins of Prefrontal Organization. *Cortex* **48**:1373–1375; discussion 1383–1377. [8]

———. 2020. How to Test Cognitive Theory with fMRI. In: New Methods in Cognitive Psychology, ed. D. Spieler and E. Schumacher. New York: Routledge. [8]

Chatham, C. H., M. J. Frank, and D. Badre. 2014. Corticostriatal Output Gating during Selection from Working Memory. *Neuron* **81**:930–942. [7, 15]

Chatham, C. H., S. A. Herd, A. M. Brant, et al. 2011. From an Executive Network to Executive Control: A Computational Model of the n-back Task. *J. Cogn. Neurosci.* **23**:3598–3619. [9]

Chau, B. K., J. Sallet, G. K. Papageorgiou, et al. 2015. Contrasting Roles for Orbitofrontal Cortex and Amygdala in Credit Assignment and Learning in Macaques. *Neuron* **87**:1106–1118. [4]

Chaudhuri, R., K. Knoblauch, M.-A. Gariel, H. Kennedy, and X.-J. Wang. 2015. A Large-Scale Circuit Mechanism for Hierarchical Dynamical Processing in the Primate Cortex. *Neuron* **88**:419–431. [12]

Chein, J. M., and W. Schneider. 2005. Neuroimaging Studies of Practice-Related Change: fMRI and Meta-Analytic Evidence of a Domain-General Control Network for Learning. *Cogn. Brain Res.* **25**:607–623. [14]

Chen, C.-H., K. Ridler, J. Suckling, et al. 2007. Brain Imaging Correlates of Depressive Symptom Severity and Predictors of Symptom Improvement after Antidepressant Treatment. *Biol. Psychiatry* **62**:407–414. [13]

Chen, G., P. Greengard, and Z. Yan. 2004. Potentiation of NMDA Receptor Currents by Dopamine D1 Receptors in Prefrontal Cortex. *PNAS* **101**:2596–2600. [6]

Chen, R., F. Gore, Q. A. Nguyen, et al. 2021. Deep Brain Optogenetics without Intracranial Surgery. *Nat. Biotechnol.* **39**:161–164. [4]

Chen, X., X. Liu, B. J. Parker, Z. Zhen, and K. S. Weiner. 2023. Functionally and Structurally Distinct Fusiform Face Area(S) in over 1000 Participants. *Neuroimage* **265**:119765. [4]

Chen, Y. C., G. Sudre, W. Sharp, et al. 2018. Neuroanatomic, Epigenetic and Genetic Differences in Monozygotic Twins Discordant for Attention Deficit Hyperactivity Disorder. *Mol. Psych.* **23**:683–690. [16]

Chi, J. G., E. C. Dooling, and F. H. Gilles. 1977. Gyral Development of the Human Brain. *Ann. Neurol.* **1**:86–93. [4]

Chiba, T., T. Kayahara, and K. Nakano. 2001. Efferent Projections of Infralimbic and Prelimbic Areas of the Medial Prefrontal Cortex in the Japanese Monkey, *Macaca Fuscata. Brain Res.* **888**:83–101. [3]

Chien, J. M., J. D. Wallis, and E. L. Rich. 2023. Abstraction of Reward Context Facilitates Relative Reward Coding in Neural Populations of the Macaque Anterior Cingulate Cortex. *J. Neurosci.* [4, 5, 8]

Chin Fatt, C. R., C. Cooper, M. K. Jha, et al. 2021. Dorsolateral Prefrontal Cortex and Subcallosal Cingulate Connectivity Show Preferential Antidepressant Response in Major Depressive Disorder. *Biol. Psych. Cogn. Neurosci. Neuroimag.* **6**:20–28. [13]

Chmielewski, W. X., and C. Beste. 2019. Stimulus-Response Recoding during Inhibitory Control Is Associated with Superior Frontal and Parahippocampal Processes. *Neuroimage* **196**:227–236. [12]

Cho, K. K. A., T. J. Davidson, G. Bouvier, et al. 2020. Cross-Hemispheric Gamma Synchrony between Prefrontal Parvalbumin Interneurons Supports Behavioral Adaptation during Rule Shift Learning. *Nat. Neurosci.* **23**:892–902. [4]

Cho, K. K. A., J. Shi, A. J. Phensy, M. L. Turner, and V. S. Sohal. 2023. Long-Range Inhibition Synchronizes and Updates Prefrontal Task Activity. *Nature* **617**:548–554. [4, 12]

Choi, E. Y., G. K. Drayna, and D. Badre. 2018. Evidence for a Functional Hierarchy of Association Networks. *J. Cogn. Neurosci.* **30**:722–736. [7]

Choi, E. Y., B. T. Yeo, and R. L. Buckner. 2012. The Organization of the Human Striatum Estimated by Intrinsic Functional Connectivity. *J. Neurophysiol.* **108**:2242–2263. [7]

Christoff, K., K. Keramatian, A. M. Gordon, R. Smith, and B. Madler. 2009. Prefrontal Organization of Cognitive Control According to Levels of Abstraction. *Brain Res.* **1286**:94–105. [7]

Christophel, T. B., P. C. Klink, B. Spitzer, P. R. Roelfsema, and J. D. Haynes. 2017. The Distributed Nature of Working Memory. *Trends Cogn. Sci.* **21**:111–124. [6]

Chudasama, Y. 2011. Animal Models of Prefrontal-Executive Function. *Behav. Neurosci.* **125**:327–343. [3]

Chung, S., and L. F. Abbott. 2021. Neural Population Geometry: An Approach for Understanding Biological and Artificial Neural Networks. *Curr. Opin. Neurobiol.* **70**:137–144. [8]

Churchland, P. S., and T. J. Sejnowski. 1988. Perspectives on Cognitive Neuroscience. *Science* **242**:741–745. [11]

Cipolotti, L., J. K. Ruffle, J. Mole, et al. 2023. Graph Lesion-Deficit Mapping of Fluid Intelligence. *Brain* **146**:167–181. [9]

Cisek, P. 2019. Resynthesizing Behavior through Phylogenetic Refinement. *Atten. Percept. Psychophys.* **81**:2265–2287. [2]

———. 2022. Evolution of Behavioural Control from Chordates to Primates. *Philos. Trans. R. Soc. Lond. B Biol. Sci.* **377**:20200522. [8]

Cisek, P., and J. F. Kalaska. 2002. Modest Gaze-Related Discharge Modulation in Monkey Dorsal Premotor Cortex during a Reaching Task Performed with Free Fixation. *J. Neurophysiol.* **88**:1064–1072. [12]

Clark, A. M., S. Bouret, A. M. Young, E. A. Murray, and B. J. Richmond. 2013. Interaction between Orbital Prefrontal and Rhinal Cortex Is Required for Normal Estimates of Expected Value. *J. Neurosci.* **33**:1833–1845. [8]

Clarke, H. F., J. W. Dalley, H. S. Crofts, T. W. Robbins, and A. C. Roberts. 2004. Cognitive Inflexibility after Prefrontal Serotonin Depletion. *Science* **304**:878–880. [4, 8, 16]

Clarke, H. F., N. K. Horst, and A. C. Roberts. 2015. Regional Inactivations of Primate Ventral Prefrontal Cortex Reveal Two Distinct Mechanisms Underlying Negative Bias in Decision Making. *PNAS* **112**:4176–4181. [8]

Clarke, H. F., T. W. Robbins, and A. C. Roberts. 2008. Lesions of the Medial Striatum in Monkeys Produce Perseverative Impairments during Reversal Learning Similar to Those Produced by Lesions of the Orbitofrontal Cortex. *J. Neurosci.* **28**:10972–10982. [4]

Clarke, H. F., S. C. Walker, H. S. Crofts, et al. 2005. Prefrontal Serotonin Depletion Affects Reversal Learning but Not Attentional Set Shifting. *J. Neurosci.* **25**:532–538. [16]

Clarke, H. F., S. C. Walker, J. W. Dalley, T. W. Robbins, and A. C. Roberts. 2007. Cognitive Inflexibility after Prefrontal Serotonin Depletion Is Behaviorally and Neurochemically Specific. *Cereb. Cortex* **17**:18–27. [8, 16]

Cocchi, L., B. J. Harrison, J. Pujol, et al. 2012. Functional Alterations of Large-Scale Brain Networks Related to Cognitive Control in Obsessive-Compulsive Disorder. *Hum. Brain Mapp.* **33**:1089–10106. [15]

Cockburn, J., V. Man, W. A. Cunningham, and J. P. O'Doherty. 2022. Novelty and Uncertainty Regulate the Balance between Exploration and Exploitation through Distinct Mechanisms in the Human Brain. *Neuron* **110**:2691–2702. [12]

Cohen, J. D., T. S. Braver, and R. C. O'Reilly. 1996. A Computational Approach to Prefrontal Cortex, Cognitive Control and Schizophrenia: Recent Developments and Current Challenges. *Philos. Trans. R. Soc. Lond. B Biol. Sci.* **351**:1515–1527. [12]

Cohen, J. R., and M. D'Esposito. 2016. The Segregation and Integration of Distinct Brain Networks and Their Relationship to Cognition. *J. Neurosci.* **36**:12083–12094. [11]

Cohen, J. R., C. L. Gallen, E. G. Jacobs, T. G. Lee, and M. D'Esposito. 2014. Quantifying the Reconfiguration of Intrinsic Networks during Working Memory. *PloS One* **9**:e106636. [11]

Cohen, M. R., and J. H. R. Maunsell. 2009. Attention Improves Performance Primarily by Reducing Interneuronal Correlations. *Nat. Neurosci.* **12**:1594–1600. [12]

Cohen, R. A., R. F. Kaplan, M. E. Meadows, and H. Wilkinson. 1994. Habituation and Sensitization of the Orienting Response Following Bilateral Anterior Cingulotomy. *Neuropsychologia* **32**:609–617. [15]

Cohen, R. A., R. F. Kaplan, D. J. Moser, M. A. Jenkins, and H. Wilkinson. 1999a. Impairments of Attention after Cingulotomy. *Neurology* **53**:819–824. [15]

Cohen, R. A., R. F. Kaplan, P. Zuffante, et al. 1999b. Alteration of Intention and Self-Initiated Action Associated with Bilateral Anterior Cingulotomy. *J Neuropsychiatry Clin Neurosci* **11**:444–453. [15]

Cohen, R. A., R. Paul, T. M. Zawacki, et al. 2001. Emotional and Personality Changes Following Cingulotomy. *Emotion* **1**:38–50. [15]

Cole, M. W., T. Ito, and T. S. Braver. 2015a. The Behavioral Relevance of Task Information in Human Prefrontal Cortex. *Cereb. Cortex* **26**:2497–2505. [6]

———. 2015b. Lateral Prefrontal Cortex Contributes to Fluid Intelligence through Multinetwork Connectivity. *Brain Connect.* **5**:497–504. [7]

Collette, F., M. Van der Linden, S. Laureys, et al. 2005. Exploring the Unity and Diversity of the Neural Substrates of Executive Functioning. *Hum. Brain Mapp.* **25**:409–423. [9]

Collins, A. G., J. F. Cavanagh, and M. J. Frank. 2014. Human EEG Uncovers Latent Generalizable Rule Structure during Learning. *J. Neurosci.* **34**:4677–4685. [7]

Collins, A. G., and M. J. Frank. 2013. Cognitive Control over Learning: Creating, Clustering, and Generalizing Task-Set Structure. *Psychol. Rev.* **120**:190–229. [7, 10]

Collins, A. G., and E. Koechlin. 2012. Reasoning, Learning, and Creativity: Frontal Lobe Function and Human Decision-Making. *PLoS Biol.* **10**:e1001293. [10]

Colom, R., X. Hua, K. Martínez, et al. 2016. Brain Structural Changes Following Adaptive Cognitive Training Assessed by Tensor-Based Morphometry (TBM). *Neuropsychologia* **91**:77–85. [14]

Colzato, L. S., B. Hommel, W. Zhang, V. Roessner, and C. Beste. 2022. The Metacontrol Hypothesis as Diagnostic Framework of OCD and ADHD: A Dimensional Approach Based on Shared Neurobiological Vulnerability. *Neurosci. Biobehav. Rev.* **137**:104677. [12]

Compte, A., N. Brunel, P. S. Goldman-Rakic, and X. J. Wang. 2000. Synaptic Mechanisms and Network Dynamics Underlying Spatial Working Memory in a Cortical Network Model. *Cereb. Cortex* **10**:910–923. [6, 10]

Compte, A., C. Constantinidis, J. Tegnér, et al. 2003. Temporally Irregular Mnemonic Persistent Activity in Prefrontal Neurons of Monkeys During a Delayed Response Task. *J. Neurophysiol.* **90**:3441–3454. [4]

Cong, F., Q.-H. Lin, L.-D. Kuang, et al. 2015. Tensor Decomposition of EEG Signals: A Brief Review. *J. Neurosci. Methods* **248**:59–69. [12]

Constantinidis, C., S. Funahashi, D. Lee, et al. 2018. Persistent Spiking Activity Underlies Working Memory. *J. Neurosci.* **38**:7020–7028. [5]

Constantinidis, C., and P. S. Goldman-Rakic. 2002. Correlated Discharges among Putative Pyramidal Neurons and Interneurons in the Primate Prefrontal Cortex. *J. Neurophysiol.* **88**:3487–3497. [6]

Constantinidis, C., and T. Klingberg. 2016. The Neuroscience of Working Memory Capacity and Training. *Nat. Rev. Neurosci.* **17**:438–449. [14]

Constantinidis, C., and X.-L. Qi. 2018. Representation of Spatial and Feature Information in the Monkey Dorsal and Ventral Prefrontal Cortex. *Front. Integr. Neurosci.* **12**:31. [6]

Constantinidis, C., and X. J. Wang. 2004. A Neural Circuit Basis for Spatial Working Memory. *Neuroscientist* **10**:553–565. [4, 6]

Constantinidis, C., G. V. Williams, and P. S. Goldman-Rakic. 2002. A Role for Inhibition in Shaping the Temporal Flow of Information in Prefrontal Cortex. *Nat. Neurosci.* **5**:175–180. [6]

Cools, R., and A. F. T. Arnsten. 2022. Neuromodulation of Prefrontal Cortex Cognitive Function in Primates: The Powerful Roles of Monoamines and Acetylcholine. *Neuropsychopharmacol.* **47**:309–328. [16]

Cools, R., R. B. Ivry, and M. D'Esposito. 2006. The Human Striatum Is Necessary for Responding to Changes in Stimulus Relevance. *J. Cogn. Neurosci.* **18**:1973–1983. [7]

Cooper, C. M., C. R. Chin Fatt, M. Jha, et al. 2019. Cerebral Blood Perfusion Predicts Response to Sertraline versus Placebo for Major Depressive Disorder in the Embarc Trial. *EClinicalMedicine* **10**:32–41. [13]

Cope, T. E., B. Wilson, H. Robson, et al. 2017. Artificial Grammar Learning in Vascular and Progressive Non-Fluent Aphasias. *Neuropsychologia* **104**:201–213. [16]

Corbetta, M., and G. L. Shulman. 2002. Control of Goal-Directed and Stimulus-Driven Attention in the Brain. *Nat. Rev. Neurosci.* **3**:201–215. [11]

Corkin, S. 1979. Hidden-Figures-Test Performance: Lasting Effects of Unilateral Penetrating Head Injury and Transient Effects of Bilateral Cingulotomy. *Neuropsychologia* **17**:585–605. [15]

Cortese, A., A. Yamamoto, M. Hashemzadeh, et al. 2021. Value Signals Guide Abstraction during Learning. *eLife* **10**:e68943. [12]

Cortese, S., C. Kelly, C. Chabernaud, et al. 2012. Toward Systems Neuroscience of ADHD: A Meta-Analysis of 55 fMRI Studies. *The American Journal of Psychiatry* **169**:1038–1055. [14]

Costa, V. D., A. R. Mitz, and B. B. Averbeck. 2019. Subcortical Substrates of Explore-Exploit Decisions in Primates. *Neuron* **103**:533–545. [8]

Costello, H., J. P. Roiser, and R. Howard. 2023. Antidepressant Medications in Dementia: Evidence and Potential Mechanisms of Treatment-Resistance. *Psychol. Med.* **53**:654–667. [16]

Course-Choi, J., H. Saville, and N. Derakshan. 2017. The Effects of Adaptive Working Memory Training and Mindfulness Meditation Training on Processing Efficiency and Worry in High Worriers. *Behav. Res. Ther.* **89**:1–13. [14]

Courtney, S. M., L. Petit, J. M. Maisog, L. G. Ungerleider, and J. V. Haxby. 1998. An Area Specialized for Spatial Working Memory in Human Frontal Cortex. *Science* **279**:1347–1351. [4]

Coutureau, E., A. S. Killcross, M. Good, et al. 2002. Acquired Equivalence and Distinctiveness of Cues: II. Neural Manipulations and Their Implications. *J. Exp. Psychol. Anim. Behav. Proc.* **28**:388–396. [7]

Coyle, J. T., G. Tsai, and D. Goff. 2003. Converging Evidence of NMDA Receptor Hypofunction in the Pathophysiology of Schizophrenia. *Ann. N. Y. Acad. Sci.* **1003**:318–327. [10]

Craig, A. D. 2009. How Do You Feel--Now? The Anterior Insula and Human Awareness. *Nat. Rev. Neurosci.* **10**:59–70. [8]

Critchley, H. D., and E. T. Rolls. 1996. Hunger and Satiety Modify the Responses of Olfactory and Visual Neurons in the Primate Orbitofrontal Cortex. *J. Neurophysiol.* **75**:1673–1686. [5]

Crittenden, B. M., and J. Duncan. 2014. Task Difficulty Manipulation Reveals Multiple Demand Activity but No Frontal Lobe Hierarchy. *Cereb. Cortex* **24**:532–540. [7]

Crone, E. A., C. Wendelken, S. E. Donohue, and S. A. Bunge. 2006. Neural Evidence for Dissociable Components of Task-Switching. *Cereb. Cortex* **16**:475–486. [9]

Cross, L., J. Cockburn, Y. Y., and J. P. O'Doherty. 2021. Using Deep Reinforcement Learning to Reveal How the Brain Encodes Abstract State-Space Representations in High-Dimensional Environments. *Neuron* **109**:724–738. [12]

Crossley, N. A., A. Mechelli, J. Scott, et al. 2014. The Hubs of the Human Connectome Are Generally Implicated in the Anatomy of Brain Disorders. *Brain* **137**:2382–2395. [11, 16]

Crosson, B., J. R. Saclek, J. A. Bobholz, et al. 1999. Activity in the Paracingulate and Cingulate Sulci during Word Generation: an fMRI Study of Functional Anatomy. *Cereb. Cortex* **9**:307–316. [4]

Csigó, K., A. Harsányi, G. Demeter, et al. 2010. Long-Term Follow-up of Patients with Obsessive-Compulsive Disorder Treated by Anterior Capsulotomy: A Neuropsychological Study. *J. Affect. Disord.* **126**:198–205. [15]

Cueva, C. J., A. Saez, E. Marcos, et al. 2020. Low-Dimensional Dynamics for Working Memory and Time Encoding. *PNAS* **117**:23021–23032. [6]

Cui, H., Y. Zhang, Y. Zhao, et al. 2023. Mechanisms Underlying Capsulotomy for Refractory Obsessive-Compulsive Disorder: Neural Correlates of Negative Affect Processing Overlap with Deep Brain Stimulation Targets. *Mol. Psych.* **28**:3063–3074. [15]

Cui, Z., H. Li, C. H. Xia, et al. 2020. Individual Variation in Functional Topography of Association Networks in Youth. *Neuron* **106**:340–353. [11]

Culbreth, A. J., E. K. Schwartz, M. J. Frank, et al. 2023. A Computational Neuroimaging Study of Reinforcement Learning and Goal-Directed Exploration in Schizophrenia Spectrum Disorders. *Psychol. Med.* 1–11. [7]

Curtis, C. E., and M. D'Esposito. 2006. Selection and Maintenance of Saccade Goals in the Human Frontal Eye Fields. *J. Neurophysiol.* **95**:3923–3927. [4]

Curtis, C. E., V. Y. Rao, and M. D'Esposito. 2004. Maintenance of Spatial and Motor Codes during Oculomotor Delayed Response Tasks. *J. Neurosci.* **24**:3944–3952. [4]

Cuthbert, B. N. 2014. The RDoC Framework: Facilitating Transition from ICD/DSM to Dimensional Approaches That Integrate Neuroscience and Psychopathology. *World Psych.* **13**:28–35. [16]

Cyr, M., D. Pagliaccio, P. Yanes-Lukin, et al. 2020. Altered Network Connectivity Predicts Response to Cognitive-Behavioral Therapy in Pediatric Obsessive-Compulsive Disorder. *Neuropsychopharmacol.* **45**:1232–1240. [16]

Czéh, B., J. I. H. Müller-Keuker, R. Rygula, et al. 2007. Chronic Social Stress Inhibits Cell Proliferation in the Adult Medial Prefrontal Cortex: Hemispheric Asymmetry and Reversal by Fluoxetine Treatment. *Neuropsychopharmacol.* **32**:1490–1503. [13]

Dagher, A., and T. W. Robbins. 2009. Personality, Addiction, Dopamine: Insights from Parkinson's Disease. *Neuron* **61**:502–510. [7]

Dahlin, E., A. S. Neely, A. Larsson, L. Backman, and L. Nyberg. 2008. Transfer of Learning after Updating Training Mediated by the Striatum. *Science* **320**:1510–1512. [14]

Dalley, J. W., B. J. Everitt, and T. W. Robbins. 2011. Impulsivity, Compulsivity, and Top-Down Cognitive Control. *Neuron* **69**:680–694. [15]

Dalley, J. W., and T. W. Robbins. 2017. Fractionating Impulsivity: Neuropsychiatric Implications. *Nat. Rev. Neurosci.* **18**:158–171. [4]

Dal Monte, O., C. C. J. Chu, N. A. Fagan, and S. W. C. Chang. 2020. Specialized Medial Prefrontal-Amygdala Coordination in Other-Regarding Decision Preference. *Nat. Neurosci.* **23**:565–574. [8]

Dalton, G. L., N. Y. Wang, A. G. Phillips, and S. B. Floresco. 2016. Multifaceted Contributions by Different Regions of the Orbitofrontal and Medial Prefrontal Cortex to Probabilistic Reversal Learning. *J. Neurosci.* **36**:1996–2006. [4]

Danet, M., S. Lapiz-Bluhm, and D. A. Morilak. 2010. A Cognitive Deficit Induced in Rats by Chronic Intermittent Cold Stress Is Reversed by Chronic Antidepressant Treatment. *Int. J. Neuropsychopharmacol.* **13**:997–1009. [16]

Dart, R. A. 1934. The Dual Structure of the Neopallium: Its History and Signficance. *J. Anat.* **69**:3–19. [5]

Dash, G. F., S. L. Karalunas, E. A. Kenyon, et al. 2023. Gene-by-Environment Interaction Effects of Social Adversity on Externalizing Behavior in ABCD Youth. *Behav. Genet.* **53**:219–231. [16]

Datta, D., and A. F. T. Arnsten. 2018. Unique Molecular Regulation of Higher-Order Prefrontal Cortical Circuits: Insights into the Neurobiology of Schizophrenia. *ACS Chem. Neurosci.* **9**:2127–2145. [13]

———. 2019. Loss of Prefrontal Cortical Higher Cognition with Uncontrollable Stress: Molecular Mechanisms, Changes with Age, and Relevance to Treatment. *Brain Sci.* **9**: [16]

Datta, D., S. T. Yang, V. C. Galvin, et al. 2019. Noradrenergic α1-Adrenoceptor Actions in the Primate Dorsolateral Prefrontal Cortex. *J. Neurosci.* **39**:2722–2734. [16]

Daugherty, A. M., C. Zwilling, E. J. Paul, et al. 2018. Multi-Modal Fitness and Cognitive Training to Enhance Fluid Intelligence. *Intelligence* **66**:32–43. [14]

David, M. C. B., M. Del Giovane, K. Y. Liu, et al. 2022. Cognitive and Neuropsychiatric Effects of Noradrenergic Treatment in Alzheimer's Disease: Systematic Review and Meta-Analysis. *J. Neurol. Neurosurg. Psych.* **93**:1080–1090. [16]

Davidson, B., C. Hamani, Y. Meng, et al. 2020a. Examining Cognitive Change in Magnetic Resonance-Guided Focused Ultrasound Capsulotomy for Psychiatric Illness. *Transl. Psych.* **10**:397. [15, 16]

Davidson, B., K. Mithani, Y. Huang, et al. 2020b. Technical and Radiographic Considerations for Magnetic Resonance Imaging-Guided Focused Ultrasound Capsulotomy. *J Neurosurg* **135**:291–299. [16]

Davidson, R. J. 1992. Anterior Cerebral Asymmetry and the Nature of Emotion. *Brain Cogn.* **20**:125–151. [8]

Davidson, R. J., W. Irwin, M. J. Anderle, and N. H. Kalin. 2003. The Neural Substrates of Affective Processing in Depressed Patients Treated with Venlafaxine. *Am. J. Psych.* **160**:64–75. [13]

Davidson, R. J., D. Pizzagalli, J. B. Nitschke, and K. Putnam. 2002. Depression: Perspectives from Affective Neuroscience. *Annu. Rev. Psychol.* **53**:545–574. [13]

Davis, A. K., F. S. Barrett, D. G. May, et al. 2021. Effects of Psilocybin-Assisted Therapy on Major Depressive Disorder: A Randomized Clinical Trial. *JAMA Psych.* **78**:481–489. [13]

Davis, K. D., K. S. Taylor, W. D. Hutchison, et al. 2005. Human Anterior Cingulate Cortex Neurons Encode Cognitive and Emotional Demands. *J. Neurosci.* **25**:8402–8406. [15]

Daw, N. D., Y. Niv, and P. Dayan. 2005. Uncertainty-Based Competition between Prefrontal and Dorsolateral Striatal Systems for Behavioral Control. *Nat. Neurosci.* **8**:1704–1711. [12]

Daw, N. D., J. P. O'Doherty, P. Dayan, B. Seymour, and R. J. Dolan. 2006. Cortical Substrates for Exploratory Decisions in Humans. *Nature* **441**:876–879. [7, 8]

Daws, R. E., C. Timmermann, B. Giribaldi, et al. 2022. Increased Global Integration in the Brain after Psilocybin Therapy for Depression. *Nat. Med.* **28**:844–851. [13]

Dayan, P. 2012. How to Set the Switches on This Thing. *Curr. Opin. Neurobiol.* **22**:1068–1074. [12]

De Baene, W., G.-J. M. Rutten, and M. M. Sitskoorn. 2019. Cognitive Functioning in Glioma Patients Is Related to Functional Connectivity Measures of the Non-Tumoural Hemisphere. *Eur. J. Neurosci.* **50**:3921–3933. [12]

Debener, S., M. Ullsperger, M. Siegel, et al. 2005. Trial-by-Trial Coupling of Concurrent Electroencephalogram and Functional Magnetic Resonance Imaging Identifies the Dynamics of Performance Monitoring. *J. Neurosci.* **25**:11730–11737. [12]

DeCasien, A. R., S. A. Williams, and J. P. Higham. 2017. Primate Brain Size Is Predicted by Diet but Not Sociality. *Nat. Ecol. Evol.* **1**:1–7. [4]

Deco, G., A. Ponce-Alvarez, P. Hagmann, et al. 2014. How Local Excitation–Inhibition Ratio Impacts the Whole Brain Dynamics. *J. Neurosci.* **34**:7886–7898. [12]

Defelipe, J., M. C. Gonzalez-Albo, M. R. Del Rio, and G. N. Elston. 1999. Distribution and Patterns of Connectivity of Interneurons Containing Calbindin, Calretinin, and Parvalbumin in Visual Areas of the Occipital and Temporal Lobes of the Macaque Monkey. *J. Comp. Neurol.* **412**:515–526. [6]

de Haan, S., E. Rietveld, M. Stokhof, and D. Denys. 2015. Effects of Deep Brain Stimulation on the Lived Experience of Obsessive-Compulsive Disorder Patients: In-Depth Interviews with 18 Patients. *PloS One* **10**:e0135524. [15]

———. 2017. Becoming More Oneself? Changes in Personality Following DBS Treatment for Psychiatric Disorders: Experiences of OCD Patients and General Considerations. *PloS One* **12**:e0175748. [15]

Deisseroth, K. 2015. Optogenetics: 10 Years of Microbial Opsins in Neuroscience. *Nat. Neurosci.* **18**:1213–1225. [2]

de la Vega, A., L. J. Chang, M. T. Banich, T. D. Wager, and T. Yarkoni. 2016. Large-Scale Meta-Analysis of Human Medial Frontal Cortex Reveals Tripartite Functional Organization. *J. Neurosci.* **36**:6553–6562. [12]

de la Vega, A., T. Yarkoni, T. D. Wager, and M. T. Banich. 2018. Large-Scale Meta-Analysis Suggests Low Regional Modularity in Lateral Frontal Cortex. *Cereb. Cortex* **28**:3414–3428. [7, 12]

Delvenne, J. F. 2005. The Capacity of Visual Short-Term Memory within and between Hemifields. *Cognition* **96**:B79–88. [8]

De Martino, B., and A. Cortese. 2023. Goals, Usefulness and Abstraction in Value-Based Choice. *Trends Cogn. Sci.* **27**:65–80. [12]

Demchenko, I., V. K. Tassone, S. H. Kennedy, K. Dunlop, and V. Bhat. 2022. Intrinsic Connectivity Networks of Glutamate-Mediated Antidepressant Response: A Neuroimaging Review. Front. Psychiatry 13:864902. [13]

Demirtas, M., J. B. Burt, M. Helmer et al. 2019. Hierarchical Heterogeneity across Human Cortex Shapes Large-Scale Neural Dynamics. *Neuron* **101**:1181–1194. [12, 16]

Denny, B. T., M. L. Jungles, P. N. Goodson, et al. 2023. Unpacking Reappraisal: A Systematic Review of fMRI Studies of distancing and reinterpretation. *Soc. Cogn. Affect. Neurosci.* **18**:nsad050. [17]

Denny, B. T., H. Kober, T. D. Wager, and K. N. Ochsner. 2012. A Meta-Analysis of Functional Neuroimaging Studies of Self- and Other Judgments Reveals a Spatial Gradient for Mentalizing in Medial Prefrontal Cortex. *J. Cogn. Neurosci.* **24**:1742–1752. [16]

den Ouden, H. E., N. D. Daw, G. Fernandez, et al. 2013. Dissociable Effects of Dopamine and Serotonin on Reversal Learning. *Neuron* **80**:1090–1100. [16]

de Oliveira Rosa, V., A. Rosa Franco, G. Abrahão Salum Júnior, et al. 2020. Effects of Computerized Cognitive Training as Add-on Treatment to Stimulants in ADHD: A Pilot fMRI Study. *Brain Imag. Behav.* **14**:1933–1944. [14]

Depue, B. E., J. M. Orr, H. R. Smolker, F. Naaz, and M. T. Banich. 2016. The Organization of Right Prefrontal Networks Reveals Common Mechanisms of Inhibitory Regulation across Cognitive, Emotional, and Motor Processes. *Cereb. Cortex* **26**:1634–1646. [12, 17]

DeRosa, J., H. Kim, J. Lewis-Peacock, and M. T. Banich. 2024. Neural Systems Underlying the Implementation of Working Memory Removal Operations. *J. Neurosci.* **44**:e0283232023. [17]

Desimone, R., and J. Duncan. 1995. Neural Mechanisms of Selective Visual Attention. *Annu. Rev. Neurosci.* **18**:193–222. [12]

D'Esposito, M., G. K. Aguirre, E. Zarahn, et al. 1998a. Functional MRI Studies of Spatial and Nonspatial Working Memory. *Cogn. Brain Res.* **7**:1–13. [11]

D'Esposito, M., D. Ballard, G. K. Aguirre, and E. Zarahn. 1998b. Human Prefrontal Cortex Is Not Specific for Working Memory: A Functional MRI Study. *Neuroimage* **8**:274–282. [16]

D'Esposito, M., and B. R. Postle. 2015. The Cognitive Neuroscience of Working Memory. *Annu. Rev. Psychol.* **66**:115–142. [6]

Desrochers, T. M., D. C. Burk, D. Badre, and D. L. Sheinberg. 2015a. The Monitoring and Control of Task Sequences in Human and Non-Human Primates. *Front. Syst. Neurosci.* **9**:185. [7]

Desrochers, T. M., C. H. Chatham, and D. Badre. 2015b. The Necessity of Rostrolateral Prefrontal Cortex for Higher-Level Sequential Behavior. *Neuron* **87**:1357–1368. [7]

Desrochers, T. M., A. G. E. Collins, and D. Badre. 2019. Sequential Control Underlies Robust Ramping Dynamics in the Rostrolateral Prefrontal Cortex. *J. Neurosci.* **39**:1471–1483. [7]

Deveau, J., S. M. Jaeggi, V. Zordan, C. Phung, and A. R. Seitz. 2015. How to Build Better Memory Training Games. *Front. Syst. Neurosci.* **8**:243. [14]

Devinsky, O., and M. D'Esposito. 2004. Executive Function and the Frontal Lobes. In: Neurology of Cognitive and Behavioral Disorders, ed. O. Devinky and M. D'Esposito, pp. 302–329. New York: Oxford Univ. Press. [7]

de Wit, S., P. R. Corlett, M. R. Aitken, A. Dickinson, and P. C. Fletcher. 2009. Differential Engagement of the Ventromedial Prefrontal Cortex by Goal-Directed and Habitual Behavior toward Food Pictures in Humans. *J. Neurosci.* **29**:11330–11338. [15]

Diamond, A. 2013. Executive Functions. *Annu. Rev. Psychol.* **64**:135–168. [14, 16]

Dias, E. C., and M. A. Segraves. 1999. Muscimol-Induced Inactivation of Monkey Frontal Eye Field: Effects on Visually and Memory-Guided Saccades. *J. Neurophysiol.* **81**:2191–2214. [4]

Dias, R., T. W. Robbins, and A. C. Roberts. 1996a. Dissociation in Prefrontal Cortex of Affective and Attentional Shifts. *Nature* **380**:69–72. [4, 8]

———. 1996b. Primate Analogue of the Wisconsin Card Sorting Test: Effects of Excitotoxic Lesions of the Prefrontal Cortex in the Marmoset. *Behav. Neurosci.* **110**:872–886. [4]

Diazgranados, N., L. Ibrahim, N. E. Brutsche, et al. 2010. A Randomized Add-on Trial of an N-Methyl-D-Aspartate Antagonist in Treatment-Resistant Bipolar Depression. *Arch. Gen. Psychiatry* **67**:793–802. [13]

Diehl, G. W., and A. D. Redish. 2023. Differential Processing of Decision Information in Subregions of Rodent Medial Prefrontal Cortex. *eLife* **12**:e82813. [3]

Diester, I., M. T. Kaufman, M. Mogri, et al. 2011. An Optogenetic Toolbox Designed for Primates. *Nat. Neurosci.* **14**:387–397. [2]

Dilcher, R., R. Jamous, A. Takacs, et al. 2021. Neurophysiology of Embedded Response Plans: Age Effects in Action Execution but Not in Feature Integration from Preadolescence to Adulthood. *J. Neurophysiol.* **125**:1382–1395. [12]

DiNicola, L. M., O. I. Ariyo, and R. L. Buckner. 2023. Functional Specialization of Parallel Distributed Networks Revealed by Analysis of Trial-to-Trial Variation in Processing Demands. *J. Neurophysiol.* **129**:17–40. [4]

DiNicola, L. M., R. M. Braga, and R. L. Buckner. 2020. Parallel Distributed Networks Dissociate Episodic and Social Functions within the Individual. *J. Neurophysiol.* **123**:1144–1179. [11, 12]

Disney, A. A., and J. S. Robert. 2019. Translational Implications of the Anatomical Nonequivalence of Functionally Equivalent Cholinergic Circuit Motifs. *PNAS* **116**:26181–26186. [13]

D'Mello, A. M., J. D. E. Gabrieli, and D. E. Nee. 2020. Evidence for Hierarchical Cognitive Control in the Human Cerebellum. *Curr. Biol.* **30**:1881–1892. [7]

Dockès, J., R. A. Poldrack, R. Primet, et al. 2020. Neuroquery, Comprehensive Meta-Analysis of Human Brain Mapping. *eLife* **9**:e53385. [8]

Dodd, S., T. R. Norman, H. A. Eyre, et al. 2022. Psilocybin in Neuropsychiatry: A Review of Its Pharmacology, Safety, and Efficacy. *CNS Spectr.* 1–11. [13]

Domenech, P., and E. Koechlin. 2015. Executive Control and Decision-Making in the Prefrontal Cortex. *Curr. Opin. Behav. Sci.* **1**:101–106. [3]

Domenech, P., S. Rheims, and E. Koechlin. 2020. Neural Mechanisms Resolving Exploitation-Exploration Dilemmas in the Medial Prefrontal Cortex. *Science* **369**:eabb0184. [1, 10, 12]

Dong, C., C. Ly, L. E. Dunlap, et al. 2021. Psychedelic-Inspired Drug Discovery Using an Engineered Biosensor. *Cell* **184**:2779–2792.e2718. [13]

Donoghue, J. P., and S. P. Wise. 1982. The Motor Cortex of the Rat: Cytoarchitecture and Microstimulation Mapping. *J. Comp. Neurol.* **212**:76–88. [4]

Donoso, M., A. G. Collins, and E. Koechlin. 2014a. Human Cognition. Foundations of Human Reasoning in the Prefrontal Cortex. *Science* **344**:1481–1486. [7, 10]

Donoso, M., A. G. E. Collins, and E. Koechlin. 2014b. Foundations of Human Reasoning in the Prefrontal Cortex. *Science* **344**:1481–1486. [12]

Dooley, J. C., J. G. Franca, A. M. Seelke, D. F. Cooke, and L. A. Krubitzer. 2014. Evolution of Mammalian Sensorimotor Cortex: Thalamic Projections to Parietal Cortical Areas in Monodelphis Domestica. *Front. Neuroanat.* **8**:163. [3]

Dorrn, A. L., K. Yuan, A. J. Barker, C. E. Schreiner, and R. C. Froemke. 2010. Developmental Sensory Experience Balances Cortical Excitation and Inhibition. *Nature* **465**:932–936. [16]

Dosenbach, N. U., D. A. Fair, A. L. Cohen, B. L. Schlaggar, and S. E. Petersen. 2008. A Dual-Networks Architecture of Top-Down Control. *Trends Cogn. Sci.* **12**:99–105. [11]

Dosenbach, N. U., K. M. Visscher, E. D. Palmer, et al. 2006. A Core System for the Implementation of Task Sets. *Neuron* **50**:799–812. [15]

Doshi-Velez, F. 2009. The Infinite Partially Observable Markov Decision Process. In: Advances in Neural Information Processing Systems 22 (Nips 2009), ed. Y. Bengio et al., pp. 477–485. Red Hook, NY: Curran Associates Inc. [10]

Dougherty, D. D., L. Baer, G. R. Cosgrove, et al. 2002. Prospective Long-Term Follow-up of 44 Patients Who Received Cingulotomy for Treatment-Refractory Obsessive-Compulsive Disorder. *Am. J. Psych.* **159**:269–275. [15]

Drevets, W. C. 2007. Orbitofrontal Cortex Function and Structure in Depression. *Ann. N. Y. Acad. Sci.* **1121**:499–527. [4]

Drevets, W. C., W. Bogers, and M. E. Raichle. 2002. Functional Anatomical Correlates of Antidepressant Drug Treatment Assessed Using PET Measures of Regional Glucose Metabolism. *Eur. Neuropsychopharmacol.* **12**:527–544. [13]

Drysdale, A. T., L. Grosenick, J. Downar, et al. 2017. Resting-State Connectivity Biomarkers Define Neurophysiological Subtypes of Depression. *Nat. Med.* **23**:28–38. [13, 16]

Duan, L. Y., N. K. Horst, S. A. W. Cranmore, et al. 2021. Controlling One's World: Identification of Sub-Regions of Primate PFC Underlying Goal-Directed Behavior. *Neuron* **109**:2485–2498. [4, 16]

Duarte, R. B., E. Patrono, A. C. Borges, et al. 2014. Consumption of a Highly Palatable Food Induces a Lasting Place-Conditioning Memory in Marmoset Monkeys. *Behavioural Processes* **107**:163–166. [4]

Duarte, R. B., E. Patrono, A. C. Borges, et al. 2015. High versus Low Fat/Sugar Food Affects the Behavioral, but Not the Cortisol Response of Marmoset Monkeys in a Conditioned-Place-Preference Task. *Physiol. Behav.* **139**:442–448. [4]

Dubreuil, A., A. Valente, M. Beiran, F. Mastrogiuseppe, and S. Ostojic. 2022. The Role of Population Structure in Computations through Neural Dynamics. *Nat. Neurosci.* **25**:783–794. [12]

Duda, B. M., and L. H. Sweet. 2020. Functional Brain Changes Associated with Cognitive Training in Healthy Older Adults: A Preliminary Ale Meta-Analysis. *Brain Imag. Behav.* **14**:1247–1262. [14]

Duffau, H. 2011. Brain Mapping: From Neural Basis of Cognition to Surgical Applications. New York: Springer. [4]

Duffau, H. 2012. The "Frontal Syndrome" Revisited: Lessons from Electrostimulation Mapping Studies. *Cortex* **48**:120–131. [15]

Dum, R. P., and P. L. Strick. 2002. Motor Areas in the Frontal Lobe of the Primate. *Physiol. Behav.* **77**:677–682. [8]

Duman, R. S., G. Sanacora, and J. H. Krystal. 2019. Altered Connectivity in Depression: GABA and Glutamate Neurotransmitter Deficits and Reversal by Novel Treatments. *Neuron* **102**:75–90. [13]

Dunbar, R. I. M., and S. Shultz. 2007. Understanding Primate Brain Evolution. *Philos. Trans. R. Soc. Lond. B Biol. Sci.* **362**:649–658. [4]

Dunbar, R. I. M., and S. Shultz. 2017. Why Are There So Many Explanations for Primate Brain Evolution? *Philos. Trans. R. Soc. Lond. B Biol. Sci.* **372**:20160244. [2]

Duncan, J. 1986. Disorganisation of Behaviour after Frontal Lobe Damage. *Cogn. Neuropsychol.* **3**:271–290. [9]

———. 1990. Goal Weighting and the Choice of Behaviour in a Complex World. *Ergonomics* **33**:1265–1279. [9]

———. 2006. EPS Mid-Career Award 2004: Brain Mechanisms of Attention. *Q. J. Exp. Psychol.* **59**:2–27. [9]

———. 2010. The Multiple-Demand (MD) System of the Primate Brain: Mental Programs for Intelligent Behaviour. *Trends Cogn. Sci.* **14**:172–179. [7, 9, 12]

———. 2013. The Structure of Cognition: Attentional Episodes in Mind and Brain. *Neuron* **80**:35–50. [7]

Duncan, J., M. Assem, and S. Shashidhara. 2020. Integrated Intelligence from Distributed Brain Activity. *Trends Cogn. Sci.* **24**:838–852. [9, 12]

Duncan, J., R. Johnson, M. Swales, and C. Freer. 1997. Frontal Lobe Deficits after Head Injury: Unity and Diversity of Function. *Cogn. Neuropsychol.* **14**:713–741. [9, 12]

Duncan, J., and A. M. Owen. 2000. Common Regions of the Human Frontal Lobe Recruited by Diverse Cognitive Demands. *Trends Neurosci.* **23**:475–483. [9, 11]

Dundon, N. M., A. D. Shapiro, V. Babenko, G. N. Okafor, and S. T. Grafton. 2021. Ventromedial Prefrontal Cortex Activity and Sympathetic Allostasis During Value-Based Ambivalence. *Front. Behav. Neurosci.* **15**:615796. [16]

Dunlop, B. W., J. K. Rajendra, W. E. Craighead, et al. 2017. Functional Connectivity of the Subcallosal Cingulate Cortex and Differential Outcomes to Treatment with Cognitive-Behavioral Therapy or Antidepressant Medication for Major Depressive Disorder. *Am. J. Psych.* **174**:533–545. [13]

Dunlop, K., B. Woodside, M. Olmsted, et al. 2016. Reductions in Cortico-Striatal Hyperconnectivity Accompany Successful Treatment of Obsessive-Compulsive Disorder with Dorsomedial Prefrontal rTMS. *Neuropsychopharmacol.* **41**:1395–1403. [15]

Dunning, D. L., and J. Holmes. 2014. Does Working Memory Training Promote the Use of Strategies on Untrained Working Memory Tasks? *Mem. Cogn.* **42**:854–862. [14]

Durstewitz, D., J. K. Seamans, and T. J. Sejnowski. 2000. Neurocomputational Models of Working Memory. *Nat. Neurosci.* **3**:1184–1191. [6]

du Toit, S. A., S. A. Kade, C. T. Danielson, et al. 2020. The Effect of Emotional Working Memory Training on Emotional and Cognitive Outcomes in Individuals with Elevated Social Anxiety. *J. Affect. Disord.* **261**:76–83. [14]

Dworetsky, A., B. A. Seitzman, B. Adeyemo, et al. 2021. Probabilistic Mapping of Human Functional Brain Networks Identifies Regions of High Group Consensus. *Neuroimage* **237**:118164. [11]

Eacott, M. J., and D. Gaffan. 1992. Inferotemporal-Frontal Disconnection: The Uncinate Fascicle and Visual Associative Learning in Monkeys. *Eur. J. Neurosci.* **4**:1320–1332. [8]

Eagle, D. M., A. Bari, and T. W. Robbins. 2008a. The Neuropsychopharmacology of Action Inhibition: Cross-Species Translation of the Stop-Signal and Go/No-Go Tasks. *Psychopharmacol.* **199**:439–456. [16]

Eagle, D. M., C. Baunez, D. M. Hutcheson, et al. 2008b. Stop-Signal Reaction-Time Task Performance: Role of Prefrontal Cortex and Subthalamic Nucleus. *Cereb. Cortex* **18**:178–188. [4]

Eckart, C., C. Stoppel, J. Kaufmann, et al. 2011. Structural Alterations in Lateral Prefrontal, Parietal and Posterior Midline Regions of Men with Chronic Posttraumatic Stress Disorder. *J. Psych. Neurosci.* **36**:176–186. [4]

Eichenbaum, A., J. M. Scimeca, and M. D'Esposito. 2020. Dissociable Neural Systems Support the Learning and Transfer of Hierarchical Control Structure. *J. Neurosci.* **40**:6624–6637. [7]

Eichenbaum, H. 2017. Prefrontal-Hippocampal Interactions in Episodic Memory. *Nat. Rev. Neurosci.* **18**:547–558. [11, 12, 17]

Eijsbouts, C., T. Zheng, N. A. Kennedy, et al. 2021. Genome-Wide Analysis of 53,400 People with Irritable Bowel Syndrome Highlights Shared Genetic Pathways with Mood and Anxiety Disorders. *Nat. Genet.* **53**:1543–1552. [16]

Eisenberg, I. W., P. G. Bissett, A. Zeynep Enkavi, et al. 2019. Uncovering the Structure of Self-Regulation through Data-Driven Ontology Discovery. *Nat. Commun.* **10**:2319. [12]

Eisenreich, B. R., B. Y. Hayden, and J. Zimmermann. 2019. Macaques Are Risk-Averse in a Freely Moving Foraging Task. *Sci. Rep.* **9**:15091. [2]

Elbau, I. G., C. J. Lynch, J. Downar, et al. 2023. Functional Connectivity Mapping for rTMS Target Selection in Depression. *Am. J. Psych.* **180**:230–240. [13, 16]

Eldar, E., G. Lievre, P. Dayan, and R. J. Dolan. 2020. The Roles of Online and Offline Replay in Planning. *eLife* **9**:e56911. [12]

Eldridge, M. A., W. Lerchner, R. C. Saunders, et al. 2016. Chemogenetic Disconnection of Monkey Orbitofrontal and Rhinal Cortex Reversibly Disrupts Reward Value. *Nat. Neurosci.* **19**:37–39. [2]

Elliott, J., A. Johnston, D. Husereau, et al. 2020. Pharmacologic Treatment of Attention Deficit Hyperactivity Disorder in Adults: A Systematic Review and Network Meta-Analysis. *PloS One* **15**:e0240584. [16]

Elliott, M. V., Esmail, S.A.S., Weiner, K.S., Johnson, S.L. 2022. Neuroanatomical Correlates of Emotion-Related Impulsivity. *Biol. Psych.* **93**:566–574. [4]

El-Shamayleh, Y., and G. D. Horwitz. 2019. Primate Optogenetics: Progress and Prognosis. *PNAS* **116**:26195–26203. [2]

Elston, G. N. 2000. Pyramidal Cells of the Frontal Lobe: All the More Spinous to Think With. *J. Neurosci.* **20**:RC95. [6, 16]

———. 2003. The Pyramidal Neuron in Occipital, Temporal and Prefrontal Cortex of the Owl Monkey (Aotus Trivirgatus): Regional Specialization in Cell Structure. *Eur. J. Neurosci.* **17**:1313–1318. [6]

Elston, G. N., R. Benavides-Piccione, A. Elston, et al. 2006. Specializations of the Granular Prefrontal Cortex of Primates: Implications for Cognitive Processing. *Anat. Rec. A Discov. Mol. Cell. Evol. Biol.* **288**:26–35. [16]

Elston, G. N., and M. C. Gonzalez-Albo. 2003. Parvalbumin-, Calbindin-, and Calretinin-Immunoreactive Neurons in the Prefrontal Cortex of the Owl Monkey (Aotus Trivirgatus): A Standardized Quantitative Comparison with Sensory and Motor Areas. *Brain Behav. Evol.* **62**:19–30. [6]

Enel, P., J. D. Wallis, and E. L. Rich. 2020. Stable and Dynamic Representations of Value in the Prefrontal Cortex. *eLife* **9**:e54313. [5]

Engelhardt, L. E., D. A. Briley, F. D. Mann, K. P. Harden, and E. M. Tucker-Drob. 2015. Genes Unite Executive Functions in Childhood. *Psychol. Sci.* **26**:1151–1163. [9]

Engelhardt, L. E., F. D. Mann, D. A. Briley, et al. 2016. Strong Genetic Overlap between Executive Functions and Intelligence. *J. Exp. Psychol. Gen.* **145**:1141–1159. [9]

Engle, R. W., S. W. Tuholski, J. E. Laughlin, and A. R. A. Conway. 1999. Working Memory, Short-Term Memory, and General Fluid Intelligence: A Latent-Variable Approach. *J. Exp. Psychol. Gen.* **128**:309–331. [9]

Eppinger, B., T. Goschke, and S. Musslick. 2021. Meta-Control: From Psychology to Computational Neuroscience. *Cogn. Affect. Behav. Neurosci.* **21**:447–452. [12]

Esterlis, I., N. DellaGioia, R. H. Pietrzak, et al. 2018. Ketamine-Induced Reduction in mGluR5 Availability Is Associated with an Antidepressant Response: an [11C] Abp688 and PET Imaging Study in Depression. *Mol. Psychiatry* **23**:824–832. [13]

Etkin, A., C. Buchel, and J. J. Gross. 2015. The Neural Bases of Emotion Regulation. *Nat. Rev. Neurosci.* **16**:693–700. [16]

Everitt, B. J., and T. W. Robbins. 2005. Neural Systems of Reinforcement for Drug Addiction: From Actions to Habits to Compulsion. *Nat. Neurosci.* **8**:1481–1489. [15]

Farashahi, S., C. H. Donahue, B. Y. Hayden, D. Lee, and A. Soltani. 2019. Flexible Combination of Reward Information across Primates. *Nat. Hum. Behav.* **3**:1215–1224. [10]

Faraza, S., J. Waldenmaier, M. Dyrba, et al. 2021. Dorsolateral Prefrontal Functional Connectivity Predicts Working Memory Training Gains. *Front. Aging Neurosci.* **13**: [14]

Fascianelli, V., L. Ferrucci, S. Tsujimoto, and A. Genovesio. 2020. Neural Correlates of Strategy Switching in the Macaque Orbital Prefrontal Cortex. *J. Neurosci.* **40**:3025–3034. [5, 8]

Fatahi, Z., A. Ghorbani, M. Ismail Zibaii, and A. Haghparast. 2020. Neural Synchronization between the Anterior Cingulate and Orbitofrontal Cortices during Effort-Based Decision Making. *Neurobiol. Learn. Mem.* **175**:107320. [4]

Fedorenko, E., M. K. Behr, and N. Kanwisher. 2011. Functional Specificity for High-Level Linguistic Processing in the Human Brain. *PNAS* **108**:16428–16433. [11]

Fedorenko, E., and I. A. Blank. 2020. Broca's Area Is Not a Natural Kind. *Trends Cogn. Sci.* **24**:270–284. [12]

Fedorenko, E., J. Duncan, and N. Kanwisher. 2013. Broad Domain Generality in Focal Regions of Frontal and Parietal Cortex. *PNAS* **110**:16616–16621. [7, 9]

Felleman, D. J., and D. C. Van Essen. 1991. Distributed Hierarchical Processing in the Primate Cerebral Cortex. *Cereb. Cortex* **1**:1–47. [6]

Fellows, L. K., and M. J. Farah. 2003. Ventromedial Frontal Cortex Mediates Affective Shifting in Humans: Evidence from a Reversal Learning Paradigm. *Brain* **126**:1830–1837. [4, 12]

Feng, W., J. S. Yokoyama, S. Yu, et al. 2015. APOE Genotype Affects Cognitive Training Response in Healthy Shanghai Community-Dwelling Elderly Individuals. *J. Alzheimers Dis.* **47**:1035–1046. [14]

Feng, Y., A. Pahor, A. R. Seitz, D. L. Barbour, and S. M. Jaeggi. 2023. Unicorn, Hare, or Tortoise? Using Machine Learning to Predict Working Memory Training Performance. *J. Cogn.* **6**:53. [14]

Ferry, B., B. Roozendaal, and J. L. McGaugh. 1999. Basolateral Amygdala Noradrenergic Influences on Memory Storage Àre Mediated by an Interaction between Beta- and Alpha1-Adrenoceptors. *J. Neurosci.* **19**:5119–5123. [16]

Festini, S. B., L. Zahodne, and P. A. Reuter-Lorenz. 2018. Theoretical Perspectives on Age Differences in Brain Activation: Harold, Pasa, Crunch—How Do They Stac Up? Oxford Research Encyclopedias. https://oxfordre.com/psychology/display/10.1093/acrefore/9780190236557.001.0001/acrefore-9780190236557-e-400. (accessed Jan. 25, 2024). [14]

Figee, M., J. Luigjes, R. Smolders, et al. 2013. Deep Brain Stimulation Restores Frontostriatal Network Activity in Obsessive-Compulsive Disorder. *Nat. Neuroscience* **16**:386–387. [15]

Filipović, D., B. Novak, J. Xiao, et al. 2022. Chronic Fluoxetine Treatment of Socially Isolated Rats Modulates Prefrontal Cortex Proteome. *Neurosci.* **501**:52–71. [13]

Finc, K., K. Bonna, X. He, et al. 2020. Dynamic Reconfiguration of Functional Brain Networks during Working Memory Training. *Nat. Commun.* **11**:2435. [14]

Findling, C., N. Chopin, and E. Koechlin. 2021. Imprecise Neural Computations as a Source of Adaptive Behaviour in Volatile Environments. *Nat. Hum. Behav.* **5**:99–112. [10]

Finlay, B. L., and R. B. Darlington. 1995. Linked Regularities in the Development and Evolution of Mammalian Brains. *Science* **268**:1578–1584. [5]

Finn, E. S., L. Huber, D. C. Jangraw, P. J. Molfese, and P. A. Bandettini. 2019. Layer-Dependent Activity in Human Prefrontal Cortex during Working Memory. *Nat. Neurosci.* **22**:1687–1695. [11]

Finn, E. S., X. Shen, D. Scheinost, et al. 2015. Functional Connectome Fingerprinting: Identifying Individuals Using Patterns of Brain Connectivity. *Nat. Neurosci.* **18**:1664–1671. [11, 12]

Fish, K. N., B. R. Rocco, and D. A. Lewis. 2018. Laminar Distribution of Subsets of GABAergic Axon Terminals in Human Prefrontal Cortex. *Front. Neuroanat.* **12**:9. [6]

Fitch, W. T., and M. D. Martins. 2014. Hierarchical Processing in Music, Language, and Action: Lashley Revisited. *Ann. N. Y. Acad. Sci.* **1316**:87–104. [12]

Fletcher, P. C., and R. N. Henson. 2001. Frontal Lobes and Human Memory: Insights from Functional Neuroimaging. *Brain* **124**:849–881. [8]

Floresco, S. B., O. Magyar, S. Ghods-Sharifi, C. Vexelman, and M. T. L. Tse. 2006. Multiple Dopamine Receptor Subtypes in the Medial Prefrontal Cortex of the Rat Regulate Set-Shifting. *Neuropsychopharmacol.* **31**:297–309. [4]

Floyd, N. S., J. L. Price, A. T. Ferry, K. A. Keay, and R. Bandler. 2001. Orbitomedial Prefrontal Cortical Projections to Hypothalamus in the Rat. *J. Comp. Neurol.* **432**:307–328. [2]

Fluckiger, C., T. Munder, A. C. Del Re, and N. Solomonov. 2023. Strength-Based Methods: A Narrative Review and Comparative Multilevel Meta-Analysis of Positive Interventions in Clinical Settings. *Psychother. Res.* **33**:856–872. [16]

Foa, E. B., and C. P. McLean. 2016. The Efficacy of Exposure Therapy for Anxiety-Related Disorders and Its Underlying Mechanisms: The Case of OCD and PTSD. *Annu. Rev. Clin. Psychol.* **12**:1–28. [16]

Folloni, D., E. Fouragnan, M. K. Wittmann, et al. 2021. Ultrasound Modulation of Macaque Prefrontal Cortex Selectively Alters Credit Assignment–Related Activity and Behavior. *Sci. Adv.* **7**:eabg7700. [4]

Folloni, D., L. Verhagen, R. B. Mars, et al. 2019. Manipulation of Subcortical and Deep Cortical Activity in the Primate Brain Using Transcranial Focused Ultrasound Stimulation. *Neuron* **101**:1109–1116. [8]

Fonzo, G. A., A. Etkin, Y. Zhang, et al. 2019. Brain Regulation of Emotional Conflict Predicts Antidepressant Treatment Response for Depression. *Nat. Hum. Behav.* **3**:1319–1331. [13]

Fornito, A., S. L. Whittle, S. J. Wood, et al. 2006. The Influence of Sulcal Variability on Morphometry of the Human Anterior Cingulate and Paracingulate Cortex. *Neuroimage* **33**:843–854. [4]

Fornito, A., M. Yücel, S. Wood, et al. 2004. Individual Differences in Anterior Cingulate/Paracingulate Morphology Are Related to Executive Functions in Healthy Males. *Cereb. Cortex* **14**:424–431. [4]

Forsberg, A., D. Fellman, M. Laine, W. Johnson, and R. H. Logie. 2020. Strategy Mediation in Working Memory Training in Younger and Older Adults. *Q. J. Exp. Psychol.* **73**:1206–1226. [14]

Forstmann, B. U., G. Dutilh, S. Brown, et al. 2008. Striatum and Pre-SMA Facilitate Decision-Making under Time Pressure. *PNAS* **105**:17538–17542. [12]

Fouragnan, E. F., B. K. H. Chau, D. Folloni, et al. 2019. The Macaque Anterior Cingulate Cortex Translates Counterfactual Choice Value into Actual Behavioral Change. *Nat. Neurosci.* **22**:797–808. [16]

Fox, K. C. R., L. Shi, S. Baek, et al. 2020. Intrinsic Network Architecture Predicts the Effects Elicited by Intracranial Electrical Stimulation of the Human Brain. *Nat. Hum. Behav.* **4**:1039–1052. [11]

Fox, M. D., R. L. Buckner, M. P. White, M. D. Greicius, and A. Pascual-Leone. 2012. Efficacy of Transcranial Magnetic Stimulation Targets for Depression Is Related to Intrinsic Functional Connectivity with the Subgenual Cingulate. *Biol. Psych.* **72**:595–603. [11, 16]

Fox, M. D., and M. E. Raichle. 2007. Spontaneous Fluctuations in Brain Activity Observed with Functional Magnetic Resonance Imaging. *Nat. Rev. Neurosci.* **8**:700–711. [11]

Fox, M. D., A. Z. Snyder, J. L. Vincent, et al. 2005. The Human Brain Is Intrinsically Organized into Dynamic, Anticorrelated Functional Networks. *PNAS* **102**:9673–9678. [11]

Frank, M. J. 2015. Linking across Levels of Computation in Model-Based Cognitive Neuroscience. In: An Introduction to Model-Based Cognitive Neuroscience, ed. B. U. Forstmann and E. Wagenmakers. New York: Springer. [8]

Frank, M. J., and D. Badre. 2012. Mechanisms of Hierarchical Reinforcement Learning in Corticostriatal Circuits 1: Computational Analysis. *Cereb. Cortex* **22**:509–526. [7, 8, 15]

———. 2015. How Cognitive Theory Guides Neuroscience. *Cognition* **135**:14–20. [7]

Frank, M. J., C. Gagne, E. Nyhus, et al. 2015. fMRI and EEG Predictors of Dynamic Decision Parameters during Human Reinforcement Learning. *J. Neurosci.* **35**:485–494. [12]

Frank, M. J., B. Loughry, and R. C. O'Reilly. 2001. Interactions between Frontal Cortex and Basal Ganglia in Working Memory: A Computational Model. *Cogn. Affect. Behav. Neurosci.* **1**:137–160. [7]

Frank, M. J., and R. C. O'Reilly. 2006. A Mechanistic Account of Striatal Dopamine Function in Human Cognition: Psychopharmacological Studies with Cabergoline and Haloperidol. *Behav. Neurosci.* **120**:497–517. [7]

Frank, M. J., L. C. Seeberger, and C. O'Reilly R. 2004. By Carrot or by Stick: Cognitive Reinforcement Learning in Parkinsonism. *Science* **306**:1940–1943. [7]

Franklin, M. E., J. Sapyta, J. B. Freeman, et al. 2011. Cognitive Behavior Therapy Augmentation of Pharmacotherapy in Pediatric Obsessive-Compulsive Disorder: The Pediatric OCD Treatment Study II (Pots II) Randomized Controlled Trial. *JAMA* **306**:1224–1232. [16]

Freedman, D. J., M. Riesenhuber, T. Poggio, and E. K. Miller. 2001. Categorical Representation of Visual Stimuli in the Primate Prefrontal Cortex. *Science* **291**:312–316. [5, 12]

———. 2002. Visual Categorization and the Primate Prefrontal Cortex: Neurophysiology and Behavior. *J. Neurophysiol.* **88**:929–941. [5]

Freedman, L. J., T. R. Insel, and Y. Smith. 2000. Subcortical Projections of Area 25 (Subgenual Cortex) of the Macaque Monkey. *J. Comp. Neurol.* **421**:172–188. [2, 13]

Freeman, W., and J. W. Watts. 1950. Psychosurgery in the Treatment of Mental Disorders and Intractable Pain. Springfield, IL: Charles C. Thomas. [15]

Freeman, W., J. W. Watts, and T. Hunt. 1942. Psychosurgery: Intelligence, Emotion, and Social Behavior Following Prefrontal Lobotomy for Mental Disorders. London: Baillière, Tindall & Cox. [15]

Freier, L., P. Gupta, D. Badre, and D. Amso. 2021. The Value of Proactive Goal Setting and Choice in 3- to 7-Year-Olds' Use of Working Memory Gating Strategies in a Naturalistic Task. *Dev Sci* **24**:e13017. [7]

Freis, S. M., C. L. Morrison, J. M. Lessem, J. K. Hewitt, and N. P. Friedman. 2022. Genetic and Environmental Influences on Executive Functions and Intelligence in Middle Childhood. *Dev. Sci.* **25**:e13150. [9]

Friedman, N. P., A. S. Hatoum, D. E. Gustavson, et al. 2020. Executive Functions and Impulsivity Are Genetically Distinct and Independently Predict Psychopathology: Results from Two Adult Twin Studies. *Clin. Psychol. Sci.* **8**:519–538. [9]

Friedman, N. P., and A. Miyake. 2017. Unity and Diversity of Executive Functions: Individual Differences as a Window on Cognitive Structure. *Cortex* **86**:186–204. [9, 11, 12, 14, 17]

Friedman, N. P., A. Miyake, L. J. Altamirano, et al. 2016. Stability and Change in Executive Function Abilities from Late Adolescence to Early Adulthood: A Longitudinal Twin Study. *Dev. Psychol.* **52**:326–340. [9]

Friedman, N. P., A. Miyake, J. L. Robinson, and J. K. Hewitt. 2011. Developmental Trajectories in Toddlers' Self-Restraint Predict Individual Differences in Executive Functions 14 Years Later: A Behavioral Genetic Analysis. *Dev Psychol* **47**:1410–1430. [9]

Friedman, N. P., A. Miyake, S. E. Young, et al. 2008. Individual Differences in Executive Functions Are Almost Entirely Genetic in Origin. *J. Exp. Psychol. Gen.* **137**:201–225. [9]

Friedman, N. P., and T. W. Robbins. 2022. The Role of Prefrontal Cortex in Cognitive Control and Executive Function. *Neuropsychopharmacol.* **47**:72–89. [1–3, 8, 9, 14]

Fries, P., J. H. Reynolds, A. E. Rorie, and R. Desimone. 2001. Modulation of Oscillatory Neuronal Synchronization by Selective Visual Attention. *Science* **291**:1560–1563. [12]

Friston, K. 2010. The Free-Energy Principle: A Unified Brain Theory? *Nat. Rev. Neurosci.* **11**:127–138. [10]

Frömer, R., and A. Shenhav. 2022. Filling the Gaps: Cognitive Control as a Critical Lens for Understanding Mechanisms of Value-Based Decision-Making. *Neurosci. Biobehav. Rev.* **134**:104483. [12]

Froudist-Walsh, S., D. P. Bliss, X. Ding, et al. 2021. A Dopamine Gradient Controls Access to Distributed Working Memory in the Large-Scale Monkey Cortex. *Neuron* **109**:3500–3520. [12, 16]

Fu, Z., D. Beam, J. M. Chung, et al. 2022. The Geometry of Domain-General Performance Monitoring in the Human Medial Frontal Cortex. *Science* **376**:eabm9922. [12]

Fu, Z., A. Sajad, S. P. Errington, J. D. Schall, and U. Rutishauser. 2023. Neurophysiological Mechanisms of Error Monitoring in Human and Non-Human Primates. *Nat. Rev. Neurosci.* **24**:153–172. [12]

Funahashi, S., C. J. Bruce, and P. S. Goldman-Rakic. 1989. Mnemonic Coding of Visual Space in the Monkey's Dorsolateral Prefrontal Cortex. *J. Neurophysiol.* **61**:331–349. [4, 5, 12]

———. 1990. Visuospatial Coding in Primate Prefrontal Neurons Revealed by Oculomotor Paradigms. *J. Neurophysiol.* **63**:814–831. [8]

———. 1993a. Dorsolateral Prefrontal Lesions and Oculomotor Delayed-Response Performance: Evidence for Mnemonic Scotomas. *J. Neurosci.* **13**:1479–1497. [4, 12]

Funahashi, S., M. V. Chafee, and P. S. Goldman-Rakic. 1993b. Prefrontal Neuronal Activity in Rhesus Monkeys Performing a Delayed Anti-Saccade Task. *Nature* **365**:753–756. [3, 16]

Fusi, S., E. K. Miller, and M. Rigotti. 2016. Why Neurons Mix: High Dimensionality for Higher Cognition. *Curr. Opin. Neurobiol.* **37**:66–74. [2, 7, 8, 12]

Fuster, J. M. 1985. The Prefrontal Cortex, Mediator of Cross-Temporal Contingencies. *Hum. Neurobiol.* **4**:169–179. [8]

———. 1989. The Prefrontal Cortex: Anatomy, Physiology and Neuropsychology of the Frontal Lobe. New York: Raven Press. [2]

———. 2001. The Prefrontal Cortex: an Update: Time Is of the Essence. *Neuron* **30**:319–333. [7, 17]

Fuster, J. M., and G. E. Alexander. 1970. Delayed Response Deficit by Cryogenic Depression of Frontal Cortex. *Brain Res.* **20**:85–90. [5]

———. 1971. Neuron Activity Related to Short-Term Memory. *Science* **173**:652–654. [3–5, 10, 12]

Fuster, J. M., and J. P. Jervey. 1982. Neuronal Firing in the Inferotemporal Cortex of the Monkey in a Visual Memory Task. *J. Neurosci.* **2**:361–375. [5]

Gabbott, P. L., and S. J. Bacon. 1997. Vasoactive Intestinal Polypeptide Containing Neurones in Monkey Medial Prefrontal Cortex (mPFC): Colocalisation with Calretinin. *Brain Res.* **744**:179–184. [6]

Gallen, C. L., P. L. Baniqued, S. B. Chapman, et al. 2016. Modular Brain Network Organization Predicts Response to Cognitive Training in Older Adults. *PloS One* **11**:e0169015. [14]

Gallen, C. L., and M. D'Esposito. 2019. Brain Modularity: A Biomarker of Intervention-Related Plasticity. *Trends Cogn. Sci.* **23**:293–304. [14, 16]

Gamo, N. J., G. Lur, M. J. Higley, et al. 2015. Stress Impairs Prefrontal Cortical Function via D1 Dopamine Receptor Interactions with Hyperpolarization-Activated Cyclic Nucleotide-Gated Channels. *Biol. Psych.* **78**:860–870. [16]

Gao, L., S. Liu, L. Gou, et al. 2022. Single-Neuron Projectome of Mouse Prefrontal Cortex. *Nat. Neurosci.* **25**:515–529. [2]

Gao, L., S. Liu, Y. Wang, et al. 2023. Single-Neuron Analysis of Dendrites and Axons Reveals the Network Organization in Mouse Prefrontal Cortex. *Nat. Neurosci.* **26**:1111–1126. [2]

Gao, P., E. Trautmann, B. Yu, et al. 2017. A Theory of Multineuronal Dimensionality, Dynamics and Measurement. https://www.biorxiv.org/content/10.1101/214262v2. [5]

Garcia-Cabezas, M. A., M. K. P. Joyce, Y. J. John, B. Zikopoulos, and H. Barbas. 2017. Mirror Trends of Plasticity and Stability Indicators in Primate Prefrontal Cortex. *Eur. J. Neurosci.* **46**:2392–2405. [6, 8]

Garin, C. M., Y. Hori, S. Everling, et al. 2022. An Evolutionary Gap in Primate Default Mode Network Organization. *Cell Rep.* **39**:110669. [4]

Garrison, J. R., C. Fernyhough, S. McCarthy-Jones, M. Haggard, and J. S. Simons. 2015. Paracingulate Sulcus Morphology Is Associated with Hallucinations in the Human Brain. *Nat. Commun.* **6**:8956. [4]

Gazzaley, A., and A. C. Nobre. 2012. Top-Down Modulation: Bridging Selective Attention and Working Memory. *Trends Cogn. Sci.* **16**:128–134. [12]

Geddes, M. R., A. Tsuchida, V. Ashley, D. Swick, and L. K. Fellows. 2014. Material-Specific Interference Control Is Dissociable and Lateralized in Human Prefrontal Cortex. *Neuropsychologia* **64**:310–319. [8]

Gehring, W. J., B. Goss, M. G. H. Coles, D. E. Meyer, and E. Donchin. 1993. A Neural System for Error Detection and Compensation. *Psychol. Sci.* **4**:385–390. [12]

Geng, J. J., and S. Vossel. 2013. Re-Evaluating the Role of Tpj in Attentional Control: Contextual Updating? *Neurosci. Biobehav. Rev.* **37**:2608–2620. [12]

Genovesio, A., S. P. Wise, and R. E. Passingham. 2014. Prefrontal–Parietal Function: From Foraging to Foresight. *Trends Cogn. Sci.* **18**:72–81. [4]

Gentil, A. F., E. N. Eskandar, C. D. Marci, K. C. Evans, and D. D. Dougherty. 2009. Physiological Responses to Brain Stimulation during Limbic Surgery: Further Evidence of Anterior Cingulate Modulation of Autonomic Arousal. *Biol. Psych.* **66**:695–701. [15]

Geraldo, A., A. R. Dores, A. Castro-Caldas, and F. Barbosa. 2023. Functional Connectivity as a Neural Correlate of Cognitive Rehabilitation Programs' Efficacy: A Systematic Review. *Curr. Psychol.* **42**:17918–17934. [14]

Gerhard, D. M., S. Pothula, R.-J. Liu, et al. 2020. GABA Interneurons Are the Cellular Trigger for Ketamine's Rapid Antidepressant Actions. *J. Clin. Invest.* **130**:1336–1349. [13]

Gerraty, R. T., J. Y. Davidow, K. Foerde, et al. 2018. Dynamic Flexibility in Striatal-Cortical Circuits Supports Reinforcement Learning. *J. Neurosci.* **38**:2442–2453. [12]

Gershman, S. J., D. M. Blei, and Y. Niv. 2010. Context Learning, and Extinction. *Psychol. Rev.* **117**:1997–1209. [10]

Gershman, S. J., A. B. Markman, and A. R. Otto. 2014. Retrospective Revaluation in Sequential Decision Making: A Tale of Two Systems. *J. Exp. Psychol. Gen.* **143**:182–194. [10]

Geschwind, N. 1965. Disconnexion Syndromes in Animals and Man. *Brain* **88**:237–294. [1, 16]

———. 1974. Disconnexion Syndromes in Animals and Man. In: Boston Studies in the Philosophy of Science, ed. R. S. Cohen and M. W. Wartofsky, pp. 105–236, Boston Studies in the Philosophy of Science, vol. 16. Boston, MA: Springer. [11]

Ghashghaei, H. T., C. C. Hilgetag, and H. Barbas. 2007. Sequence of Information Processing for Emotions Based on the Anatomic Dialogue between Prefrontal Cortex and Amygdala. *Neuroimage* **34**:905–923. [3]

Ghosh, B. C., A. J. Calder, P. V. Peers, et al. 2012. Social Cognitive Deficits and Their Neural Correlates in Progressive Supranuclear Palsy. *Brain* **135**:2089–2102. [16]

Giarrocco, F., and B. B. Averbeck. 2021. Organization of Parietoprefrontal and Temporoprefrontal Networks in the Macaque. *J. Neurophysiol.* **126**:1289–1309. [5, 8, 11]

———. 2023. Anatomical Organization of Forebrain Circuits in the Primate. *Brain Struct. Funct.* **228**:393–411. [5, 8, 11, 16]

Giedd, J. N., A. C. Vaituzis, S. D. Hamburger, et al. 1996. Quantitative MRI of the Temporal Lobe, Amygdala, and Hippocampus in Normal Human Development: Ages 4–18 Years. *J. Comp. Neurol.* **366**:223–230. [16]

Giguere, M., and P. S. Goldman-Rakic. 1988. Mediodorsal Nucleus: Areal, Laminar, and Tangential Distribution of Afferents and Efferents in the Frontal Lobe of Rhesus Monkeys. *J. Comp. Neurol.* **277**:195–213. [6, 8]

Gillan, C. M., M. Papmeyer, S. Morein-Zamir, et al. 2011. Disruption in the Balance between Goal-Directed Behavior and Habit Learning in Obsessive-Compulsive Disorder. *Am. J. Psych.* **168**:718–726. [15]

Giller, F., W. Bensmann, M. Mückschel, A.-K. Stock, and C. Beste. 2020. Evidence for a Causal Role of Superior Frontal Cortex Theta Oscillations during the Processing of Joint Subliminal and Conscious Conflicts. *Cortex* **132**:15–28. [12]

Gilman, J. P., M. Medalla, and J. I. Luebke. 2017. Area-Specific Features of Pyramidal Neurons-a Comparative Study in Mouse and Rhesus Monkey. *Cereb. Cortex* **27**:2078–2094. [16]

Gilzenrat, M. S., S. Nieuwenhuis, M. Jepma, and J. D. Cohen. 2010. Pupil Diameter Tracks Changes in Control State Predicted by the Adaptive Gain Theory of Locus Coeruleus Function. *Cogn. Affect. Behav. Neurosci.* **10**:252–269. [12]

Gläscher, J., R. Adolphs, H. Damasio, et al. 2012. Lesion Mapping of Cognitive Control and Value-Based Decision Making in the Prefrontal Cortex. *PNAS* **109**:14681–14686. [15]

Gläscher, J., D. Tranel, L. K. Paul, et al. 2009. Lesion Mapping of Cognitive Abilities Linked to Intelligence. *Neuron* **61**:681–691. [8]

Glasser, M. F., T. S. Coalson, E. C. Robinson, et al. 2016a. A Multi-Modal Parcellation of Human Cerebral Cortex. *Nature* **536**:171–178. [9]

Glasser, M. F., S. M. Smith, D. S. Marcus, et al. 2016b. The Human Connectome Project's Neuroimaging Approach. *Nat. Neurosci.* **19**:1175–1187. [9]

Glasser, M. F., and D. C. Van Essen. 2011. Mapping Human Cortical Areas *in Vivo* Based on Myelin Content as Revealed by T1- and T2-Weighted MRI. *J. Neurosci.* **31**:11597–11616. [6]

Gnadt, J. W., and R. A. Andersen. 1988. Memory Related Motor Planning Activity in Posterior Parietal Cortex of Macaque. *Exp. Brain Res.* **70**:216–220. [5]

Godefroy, O., A. Aarabi, F. Dorchies, et al. 2023. Functional Architecture of Executive Processes: Evidence from Verbal Fluency and Lesion Mapping in Stroke Patients. *Cortex* **164**:129–143. [12]

Godfrey, J. R., B. R. Howell, A. Mummert, et al. 2023. Effects of Social Rank and Pubertal Delay on Brain Structure in Female Rhesus Macaques. *Psychoneuroendocrin.* **149**:105987. [16]

Godlewska, B. R., M. Browning, R. Norbury, P. J. Cowen, and C. J. Harmer. 2016. Early Changes in Emotional Processing as a Marker of Clinical Response to SSRI Treatment in Depression. *Transl. Psychiatry* **6**:e957. [13]

Godlewska, B. R., M. Browning, R. Norbury, et al. 2018. Predicting Treatment Response in Depression: The Role of Anterior Cingulate Cortex. *Int. J. Neuropsychopharmacol.* **21**:988–996. [13]

Gogtay, N., J. N. Giedd, L. Lusk, et al. 2004. Dynamic Mapping of Human Cortical Development during Childhood through Early Adulthood. *PNAS* **101**:8174–8179. [16]

Goldapple, K., Z. Segal, C. Garson, et al. 2004. Modulation of Cortical-Limbic Pathways in Major Depression: Treatment-Specific Effects of Cognitive Behavior Therapy. *Arch. Gen. Psychiatry* **61**:34–41. [13]

Goldfarb, E. V., M. D. Rosenberg, D. Seo, R. T. Constable, and R. Sinha. 2020. Hippocampal Seed Connectome-Based Modeling Predicts the Feeling of Stress. *Nat. Commun.* **11**:2650. [16]

Goldman-Rakic, P. S. 1984. Modular Organization of Prefrontal Cortex. *Trends Neurosci.* **7**:419–424. [6]

———. 1987. Circuitry of the Prefrontal Cortex and the Regulation of Behavior by Representational Memory. In: Handbook of Physiology, ed. F. Plum and V. Mountcastle, pp. 373–417. Bethesda: American Physiological Society. [5]

———. 1988. Topography of Cognition: Parallel Distributed Networks in Primate Association Cortex. *Annu. Rev. Neurosci.* **11**:137–156. [11]

———. 1995. Cellular Basis of Working Memory. *Neuron* **14**:477–485. [5, 10, 12]

Goldman-Rakic, P. S., and L. J. Porrino. 1985. The Primate Mediodorsal (MD) Nucleus and Its Projection to the Frontal Lobe. *J. Comp. Neurol.* **242**:535–560. [8]

Goldman, P. S., and H. E. Rosvold. 1970. Localization of Function within the Dorsolateral Prefrontal Cortex of the Rhesus Monkey. *Exp Neurol* **27**:291–304. [4]

Goldman, P. S., H. E. Rosvold, B. Vest, and T. W. Galkin. 1971. Analysis of the Delayed-Alternation Deficit Produced by Dorsolateral Prefrontal Lesions in the Rhesus Monkey. *J. Comp. Physiol. Psychol.* **77**:212–220. [8]

Goldstein-Piekarski, A. N., T. M. Ball, Z. Samara, et al. 2022. Mapping Neural Circuit Biotypes to Symptoms and Behavioral Dimensions of Depression and Anxiety. *Biol. Psychiatry* **91**:561–571. [13]

Goldstein-Piekarski, A. N., B. R. Staveland, T. M. Ball, et al. 2018. Intrinsic Functional Connectivity Predicts Remission on Antidepressants: A Randomized Controlled Trial to Identify Clinically Applicable Imaging Biomarkers. *Transl. Psychiatry* **8**:1–11. [13]

Gonzalez-Burgos, G., T. Miyamae, Y. Krimer, et al. 2019. Distinct Properties of Layer 3 Pyramidal Neurons from Prefrontal and Parietal Areas of the Monkey Neocortex. *J. Neurosci.* **39**:7277–7290. [16]

González, V. V., A. Izquierdo, and A. P. Blaisdell. 2023. Theoretical Mechanisms of Paradoxical Choices Involving Information. *Comp. Cogn. Behav. Rev.* **18**:11–31. [2]

Goodfellow, I., J. Pouget-Abadie, M. Mirza, et al. 2014. Generative Adversarial Nets. 27. https://proceedings.neurips.cc/paper_files/paper/2014/hash/5ca3e9b122f61f8f06494c97b1afccf3-Abstract.html. (accessed Jan. 25, 2024). [12]

Goodkind, M., S. B. Eickhoff, D. J. Oathes, et al. 2015. Identification of a Common Neurobiological Substrate for Mental Illness. *JAMA Psych.* **72**: 305–315. [13, 16]

Goodman, W. K., and R. L. Alterman. 2012. Deep Brain Stimulation for Intractable Psychiatric Disorders. *Annu. Rev. Med.* **63**:511–524. [15]

Goodman, W. K., K. D. Foote, B. D. Greenberg, et al. 2010. Deep Brain Stimulation for Intractable Obsessive Compulsive Disorder: Pilot Study Using a Blinded, Staggered-Onset Design. *Biol. Psych.* **67**:535–542. [15]

Goodwin, S. J., R. K. Blackman, S. Sakellaridi, and M. V. Chafee. 2012. Executive Control over Cognition: Stronger and Earlier Rule-Based Modulation of Spatial Category Signals in Prefrontal Cortex Relative to Parietal Cortex. *J. Neurosci.* **32**:3499–3515. [9]

Gordon, E. M., T. O. Laumann, A. W. Gilmore, et al. 2017. Precision Functional Mapping of Individual Human Brains. *Neuron* **95**:791–807. [7, 11, 12]

Gordon, E. M., and S. M. Nelson. 2021. Three Types of Individual Variation in Brain Networks Revealed by Single-Subject Functional Connectivity Analyses. *Curr. Opin. Behav. Sci.* **40**:79–86. [12]

Gottfried, J. A., J. O'Doherty, and R. J. Dolan. 2003. Encoding Predictive Reward Value in Human Amygdala and Orbitofrontal Cortex. *Science* **301**:1104–1107. [4]

Gottlieb, J. 2007. From Thought to Action: The Parietal Cortex as a Bridge between Perception, Action, and Cognition. *Neuron* **53**:9–16. [12]

Goudar, V., B. Peysakhovich, D. J. Freedman, E. A. Buffalo, and X.-J. Wang. 2023. Schema Formation in a Neural Population Subspace Underlies Learning-to-Learn in Flexible Sensorimotor Problem-Solving. *Nat. Neurosci.* **26**:879–890. [12]

Goulas, A., J.-P. Changeux, K. Wagstyl, et al. 2021. The Natural Axis of Transmitter Receptor Distribution in the Human Cerebral Cortex. *PNAS* **118**:e2020574118. [13]

Goulas, A., H. B. Uylings, and P. Stiers. 2014. Mapping the Hierarchical Layout of the Structural Network of the Macaque Prefrontal Cortex. *Cereb. Cortex* **24**:1178–1194. [4, 7, 8]

Goulas, A., K. Zilles, and C. C. Hilgetag. 2018. Cortical Gradients and Laminar Projections in Mammals. *Trends Neurosci.* **41**:775–788. [16]

Gouwens, N. W., S. A. Sorensen, F. Baftizadeh, et al. 2020. Integrated Morphoelectric and Transcriptomic Classification of Cortical GABAergic Cells. *Cell* **183**:935–953. [8]

Granon, S., and B. Poucet. 2000. Involvement of the Rat Prefrontal Cortex in Cognitive Functions: A Central Role for the Prelimbic Area. *Psychobiol.* **28**:229–237. [3]

Grattan, L. E., and P. W. Glimcher. 2014. Absence of Spatial Tuning in the Orbitofrontal Cortex. *PloS One* **9**:e112750. [5]

Gratton, C., A. Dworetsky, B. Adeyemo, et al. 2022. The Cingulo-Opercular Network Is Composed of Two Distinct Sub-Systems. 2022.9. [11]

Gratton, C., T. O. Laumann, E. M. Gordon, B. Adeyemo, and S. E. Petersen. 2016. Evidence for Two Independent Factors That Modify Brain Networks to Meet Task Goals. *Cell Rep.* **17**:1276–1288. [11]

Gratton, C., T. O. Laumann, A. N. Nielsen, et al. 2018a. Functional Brain Networks Are Dominated by Stable Group and Individual Factors, Not Cognitive or Daily Variation. *Neuron* **98**:439–452. [11]

Gratton, C., T. G. Lee, E. M. Nomura, and M. D'Esposito. 2014. Perfusion MRI Indexes Variability in the Functional Brain Effects of Theta-Burst Transcranial Magnetic Stimulation. *PloS One* **9**:e101430. [16]

Gratton, C., E. M. Nomura, F. Pérez, and M. D'Esposito. 2012. Focal Brain Lesions to Critical Locations Cause Widespread Disruption of the Modular Organization of the Brain. *J. Cog. Neuro.* **24**:1275–1285. [11, 12]

Gratton, C., H. Sun, and S. E. Petersen. 2018b. Control Networks and Hubs. *Psychophysiol.* **55**:e13032. [7, 11]

Grayson, D. S., E. Bliss-Moreau, C. J. Machado, et al. 2016. The Rhesus Monkey Connectome Predicts Disrupted Functional Networks Resulting from Pharmacogenetic Inactivation of the Amygdala. *Neuron* **91**:453–466. [2]

Graziano, M. S., C. S. Taylor, and T. Moore. 2002. Complex Movements Evoked by Microstimulation of Precentral Cortex. *Neuron* **34**:841–851. [8]

Green, C. S., D. Bavelier, A. F. Kramer, et al. 2019. Improving Methodological Standards in Behavioral Interventions for Cognitive Enhancement. *J. Cogn. Enhanc.* **3**:2–29. [14]

Greenberg, B. D., L. A. Gabriels, D. A. Malone, Jr., et al. 2010. Deep Brain Stimulation of the Ventral Internal Capsule/Ventral Striatum for Obsessive-Compulsive Disorder: Worldwide Experience. *Mol. Psych.* **15**:64–79. [15]

Greenberg, B. D., L. H. Price, S. L. Rauch, et al. 2003. Neurosurgery for Intractable Obsessive-Compulsive Disorder and Depression: Critical Issues. *Neurosurg Clin N Am* **14**:199–212. [16]

Greicius, M. D., B. H. Flores, V. Menon, et al. 2007. Resting-State Functional Connectivity in Major Depression: Abnormally Increased Contributions from Subgenual Cingulate Cortex and Thalamus. *Biol. Psychiatry* **62**:429–437. [13]

Griffiths, R. R., M. W. Johnson, M. A. Carducci, et al. 2016. Psilocybin Produces Substantial and Sustained Decreases in Depression and Anxiety in Patients with Life-Threatening Cancer: A Randomized Double-Blind Trial. *J. Psychopharmacol.* **30**:1181–1197. [13]

Grimm, S., P. Boesiger, J. Beck, et al. 2009. Altered Negative BOLD Responses in the Default-Mode Network during Emotion Processing in Depressed Subjects. *Neuropsychopharmacol.* **34**:932–943. [13]

Grisot, G., S. N. Haber, and A. Yendiki. 2021. Diffusion MRI and Anatomic Tracing in the Same Brain Reveal Common Failure Modes of Tractography. *Neuroimage* **239**:118300. [11]

Groenewegen, H. J. 1988. Organization of the Afferent Connections of the Mediodorsal Thalamic Nucleus in the Rat, Related to the Mediodorsal-Prefrontal Topography. *Neurosci.* **24**:379–431. [3]

Groman, S. M., C. Keistler, A. J. Keip, et al. 2019. Orbitofrontal Circuits Control Multiple Reinforcement-Learning Processes. *Neuron* **103**:734–746.e733. [4]

Gross, C. G. 1993. Huxley versus Owen: The Hippocampus Minor and Evolution. *Trends in Neurosciences* **16**:493–498. [4]

Gross, C. G. 1994. How Inferior Temporal Cortex Became a Visual Area. *Cerebral Cortex* **4**:455–469. [4]

Gu, S., F. Pasqualetti, M. Cieslak, et al. 2015. Controllability of Structural Brain Networks. *Nat. Commun.* **6**:8414. [12]

Gunning, F. M., J. Cheng, C. F. Murphy, et al. 2009. Anterior Cingulate Cortical Volumes and Treatment Remission of Geriatric Depression. *Int. J. Geriatr. Psychiatry* **24**:829–836. [13]

Gurnani, H., and N. A. Cayco Gajic. 2023. Signatures of Task Learning in Neural Representations. *Curr. Opin. Neurobiol.* **83**:102759. [4]

Gustavson, D. E., M. S. Panizzon, C. E. Franz, et al. 2018. Genetic and Environmental Architecture of Executive Functions in Midlife. *Neuropsychol.* **32**:18–30. [9]

Gustavson, D. E., C. A. Reynolds, R. P. Corley, et al. 2022. Genetic Associations between Executive Functions and Intelligence: A Combined Twin and Adoption Study. *J. Exp. Psychol. Gen.* **151**:1745–1761. [9]

Gyurak, A., B. Patenaude, M. S. Korgaonkar, et al. 2016. Frontoparietal Activation During Response Inhibition Predicts Remission to Antidepressants in Patients with Major Depression. *Biol. Psychiatry* **79**:274–281. [13]

Haber, S. N. 2003. The Primate Basal Ganglia: Parallel and Integrative Networks. *J. Chem. Neuroanat.* **26**:317–330. [7]

———. 2016. Corticostriatal Circuitry. *Dialogues Clin. Neurosci.* **18**:7–21. [8, 14]

Haber, S. N., and T. E. Behrens. 2014. The Neural Network Underlying Incentive-Based Learning: Implications for Interpreting Circuit Disruptions in Psychiatric Disorders. *Neuron* **83**:1019–1039. [1, 16]

Haber, S. N., and R. Calzavara. 2009. The Cortico-Basal Ganglia Integrative Network: The Role of the Thalamus. *Brain Res. Bull.* **78**:69–74. [15]

Haber, S. N., K. S. Kim, P. Mailly, and R. Calzavara. 2006. Reward-Related Cortical Inputs Define a Large Striatal Region in Primates That Interface with Associative Cortical Connections, Providing a Substrate for Incentive-Based Learning. *J. Neurosci.* **26**:8368–8376. [8, 11, 15]

Haber, S. N., H. Liu, J. Seidlitz, and E. Bullmore. 2022. Prefrontal Connectomics: From Anatomy to Human Imaging. *Neuropsychopharmacol.* **47**:20–40. [1, 16]

Haber, S. N., and T. Robbins. 2022. The Prefrontal Cortex. *Neuropsychopharmacol.* **47**:1–2. [3]

Haber, S. N., W. Tang, E. Y. Choi, et al. 2020. Circuits, Networks, and Neuropsychiatric Disease: Transitioning from Anatomy to Imaging. *Biol. Psych.* **87**:318–327. [7, 15]

Haber, S. N., A. Yendiki, and S. Jbabdi. 2021. Four Deep Brain Stimulation Targets for Obsessive-Compulsive Disorder: Are They Different? *Biol. Psych.* **90**:667–677. [15]

Haghir, H., A. Kuckertz, L. Zhao, J. Hami, and N. Palomero-Gallagher. 2023. A New Map of the Rat Isocortex and Proisocortex: Cytoarchitecture and M2 Receptor Distribution Patterns. *Brain Struct. Funct.* **5**: [3, 4]

Hains, A. B., M. A. Vu, P. K. Maciejewski, et al. 2009. Inhibition of Protein Kinase C Signaling Protects Prefrontal Cortex Dendritic Spines and Cognition from the Effects of Chronic Stress. *PNAS* **106**:17957–17962. [16]

Hains, A. B., Y. Yabe, and A. F. Arnsten. 2015. Chronic Stimulation of Alpha-2a-Adrenoceptors with Guanfacine Protects Rodent Prefrontal Cortex Dendritic Spines and Cognition from the Effects of Chronic Stress. *Neurobiol Stress* **2**:1–9. [16]

Halassa, M. M., and S. M. Sherman. 2019. Thalamocortical Circuit Motifs: A General Framework. *Neuron* **103**:762–770. [8]

Hallenbeck, G. E., T. C. Sprague, M. Rahmati, K. K. Sreenivasan, and C. E. Curtis. 2021. Working Memory Representations in Visual Cortex Mediate Distraction Effects. *Nat. Commun.* **12**:4714. [4]

Halley, A. C., M. K. L. Baldwin, D. F. Cooke, M. Englund, and L. Krubitzer. 2020. Distributed Motor Control of Limb Movements in Rat Motor and Somatosensory Cortex: The Sensorimotor Amalgam Revisited. *Cereb. Cortex* **30**:6296–6312. [8]

Hamilton, J. P., D. J. Furman, C. Chang, et al. 2011. Default-Mode and Task-Positive Network Activity in Major Depressive Disorder: Implications for Adaptive and Maladaptive Rumination. *Biol. Psychiatry* **70**:327–333. [13]

Hampton, A. N., P. Bossaerts, and J. P. O'Doherty. 2006. The Role of the Ventromedial Prefrontal Cortex in Abstract State-Based Inference during Decision Making in Humans. *J. Neurosci.* **26**:8360–8367. [12]

Han, L., X. Wei, C. Liu, et al. 2022. Cell Transcriptomic Atlas of the Non-Human Primate Macaca fascicularis. *Nature* **604**:723–731. [2]

Han, L. K. M., R. Dinga, T. Hahn, et al. 2021. Brain Aging in Major Depressive Disorder: Results from the Enigma Major Depressive Disorder Working Group. *Mol. Psych.* **26**:5124–5139. [16]

Hansen, J. Y., G. Shafiei, R. D. Markello, et al. 2022. Mapping Neurotransmitter Systems to the Structural and Functional Organization of the Human Neocortex. *Nat. Neurosci.* **25**:1569–1581. [13]

Hanson, J. L., M. K. Chung, B. B. Avants, et al. 2012. Structural Variations in Prefrontal Cortex Mediate the Relationship between Early Childhood Stress and Spatial Working Memory. *J. Neurosci.* **32**:7917-7925. [16]

Hanson, J. L., A. V. Williams, D. A. Bangasser, and C. J. Peña. 2021. Impact of Early Life Stress on Reward Circuit Function and Regulation. *Front. Psych.* **12**:744690. [16]

Hardung, S., R. Epple, Z. Jackel, et al. 2017. A Functional Gradient in the Rodent Prefrontal Cortex Supports Behavioral Inhibition. *Curr. Biol.* **27**:549–555. [3, 12]

Harhen, N. C., and A. M. Bornstein. 2023. Overharvesting in Human Patch Foraging Reflects Rational Structure Learning and Adaptive Planning. *PNAS* **120**:e2216524120. [2]

Hart, E., and A. C. Huk. 2020. Recurrent Circuit Dynamics Underlie Persistent Activity in the Macaque Frontoparietal Network. *eLife* **9**:e52460. [6, 8]

Hart, E. E., G. J. Blair, T. J. O'Dell, H. T, Blair, and A. Izquierdo. 2020. Chemogenetic Modulation and Single-Photon Calcium Imaging in Anterior Cingulate Cortex Reveal a Mechanism for Effort-Based Decisions. *J. Neurosci.* **40**:5628–5643. [2, 4]

Hartshorne, J. K., and L. T. Germine. 2015. When Does Cognitive Functioning Peak? The Asynchronous Rise and Fall of Different Cognitive Abilities across the Life Span. *Psychol. Sci.* **26**:433–443. [14]

Hasler, G., J. W. van der Veen, T. Tumonis, et al. 2007. Reduced Prefrontal Glutamate/Glutamine and Gamma-Aminobutyric Acid Levels in Major Depression Determined Using Proton Magnetic Resonance Spectroscopy. *Arch. Gen. Psychiatry* **64**:193–200. [13]

Hathaway, C. B., W. I. Voorhies, N. Sathishkumar, et al. 2023. Defining Tertiary Sulci in Lateral Prefrontal Cortex in Chimpanzees Using Human Predictions. *Brain Struct. Funct.*, in press. [4]

Hatoum, A. S., C. L. Morrison, E. C. Mitchell, et al. 2023. Genome-Wide Association Study Shows That Executive Functioning Is Influenced by GABAergic Processes and Is a Neurocognitive Genetic Correlate of Psychiatric Disorders. *Biol. Psych.* **93**:59–70. [9]

Hauser, T. U., R. Iannaccone, P. Stämpfli, et al. 2014. The Feedback-Related Negativity (FRN) Revisited: New Insights into the Localization, Meaning and Network Organization. *Neuroimage* **1**:159–168. [12]

Hayashi-Takagi, A., S. Yagishita, M. Nakamura, et al. 2015. Labelling and Optical Erasure of Synaptic Memory Traces in the Motor Cortex. *Nature* **525**:333–338. [13]

Hayden, B. Y. 2023. The Dangers of Cortical Brain Maps. *J. Cogn. Neurosci.* **35**:372–375. [8]

Hayden, B. Y., H. S. Park, and J. Zimmermann. 2022. Automated Pose Estimation in Primates. *Am. J. Primatol.* **84**:e23348. [8]

Hayden, B. Y., J. M. Pearson, and M. L. Platt. 2011. Neuronal Basis of Sequential Foraging Decisions in a Patchy Environment. *Nat. Neurosci.* **14**:933–939. [4]

Hayes, S. C. 2019. Acceptance and Commitment Therapy: Towards a Unified Model of Behavior Change. *World Psych.* **18**:226–227. [16]

Hazy, T. E., M. J. Frank, and C. O'Reilly R. 2007. Towards an Executive without a Homunculus: Computational Models of the Prefrontal Cortex/Basal Ganglia System. *Philos. Trans. R. Soc. Lond. B Biol. Sci.* **362**:1601–1613. [17]

Hazy, T. E., M. J. Frank, and R. C. O'Reilly. 2006. Banishing the Homunculus: Making Working Memory Work. *Neurosci.* **139**:105–118. [7]

Heidbreder, C. A., and H. J. Groenewegen. 2003. The Medial Prefrontal Cortex in the Rat: Evidence for a Dorso-Ventral Distinction Based Upon Functional and Anatomical Characteristics. *Neurosci. Biobehav. Rev.* **27**:555–579. [5]

Heilbronner, S. R., and B. Y. Hayden. 2016. Dorsal Anterior Cingulate Cortex: A Bottom-Up View. *Annu. Rev. Neurosci.* **39**:149–170. [15]

Heilbronner, S. R., J. Rodriguez-Romaguera, G. J. Quirk, H. J. Groenewegen, and S. N. Haber. 2016. Circuit-Based Corticostriatal Homologies between Rat and Primate. *Biol. Psych.* **80**:509–521. [2–4]

Heine, L., A. Soddu, F. Gómez, et al. 2012. Resting State Networks and Consciousness: Alterations of Multiple Resting State Network Connectivity in Physiological, Pharmacological, and Pathological Consciousness States. *Front. Psychol.* **3**:295. [11]

Heller, A. S., T. C. Shi, C. E. C. Ezie, et al. 2020. Association between Real-World Experiential Diversity and Positive Affect Relates to Hippocampal-Striatal Functional Connectivity. *Nat. Neurosci.* **23**:800–804. [12]

Hempel, C. M., K. H. Hartman, X. J. Wang, G. G. Turrigiano, and S. B. Nelson. 2000. Multiple Forms of Short-Term Plasticity at Excitatory Synapses in Rat Medial Prefrontal Cortex. *J. Neurophysiol.* **83**:3031–3041. [8]

Hensch, T. K., and M. Fagiolini. 2005. Excitatory-Inhibitory Balance and Critical Period Plasticity in Developing Visual Cortex. *Prog. Brain Res.* **147**:115–124. [16]

Henson, M. A., A. C. Roberts, K. Salimi, et al. 2008. Developmental Regulation of the NMDA Receptor Subunits, NR3A and NR1, in Human Prefrontal Cortex. *Cereb. Cortex* **18**:2560–2573. [16]

Henssen, A., K. Zilles, N. Palomero-Gallagher, et al. 2016. Cytoarchitecture and Probability Maps of the Human Medial Orbitofrontal Cortex. *Cortex* **75**:87–112. [4]

Herculano-Houzel, S. 2009. The Human Brain in Numbers: A Linearly Scaled-up Primate Brain. *Front. Hum. Neurosci.* **3**:31. [2, 4]

Herd, S. A., R. C. O'Reilly, T. E. Hazy, et al. 2014. A Neural Network Model of Individual Differences in Task Switching Abilities. *Neuropsychologia* **62**:375–389. [9]

Hernes, S. S., M. M. Flak, G. C. C. Løhaugen, et al. 2021. Working Memory Training in Amnestic and Non-Amnestic Patients with Mild Cognitive Impairment: Preliminary Findings from Genotype Variants on Training Effects. *Front. Aging Neurosci.* **13**: [14]

Hervig, M. E., L. Fiddian, L. Piilgaard, et al. 2019. Dissociable and Paradoxical Roles of Rat Medial and Lateral Orbitofrontal Cortex in Visual Serial Reversal Learning. *Cerebral Cortex* **30**:1016–1029. [4]

Hesselgrave, N., T. A. Troppoli, A. B. Wulff, A. B. Cole, and S. M. Thompson. 2021. Harnessing Psilocybin: Antidepressant-Like Behavioral and Synaptic Actions of Psilocybin Are Independent of 5-Ht2r Activation in Mice. *PNAS* **118**: [13]

Hilgetag, C. C., G. A. P. C. Burns, M. A. O'Neill, J. W. Scannell, and M. P. Young. 2000. Anatomical Connectivity Defines the Organization of Clusters of Cortical Areas in the Macaque and the Cat. *Philos. Trans. R. Soc. Lond. B Biol. Sci.* **355**:91–110. [11]

Hinds, A. L., E. Z. Woody, A. Drandic, et al. 2010. The Psychology of Potential Threat: Properties of the Security Motivation System. *Biol. Psychol.* **85**:331–337. [15]

Hinds, A. L., E. Z. Woody, M. Van Ameringen, L. A. Schmidt, and H. Szechtman. 2012. When Too Much Is Not Enough: Obsessive-Compulsive Disorder as a Pathology of Stopping, Rather Than Starting. *PloS One* **7**:e30586. [15]

His, W. 1895. Die Anatomische Nomenclatur: Nomina Anatomica, Verzeichniss der von der Anatomischen Gesellschaft auf Ihrer IX. Versammlung in Basel Angenommenen Namen. Leipzig: Verlag von Veit. [4]

Hochman, E. Y., S. Wang, T. E. Milner, and L. K. Fellows. 2015. Double Dissociation of Error Inhibition and Correction Deficits after Basal Ganglia or Dorsomedial Frontal Damage in Humans. *Neuropsychologia* **69**:130–139. [12]

Hochreiter, S., and J. Schmidhuber. 1997. Long Short-Term Memory. *Neural Comput.* **9**:1735–1780. [7, 12]

Hodge, R. D., T. E. Bakken, J. A. Miller, et al. 2019. Conserved Cell Types with Divergent Features in Human versus Mouse Cortex. *Nature* **573**:61–68. [2]

Hoekzema, E., S. Carmona, J. A. Ramos-Quiroga, et al. 2011. Training-Induced Neuroanatomical Plasticity in ADHD: A Tensor-Based Morphometric Study. *Hum. Brain Mapp.* **32**:1741–1749. [14]

Hoekzema, E., S. Carmona, V. Tremols, et al. 2010. Enhanced Neural Activity in Frontal and Cerebellar Circuits after Cognitive Training in Children with Attention-Deficit/Hyperactivity Disorder. *Hum. Brain Mapp.* **31**:1942–1950. [14]

Hofmann, S. G., A. Asnaani, I. J. Vonk, A. T. Sawyer, and A. Fang. 2012a. The Efficacy of Cognitive Behavioral Therapy: A Review of Meta-Analyses. *Cognit. Ther. Res.* **36**:427–440. [16]

Hofmann, W., R. F. F. Baumeister, G. Förster, and K. D. Vohs. 2012b. Everyday Temptations: An Experience Sampling Study of Desire, Conflict, and Self-Control. *J. Pers. Soc. Psychol.* **102**:1318–1335. [12]

Hoftman, G. D., S. J. Dienel, H. H. Bazmi, et al. 2018. Altered Gradients of Glutamate and Gamma-Aminobutyric Acid Transcripts in the Cortical Visuospatial Working Memory Network in Schizophrenia. *Biol. Psych.* **83**:670–679. [16]

Holland, N., P. S. Jones, G. Savulich, et al. 2023. Longitudinal Synaptic Loss in Primary Tauopathies: An *in Vivo* [(11) C]Ucb-J Positron Emission Tomography Study. *Mov. Disord.* **38**:1316–1326. [16]

Holland, N., T. W. Robbins, and J. B. Rowe. 2021. The Role of Noradrenaline in Cognition and Cognitive Disorders. *Brain* **144**:2243–2256. [16]

Holland, P. C. 2008. Cognitive versus Stimulus-Response Theories of Learning. *Learn. Behav.* **36**:227–241. [2]

Holland, P. C., and J. J. Straub. 1979. Differential Effects of Two Ways of Devaluing the Unconditioned Stimulus after Pavlovian Appetitive Conditioning. *J. Exp. Psychol. Anim. Behav. Proc.* **5**:65–78. [4]

Hollunder, B., J. L. Ostrem, I. A. Sahin, et al. 2023. Mapping Dysfunctional Circuits in the Frontal Cortex Using Deep Brain Stimulation. medRxiv. https://www.medrxiv.org/content/10.1101/2023.03.07.23286766v2. [16]

Hollunder, B., N. Rajamani, S. H. Siddiqi, et al. 2022. Toward Personalized Medicine in Connectomic Deep Brain Stimulation. *Prog. Neurobiol.* **210**:102211. [16]

Holmes, S. E., S. J. Finnema, M. Naganawa, et al. 2022. Imaging the Effect of Ketamine on Synaptic Density (Sv2a) in the Living Brain. *Mol. Psychiatry* **27**:2273–2281. [13]

Holmes, S. E., D. Scheinost, S. J. Finnema, et al. 2019. Lower Synaptic Density Is Associated with Depression Severity and Network Alterations. *Nat. Commun.* **10**:1–10. [13]

Holroyd, C. B., and M. G. H. Coles. 2002. The Neural Basis of Human Error Processing: Reinforcement Learning, Dopamine, and the Error-Related Negativity. *Psychol. Rev.* **109**:679–709. [12]

Holroyd, C. B., and T. Verguts. 2021. The Best Laid Plans: Computational Principles of Anterior Cingulate Cortex. *Trends Cogn. Sci.* **25**:316–329. [15]

Holroyd, C. B., and N. Yeung. 2012. Motivation of Extended Behaviors by Anterior Cingulate Cortex. *Trends Cogn. Sci.* **16**:122–128. [15]

Holtmaat, A., and K. Svoboda. 2009. Experience-Dependent Structural Synaptic Plasticity in the Mammalian Brain. *Nat. Rev. Neurosci.* **10**:647–658. [13]

Hommel, B. 2004. Event Files: Feature Binding in and across Perception and Action. *Trends Cogn. Sci.* **8**:494–500. [12]

———. 2009. Action Control According to Tec (Theory of Event Coding). *Psychol. Res.* **73**:512–526. [12]

Hommel, B., J. Müsseler, G. Aschersleben, and W. Prinz. 2001. The Theory of Event Coding (Tec): A Framework for Perception and Action Planning. *Behav. Brain Sci.* **24**:849–878; discussion 878–937. [12]

Hommel, B., and R. W. Wiers. 2017. Towards a Unitary Approach to Human Action Control. *Trends Cogn. Sci.* **21**:940–949. [12]

Honey, C. J., and O. Sporns. 2008. Dynamical Consequences of Lesions in Cortical Networks. *Hum. Brain Mapp.* **29**:802–809. [12]

Honey, C. J., O. Sporns, L. Cammoun, et al. 2009. Predicting Human Resting-State Functional Connectivity from Structural Connectivity. *PNAS* **106**:2035–2040. [11]

Hoover, W. B., and R. P. Vertes. 2007. Anatomical Analysis of Afferent Projections to the Medial Prefrontal Cortex in the Rat. *Brain Struct. Funct.* **212**:149–179. [3]

Hornak, J., J. P. O'Doherty, J. Bramham, et al. 2004. Reward-Related Reversal Learning after Surgical Excisions in Orbito-Frontal or Dorsolateral Prefrontal Cortex in Humans. *J. Cogn. Neurosci.* **16**:463–478. [12]

Horrillo, I., J. E. Ortega, R. Diez-Alarcia, L. Urigüen, and J. J. Meana. 2019. Chronic Fluoxetine Reverses the Effects of Chronic Corticosterone Treatment on A2-Adrenoceptors in the Rat Frontal Cortex but Not Locus Coeruleus. *Neuropharmacol.* **158**:107731. [13]

Hosokawa, T., K. Kato, M. Inoue, and A. Mikami. 2005. Correspondence of Cue Activity to Reward Activity in the Macaque Orbitofrontal Cortex. *Neurosci. Lett.* **389**:146–151. [5]

———. 2007. Neurons in the Macaque Orbitofrontal Cortex Code Relative Preference of Both Rewarding and Aversive Outcomes. *Neurosci. Res.* **57**:434–445. [5]

Hosokawa, T., S. W. Kennerley, J. Sloan, and J. D. Wallis. 2013. Single-Neuron Mechanisms Underlying Cost-Benefit Analysis in Frontal Cortex. *J. Neurosci.* **33**:17385–17397. [4, 5]

Howard, J. D., and T. Kahnt. 2017. Identity-Specific Reward Representations in Orbitofrontal Cortex Are Modulated by Selective Devaluation. *J. Neurosci.* **37**:2627–2638. [8]

Hrdy, S. B., and J. M. Burkart. 2022. How Reliance on Allomaternal Care Shapes Primate Development with Special Reference to the Genus Homo. In Evolutionary Perspectives on Infancy: Springer. [16]

Hsu, N. S., and S. M. Jaeggi. 2014. The Emergence of Cognitive Control Abilities in Childhood. *Curr. Top. Behav. Neurosci.* **16**:149–166. [14]

Hsu, N. S., J. M. Novick, and S. M. Jaeggi. 2014. The Development and Malleability of Executive Control Abilities. *Front. Behav. Neurosci.* **8**:221. [14]

Hubel, D. H., and T. N. Wiesel. 1977. Ferrier Lecture: Functional Architecture of Macaque Monkey Visual Cortex. *Proc. R. Soc. Lond. B Biol. Sci.* **198**:1–59. [4]

Huber, L., E. S. Finn, Y. Chai, et al. 2021. Layer-Dependent Functional Connectivity Methods. *Prog. Neurobiol.* **207**:101835. [11]

Hughes, L. E., T. Rittman, R. Regenthal, T. W. Robbins, and J. B. Rowe. 2015. Improving Response Inhibition Systems in Frontotemporal Dementia with Citalopram. *Brain* **138**:1961–1975. [16]

Hughes, L. E., T. Rittman, T. W. Robbins, and J. B. Rowe. 2018. Reorganization of Cortical Oscillatory Dynamics Underlying Disinhibition in Frontotemporal Dementia. *Brain* **141**:2486–2499. [16]

Hulbert, J. C., R. N. Henson, and M. C. Anderson. 2016. Inducing Amnesia through Systemic Suppression. *Nat. Commun.* **7**:11003. [12]

Humphrey, N. K. 1976. The Social Function of Intellect. In: Growing Points in Ethology, ed. P. P. G. Bateson and R. A. Hinde, pp. 303–317. Cambridge: Cambridge Univ. Press. [4]

Hunt, L. T., T. E. Behrens, T. Hosokawa, J. D. Wallis, and S. W. Kennerley. 2015. Capturing the Temporal Evolution of Choice across Prefrontal Cortex. *eLife* **4**:e11945. [5]

Hunt, L. T., W. M. N. Malalasekera, A. O. de Berker, et al. 2018. Triple Dissociation of Attention and Decision Computations across Prefrontal Cortex. *Nat. Neurosci.* **21**:1471–1481. [4, 5]

Huntenburg, J. M., P. L. Bazin, A. Goulas, et al. 2017. A Systematic Relationship between Functional Connectivity and Intracortical Myelin in the Human Cerebral Cortex. *Cereb. Cortex* **27**:981–997. [6]

Huntenburg, J. M., P. L. Bazin, and D. S. Margulies. 2018. Large-Scale Gradients in Human Cortical Organization. *Trends Cogn. Sci.* **22**:21–31. [4, 7]

Huth, A. G., W. A. de Heer, T. L. Griffiths, F. E. Theunissen, and J. L. Gallant. 2016. Natural Speech Reveals the Semantic Maps That Tile Human Cerebral Cortex. *Nature* **532**:453–458. [8]

Huttenlocher, P. R. 1990. Morphometric Study of Human Cerebral Cortex Development. *Neuropsychologia* **28**:517–527. [16]

Huys, Q. J. M., N. Eshel, E. O'Nions, et al. 2012. Bonsai Trees in Your Head: How the Pavlovian System Sculpts Goal-Directed Choices by Pruning Decision Trees. *PLoS Comput. Biol.* **8**:e1002410. [12]

Hwang, K., J. Bruss, D. Tranel, and A. D. Boes. 2020. Network Localization of Executive Function Deficits in Patients with Focal Thalamic Lesions. *J. Cog. Neuro.* **32**:2303–2319. [17]

Hwang, K., M. N. Hallquist, and B. Luna. 2013. The Development of Hub Architecture in the Human Functional Brain Network. *Cereb. Cortex* **23**:2380–2393. [16]

Hwang, K., J. M. Shine, J. Bruss, D. Tranel, and A. Boes. 2021. Neuropsychological Evidence of Multi-Domain Network Hubs in the Human Thalamus. *eLife* **10**:e69480. [8]

Ichihara-Takeda, S., and S. Funahashi. 2007. Activity of Primate Orbitofrontal and Dorsolateral Prefrontal Neurons: Task-Related Activity during an Oculomotor Delayed-Response Task. *Exp. Brain Res.* **181**:409–425. [5]

Iigaya, K., S. Yi, I. A. Wahle, et al. 2023. Neural Mechanisms Underlying the Hierarchical Construction of Perceived Aesthetic Value. *Nat. Commun.* **14**:127. [12]

Iordan, A. D., K. A. Cooke, K. D. Moored, et al. 2018. Aging and Network Properties: Stability over Time and Links with Learning during Working Memory Training. *Front. Aging Neurosci.* **9**:419. [14]

Iordan, A. D., K. A. Cooke, K. D. Moored, et al. 2020. Neural Correlates of Working Memory Training: Evidence for Plasticity in Older Adults. *Neuroimage* **217**:116887. [14]

Iordan, A. D., K. D. Moored, B. Katz, et al. 2021. Age Differences in Functional Network Reconfiguration with Working Memory Training. *Hum. Brain Mapp.* **42**:1888–1909. [14]

Iordanova, M. D., A. S. Killcross, and R. C. Honey. 2007. Role of the Medial Prefrontal Cortex in Acquired Distinctiveness and Equivalence of Cues. *Behav. Neurosci.* **121**:1431–1436. [7]

Iversen, S. D., and M. Mishkin. 1970. Perseverative Interference in Monkeys Following Selective Lesions of the Inferior Prefrontal Convexity. *Exp. Brain Res.* **11**:376–386. [4]

Iyadurai, L., S. E. Blackwell, R. Meiser-Stedman, et al. 2018. Preventing Intrusive Memories after Trauma via a Brief Intervention Involving Tetris Computer Game Play in the Emergency Department: A Proof-of-Concept Randomized Controlled Trial. *Mol. Psych.* **23**:674–682. [16]

Izquierdo, A. 2017. Functional Heterogeneity within Rat Orbitofrontal Cortex in Reward Learning and Decision Making. *J. Neurosci.* **37**:10529–10540. [2]

———. 2021. Touchscreen Response Technology and the Power of Stimulus-Based Approaches in Freely Behaving Animals. *Genes Brain Behav.* **20**:e12720. [2]

Izquierdo, A., J. L. Brigman, A. K. Radke, P. H. Rudebeck, and A. Holmes. 2017. The Neural Basis of Reversal Learning: An Updated Perspective. *Neurosci.* **345**:12–26. [4]

Izquierdo, A., C. Darling, N. Manos, et al. 2013. Basolateral Amygdala Lesions Facilitate Reward Choices after Negative Feedback in Rats. *J. Neurosci.* **33**:4105–4109. [4]

Izquierdo, A., R. K. Suda, and E. A. Murray. 2004. Bilateral Orbital Prefrontal Cortex Lesions in Rhesus Monkeys Disrupt Choices Guided by Both Reward Value and Reward Contingency. *J. Neurosci.* **24**:7540–7548. [4]

Izquierdo, A., C. L. Wellman, and A. Holmes. 2006. Brief Uncontrollable Stress Causes Dendritic Retraction in Infralimbic Cortex and Resistance to Fear Extinction in Mice. *J. Neurosci.* **26**:5733–5738. [16]

Jackson, S. A. W., N. K. Horst, A. Pears, T. W. Robbins, and A. C. Roberts. 2016. Role of the Perigenual Anterior Cingulate and Orbitofrontal Cortex in Contingency Learning in the Marmoset. *Cerebral Cortex* **26**:3273–3284. [4]

Jacob, S. N., M. Stalter, and A. Nieder. 2016. Cell-Type-Specific Modulation of Targets and Distractors by Dopamine D1 Receptors in Primate Prefrontal Cortex. *Nat. Commun.* **7**:13218. [6]

Jacobs, B., M. Schall, M. Prather, et al. 2001. Regional Dendritic and Spine Variation in Human Cerebral Cortex: A Quantitative Golgi Study. *Cereb. Cortex* **11**:558–571. [7]

Jaeggi, S. M., M. Buschkuehl, J. Jonides, and P. Shah. 2011. Short- and Long-Term Benefits of Cognitive Training. *PNAS* **108**:10081–10086. [14]

Jaeggi, S. M., M. Buschkuehl, P. Shah, and J. Jonides. 2014. The Role of Individual Differences in Cognitive Training and Transfer. *Mem. Cogn.* **42**:464–480. [14]

Jaeggi, S. M., A. N. Weaver, E. Carbone, et al. 2023. Engage: A Metacognitive Intervention to Supplement Working Memory Training: A Feasibility Study in Older Adults. *Aging Brain* **4**:100083. [14, 16]

Jaffard, M., M. Longcamp, J.-L. Velay, et al. 2008. Proactive Inhibitory Control of Movement Assessed by Event-Related fMRI. *Neuroimage* **42**:1196–1206. [12]

Jaffe, R. J., and C. Constantinidis. 2021. Working Memory: From Neural Activity to the Sentient Mind. *Compr. Physiol.* **11**:1–41. [6]

Jahn, A., D. E. Nee, W. H. Alexander, and J. W. Brown. 2014. Distinct Regions of Anterior Cingulate Cortex Signal Prediction and Outcome Evaluation. *Neuroimage* **95**:80–89. [2]

Jalbrzikowski, M., B. Larsen, M. N. Hallquist, et al. 2017. Development of White Matter Microstructure and Intrinsic Functional Connectivity between the Amygdala and Ventromedial Prefrontal Cortex: Associations with Anxiety and Depression. *Biol. Psych.* **82**:511–521. [16]

James, G. M., G. Gryglewski, T. Vanicek, et al. 2019. Parcellation of the Human Cerebral Cortex Based on Molecular Targets in the Serotonin System Quantified by Positron Emission Tomography *in Vivo*. *Cereb. Cortex* **29**:372–382. [13]

Jazayeri, M., and A. Afraz. 2017. Navigating the Neural Space in Search of the Neural Code. *Neuron* **93**:1003–1014. [4]

Jbabdi, S., J. F. Lehman, S. N. Haber, and T. E. Behrens. 2013. Human and Monkey Ventral Prefrontal Fibers Use the Same Organizational Principles to Reach Their Targets: Tracing versus Tractography. *J. Neurosci.* **33**:3190–3201. [15]

Jenike, M. A., L. Baer, T. Ballantine, et al. 1991. Cingulotomy for Refractory Obsessive-Compulsive Disorder. A Long-Term Follow-up of 33 Patients. *Arch. Gen. Psych.* **48**:548–555. [15]

Jeon, H. A., A. Anwander, and A. D. Friederici. 2014. Functional Network Mirrored in the Prefrontal Cortex, Caudate Nucleus, and Thalamus: High-Resolution Functional Imaging and Structural Connectivity. *J. Neurosci.* **34**:9202–9212. [7]

Jerde, T. A., E. P. Merriam, A. C. Riggall, J. H. Hedges, and C. E. Curtis. 2012. Prioritized Maps of Space in Human Frontoparietal Cortex. *J. Neurosci.* **32**:17382–17390. [4]

Jezzini, A., E. S. Bromberg-Martin, L. R. Trambaiolli, S. N. Haber, and I. E. Monosov. 2021. A Prefrontal Network Integrates Preferences for Advance Information About Uncertain Rewards and Punishments. *Neuron* **109**:2339–2352. [2]

Ji, J. L., M. Spronk, K. Kulkarni, et al. 2019. Mapping the Human Brain's Cortical-Subcortical Functional Network Organization. *Neuroimage* **185**:35–57. [7, 9]

Jin, L. E., M. Wang, V. C. Galvin, et al. 2018. Mglur2 versus mGluR3 Metabotropic Glutamate Receptors in Primate Dorsolateral Prefrontal Cortex: Postsynaptic mGluR3 Strengthen Working Memory Networks. *Cereb. Cortex* **28**:974–987. [16]

Jin, L. E., M. Wang, S. T. Yang, et al. 2017. Mglur2/3 Mechanisms in Primate Dorsolateral Prefrontal Cortex: Evidence for Both Presynaptic and Postsynaptic Actions. *Mol. Psych.* **22**:1615–1625. [16]

Jin, X., F. Tecuapetla, and R. M. Costa. 2014. Basal Ganglia Subcircuits Distinctively Encode the Parsing and Concatenation of Action Sequences. *Nat. Neurosci.* **17**:423–430. [7]

Jocham, G., L. T. Hunt, J. Near, and T. E. Behrens. 2012. A Mechanism for Value-Guided Choice Based on the Excitation-Inhibition Balance in Prefrontal Cortex. *Nat. Neurosci.* **15**:960–961. [16]

John, C. S., K. L. Smith, A. Van'T Veer, et al. 2012. Blockade of Astrocytic Glutamate Uptake in the Prefrontal Cortex Induces Anhedonia. *Neuropsychopharmacol.* **37**:2467–2475. [4]

Johnson, E. L., J. J. Lin, D. King-Stephens, et al. 2023. A Rapid Theta Network Mechanism for Flexible Information Encoding. *Nat. Commun.* **14**:2872. [7]

Jones, D. T., and J. Graff-Radford. 2021. Executive Dysfunction and the Prefrontal Cortex. *Continuum* **27**:1586–1601. [14]

Jones, J. L., G. R. Esber, M. A. McDannald, et al. 2012. Orbitofrontal Cortex Supports Behavior and Learning Using Inferred but Not Cached Values. *Science* **338**:953–956. [7, 10]

Jonides, J., E. E. Smith, R. A. Koeppe, et al. 1993. Spatial Working Memory in Humans as Revealed by PET. *Nature* **363**:623–625. [4]

Joyce, M. K. P., and H. Barbas. 2018. Cortical Connections Position Primate Area 25 as a Keystone for Interoception, Emotion, and Memory. *J. Neurosci.* **38**:1677–1698. [3]

Juavinett, A. L., G. Bekheet, and A. K. Churchland. 2019. Chronically Implanted Neuropixels Probes Enable High-Yield Recordings in Freely Moving Mice. *eLife* **8**:e47188. [4]

Jun, J. J., N. A. Steinmetz, J. H. Siegle, et al. 2017. Fully Integrated Silicon Probes for High-Density Recording of Neural Activity. *Nature* **551**:232–236. [4]

Jung, H. H., C. H. Kim, J. H. Chang, et al. 2006. Bilateral Anterior Cingulotomy for Refractory Obsessive-Compulsive Disorder: Long-Term Follow-up Results. *Stereotact Funct Neurosurg* **84**:184–189. [15]

Kaalund, S. S., L. Passamonti, K. S. J. Allinson, et al. 2020. Locus Coeruleus Pathology in Progressive Supranuclear Palsy, and Its Relation to Disease Severity. *Acta Neuropathol. Commun.* **8**:11. [16]

Kaas, J. H. 1987. The Organization of Neocortex in Mammals: Implications for Theories of Brain Function. *Annu. Rev. Psychol.* **38**:129–151. [5]

———. 2004. Evolution of Somatosensory and Motor Cortex in Primates. *Anat. Rec. A Discov. Mol. Cell. Evol. Biol.* **281**:1148–1156. [3]

Kaiser, R. H., J. R. Andrews-Hanna, T. D. Wager, and D. A. Pizzagalli. 2015. Large-Scale Network Dysfunction in Major Depressive Disorder: A Meta-Analysis of Resting-State Functional Connectivity. *JAMA Psych.* **72**:603–611. [11, 13]

Kamiński, J., S. Sullivan, J. M. Chung, et al. 2017. Persistently Active Neurons in Human Medial Frontal and Medial Temporal Lobe Support Working Memory. *Nat. Neurosci.* **20**:590–601. [5]

Kampman, O., M. Viikki, K. Järventausta, and E. Leinonen. 2014. Meta-Analysis of Anxiety Disorders and Temperament. *Neuropsychobiol.* **69**:175–186. [15]

Kane, G. A., A. M. Bornstein, A. Shenhav, et al. 2019. Rats Exhibit Similar Biases in Foraging and Intertemporal Choice Tasks. *eLife* **8**:e48429. [2]

Kang, H. J., B. Voleti, T. Hajszan, et al. 2012. Decreased Expression of Synapse-Related Genes and Loss of Synapses in Major Depressive Disorder. *Nat. Med.* **18**:1413–1417. [13]

Karbach, J., T. Könen, and M. Spengler. 2017. Who Benefits the Most? Individual Differences in the Transfer of Executive Control Training across the Lifespan. *J. Cogn. Enhanc.* **1**:394–405. [14, 16]

Karbach, J., T. Strobach, and T. Schubert. 2015. Adaptive Working-Memory Training Benefits Reading, but Not Mathematics in Middle Childhood. *Child Neuropsych.* **21**:285–301. [16]

Karbach, J., and P. Verhaeghen. 2014. Making Working Memory Work: A Meta-Analysis of Executive-Control and Working Memory Training in Older Adults. *Psychol. Sci.* **25**:2027–2037. [14]

Karlen, S. J., and L. Krubitzer. 2007. The Functional and Anatomical Organization of Marsupial Neocortex: Evidence for Parallel Evolution across Mammals. *Prog. Neurobiol.* **82**:122–141. [3]

Karr, J. E., C. N. Areshenkoff, P. Rast, et al. 2018. The Unity and Diversity of Executive Functions: A Systematic Review and Re-Analysis of Latent Variable Studies. *Psychol. Bull.* **144**:1147–1185. [9, 14]

Karssemeijer, E. G. A., J. A. Aaronson, W. J. Bossers, et al. 2017. Positive Effects of Combined Cognitive and Physical Exercise Training on Cognitive Function in Older Adults with Mild Cognitive Impairment or Dementia: A Meta-Analysis. *Ageing Res. Rev.* **40**:75–83. [14]

Kastner, S., K. DeSimone, C. S. Konen, et al. 2007. Topographic Maps in Human Frontal Cortex Revealed in Memory-Guided Saccade and Spatial Working-Memory Tasks. *J. Neurophysiol.* **97**:3494–3507. [8]

Kastner, S., and L. G. Ungerleider. 2001. The Neural Basis of Biased Competition in Human Visual Cortex. *Neuropsychologia* **39**:1263–1276. [12]

Katsuki, F., X. L. Qi, T. Meyer, et al. 2014. Differences in Intrinsic Functional Organization between Dorsolateral Prefrontal and Posterior Parietal Cortex. *Cereb. Cortex* **24**:2334–2349. [6]

Katz, B., M. R. Jones, P. Shah, M. Buschkuehl, and S. M. Jaeggi. 2021. Individual Differences in Cognitive Training Research. In: Cognitive Training: An Overview of Features and Applications, ed. T. Strobach et al., pp. 107–123. Cham: Springer. [14]

Kay, K. N., T. Naselaris, R. J. Prenger, and J. L. Gallant. 2008. Identifying Natural Images from Human Brain Activity. *Nature* **452**:352–355. [8]

Kebschull, J. M., P. Garcia da Silva, A. P. Reid, et al. 2016. High-Throughput Mapping of Single-Neuron Projections by Sequencing of Barcoded RNA. *Neuron* **91**:975–987. [2, 4]

Kehagia, A. A., R. Ye, D. W. Joyce, et al. 2017. Parsing the Roles of the Frontal Lobes and Basal Ganglia in Task Control Using Multivoxel Pattern Analysis. *J. Cogn. Neurosci.* **29**:1390–1401. [16]

Kennedy, K. M., M. A. Boylan, J. R. Rieck, C. M. Foster, and K. M. Rodrigue. 2017. Dynamic Range in BOLD Modulation: Lifespan Aging Trajectories and Association with Performance. *Neurobiol. Aging* **60**:153–163. [14]

Kennedy, S. H., J. Z. Konarski, Z. V. Segal, et al. 2007. Differences in Brain Glucose Metabolism between Responders to CBT and Venlafaxine in a 16-Week Randomized Controlled Trial. *Am. J. Psych.* **164**:778–788. [13]

Kennerley, S. W., A. F. Dahmubed, A. H. Lara, and J. D. Wallis. 2009. Neurons in the Frontal Lobe Encode the Value of Multiple Decision Variables. *J. Cogn. Neurosci.* **21**:1162–1178. [5]

Kennerley, S. W., and J. D. Wallis. 2009a. Encoding of Reward and Space during a Working Memory Task in the Orbitofrontal Cortex and Anterior Cingulate Sulcus. *J. Neurophysiol.* **102**:3352–3364. [5, 8]

———. 2009b. Evaluating Choices by Single Neurons in the Frontal Lobe: Outcome Value Encoded across Multiple Decision Variables. *Eur. J. Neurosci.* **29**:2061–2073. [5]

Kennerley, S. W., M. E. Walton, T. E. J. Behrens, M. J. Buckley, and M. F. S. Rushworth. 2006. Optimal Decision Making and the Anterior Cingulate Cortex. *Nat. Neurosci.* **9**:940–947. [4]

Kenwood, M. M., N. H. Kalin, and H. Barbas. 2022. The Prefrontal Cortex, Pathological Anxiety, and Anxiety Disorders. *Neuropsychopharmacol.* **47**:260–275. [12]

Kepecs, A., and G. Fishell. 2014. Interneuron Cell Types Are Fit to Function. *Nature* **505**:318–326. [2]

Kesner, R. P., and M. E. Ragozzino. 2003. The Role of the Prefrontal Cortex in Object–Place Learning: A Test of the Attribute Specificity Model. *Behav. Brain Res.* **146**:159–165. [4]

Kessler, R. C., J. Ormel, O. Demler, and P. E. Stang. 2003. Comorbid Mental Disorders Account for the Role Impairment of Commonly Occurring Chronic Physical Disorders: Results from the National Comorbidity Survey. *J. Occup. Environ. Med.* **45**:1257–1266. [16]

Khrameeva, E., I. Kurochkin, D. Han, et al. 2020. Single-Cell-Resolution Transcriptome Map of Human, Chimpanzee, Bonobo, and Macaque Brains. *Genome Res.* **30**:776–789. [2]

Kie, J. G. 1999. Optimal Foraging and Risk of Predation: Effects on Behavior and Social Structure in Ungulates. *J. Mammal.* **80**:1114–1129. [2]

Kievit, R. A., U. Lindenberger, I. M. Goodyer, et al. 2017. Mutualistic Coupling between Vocabulary and Reasoning Supports Cognitive Development During Late Adolescence and Early Adulthood. *Psychol. Sci.* **28**:1419–1431. [9]

Kikumoto, A., and U. Mayr. 2020. Conjunctive Representations That Integrate Stimuli, Responses, and Rules Are Critical for Action Selection. *PNAS* **117**:10603–10608. [7]

Kikumoto, A., U. Mayr, and D. Badre. 2022. The Role of Conjunctive Representations in Prioritizing and Selecting Planned Actions. *eLife* **11**:e80153. [7]

Kim, C. K., A. Adhikari, and K. Deisseroth. 2017. Integration of Optogenetics with Complementary Methodologies in Systems Neuroscience. *Nat. Rev. Neurosci.* **18**:222–235. [3]

Kim, H. 2011. Neural Activity That Predicts Subsequent Memory and Forgetting: A Meta-Analysis of 74 fMRI Studies. *Neuroimage* **54**:2446–2461. [17]

Kim, J.-W., A. E. Autry, E. S. Na, et al. 2021. Sustained Effects of Rapidly Acting Antidepressants Require BDNF-Dependent MeCP2 Phosphorylation. *Nat. Neurosci.* **24**:1100–1109. [13]

Kim, R., and T. J. Sejnowski. 2021. Strong Inhibitory Signaling Underlies Stable Temporal Dynamics and Working Memory in Spiking Neural Networks. *Nat. Neurosci.* **24**:129–139. [6]

Kim, S. J., D. Roh, H. H. Jung, et al. 2018. A Study of Novel Bilateral Thermal Capsulotomy with Focused Ultrasound for Treatment-Refractory Obsessive-Compulsive Disorder: 2-Year Follow-Up. *J. Psych. Neurosci.* **43**:170188. [15]

Kimmel, D. L., G. F. Elsayed, J. P. Cunningham, and W. T. Newsome. 2020. Value and Choice as Separable and Stable Representations in Orbitofrontal Cortex. *Nat. Commun.* **11**:3466. [5]

King, D., K. Holt, J. Toombs, et al. 2023. Synaptic Resilience Is Associated with Maintained Cognition during Ageing. *Alzheimers Dement.* **19**:2560–2574. [16]

King, G., and G. N. Bailey. 2006. Tectonics and Human Evolution. *Antiquity* **80**:265–286. [4]

Kirkland, A. E., and K. F. Holton. 2019. Measuring Treatment Response in Pharmacological and Lifestyle Interventions Using Electroencephalography in ADHD: A Review. *Clin. EEG Neurosci.* **50**:256–266. [14]

Kleimaker, M., A. Takacs, G. Conte, et al. 2020. Increased Perception-Action Binding in Tourette Syndrome. *Brain* **143**:1934–1945. [12]

Klein-Flugge, M. C., H. C. Barron, K. H. Brodersen, R. J. Dolan, and T. E. Behrens. 2013. Segregated Encoding of Reward-Identity and Stimulus-Reward Associations in Human Orbitofrontal Cortex. *J. Neurosci.* **33**:3202–3211. [8]

Klein, M. E., J. Chandra, S. Sheriff, and R. Malinow. 2020. Opioid System Is Necessary but Not Sufficient for Antidepressive Actions of Ketamine in Rodents. *PNAS* **117**:2656–2662. [13]

Kleinsorge, T., and H. Heuer. 1999. Hierarchical Swithcing in a Multi-Dimensional Task Space. *Psychol. Res.* **62**:300–312. [7]

Klingberg, T. 2010. Training and Plasticity of Working Memory. *Trends Cogn. Sci.* **14**:317–324. [14]

Knoblich, U., L. Huang, H. Zeng, and L. Li. 2019. Neuronal Cell-Subtype Specificity of Neural Synchronization in Mouse Primary Visual Cortex. *Nat. Commun.* **10**:2533. [2]

Knott, G. W., A. Holtmaat, L. Wilbrecht, E. Welker, and K. Svoboda. 2006. Spine Growth Precedes Synapse Formation in the Adult Neocortex *in Vivo*. *Nat. Neurosci.* **9**:1117–1124. [13]

Knudsen, E. B., and J. D. Wallis. 2020. Closed-Loop Theta Stimulation in the Orbitofrontal Cortex Prevents Reward-Based Learning. *Neuron* **106**:537–547. [4]

———. 2022. Taking Stock of Value in the Orbitofrontal Cortex. *Nat. Rev. Neurosci.* **23**:428–438. [4]

Kocagoncu, E., A. Klimovich-Gray, L. E. Hughes, and J. B. Rowe. 2021. Evidence and Implications of Abnormal Predictive Coding in Dementia. *Brain* **144**:3311–3321. [16]

Koechlin, E. 2014. An Evolutionary Computational Theory of Prefrontal Executive Function in Decision-Making. *Philos. Trans. R. Soc. Lond. B Biol. Sci.* **369**:20130474. [10]

———. 2020. Executive Control and Decision-Making: A Neural Theory of Prefrontal Function. In: The Cognitive Neurosciences, 6th edition, ed. D. Poeppel et al., pp. 451–468. Cambridge, MA: The MIT Press. [10]

Koechlin, E., G. Basso, P. Pietrini, S. Panzer, and J. Grafman. 1999. The Role of the Anterior Prefrontal Cortex in Human Cognition. *Nature* **399**:148–151. [7, 15]

Koechlin, E., and A. Hyafil. 2007. Anterior Prefrontal Function and the Limits of Human Decision-Making. *Science* **318**:594–598. [7, 8]

Koechlin, E., C. Ody, and F. Kouneiher. 2003. The Architecture of Cognitive Control in the Human Prefrontal Cortex. *Science* **302**:1181–1185. [6–8, 10, 11, 12, 15]

Koechlin, E., and C. Summerfield. 2007. An Information Theoretical Approach to Prefrontal Executive Function. *Trends Cogn. Sci.* **11**:229–235. [16]

Kolling, N., T. Behrens, M. K. Wittmann, and M. Rushworth. 2016a. Multiple Signals in Anterior Cingulate Cortex. *Curr. Opin. Neurobiol.* **37**:36–43. [5]

Kolling, N., T. E. J. Behrens, R. B. Mars, and M. F. S. Rushworth. 2012. Neural Mechanisms of Foraging. *Science* **6**:95–98. [4]

Kolling, N., J. Scholl, A. Chekroud, H. A. Trier, and M. F. S. Rushworth. 2018. Prospection, Perseverance, and Insight in Sequential Behavior. *Neuron* **99**:1069–1082. [16]

Kolling, N., M. K. Wittmann, T. E. Behrens, et al. 2016b. Value, Search, Persistence and Model Updating in Anterior Cingulate Cortex. *Nat. Neurosci.* **19**:1280–1285. [16]

Konarski, J. Z., S. H. Kennedy, Z. V. Segal, et al. 2009. Predictors of Nonresponse to Cognitive Behavioural Therapy or Venlafaxine Using Glucose Metabolism in Major Depressive Disorder. *J. Psychiatry Neurosci.* **34**:175–180. [13]

Kong, R., J. Li, C. Orban, et al. 2019. Spatial Topography of Individual-Specific Cortical Networks Predicts Human Cognition, Personality, and Emotion. *Cereb. Cortex* **29**:2533–2551. [11, 12]

Konidaris, G. 2019. On the Necessity of Abstraction. *Curr. Opin. Behav. Sci.* **29**:1–7. [7]

Korgaonkar, M. S., W. Rekshan, E. Gordon, et al. 2015. Magnetic Resonance Imaging Measures of Brain Structure to Predict Antidepressant Treatment Outcome in Major Depressive Disorder. *EBioMedicine* **2**:37–45. [13]

Kornblith, S., T. J. Buschman, and E. K. Miller. 2016. Stimulus Load and Oscillatory Activity in Higher Cortex. *Cereb. Cortex* **26**:3772–3784. [8]

Kornblith, S., R. Quian Quiroga, C. Koch, I. Fried, and F. Mormann. 2017. Persistent Single-Neuron Activity during Working Memory in the Human Medial Temporal Lobe. *Curr. Biol.* **27**:1026–1032. [5]

Kotter, R. 2004. Online Retrieval, Processing, and Visualization of Primate Connectivity Data from the CoCoMac Database. *Neuroinformatics* **2**:127–144. [7]

Kouneiher, F., S. Charron, and E. Koechlin. 2009. Motivation and Cognitive Control in the Human Prefrontal Cortex. *Nat. Neurosci.* **12**:939–945. [16]

Kovacs, K., and A. R. A. Conway. 2016. Process Overlap Theory: A Unified Account of the General Factor of Intelligence. *Psychological Inquiry* **27**:151–177. [9]

Kozak, M. J., and B. N. Cuthbert. 2016. The NIMH Research Domain Criteria Initiative: Background, Issues, and Pragmatics. *Psychophysiol.* **53**:286–297. [17]

Kozlenkov, A., M. W. Vermunt, P. Apontes, et al. 2020. Evolution of Regulatory Signatures in Primate Cortical Neurons at Cell-Type Resolution. *PNAS* **117**:28422–28432. [2]

Krámská, L., D. Urgošík, R. Liščák, L. Hrešková, and J. Skopová. 2021. Neuropsychological Outcome in Refractory Obsessive-Compulsive Disorder Treated with Anterior Capsulotomy Including Repeated Surgery. *Psychiatry Clin Neurosci* **75**:101–107. [15]

Kriegeskorte, N. 2015. Deep Neural Networks: A New Framework for Modeling Biological Vision and Brain Information Processing. *Annu. Rev. Vis. Sci.* **1**:417–446. [12]

Kriegstein, A., S. Noctor, and V. Martinez-Cerdeno. 2006. Patterns of Neural Stem and Progenitor Cell Division May Underlie Evolutionary Cortical Expansion. *Nat. Rev. Neurosci.* **7**:883–890. [2]

Krienen, F. M., M. Goldman, Q. Zhang, et al. 2020. Innovations Present in the Primate Interneuron Repertoire. *Nature* **586**:262–269. [2]

Kringelbach, M. L. 2005. The Human Orbitofrontal Cortex: Linking Reward to Hedonic Experience. *Nat. Rev. Neurosci.* **6**:691–702. [4]

Kringelbach, M. L., and E. T. Rolls. 2004. The Functional Neuroanatomy of the Human Orbitofrontal Cortex: Evidence from Neuroimaging and Neuropsychology. *Prog. Neurobiol.* **72**:341–372. [4, 16]

Krishnan, V., M. H. Han, D. L. Graham, et al. 2007. Molecular Adaptations Underlying Susceptibility and Resistance to Social Defeat in Brain Reward Regions. *Cell* **131**:391–404. [16]

Kritzer, M. F., and P. S. Goldman-Rakic. 1995. Intrinsic Circuit Organization of the Major Layers and Sublayers of the Dorsolateral Prefrontal Cortex in the Rhesus Monkey. *J. Comp. Neurol.* **359**:131–143. [6]

Krubitzer, L., H. Kunzle, and J. Kaas. 1997. Organization of Sensory Cortex in a Madagascan Insectivore, the Tenrec (Echinops telfairi). *J. Comp. Neurol.* **379**:399–414. [3]

Krystal, J. H., and M. W. State. 2014. Psychiatric Disorders: Diagnosis to Therapy. *Cell* **157**:201–214. [13]

Kubota, K., and H. Niki. 1971. Prefrontal Cortical Unit Activity and Delayed Alternation Performance in Monkeys. *J. Neurophysiol.* **34**:337–347. [4, 5]

Kucyi, A., J. Schrouff, S. Bickel, et al. 2018. Intracranial Electrophysiology Reveals Reproducible Intrinsic Functional Connectivity within Human Brain Networks. *J. Neurosci.* **38**:4230–4242. [11]

Kühn, S., F. Schmiedek, H. Noack, et al. 2013. The Dynamics of Change in Striatal Activity Following Updating Training. *Hum. Brain Mapp.* **34**:1530–1541. [14]

Kühn, S., F. Schmiedek, B. Schott, et al. 2011. Brain Areas Consistently Linked to Individual Differences in Perceptual Decision-Making in Younger as Well as Older Adults before and after Training. *J. Cog. Neuro.* **23**:2147–2158. [14]

Kundu, B., D. W. Sutterer, S. M. Emrich, and B. R. Postle. 2013. Strengthened Effective Connectivity Underlies Transfer of Working Memory Training to Tests of Short-Term Memory and Attention. *J. Neurosci.* **33**:8705–8715. [14]

Kurth-Nelson, Z., M. Economides, R. J. Dolan, and P. Dayan. 2016. Fast Sequences of Non-Spatial State Representations in Humans. *Neuron* **91**:194–204. [12]

Kwan, A. C., D. E. Olson, K. H. Preller, and B. L. Roth. 2022. The Neural Basis of Psychedelic Action. *Nat. Neurosci.* **25**:1407–1419. [13]

Kwon, Y. H., J. J. Salvo, N. Anderson, et al. 2023. Situating the Parietal Memory Network in the Context of Multiple Parallel Distributed Networks Using High-Resolution Functional Connectivity. 2023.08. bioRxiv. [11]

Kyllonen, P. C., and R. E. Christal. 1990. Reasoning Ability Is (Little More Than) Working-Memory Capacity?! *Intelligence* **14**:389–433. [9]

Ladwig, Z., B. A. Seitzman, A. Dworetsky, et al. 2022. BOLD Cofluctuation "Events" Are Predicted from Static Functional Connectivity. *Neuroimage* **260**:119476. [11]

Laine, M., D. Fellman, O. Waris, and T. J. Nyman. 2018. The Early Effects of External and Internal Strategies on Working Memory Updating Training. *Sci. Rep.* **8**:4045. [14]

Lak, A., G. M. Costa, E. Romberg, et al. 2014. Orbitofrontal Cortex Is Required for Optimal Waiting Based on Decision Confidence. *Neuron* **84**:190–201. [2, 4]

Lambez, B., A. Harwood-Gross, E. Z. Golumbic, and Y. Rassovsky. 2020. Non-Pharmacological Interventions for Cognitive Difficulties in ADHD: A Systematic Review and Meta-Analysis. *J. Psychiatr. Res.* **120**:40–55. [14]

Lang, S., A. Kroll, S. J. Lipinski, et al. 2009. Context Conditioning and Extinction in Humans: Differential Contribution of the Hippocampus, Amygdala and Prefrontal Cortex. *Eur. J. Neurosci.* **29**:823–832. [4]

Langenecker, S. A., S. E. Kennedy, L. M. Guidotti, et al. 2007. Frontal and Limbic Activation during Inhibitory Control Predicts Treatment Response in Major Depressive Disorder. *Biol. Psychiatry* **62**:1272–1280. [13]

Lansdall, C. J., I. T. S. Coyle-Gilchrist, P. S. Jones, et al. 2017. Apathy and Impulsivity in Frontotemporal Lobar Degeneration Syndromes. *Brain* **140**:1792–1807. [16]

———. 2018. White Matter Change with Apathy and Impulsivity in Frontotemporal Lobar Degeneration Syndromes. *Neurology* **90**:e1066–e1076. [16]

Lapish, C. C., D. Durstewitz, L. J. Chandler, and J. K. Seamans. 2008. Successful Choice Behavior Is Associated with Distinct and Coherent Network States in Anterior Cingulate Cortex. *PNAS* **105**:11963–11968. [4]

Lara, A. H., S. W. Kennerley, and J. D. Wallis. 2009. Encoding of Gustatory Working Memory by Orbitofrontal Neurons. *J. Neurosci.* **29**:765–774. [5]

Lara, A. H., and J. D. Wallis. 2014. Executive Control Processes Underlying Multi-Item Working Memory. *Nat. Neurosci.* **17**:876–883. [5]

Laroche, S., S. Davis, and T. M. Jay. 2000. Plasticity at Hippocampal to Prefrontal Cortex Synapses: Dual Roles in Working Memory and Consolidation. *Hippocampus* **10**:438–446. [6, 8]

Larsen, B., and B. Luna. 2018. Adolescence as a Neurobiological Critical Period for the Development of Higher-Order Cognition. *Neurosci. Biobehav. Rev.* **94**:179–195. [16]

Lasaponara, S., F. Marson, F. Doricchi, and M. Cavallo. 2021. A Scoping Review of Cognitive Training in Neurodegenerative Diseases via Computerized and Virtual Reality Tools: What We Know So Far. *Brain Sci.* **11**:528. [14]

Lashley, K. S., and G. Clark. 1946. The Cytoarchitecture of the Cerebral Cortex of Ateles; a Critical Examination of Architectonic Studies. *J. Comp. Neurol.* **85**:223–305. [5]

Laubach, M., L. M. Amarante, K. Swanson, and S. R. White. 2018. What, If Anything, Is Rodent Prefrontal Cortex? *eNeuro* **5**:ENEURO.0315–0318.2018. [2, 4, 5]

Lauer, J., M. Zhou, S. Ye, et al. 2022. Multi-Animal Pose Estimation, Identification and Tracking with DeepLabCut. *Nat. Methods* **19**:496–504. [2]

Laumann, T. O., A. Z. Snyder, A. Mitra, et al. 2017. On the Stability of BOLD fMRI Correlations. *Cereb. Cortex* **27**:4719–4732. [11]

Laurent, V., and B. W. Balleine. 2021. How Predictive Learning Influences Choice: Evidence for a GPCR-based Memory Process Necessary for Pavlovian-Instrumental Transfer. *J. Neurochem.* **157**:1436–1449. [4]

Lawhern, V. J., A. J. Solon, N. R. Waytowich, et al. 2018. Eegnet: A Compact Convolutional Neural Network for EEG-Based Brain–Computer Interfaces. *J. Neural Eng.* **15**:056013. [12]

Leavitt, M. L., D. Mendoza-Halliday, and J. C. Martinez-Trujillo. 2017. Sustained Activity Encoding Working Memories: Not Fully Distributed. *Trends Neurosci.* **40**:328–346. [6, 10, 16]

Lebedev, M. A., A. Messinger, J. D. Kralik, and S. P. Wise. 2004. Representation of Attended versus Remembered Locations in Prefrontal Cortex. *PLoS Biol.* **2**:e365. [4]

Lebel, C., and C. Beaulieu. 2011. Longitudinal Development of Human Brain Wiring Continues from Childhood into Adulthood. *J. Neurosci.* **31**:10937–10947. [16]

Le Bouc, R., and M. Pessiglione. 2022. A Neuro-Computational Account of Procrastination Behavior. *Nat. Commun.* **13**:5639. [12]

Lee, A. M., L. H. Tai, A. Zador, and L. Wilbrecht. 2015. Between the Primate and 'Reptilian' Brain: Rodent Models Demonstrate the Role of Corticostriatal Circuits in Decision Making. *Neurosci.* **296**:66–74. [7]

Lee, S. W., S. Shimojo, and J. P. O'Doherty. 2014. Neural Computations Underlying Arbitration between Model-Based and Model-Free Learning. *Neuron* **81**:687–699. [12]

Lee, T. G., and M. D'Esposito. 2012. The Dynamic Nature of Top-Down Signals Originating from Prefrontal Cortex: A Combined fMRI-TMS Study. *J. Neurosci.* **32**:15458–15466. [4, 8]

Lefaucheur, J. P., N. Andre-Obadia, A. Antal, et al. 2014. Evidence-Based Guidelines on the Therapeutic Use of Repetitive Transcranial Magnetic Stimulation (rTMS). *J. Clin. Neurophysiol.* **125**:2150–2206. [16]

Lehman, J. F., B. D. Greenberg, C. C. McIntyre, S. A. Rasmussen, and S. N. Haber. 2011. Rules Ventral Prefrontal Cortical Axons Use to Reach Their Targets: Implications for Diffusion Tensor Imaging Tractography and Deep Brain Stimulation for Psychiatric Illness. *J. Neurosci.* **31**:10392–10402. [4]

Le Merre, P., S. Ahrlund-Richter, and M. Carlen. 2021. The Mouse Prefrontal Cortex: Unity in Diversity. *Neuron* **109**:1925–1944. [2, 4]

Leng, X., D. Yee, H. Ritz, and A. Shenhav. 2021. Dissociable Influences of Reward and Punishment on Adaptive Cognitive Control. *PLoS Comput. Biol.* **17**:e1009737. [12]

Lenze, E. J., G. E. Nicol, D. L. Barbour, et al. 2021. Precision Clinical Trials: A Framework for Getting to Precision Medicine for Neurobehavioural Disorders. *J. Psych. Neurosci.* **46**:E97–E110. [14]

Leon, M. I., and M. N. Shadlen. 1999. Effect of Expected Reward Magnitude on the Response of Neurons in the Dorsolateral Prefrontal Cortex of the Macaque. *Neuron* **24**:415–425. [5, 8]

Leung, B. K., and B. W. Balleine. 2015. Ventral Pallidal Projections to Mediodorsal Thalamus and Ventral Tegmental Area Play Distinct Roles in Outcome-Specific Pavlovian-Instrumental Transfer. *J. Neurosci.* **35**:4953–4964. [4]

Levin, F. R., J. J. Mariani, A. Secora, et al. 2009. Atomoxetine Treatment for Cocaine Abuse and Adult Attention-Deficit Hyperactivity Disorder (ADHD): A Preliminary Open Trial. *J. Dual Diagn.* **5**:41–56. [16]

Levitt, J. B., D. A. Lewis, T. Yoshioka, and J. S. Lund. 1993. Topography of Pyramidal Neuron Intrinsic Connections in Macaque Monkey Prefrontal Cortex (Areas 9 and 46). *J. Comp. Neurol.* **338**:360–376. [6]

Levitt, P., P. Rakic, and P. Goldman-Rakic. 1984. Region-Specific Distribution of Catecholamine Afferents in Primate Cerebral Cortex: A Fluorescence Histochemical Analysis. *J. Comp. Neurol.* **227**:23–36. [6]

Levy, R. 2023. The Prefrontal Cortex: From Monkey to Man. https://doi.org/10.1093/brain/awad389. (accessed 2/12/2024). [17]

Lhermitte, F. 1986. Human Autonomy and the Frontal Lobes. Part II: Patient Behavior in Complex and Social Situations: The "Environmental Dependency Syndrome." *Ann. Neurol.* **19**:335–343. [12]

Lhermitte, F., B. Pillon, and M. Serdaru. 1986. Human Autonomy and the Frontal Lobes. Part I: Imitation and Utilization Behavior: A Neuropsychological Study of 75 Patients. *Ann. Neurol.* **19**:326–334. [12]

Li, F., H. Bai, W. Zhang, et al. 2015. Cloning of Genomic Sequences of Three Crustacean Hyperglycemic Hormone Superfamily Genes and Elucidation of Their Roles of Regulating Insulin-Like Androgenic Gland Hormone Gene. *Gene* **561**:68–75. [4]

Li, J. Y., H. Wu, S. Yuan, et al. 2022. A Meta-Analysis on Neural Changes of Cognitive Training for Mental Disorders in Executive Function Tasks: Increase or Decrease Brain Activation? *BMC Psych.* **22**:155. [14]

Li, N., J. C. Baldermann, A. Kibleur, et al. 2020a. A Unified Connectomic Target for Deep Brain Stimulation in Obsessive-Compulsive Disorder. *Nat. Commun.* **11**:3364. [15]

Li, N., B. Lee, R.-J. Liu, et al. 2010a. mTOR-Dependent Synapse Formation Underlies the Rapid Antidepressant Effects of NMDA Antagonists. *Science* **329**:959–964. [13]

Li, S., X. Zhou, C. Constantinidis, and X. L. Qi. 2020b. Plasticity of Persistent Activity and Its Constraints. *Front. Neural Circuits* **14**:15. [6]

Li, W.-Y., T.-S. Shi, J. Huang, et al. 2023. Activation of the mTORC1 Signaling Cascade in the Hippocampus and Medial Prefrontal Cortex Is Required for the Antidepressant Actions of Vortioxetine in Mice. *Int. J. Neuropsychopharmacol.* [13]

Li, Y. C., G. Liu, J. L. Hu, W. J. Gao, and Y. Q. Huang. 2010b. Dopamine D(1) Receptor-Mediated Enhancement of NMDA Receptor Trafficking Requires Rapid PKC-Dependent Synaptic Insertion in the Prefrontal Neurons. *J. Neurochem.* **114**:62–73. [16]

Libby, A., and T. J. Buschman. 2021. Rotational Dynamics Reduce Interference between Sensory and Memory Representations. *Nat. Neurosci.* **24**:715–726. [12]

Liegeois, R., T. O. Laumann, A. Z. Snyder, J. Zhou, and B. T. T. Yeo. 2017. Interpreting Temporal Fluctuations in Resting-State Functional Connectivity MRI. *Neuroimage* **163**:437–455. [11]

Liljeholm, M., E. Tricomi, J. P. O'Doherty, and B. W. Balleine. 2011. Neural Correlates of Instrumental Contingency Learning: Differential Effects of Action–Reward Conjunction and Disjunction. *J. Neurosci.* **31**:2474–2480. [4]

Lin, J. P., H. M. Kelly, Y. Song, et al. 2022. Transcriptomic Architecture of Nuclei in the Marmoset CNS. *Nat. Commun.* **13**:5531. [2]

Lin, P.-Y., Z. Z. Ma, M. Mahgoub, E. T. Kavalali, and L. M. Monteggia. 2021. A Synaptic Locus for Trkb Signaling Underlying Ketamine Rapid Antidepressant Action. *Cell Rep.* **36**:109513. [13]

Lin, Y., Q. Li, and A. Chen. 2023. The Causal Mechanisms Underlying Analogical Reasoning Performance Improvement by Executive Attention Intervention. *Hum. Brain Mapp.* **44**:3241–3253. [17]

Lin, Z., C. Nie, Y. Zhang, Y. Chen, and T. Yang. 2020. Evidence Accumulation for Value Computation in the Prefrontal Cortex during Decision Making. *PNAS* **117**:30728–30737. [4]

Liston, C., J. M. Cichon, F. Jeanneteau, et al. 2013. Circadian Glucocorticoid Oscillations Promote Learning-Dependent Synapse Formation and Maintenance. *Nat. Neurosci.* **16**:698–705. [16]

Liston, C., and W. B. Gan. 2011. Glucocorticoids Are Critical Regulators of Dendritic Spine Development and Plasticity *in Vivo*. *PNAS* **108**:16074–16079. [16]

Liston, C., M. Malter Cohen, T. Teslovich, D. Levenson, and B. J. Casey. 2011. Atypical Prefrontal Connectivity in Attention-Deficit/Hyperactivity Disorder: Pathway to Disease or Pathological End Point? *Biol. Psych.* **69**:1168–1177. [3, 16]

Liston, C., B. S. McEwen, and B. J. Casey. 2009. Psychosocial Stress Reversibly Disrupts Prefrontal Processing and Attentional Control. *PNAS* **106**:912–917. [16]

Liu, Y., M. G. Mattar, T. E. J. Behrens, N. D. Daw, and R. J. Dolan. 2021. Experience Replay Is Associated with Efficient Nonlocal Learning. *Science* **372**:eabf1357. [10]

Livneh, Y., A. U. Sugden, J. C. Madara, et al. 2020. Estimation of Current and Future Physiological States in Insular Cortex. *Neuron* **105**:1094–1111. [10]

Logan, G. D., T. Van Zandt, F. Verbruggen, and E. J. Wagenmakers. 2014. On the Ability to Inhibit Thought and Action: General and Special Theories of an Act of Control. *Psychol. Rev.* **121**:66–95. [4]

Lopez-Persem, A., L. Verhagen, C. Amiez, M. Petrides, and J. Sallet. 2019. The Human Ventromedial Prefrontal Cortex: Sulcal Morphology and Its Influence on Functional Organization. *J. Neurosci.* **39**:3627–3639. [4]

Lorsch, Z. S., A. Ambesi-Impiombato, R. Zenowich, et al. 2021. Computational Analysis of Multidimensional Behavioral Alterations after Chronic Social Defeat Stress. *Biol. Psych.* **89**:920–928. [16]

Lövdén, M., Y. Brehmer, S. C. Li, and U. Lindenberger. 2012. Training-Induced Compensation versus Magnification of Individual Differences in Memory Performance. *Front. Hum. Neurosci.* **6**:141. [14]

Lovett, M. C., and J. R. Anderson. 2005. Thinking as a Production System. In: The Cambridge Handbook of Thinking and Reasoning, ed. K. J. Holyoak and R. G. Morrison, pp. 401–429. Cambridge Cambridge University Press. [7]

Løvstad, M., I. Funderud, T. Endestad, et al. 2012. Executive Functions after Orbital or Lateral Prefrontal Lesions: Neuropsychological Profiles and Self-Reported Executive Functions in Everyday Living. *Brain Inj.* **26**:1586–1598. [12]

Low, R. J., Y. Gu, and D. W. Tank. 2014. Cellular Resolution Optical Access to Brain Regions in Fissures: Imaging Medial Prefrontal Cortex and Grid Cells in Entorhinal Cortex. *PNAS* **111**:18739–18744. [13]

Luebke, J. I. 2017. Pyramidal Neurons Are Not Generalizable Building Blocks of Cortical Networks. *Front. Neuroanat.* **11**:11. [2]

Lui, J. H., D. V. Hansen, and A. R. Kriegstein. 2011. Development and Evolution of the Human Neocortex. *Cell* **146**:18–36. [2]

Lui, J. H., N. D. Nguyen, S. M. Grutzner, et al. 2021. Differential Encoding in Prefrontal Cortex Projection Neuron Classes across Cognitive Tasks. *Cell* **184**:489–506. [2]

Luk, C. H., and J. D. Wallis. 2013. Choice Coding in Frontal Cortex during Stimulus-Guided or Action-Guided Decision-Making. *J. Neurosci.* **33**:1864–1871. [5, 8]

Luna, B., S. Marek, B. Larsen, B. Tervo-Clemmens, and R. Chahal. 2015. An Integrative Model of the Maturation of Cognitive Control. *Annu. Rev. Neurosci.* **38**:151–170. [16]

Lundqvist, M., P. Herman, and E. K. Miller. 2018. Working Memory: Delay Activity, Yes! Persistent Activity? Maybe Not. *J. Neurosci.* **38**:7013–7019. [5]

Lundqvist, M., J. Rose, P. Herman, et al. 2016. Gamma and Beta Bursts Underlie Working Memory. *Neuron* **90**:152–164. [3]

Luo, T. Z., A. G. Bondy, D. Gupta, et al. 2020. An Approach for Long-Term, Multi-Probe Neuropixels Recordings in Unrestrained Rats. *eLife* **9**:e59716. [4]

Luo, T. Z., and J. H. R. Maunsell. 2015. Neuronal Modulations in Visual Cortex Are Associated with Only One of Multiple Components of Attention. *Neuron* **86**:1182–1188. [12]

———. 2018. Attentional Changes in Either Criterion or Sensitivity Are Associated with Robust Modulations in Lateral Prefrontal Cortex. *Neuron* **97**:1382–1393. [12]

Luo, Y., G. Fernandez, E. Hermans, et al. 2018. How Acute Stress May Enhance Subsequent Memory for Threat Stimuli Outside the Focus of Attention: DLPFC-Amygdala Decoupling. *Neuroimage* **171**:311–322. [16]

Luppino, G., M. Matelli, R. M. Camarda, V. Gallese, and G. Rizzolatti. 1991. Multiple Representations of Body Movements in Mesial Area 6 and the Adjacent Cingulate Cortex: an Intracortical Microstimulation Study in the Macaque Monkey. *J. Comp. Neurol.* **311**:463–482. [8]

Luria, A. R. 1966a. Higher Cortical Functions in Man. Oxford: Basic Books. [9]

———. 1966b. Human Brain and Psychological Processes. New York: Harper and Row. [2]

Luria, R., H. Balaban, E. Awh, and E. K. Vogel. 2016. The Contralateral Delay Activity as a Neural Measure of Visual Working Memory. *Neurosci. Biobehav. Rev.* **62**:100–108. [8]

Lurie, D. J., D. Kessler, D. S. Bassett, et al. 2020. Questions and Controversies in the Study of Time-Varying Functional Connectivity in Resting fMRI. *Netw. Neurosci.* **4**:30–69. [4]

Lustig, C., P. Shah, R. Seidler, and P. A. Reuter-Lorenz. 2009. Aging, Training, and the Brain: A Review and Future Directions. *Neuropsychol. Rev.* **19**:504–522. [14]

Luyten, L., S. Hendrickx, S. Raymaekers, L. Gabriëls, and B. Nuttin. 2016. Electrical Stimulation in the Bed Nucleus of the Stria Terminalis Alleviates Severe Obsessive-Compulsive Disorder. *Mol. Psych.* **21**:1272–1280. [15]

Lynch, C. J., I. G. Elbau, T. H. Ng, et al. 2022. Automated Optimization of TMS Coil Placement for Personalized Functional Network Engagement. *Neuron* **110**:3263–3277. [16]

MacDowell, C. J., and T. J. Buschman. 2020. Low-Dimensional Spatiotemporal Dynamics Underlie Cortex-Wide Neural Activity. *Curr. Biol.* **30**:2665–2680. [12]

MacDowell, C. J., A. Libby, C. I. Jahn, S. Tafazoli, and T. J. Buschman. 2023. Multiplexed Subspaces Route Neural Activity across Brain-Wide Networks. bioRxiv. https://www.ncbi.nlm.nih.gov/pubmed/36798411. (accessed 25. Jan, 2024). [8, 12]

Mack, M. L., A. R. Preston, and B. C. Love. 2017. Medial Prefrontal Cortex Compresses Concept Representations through Learning. In: International Workshop on Pattern Recognition in Neuroimaging (PRNI). Toronto, ON, Canada. [8]

Mack, M. L., A. R. Preston, and B. C. Love. 2020. Ventromedial Prefrontal Cortex Compression during Concept Learning. *Nat. Commun.* **11**:46. [12]

Mackey, A. P., R. D. S. Raizada, and S. A. Bunge. 2013. Environmental Influences on Prefrontal Development. In: Principles of Frontal Lobe Function, ed. D. T. Stuss and R. T. Knight. Oxford: Oxford Univ. Press. [14]

Mackey, S., and M. Petrides. 2009. Architectonic Mapping of the Medial Region of the Human Orbitofrontal Cortex by Density Profiles. *Neurosci.* **159**:1089–1107. [4]

Mackey, W. E., and C. E. Curtis. 2017. Distinct Contributions by Frontal and Parietal Cortices Support Working Memory. *Sci. Rep.* **7**:6188. [4]

Mackey, W. E., O. Devinsky, W. K. Doyle, J. G. Golfinos, and C. E. Curtis. 2016. Human Parietal Cortex Lesions Impact the Precision of Spatial Working Memory. *J. Neurophysiol.* **116**:1049–1054. [4]

Mackey, W. E., J. Winawer, and C. E. Curtis. 2017. Visual Field Map Clusters in Human Frontoparietal Cortex. *eLife* **6**:e22974. [8]

Maclean, P. D. 1958. Contrasting Functions of Limbic and Neocortical Systems of the Brain and Their Relevance to Psychophysiological Aspects of Medicine. *Am. J. Med.* **25**:611–626. [15]

Macmillan, M., and M. L. Lena. 2010. Rehabilitating Phineas Gage. *Neuropsychol. Rehabil.* **20**:641–658. [12]

Maisson, D. J., R. L. Cervera, B. Voloh, et al. 2023. Widespread Coding of Navigational Variables in Prefrontal Cortex. *Curr. Biol.* **33**:3478–3488. [8]

Malejko, K., B. Abler, P. L. Plener, and J. Straub. 2017. Neural Correlates of Psychotherapeutic Treatment of Post-Traumatic Stress Disorder: A Systematic Literature Review. *Front. Psych.* **8**:85. [16]

Málková, L., D. Gaffan, and E. A. Murray. 1997. Excitotoxic Lesions of the Amygdala Fail to Produce Impairment in Visual Learning for Auditory Secondary Reinforcement but Interfere with Reinforcer Devaluation Effects in Rhesus Monkeys. *J. Neurosci.* **17**:6011–6020. [4]

Malvaez, M., C. Shieh, M. D. Murphy, V. Y. Greenfield, and K. M. Wassum. 2019. Distinct Cortical–Amygdala Projections Drive Reward Value Encoding and Retrieval. *Nat. Neurosci.* **22**:762–769. [4]

Mansouri, F. A., and M. J. Buckley. 2018. Context-Dependent Adjustments in Executive Control of Goal-Directed Behaviour: Contribution of Frontal Brain Areas to Conflict-Induced Behavioural Adjustments in Primates. *Adv. Neurobiol.* **21**:71–83. [4]

Mansouri, F. A., D. J. Freedman, and M. J. Buckley. 2020. Emergence of Abstract Rules in the Primate Brain. *Nat. Rev. Neurosci.* **21**:595–610. [7]

Mansouri, F. A., E. Koechlin, M. G. P. Rosa, and M. J. Buckley. 2017. Managing Competing Goals: A Key Role for the Frontopolar Cortex. *Nat. Rev. Neurosci.* **18**:645–657. [10, 15, 16]

Mansouri, F. A., K. Matsumoto, and K. Tanaka. 2006. Prefrontal Cell Activities Related to Monkeys' Success and Failure in Adapting to Rule Changes in a Wisconsin Card Sorting Test Analog. *J. Neurosci.* **26**:2745–2756. [10]

Mante, V., D. Sussillo, K. V. Shenoy, and W. T. Newsome. 2013. Context-Dependent Computation by Recurrent Dynamics in Prefrontal Cortex. *Nature* **503**:78–84. [5, 12]

Marek, S., K. Hwang, W. Foran, M. N. Hallquist, and B. Luna. 2015. The Contribution of Network Organization and Integration to the Development of Cognitive Control. *PLoS Biol.* **13**:e1002328. [16]

Marek, S., B. Tervo-Clemmens, F. J. Calabro, et al. 2022. Reproducible Brain-Wide Association Studies Require Thousands of Individuals. *Nature* **603**:654–660. [9, 11, 13]

Margulies, D. S., S. S. Ghosh, A. Goulas, et al. 2016. Situating the Default-Mode Network Along a Principal Gradient of Macroscale Cortical Organization. *PNAS* **113**:12574–12579. [4, 7]

Markov, N. T., M. Ercsey-Ravasz, D. C. Van Essen, et al. 2013. Cortical High-Density Counterstream Architectures. *Science* **342**:1238406. [11]

Markov, N. T., M. M. Ercsey-Ravasz, A. R. Ribeiro Gomes, et al. 2014. A Weighted and Directed Interareal Connectivity Matrix for Macaque Cerebral Cortex. *Cereb. Cortex* **24**:17–36. [10]

Markowitz, J. E., W. F. Gillis, C. C. Beron, et al. 2018. The Striatum Organizes 3D Behavior via Moment-to-Moment Action Selection. *Cell* **174**:44–58. [12]

Markowitz, J. E., W. F. Gillis, M. Jay, et al. 2023. Spontaneous Behaviour Is Structured by Reinforcement without Explicit Reward. *Nature* **614**:108–117. [12]

Marquand, A. F., K. V. Haak, and C. F. Beckmann. 2017. Functional Corticostriatal Connection Topographies Predict Goal Directed Behaviour in Humans. *Nat. Hum. Behav.* **1**:0146. [8]

Marquardt, K., R. Sigdel, and J. L. Brigman. 2017. Touch-Screen Visual Reversal Learning Is Mediated by Value Encoding and Signal Propagation in the Orbitofrontal Cortex. *Neurobiol. Learn. Mem.* **139**:179–188. [4]

Marr, D. 1982. Vision: A Computational Approach. San Francisco: Freeman & Co. [10, 12]

Mars, R. B., and M. J. Grol. 2007. Dorsolateral Prefrontal Cortex, Working Memory, and Prospective Coding for Action. *J. Neurosci.* **27**:1801–1802. [12]

Mars, R. B., L. Verhagen, T. E. Gladwin, et al. 2016. Comparing Brains by Matching Connectivity Profiles. *Neurosci. Biobehav. Rev.* **60**:90–97. [8]

Martinez-Fernandez, R., and J. A. Pineda-Pardo. 2020. Magnetic Resonance-Guided Focused Ultrasound for Movement Disorders: Clinical and Neuroimaging Advances. *Curr Opin Neurol* **33**:488–497. [16]

Martinez-Rivera, F. J., J. Perez-Torres, C. D. Velazquez-Diaz, et al. 2023. A Novel Insular/Orbital-Prelimbic Circuit That Prevents Persistent Avoidance in a Rodent Model of Compulsive Behavior. *Biol. Psych.* **93**:1000–1009. [16]

Mashhoori, A., S. Hashemnia, B. L. McNaughton, D. R. Euston, and A. J. Gruber. 2018. Rat Anterior Cingulate Cortex Recalls Features of Remote Reward Locations after Disfavoured Reinforcements. *eLife* **7**:e29793. [4]

Masís, J., T. Chapman, J. Y. Rhee, D. D. Cox, and A. M. Saxe. 2023. Strategically Managing Learning during Perceptual Decision Making. *eLife* **12**:e64978. [12]

Masland, R. H. 2004. Neuronal Cell Types. *Curr. Biol.* **14**:R497–R500. [2]

Masse, N. Y., G. R. Yang, H. F. Song, X.-J. Wang, and D. J. Freedman. 2019. Circuit Mechanisms for the Maintenance and Manipulation of Information in Working Memory. *Nat. Neurosci.* **22**:1159–1167. [12]

Massi, B., C. H. Donahue, and D. Lee. 2018. Volatility Facilitates Value Updating in the Prefrontal Cortex. *Neuron* **99**:598–608. [8]

Matelli, M., G. Luppino, and G. Rizzolatti. 1985. Patterns of Cytochrome Oxidase Activity in the Frontal Agranular Cortex of the Macaque Monkey. *Behav. Brain Res.* **18**:125–136. [8]

———. 1991. Architecture of Superior and Mesial Area 6 and the Adjacent Cingulate Cortex in the Macaque Monkey. *J. Comp. Neurol.* **311**:445–462. [8]

Mathis, A., P. Mamidanna, K. M. Cury, et al. 2018. DeepLabCut: Markerless Pose Estimation of User-Defined Body Parts with Deep Learning. *Nat. Neurosci.* **21**:1281–1289. [5]

Mathis, A., S. Schneider, J. Lauer, and M. W. Mathis. 2020. A Primer on Motion Capture with Deep Learning: Principles, Pitfalls, and Perspectives. *Neuron* **108**:44–65. [2]

Matsumoto, K., W. Suzuki, and K. Tanaka. 2003. Neuronal Correlates of Goal-Based Motor Selection in the Prefrontal Cortex. *Science* **301**:229–232. [8]

Mattar, M. G., and N. D. Daw. 2018. Prioritized Memory Access Explains Planning and Hippocampal Replay. *Nat. Neurosci.* **21**:1609–1617. [12]

Mattay, V. S., F. Fera, A. Tessitore, et al. 2006. Neurophysiological Correlates of Age-Related Changes in Working Memory Capacity. *Neurosci. Lett.* **392**:32–37. [9]

Matthews, K., and T. W. Robbins. 2003. Early Experience as a Determinant of Adult Behavioural Responses to Reward: The Effects of Repeated Maternal Separation in the Rat. *Neurosci. Biobehav. Rev.* **27**:45–55. [16]

Mayberg, H. S., M. Liotti, S. K. Brannan, et al. 1999. Reciprocal Limbic-Cortical Function and Negative Mood: Converging PET Findings in Depression and Normal Sadness. *Am. J. Psych.* **156**:675–682. [13]

Mayberg, H. S., A. M. Lozano, V. Voon, et al. 2005. Deep Brain Stimulation for Treatment-Resistant Depression. *Neuron* **45**:651–660. [4, 13]

McAlonan, K., J. Cavanaugh, and R. H. Wurtz. 2006. Attentional Modulation of Thalamic Reticular Neurons. *J. Neurosci.* **26**:4444–4450. [4]

McEwen, B. S. 2013. The Brain on Stress: Toward an Integrative Approach to Brain, Body, and Behavior. *Perspect. Psychol. Sci.* **8**:673–675. [16]

McEwen, B. S., N. P. Bowles, J. D. Gray, et al. 2015. Mechanisms of Stress in the Brain. *Nat. Neurosci.* **18**:1353–1363. [16]

McEwen, B. S., and J. H. Morrison. 2013. The Brain on Stress: Vulnerability and Plasticity of the Prefrontal Cortex over the Life Course. *Neuron* **79**:16–29. [6, 8]

McGaugh, J. L. 2004. The Amygdala Modulates the Consolidation of Memories of Emotionally Arousing Experiences. *Annu. Rev. Neurosci.* **27**:1–28. [16]

McGrath, C. L., M. E. Kelley, B. W. Dunlop, et al. 2014. Pretreatment Brain States Identify Likely Nonresponse to Standard Treatments for Depression. *Biol. Psych.* **76**:527–535. [13, 16]

McGuire, J. T., M. R. Nassar, J. I. Gold, and J. W. Kable. 2014. Functionally Dissociable Influences on Learning Rate in a Dynamic Environment. *Neuron* **84**:870–881. [12]

McKiernan, K. A., J. N. Kaufman, J. Kucera-Thompson, and J. R. Binder. 2003. A Parametric Manipulation of Factors Affecting Task-Induced Deactivation in Functional Neuroimaging. *J. Cog. Neuro.* **15**:394–408. [11]

McLardy, T. 1950. Thalamic Projection to Frontal Cortex in Man. *J. Neurol. Neurosurg. Psych.* **13**:198–202. [15]

McNab, F., and T. Klingberg. 2008. Prefrontal Cortex and Basal Ganglia Control Access to Working Memory. *Nat. Neurosci.* **11**:103–107. [7]

McNamara, J. M., T. W. Fawcett, and A. I. Houston. 2013. An Adaptive Response to Uncertainty Generates Positive and Negative Contrast Effects. *Science* **340**:1084–1086. [2]

Meier, E. L., C. R. Kelly, E. B. Goldberg, and A. E. Hillis. 2022. Executive Control Deficits and Lesion Correlates in Acute Left Hemisphere Stroke Survivors with and without Aphasia. *Brain Imag. Behav.* **16**:868–877. [12]

Mejias, J. F., and X. J. Wang. 2022. Mechanisms of Distributed Working Memory in a Large-Scale Network of Macaque Neocortex. *eLife* **11**:e72136. [6, 10, 12]

Melby-Lervåg, M., T. S. Redick, and C. Hulme. 2016. Working Memory Training Does Not Improve Performance on Measures of Intelligence or Other Measures of "Far Transfer": Evidence from a Meta-Analytic Review. *Perspect. Psychol. Sci.* **11**:512–534. [14]

Melchitzky, D. S., and D. A. Lewis. 2008. Dendritic-Targeting GABA Neurons in Monkey Prefrontal Cortex: Comparison of Somatostatin- and Calretinin-Immunoreactive Axon Terminals. *Synapse* **62**:456–465. [6]

Menon, V., and M. D'Esposito. 2022. The Role of PFC Networks in Cognitive Control and Executive Function. *Neuropsychopharmacol.* **47**:90–103. [8]

Menon, V., and L. Q. Uddin. 2010. Saliency, Switching, Attention and Control: A Network Model of Insula Function. *Brain Struct. Funct.* **214**:655–667. [4, 11]

Meskenaite, V. 1997. Calretinin-Immunoreactive Local Circuit Neurons in Area 17 of the Cynomolgus Monkey, Macaca fascicularis. *J. Comp. Neurol.* **379**:113–132. [6]

Mesulam, M. M. 1981. A Cortical Network for Directed Attention and Unilateral Neglect. *Annals of Neurology: Official Journal of the American Neurological Association and the Child Neurology Society* **10**:309–325. [11]

———. 1990. Large-Scale Neurocognitive Networks and Distributed Processing for Attention, Language, and Memory. *Annals of Neurology: Official Journal of the American Neurological Association and the Child Neurology Society* **28**:597–613. [11]

———. 1998. From Sensation to Cognition. *Brain* **121 (Pt. 6)**:1013–1052. [1]

Michalka, S. W., L. Kong, M. L. Rosen, B. G. Shinn-Cunningham, and D. C. Somers. 2015. Short-Term Memory for Space and Time Flexibly Recruit Complementary Sensory-Biased Frontal Lobe Attention Networks. *Neuron* **87**:882–892. [11]

Middleton, F. A., and P. L. Strick. 2002. Basal-Ganglia "Projections" to the Prefrontal Cortex of the Primate. *Cereb. Cortex* **12**:926–935. [6]

Milad, M. R., and G. J. Quirk. 2012. Fear Extinction as a Model for Translational Neuroscience: Ten Years of Progress. *Annu. Rev. Psychol.* **63**:129–151. [4]

Milad, M. R., and S. L. Rauch. 2012. Obsessive-Compulsive Disorder: Beyond Segregated Cortico-Striatal Pathways. *Trends Cogn. Sci.* **16**:43–51. [15]

Milak, M. S., C. J. Proper, S. T. Mulhern, et al. 2016. A Pilot *in Vivo* Proton Magnetic Resonance Spectroscopy Study of Amino Acid Neurotransmitter Response to Ketamine Treatment of Major Depressive Disorder. *Mol. Psychiatry* **21**:320–327. [13]

Milak, M. S., R. Rashid, Z. Dong, et al. 2020. Assessment of Relationship of Ketamine Dose with Magnetic Resonance Spectroscopy of Glx and GABA Responses in Adults with Major Depression: A Randomized Clinical Trial. *JAMA Netw. Open* **3**:e2013211–e2013211. [13]

Miller, E. K., and J. D. Cohen. 2001. An Integrative Theory of Prefrontal Cortex Function. *Annu. Rev. Neurosci.* **24**:167–202. [3, 7–10, 12]

Miller, E. N., P. R. Hof, C. C. Sherwood, and W. D. Hopkins. 2021a. The Paracingulate Sulcus Is a Unique Feature of the Medial Frontal Cortex Shared by Great Apes and Humans. *Brain Behav. Evol.* **96**:26–36. [4]

Miller, J. A., M. D'Esposito, and K. S. Weiner. 2021b. Using Tertiary Sulci to Map the "Cognitive Globe" of Prefrontal Cortex. *J. Cogn. Neurosci.* **33**:1698–1715. [4]

Miller, J. A., A. Tambini, A. Kiyonaga, and M. D'Esposito. 2022. Long-Term Learning Transforms Prefrontal Cortex Representations during Working Memory. *Neuron* **110**:3805–3819. [12]

Miller, S., L. M. McTeague, A. Gyurak, et al. 2015. Cognition-Childhood Maltreatment Interactions in the Prediction of Antidepressant Outcomes in Major Depressive Disorder Patients: Results from the Ispot-D Trial. *Depress. Anxiety* **32**:594–604. [13]

Milton, R., N. Shahidi, and V. Dragoi. 2020. Dynamic States of Population Activity in Prefrontal Cortical Networks of Freely-Moving Macaque. *Nat. Commun.* **11**:1948. [8]

Mindus, P., G. Edman, and S. Andréewitch. 1999. A Prospective, Long-Term Study of Personality Traits in Patients with Intractable Obsessional Illness Treated by Capsulotomy. *Acta Psychiatr Scand* **99**:40–50. [15]

Minxha, J., R. Adolphs, S. Fusi, A. N. Mamelak, and U. Rutishauser. 2020. Flexible Recruitment of Memory-Based Choice Representations by the Human Medial Frontal Cortex. *Science* **368**:eaba3313. [6]

Mion, G., and T. Villevieille. 2013. Ketamine Pharmacology: an Update (Pharmacodynamics and Molecular Aspects, Recent Findings). *CNS Neurosci. Ther.* **19**:370–380. [13]

Misztak, P., M. Sowa-Kućma, P. Pańczyszyn-Trzewik, B. Szewczyk, and G. Nowak. 2021. Antidepressant-Like Effects of Combined Fluoxetine and Zinc Treatment in Mice Exposed to Chronic Restraint Stress Are Related to Modulation of Histone Deacetylase. *Molecules* **27**: [13]

Mitchell, A. S. 2015. The Mediodorsal Thalamus as a Higher Order Thalamic Relay Nucleus Important for Learning and Decision-Making. *Neurosci. Biobehav. Rev.* **54**:76–88. [8]

Mitchell, D. J., A. L. S. Mousley, M. A. Shafto, Cam-CAN, and J. Duncan. 2023. Neural Contributions to Reduced Fluid Intelligence across the Adult Lfespan. *J. Neurosci.* **43**:293–307. [9]

Mitz, A. R., R. Bartolo, R. C. Saunders, et al. 2017. High Channel Count Single-Unit Recordings from Nonhuman Primate Frontal Cortex. *J. Neurosci. Methods* **289**:39–47. [2]

Mitz, A. R., and S. P. Wise. 1987. The Somatotopic Organization of the Supplementary Motor Area: Intracortical Microstimulation Mapping. *J. Neurosci.* **7**:1010–1021. [8]

Miyake, A., N. P. Friedman, M. J. Emerson, et al. 2000. The Unity and Diversity of Executive Functions and Their Contributions to Complex "Frontal Lobe" Tasks: A Latent Variable Analysis. *Cogn. Psychol.* **41**:49–100. [4, 9, 12]

Mnih, V., K. Kavukcuoglu, D. Silver, et al. 2015. Human-Level Control through Deep Reinforcement Learning. *Nature* **518**:529–533. [11, 12]

Mobbs, D., D. B. Headley, W. Ding, and P. Dayan. 2020. Space, Time, and Fear: Survival Computations Along Defensive Circuits. *Trends Cogn. Sci.* **24**:228–241. [8]

Moda-Sava, R. N., M. H. Murdock, P. K. Parekh, et al. 2019. Sustained Rescue of Prefrontal Circuit Dysfunction by Antidepressant-Induced Spine Formation. *Science* **364**:eaat8078. [13, 16]

Monosov, I. E., S. N. Haber, E. C. Leuthardt, and A. Jezzini. 2020. Anterior Cingulate Cortex and the Control of Dynamic Behavior in Primates. *Curr. Biol.* **30**:R1442–R1454. [15]

Monosov, I. E., and M. F. S. Rushworth. 2022. Interactions between Ventrolateral Prefrontal and Anterior Cingulate Cortex during Learning and Behavioural Change. *Neuropsychopharmacol.* **47**:196–210. [1, 12]

Moolchand, P., S. R. Jones, and M. J. Frank. 2022. Biophysical and Architectural Mechanisms of Subthalamic Theta under Response Conflict. *J. Neurosci.* **42**:4470–4487. [8]

Moore, T., and K. M. Armstrong. 2003. Selective Gating of Visual Signals by Microstimulation of Frontal Cortex. *Nature* **421**:370–373. [12]

Morecraft, R. J., K. S. Stilwell-Morecraft, P. B. Cipolloni, et al. 2012. Cytoarchitecture and Cortical Connections of the Anterior Cingulate and Adjacent Somatomotor Fields in the Rhesus Monkey. *Brain Res. Bull.* **87**:457–497. [4, 16]

Moriguchi, S., A. Takamiya, Y. Noda, et al. 2019. Glutamatergic Neurometabolite Levels in Major Depressive Disorder: A Systematic Review and Meta-Analysis of Proton Magnetic Resonance Spectroscopy Studies. *Mol. Psychiatry* **24**:952–964. [13]

Morin, E. L., B. R. Howell, E. Feczko, et al. 2020. Developmental Outcomes of Early Adverse Care on Amygdala Functional Connectivity in Nonhuman Primates. *Dev. Psychopathol.* **32**:1579–1596. [16]

Morris, L. S., S. Costi, A. Tan, et al. 2020. Ketamine Normalizes Subgenual Cingulate Cortex Hyper-Activity in Depression. *Neuropsychopharmacol.* **45**:975–981. [13]

Morris, R. W., A. Dezfouli, K. R. Griffiths, and B. W. Balleine. 2014. Action-Value Comparisons in the Dorsolateral Prefrontal Cortex Control Choice between Goal-Directed Actions. *Nat. Commun.* **5**:4390. [4]

Morris, R. W., A. Dezfouli, K. R. Griffiths, M. E. Le Pelley, and B. W. Balleine. 2022. The Neural Bases of Action-Outcome Learning in Humans. *J. Neurosci.* **42**:3636–3647. [4]

Morrison, S. E., and C. D. Salzman. 2009. The Convergence of Information About Rewarding and Aversive Stimuli in Single Neurons. *J. Neurosci.* **29**:11471–11483. [5, 8]

Morse, A. K., B. K. Leung, E. Heath, et al. 2020. Basolateral Amygdala Drives a Gpcr-Mediated Striatal Memory Necessary for Predictive Learning to Influence Choice. *Neuron* **106**:855–869.e858. [4]

Morton, N. W., and A. R. Preston. 2021. Concept Formation as a Computational Cognitive Process. *Curr. Opin. Behav. Sci.* **38**:83–89. [17]

Mostame, P., and S. Sadaghiani. 2021. Oscillation-Based Connectivity Architecture Is Dominated by an Intrinsic Spatial Organization, Not Cognitive State or Frequency. *J. Neurosci.* **41**:179–192. [11]

Mothersill, D., and G. Donohoe. 2019. Neural Effects of Cognitive Training in Schizophrenia: A Systematic Review and Activation Likelihood Estimation Meta-Analysis. *Biol. Psych. Cogn. Neurosci. Neuroimag.* **4**:688–696. [14]

Mukherjee, A., N. Bajwa, N. H. Lam, et al. 2020. Variation of Connectivity across Exemplar Sensory and Associative Thalamocortical Loops in the Mouse. *eLife* **9**:e62554. [3]

Muldoon, S. F., and D. S. Bassett. 2016. Network and Multilayer Network Approaches to Understanding Human Brain Dynamics. *Philos. Sci.* **83**:710–720. [12]

Munakata, Y., S. A. Herd, C. H. Chatham, et al. 2011. A Unified Framework for Inhibitory Control. *Trends Cogn. Sci.* **15**:453–459. [12, 17]

Murley, A. G., I. Coyle-Gilchrist, M. A. Rouse, et al. 2020. Redefining the Multidimensional Clinical Phenotypes of Frontotemporal Lobar Degeneration Syndromes. *Brain* **143**:1555–1571. [16]

Murley, A. G., and J. B. Rowe. 2018. Neurotransmitter Deficits from Frontotemporal Lobar Degeneration. *Brain* **141**:1263–1285. [16]

Murray, E. A., and M. G. Baxter. 2006. Cognitive Neuroscience and Nonhuman Primates: Lesion Studies. In: Methods in Mind, ed. C. Senior et al., pp. 43–69. Cambridge, MA: The MIT Press. [8]

Murray, E. A., and L. K. Fellows. 2022. Prefrontal Cortex Interactions with the Amygdala in Primates. *Neuropsychopharmacol.* **47**:163–179. [8]

Murray, E. A., E. J. Moylan, K. S. Saleem, B. M. Basile, and J. Turchi. 2015. Specialized Areas for Value Updating and Goal Selection in the Primate Orbitofrontal Cortex. *eLife* **4**:e11695. [8]

Murray, E. A., and P. H. Rudebeck. 2013. The Drive to Strive: Goal Generation Based on Current Needs. *Front. Neurosci.* **7**:112. [8]

———. 2018. Specializations for Reward-Guided Decision-Making in the Primate Ventral Prefrontal Cortex. *Nat. Rev. Neurosci.* **19**:404–417. [2, 8, 16]

Murray, E. A., S. P. Wise, and K. S. Graham. 2017. The Evolution of Memory Systems: Ancestors, Anatomy, and Adaptations. Oxford: Oxford Univ. Press. [4, 8]

Murray, E. A., S. P. Wise, and S. E. V. Rhodes. 2011. What Can Different Brains Do with Reward? In: Neurobiology of Sensation and Reward, ed. J. A. Gottfried. Boca Raton: CRC Press/Taylor & Francis. [2]

Murray, J. D., A. Bernacchia, D. J. Freedman, et al. 2014. A Hierarchy of Intrinsic Timescales across Primate Cortex. *Nat. Neurosci.* **17**:1661–1663. [4, 6, 12, 16]

Murrough, J. W., D. V. Iosifescu, L. C. Chang, et al. 2013. Antidepressant Efficacy of Ketamine in Treatment-Resistant Major Depression: A Two-Site Randomized Controlled Trial. *Am. J. Psych.* **170**:1134–1142. [13]

Murty, V. P., H. Shah, D. Montez, et al. 2018. Age-Related Trajectories of Functional Coupling between the VTA and Nucleus Accumbens Depend on Motivational State. *J. Neurosci.* **38**:7420–7427. [16]

Musslick, S., and J. D. Cohen. 2021. Rationalizing Constraints on the Capacity for Cognitive Control. *Trends Cogn. Sci.* **25**:757–775. [11, 12]

Musslick, S., J. D. Cohen, and A. Shenhav. 2019. Decomposing Individual Differences in Cognitive Control: A Model-Based Approach. In: Proceedings of the 41st Annual Meeting of the Cognitive Science Society: Creativity + Cognition + Computation, Cogsci 2019. Montrel: Cognitive Science Society. [12]

Nagai, Y., N. Miyakawa, H. Takuwa, et al. 2020. Deschloroclozapine, a Potent and Selective Chemogenetic Actuator Enables Rapid Neuronal and Behavioral Modulations in Mice and Monkeys. *Nat. Neurosci.* **23**:1157–1167. [2]

Najafi, F., G. F. Elsayed, R. Cao, et al. 2020. Excitatory and Inhibitory Subnetworks Are Equally Selective during Decision-Making and Emerge Simultaneously during Learning. *Neuron* **105**:165–179. [2]

Nakajima, M., L. I. Schmitt, and M. M. Halassa. 2019. Prefrontal Cortex Regulates Sensory Filtering through a Basal Ganglia-to-Thalamus Pathway. *Neuron* **103**:445–458. [3]

Nakamura, K., and K. Kubota. 1995. Mnemonic Firing of Neurons in the Monkey Temporal Pole during a Visual Recognition Memory Task. *J. Neurophysiol.* **74**:162–178. [5]

Nakamura, M., P. G. Nestor, and M. E. Shenton. 2020. Orbitofrontal Sulcogyral Pattern as a Transdiagnostic Trait Marker of Early Neurodevelopment in the Social Brain. *Clin. EEG Neurosci.* **51**:275–284. [4]

Nakayama, H., I. Ibanez-Tallon, and N. Heintz. 2018. Cell-Type-Specific Contributions of Medial Prefrontal Neurons to Flexible Behaviors. *J. Neurosci.* **38**:4490–4504. [2]

Napoli, J. L., C. R. Camalier, A. L. Brown, et al. 2021. Correlates of Auditory Decision-Making in Prefrontal, Auditory, and Basal Lateral Amygdala Cortical Areas. *J. Neurosci.* **41**:1301–1316. [5]

Narayanan, N. S., J. F. Cavanagh, M. J. Frank, and M. Laubach. 2013. Common Medial Frontal Mechanisms of Adaptive Control in Humans and Rodents. *Nat. Neurosci.* **16**:1888–1895. [12]

Nasca, C., C. Menard, G. Hodes, et al. 2019. Multidimensional Predictors of Susceptibility and Resilience to Social Defeat Stress. *Biol. Psych.* **86**:483–491. [16]

Naselaris, T., E. Allen, and K. Kay. 2021. Extensive Sampling for Complete Models of Individual Brains. *Curr. Opin. Behav. Sci.* **40**:45–51. [12]

Nauczyciel, C., F. Le Jeune, F. Naudet, et al. 2014. Repetitive Transcranial Magnetic Stimulation over the Orbitofrontal Cortex for Obsessive-Compulsive Disorder: A Double-Blind, Crossover Study. *Transl. Psych.* **4**:e436. [15]

Neafsey, E. J., E. L. Bold, G. Haas, et al. 1986. The Organization of the Rat Motor Cortex: A Microstimulation Mapping Study. *Brain Research Reviews* **11**:77–96. [4]

Neale, J. H., R. T. Olszewski, D. Zuo, et al. 2011. Advances in Understanding the Peptide Neurotransmitter NAAG and Appearance of a new Member of the NAAG Neuropeptide Family. *J. Neurochem.* **118**:490–498. [16]

Nee, D. E. 2021. Integrative Frontal-Parietal Dynamics Supporting Cognitive Control. *eLife* **10**:e57244. [7]

Nee, D. E., and J. W. Brown. 2012. Rostral-Caudal Gradients of Abstraction Revealed by Multi-Variate Pattern Analysis of Working Memory. *Neuroimage* **63**:1285–1294. [7]

Nee, D. E., and J. W. Brown. 2013. Dissociable Frontal-Striatal and Frontal-Parietal Networks Involved in Updating Hierarchical Contexts in Working Memory. *Cereb. Cortex* **23**:2146–2158. [7]

Nee, D. E., and M. D'Esposito. 2016. The Hierarchical Organization of the Lateral Prefrontal Cortex. *eLife* **21**:e12112. [4, 7, 8, 10]

———. 2017. Causal Evidence for Lateral Prefrontal Cortex Dynamics Supporting Cognitive Control. *eLife* **6**:e28040. [4, 7, 8, 10]

Nelson, S. M., N. U. F. Dosenbach, A. L. Cohen, et al. 2010. Role of the Anterior Insula in Task-Level Control and Focal Attention. *Brain Struct. Funct.* **214**:669–680. [11]

Nemati, S., T. J. Akiki, J. Roscoe, et al. 2020. A Unique Brain Connectome Fingerprint Predates and Predicts Response to Antidepressants. *iScience* **23**:100800. [13]

Neubauer, A. C., and A. Fink. 2009. Intelligence and Neural Efficiency. *Neuroscience and Biobehavioral Reviews* **33**:1004–1023. [14]

Neubert, F. X., R. B. Mars, J. Sallet, and M. F. Rushworth. 2015. Connectivity Reveals Relationship of Brain Areas for Reward-Guided Learning and Decision Making in Human and Monkey Frontal Cortex. *PNAS* **112**:E2695–2704. [4, 8]

Newman, M. E. J. 2006. Modularity and Community Structure in Networks. *PNAS* **103**:8577–8582. [11]

Neymotin, S. A., D. S. Daniels, B. Caldwell, et al. 2020. Human Neocortical Neurosolver (Hnn), a new Software Tool for Interpreting the Cellular and Network Origin of Human MEG/EEG Data. *eLife* **9**:e51214. [8]

Nguyen, K. P., C. Chin Fatt, A. Treacher, et al. 2022. Patterns of Pretreatment Reward Task Brain Activation Predict Individual Antidepressant Response: Key Results from the Embarc Randomized Clinical Trial. *Biol. Psychiatry* **91**:550–560. [13]

Niendam, T. A., A. R. Laird, K. L. Ray, et al. 2012. Meta-Analytic Evidence for a Superordinate Cognitive Control Network Subserving Diverse Executive Functions. *Cogn. Affect. Behav. Neurosci.* **12**:241–268. [9, 16]

Nigg, J. T., M. H. Sibley, A. Thapar, and S. L. Karalunas. 2020. Development of ADHD: Etiology, Heterogeneity, and Early Life Course. *Annu. Rev. Dev. Psychol.* **2**:559–583. [16]

Niki, H. 1974. Differential Activity of Prefrontal Units during Right and Left Delayed Response Trials. *Brain Res.* **70**:346–349. [5]

Niki, H., and M. Watanabe. 1976. Prefrontal Unit Activity and Delayed Response: Relation to Cue Location versus Direction of Response. *Brain Res.* **105**:79–88. [5]

Niv, Y. 2019. Learning Task-State Representations. *Nat. Neurosci.* **22**:1544–1553. [12]

Nomura, E. M., C. Gratton, R. M. Visser, et al. 2010. Double Dissociation of Two Cognitive Control Networks in Patients with Focal Brain Lesions. *PNAS* **107**:12017–12022. [11]

Noonan, M. P., M. E. Walton, T. E. J. Behrens, et al. 2010. Separate Value Comparison and Learning Mechanisms in Macaque Medial and Lateral Orbitofrontal Cortex. *PNAS* **107**:20547–20552. [4, 12]

Norman, D. A., and T. Shallice. 1986. Attention to Action: Willed and Automatic Control of Behaviour. In: Consciousness and Self-Regulation: Advances in Research and Theory, ed. R. J. Davidson et al., pp. 1–18, vol. 4. New York: Plenum Press. Also available over Springer Book Archive at https://doi.org/10.1007/978-1-4757-0629-1 (accessed 29.3.24). [9, 14]

Novick, A. M., M. Mears, G. L. Forster, et al. 2016. Adolescent Social Defeat Alters N-Methyl-D-Aspartic Acid Receptor Expression and Impairs Fear Learning in Adulthood. *Behav. Brain Res.* **304**:51–59. [16]

Nuttin, B., P. Cosyns, H. Demeulemeester, J. Gybels, and B. Meyerson. 1999. Electrical Stimulation in Anterior Limbs of Internal Capsules in Patients with Obsessive-Compulsive Disorder. *Lancet* **354**:1526. [15]

Nyman, H., S. Andréewitch, E. Lundbäck, and P. Mindus. 2001. Executive and Cognitive Functions in Patients with Extreme Obsessive-Compulsive Disorder Treated by Capsulotomy. *Appl Neuropsychol* **8**:91–98. [15]

Nyman, H., and P. Mindus. 1995. Neuropsychological Correlates of Intractable Anxiety Disorder before and after Capsulotomy. *Acta Psychiatr Scand* **91**:23–31. [15]

O'Callaghan, C., M. Bertoux, M. Irish, et al. 2016. Fair Play: Social Norm Compliance Failures in Behavioural Variant Frontotemporal Dementia. *Brain* **139**:204–216. [16]

O'Doherty, J. P. 2007. Lights, Camembert, Action! The Role of Human Orbitofrontal Cortex in Encoding Stimuli, Rewards, and Choices. *Ann. N. Y. Acad. Sci.* **1121**:254–272. [1]

O'Doherty, J. P., S. W. Lee, R. Tadayonnejad, et al. 2021. Why and How the Brain Weights Contributions from a Mixture of Experts. *Neurosci. Biobehav. Rev.* **123**:14–23. [12]

Oguchi, M., J. Jiasen, T. W. Yoshioka, et al. 2021a. Microendoscopic Calcium Imaging of the Primary Visual Cortex of Behaving Macaques. *Sci. Rep.* **11**:17021. [4]

Oguchi, M., S. Tanaka, X. Pan, et al. 2021b. Chemogenetic Inactivation Reveals the Inhibitory Control Function of the Prefronto-Striatal Pathway in the Macaque Brain. *Commun. Biol.* **4**:1088. [2, 4]

Ohira, K., R. Takeuchi, T. Iwanaga, and T. Miyakawa. 2013. Chronic Fluoxetine Treatment Reduces Parvalbumin Expression and Perineuronal Nets in Gamma-Aminobutyric Acidergic Interneurons of the Frontal Cortex in Adult Mice. *Mol. Brain* **6**:43. [13]

Okazawa, G., C. E. Hatch, A. Mancoo, C. K. Machens, and R. Kiani. 2021. Representational Geometry of Perceptual Decisions in the Monkey Parietal Cortex. *Cell* **184**:3748–3761. [6]

Olivers, C. N., J. Peters, R. Houtkamp, and P. R. Roelfsema. 2011. Different States in Visual Working Memory: When It Guides Attention and When It Does Not. *Trends Cogn. Sci.* **15**:327–334. [12]

O'Neill, M., and W. Schultz. 2010. Coding of Reward Risk by Orbitofrontal Neurons Is Mostly Distinct from Coding of Reward Value. *Neuron* **68**:789–800. [5]

Öngür, D., A. T. Ferry, and J. L. Price. 2003. Architectonic Subdivision of the Human Orbital and Medial Prefrontal Cortex. *J. Comp. Neurol.* **460**:425–449. [4, 8]

Öngür, D., and J. L. Price. 2000. The Organization of Networks within the Orbital and Medial Prefrontal Cortex of Rats, Monkeys and Humans. *Cereb. Cortex* **10**:206–219. [4, 13, 16]

Ophey, A., M. Roheger, A.-K. Folkerts, N. Skoetz, and E. Kalbe. 2020. A Systematic Review on Predictors of Working Memory Training Responsiveness in Healthy Older Adults: Methodological Challenges and Future Directions. *Front. Aging Neurosci.* **12**: [14]

Opler, L. A., M. G. A. Opler, and A. F. T. Arnsten. 2016. Ameliorating Treatment-Refractory Depression with Intranasal Ketamine: Potential NMDA Receptor Actions in the Pain Circuitry Representing Mental Anguish. *CNS Spectr.* **21**:12–22. [13]

Oppenheimer, S. M., A. Gelb, J. P. Girvin, and V. C. Hachinski. 1992. Cardiovascular Effects of Human Insular Cortex Stimulation. *Neurology* **42**:1727–1732. [8]

Ordaz, S. J., W. Foran, K. Velanova, and B. Luna. 2013. Longitudinal Growth Curves of Brain Function Underlying Inhibitory Control through Adolescence. *J. Neurosci.* **33**:18109–18124. [16]

O'Reilly, R. C. 2010. The What and How of Prefrontal Cortical Organization. *Trends Neurosci.* **33**:355–361. [5, 8]

O'Reilly, R. C., and M. J. Frank. 2006. Making Working Memory Work: A Computational Model of Learning in the Prefrontal Cortex and Basal Ganglia. *Neural Comput.* **18**:283–328. [7, 12]

Orr, J. M., and M. T. Banich. 2014. The Neural Mechanisms Underlying Internally and Externally Guided Task Selection. *Neuroimage* **84**:191–205. [12]

Ørskov, P. T., A. Norup, E. L. Beatty, and S. M. Jaeggi. 2021. Exploring Individual Differences as Predictors of Performance Change During Dual-N-Back Training. *J. Cogn. Enhanc.* **5**:480–498. [14]

Ostlund, S. B., and B. W. Balleine. 2007. Orbitofrontal Cortex Mediates Outcome Encoding in Pavlovian but Not Instrumental Conditioning. *J. Neurosci.* **27**:4819–4825. [4]

Ott, T., S. N. Jacob, and A. Nieder. 2014. Dopamine Receptors Differentially Enhance Rule Coding in Primate Prefrontal Cortex Neurons. *Neuron* **84**:1317–1328. [6]

Owen, A. M., J. J. Downes, B. J. Sahakian, C. E. Polkey, and T. W. Robbins. 1990. Planning and Spatial Working Memory Following Frontal Lobe Lesions in Man. *Neuropsychologia* **28**:1021–1034. [12]

Oyama, K., Y. Hori, K. Mimura, et al. 2022. Chemogenetic Disconnection between the Orbitofrontal Cortex and the Rostromedial Caudate Nucleus Disrupts Motivational Control of Goal-Directed Action. *J. Neurosci.* **42**:6267–6275. [2, 4]

Packard, M. G., and B. J. Knowlton. 2002. Learning and Memory Functions of the Basal Ganglia. *Annu. Rev. Neurosci.* **25**:563–593. [14]

Padmanabhan, J. L., D. Cooke, J. Joutsa, et al. 2019. A Human Depression Circuit Derived from Focal Brain Lesions. *Biol. Psych.* **86**:749–758. [11]

Padoa-Schioppa, C. 2009. Range-Adapting Representation of Economic Value in the Orbitofrontal Cortex. *J. Neurosci.* **29**:14004–14014. [4]

Padoa-Schioppa, C. 2011. Neurobiology of Economic Choice: A Good-Based Model. *Annu. Rev. Neurosci.* **34**:333–359. [8]

Padoa-Schioppa, C., and J. A. Assad. 2006. Neurons in the Orbitofrontal Cortex Encode Economic Value. *Nature* **441**:223–226. [5]

———. 2008. The Representation of Economic Value in the Orbitofrontal Cortex Is Invariant for Changes of Menu. *Nat. Neurosci.* **11**:95–102. [5]

Padoa-Schioppa, C., and K. E. Conen. 2017. Orbitofrontal Cortex: A Neural Circuit for Economic Decisions. *Neuron* **96**:736–754. [5]

Pahor, A., S. M. Jaeggi, and A. R. Seitz. 2018. Brain Training. In: Encyclopedia of Life Sciences, pp. 1–9. Online Library: John Wiley & Sons. [14]

Pahor, A., A. R. Seitz, and S. M. Jaeggi. 2022. Near Transfer to an Unrelated N-Back Task Mediates the Effect of N-Back Working Memory Training on Matrix Reasoning. *Nat. Hum. Behav.* 1–14. [14]

Palanca, B. J. A., A. Mitra, L. Larson-Prior, et al. 2015. Resting-State Functional Magnetic Resonance Imaging Correlates of Sevoflurane-Induced Unconsciousness. *Anesthesiology* **123**:346–356. [11]

Pallier, C., A.-D. Devauchelle, and S. Dehaene. 2011. Cortical Representation of the Constituent Structure of Sentences. *PNAS* **108**:2522–2527. [12]

Palomero-Gallagher, N., H. J. Bidmon, M. Cremer, et al. 2009a. Neurotransmitter Receptor Imbalances in Motor Cortex and Basal Ganglia in Hepatic Encephalopathy. *Cell Physiol. Biochem.* **24**:291–306. [4]

Palomero-Gallagher, N., S. B. Eickhoff, F. Hoffstaedter, et al. 2015. Functional Organization of Human Subgenual Cortical Areas: Relationship between Architectonical Segregation and Connectional Heterogeneity. *Neuroimage* **115**:177–190. [3]

Palomero-Gallagher, N., B. A. Vogt, A. Schleicher, H. S. Mayberg, and K. Zilles. 2009b. Receptor Architecture of Human Cingulate Cortex: Evaluation of the Four-Region Neurobiological Model. *Hum. Brain Mapp.* **30**:2336–2355. [13]

Palomero-Gallagher, N., and K. Zilles. 2018. Cyto- and Receptor Architectonic Mapping of the Human Brain. *Handb. Clin. Neurol.* **150**:355–387. [4]

Pandya, D. N., and B. Seltzer. 1982. Intrinsic Connections and Architectonics of Posterior Parietal Cortex in the Rhesus Monkey. *J. Comp. Neurol.* **204**:196–210. [8]

Panichello, M. F., and T. J. Buschman. 2021. Shared Mechanisms Underlie the Control of Working Memory and Attention. *Nature* **592**:601–605. [12]

Park, D. C., and P. Reuter-Lorenz. 2009. The Adaptive Brain: Aging and Neurocognitive Scaffolding. *Annu. Rev. Psychol.* **60**:173–196. [14]

Parker, B. J., W. I. Voorhies, G. Jiahui, et al. 2023. Hominoid-Specific Sulcal Variability Is Related to Face Perception Ability. *Brain Struct. Funct.* **228**:677–685. [4]

Parker, N. F., A. Baidya, J. Cox, et al. 2022. Choice-Selective Sequences Dominate in Cortical Relative to Thalamic Inputs to Nac to Support Reinforcement Learning. *Cell Rep.* **39**:110756. [8]

Parong, J., A. R. Seitz, S. M. Jaeggi, and C. S. Green. 2022. Expectation Effects in Working Memory Training. *PNAS* **119**:e2209308119. [14]

Parr, A. C., F. Calabro, B. Larsen, et al. 2021. Dopamine-Related Striatal Neurophysiology Is Associated with Specialization of Frontostriatal Reward Circuitry through Adolescence. *Prog. Neurobiol.* **201**:101997. [16]

Parr, A. C., F. Calabro, B. Tervo-Clemmens, et al. 2022. Contributions of Dopamine-Related Basal Ganglia Neurophysiology to the Developmental Effects of Incentives on Inhibitory Control. *Dev. Cogn. Neurosci.* **54**:101100. [17]

Parro, C., M. L. Dixon, and K. Christoff. 2018. The Neural Basis of Motivational Influences on Cognitive Control. *Hum. Brain Mapp.* **39**:5097–5111. [12]

Pasqualotto, A., I. Altarelli, A. De Angeli, et al. 2022. Enhancing Reading Skills through a Video Game Mixing Action Mechanics and Cognitive Training. *Nat. Hum. Behav.* **6**:545–554. [14]

Passamonti, L., C. J. Lansdall, and J. B. Rowe. 2018. The Neuroanatomical and Neurochemical Basis of Apathy and Impulsivity in Frontotemporal Lobar Degeneration. *Curr. Opin. Behav. Sci.* **22**:14–20. [16]

Passingham, R. E. 1975. Changes in the Size and Organisation of the Brain in Man and His Ancestors. *Brain Behav. Evol.* **11**:73–90. [4]

———. 2021. Understanding the Prefrontal Cortex: Selective Advantage, Connectivity, and Neural Operations. Oxford: Oxford Univ. Press. [8]

Passingham, R. E., and J. B. Smaers. 2014. Is the Prefrontal Cortex Especially Enlarged in the Human Brain Allometric Relations and Remapping Factors. *Brain Behav. Evol.* **84**:156–166. [4]

Passingham, R. E., J. B. Smaers, and C. C. Sherwood. 2017. Evolutionary Specializations of the Human Prefrontal Cortex. In: Evolution of Nervous Systems: Second Edition, ed. J. H. Kaas. Amsterdam: Elsevier. [4]

Passingham, R. E., and S. P. Wise. 2012. The Neurobiology of the Prefrontal Cortex. Oxford: Oxford Univ. Press. [8]

Pasupathy, A., and E. K. Miller. 2005. Different Time Courses of Learning-Related Activity in the Prefrontal Cortex and Striatum. *Nature* **433**:873–876. [8]

Paton, J. J., M. A. Belova, S. E. Morrison, and C. D. Salzman. 2006. The Primate Amygdala Represents the Positive and Negative Value of Visual Stimuli during Learning. *Nature* **439**:865–870. [8]

Patti, M. A., and V. Troiani. 2017. Orbitofrontal Sulcogyral Morphology Is a Transdiagnostic Indicator of Brain Dysfunction. *Neuroimage Clin.* **17**:910–917. [4]

Paus, T. 1996. Location and Function of the Human Frontal Eye-Field: A Selective Review. *Neuropsychologia* **34**:475–483. [4]

Paus, T., M. Keshavan, and J. N. Giedd. 2008. Why Do Many Psychiatric Disorders Emerge during Adolescence? *Nat. Rev. Neurosci.* **9**:947–957. [16]

Paxinos, G., C. Watson, M. Petrides, R. Marcello, and H. Tokuno. 2012. The Marmoset Brain in Stereotaxic Coordinates. San Diego: Elsevier. [4]

Payzan-LeNestour, E., and P. Bossaerts. 2011. Risk, Unexpected Uncertainty, and Estimation Uncertainty: Bayesian Learning in Unstable Settings. *PLoS Comput. Biol.* **7**:e1001048. [10]

Penfield, W. 1954. Mechanisms of Voluntary Movement. *Brain* **77**:1–17. [8]

———. 1965. Conditioning the Uncommitted Cortex for Language Learning. *Brain* **88**:787–798. [5]

Pereyra, A. E., C. J. Mininni, and B. S. Zanutto. 2021. Information Capacity and Robustness of Encoding in the Medial Prefrontal Cortex Are Modulated by the Bioavailability of Serotonin and the Time Elapsed from the Cue during a Reward-Driven Task. *Sci. Rep.* **11**:1–13. [13]

Pergher, V., B. Schoenmakers, P. Demaerel, J. Tournoy, and M. M. Van Hulle. 2020a. Differential Impact of Cognitive Impairment in MCI Patients: A Case-Based Report. *Case Rep. Neurol.* **12**:222–231. [16]

Pergher, V., M. A. Shalchy, A. Pahor, et al. 2020b. Divergent Research Methods Limit Understanding of Working Memory Training. *J. Cogn. Enhanc.* **4**:100–120. [14, 16]

Perica, M. I., F. J. Calabro, B. Larsen, et al. 2022. Development of Frontal GABA and Glutamate Supports Excitation/Inhibition Balance from Adolescence into Adulthood. *Prog. Neurobiol.* **219**:102370. [16]

Perich, M. G., J. A. Gallego, and L. E. Miller. 2018. A Neural Population Mechanism for Rapid Learning. *Neuron* **100**:964–976. [4]

Perich, M. G., and K. Rajan. 2020. Rethinking Brain-Wide Interactions through Multi-Region 'Network of Networks' Models. *Curr. Opin. Neurobiol.* **65**:146–151. [12]

Perkes, I. E., R. W. Morris, K. R. Griffiths, et al. 2023. The Motivational Determinants of Human Action, Their Neural Bases and Functional Impact in Adolescents with Obsessive-Compulsive Disorder. *Biol. Psychiatry Glob. Open Sci.* **3**:1062–1072. [4]

Perry, B. A. L., E. Lomi, and A. S. Mitchell. 2021. Thalamocortical Interactions in Cognition and Disease: The Mediodorsal and Anterior Thalamic Nuclei. *Neurosci. Biobehav. Rev.* **130**:162–177. [8]

Petanjek, Z., I. Banovac, D. Sedmak, and A. Hladnik. 2023. Dendritic Spines: Synaptogenesis and Synaptic Pruning for the Developmental Organization of Brain Circuits. *Adv. Neurobiol.* **34**:143–221. [16]

Petanjek, Z., M. Judas, G. Simic, et al. 2011. Extraordinary Neoteny of Synaptic Spines in the Human Prefrontal Cortex. *PNAS* **108**:13281–13286. [16]

Petersen, S. E., and O. Sporns. 2015. Brain Networks and Cognitive Architectures. *Neuron* **88**:207–219. [11]

Petrides, M. 1991. Functional Specialization within the Dorsolateral Frontal Cortex for Serial Order Memory. *Proc. Biol. Sci.* **246**:299–306. [8]

———. 1994. Frontal Lobes and Behaviour. *Curr. Opin. Neurobiol.* **4**:207–211. [4, 5, 11]

———. 1996. Specialized Systems for the Processing of Mnemonic Information within the Primate Frontal Cortex. *Philos. Trans. R. Soc. Lond. B Biol. Sci.* **351**:1455–1462. [4]

———. 2005. Lateral Prefrontal Cortex: Architectonic and Functional Organization. *Philos. Trans. R. Soc. Lond. B Biol. Sci.* **360**:781795. [4, 6]

———. 2019. Atlas of the Morphology of the Human Cerebral Cortex on the Average MNI brain. London: Elsevier. [4]

Petrides, M., B. Alivisatos, and S. Frey. 2002. Differential Activation of the Human Orbital, Mid-Ventrolateral, and Mid-Dorsolateral Prefrontal Cortex during the Processing of Visual Stimuli. *PNAS* **99**:5649–5654. [4]

Petrides, M., G. Cadoret, and S. Mackey. 2005. Orofacial Somatomotor Responses in the Macaque Monkey Homologue of Broca's Area. *Nature* **435**:1235–1238. [4]

Petrides, M., and D. N. Pandya. 1999. Dorsolateral Prefrontal Cortex: Comparative Cytoarchitectonic Analysis in the Human and the Macaque Brain and Corticocortical Connection Patterns. *Eur. J. Neurosci.* **11**:1011–1036. [8]

———. 2002. Comparative Cytoarchitectonic Analysis of the Human and the Macaque Ventrolateral Prefrontal Cortex and Corticocortical Connection Patterns in the Monkey. *Eur. J. Neurosci.* **16**:291–310. [4, 8]

———. 2007. Efferent Association Pathways from the Rostral Prefrontal Cortex in the Macaque Monkey. *J. Neurosci.* **27**:11573–11586. [8]

Petrides, M., F. Tomaiuolo, E. H. Yeterian, and D. N. Pandya. 2012. The Prefrontal Cortex: Comparative Architectonic Organization in the Human and the Macaque Monkey Brains. *Cortex* **48**:46–57. [13]

Philipp, R., L. Kriston, J. Lanio, et al. 2019. Effectiveness of Metacognitive Interventions for Mental Disorders in Adults—a Systematic Review and Meta-Analysis (Metacog). *Clin. Psych. Psychother.* **26**:227–240. [14]

Phillips, J. M., and S. Everling. 2014. Event-Related Potentials Associated with Performance Monitoring in Non-Human Primates. *Neuroimage* **97**:308–320. [12]

Phillips, J. M., N. A. Kambi, M. J. Redinbaugh, S. Mohanta, and Y. B. Saalmann. 2021. Disentangling the Influences of Multiple Thalamic Nuclei on Prefrontal Cortex and Cognitive Control. *Neurosci. Biobehav. Rev.* **128**:487–510. [7]

Phoumthipphavong, V., F. Barthas, S. Hassett, and A. C. Kwan. 2016. Longitudinal Effects of Ketamine on Dendritic Architecture *in Vivo* in the Mouse Medial Frontal Cortex. *eNeuro* **3**: [13]

Piantadosi, S. 2023. Modern Language Models Refute Chomsky's Approach to Language. University of Tromsø. file:///C:/Users/stephen/Downloads/piantadosi_23_Modern-lang.7.pdf. (accessed Feb. 28, 2024). [12]

Picard, N., and P. L. Strick. 1996. Motor Areas of the Medial Wall: A Review of Their Location and Functional Activation. *Cereb. Cortex* **6**:342–353. [16]

Piette, C. E., M. A. Baez-Santiago, E. E. Reid, D. B. Katz, and A. Moran. 2012. Inactivation of Basolateral Amygdala Specifically Eliminates Palatability-Related Information in Cortical Sensory Responses. *J. Neurosci.* **32**:9981–9991. [8]

Pinto, L., and Y. Dan. 2015. Cell-Type-Specific Activity in Prefrontal Cortex during Goal-Directed Behavior. *Neuron* **87**:437–450. [2]

Pitts, M., and D. E. Nee. 2022. Generalizing the Control Architecture of the Lateral Prefrontal Cortex. *Neurobiol. Learn. Mem.* **195**:107688. [7]

Pizzagalli, D. A. 2014. Depression, Stress, and Anhedonia: Toward a Synthesis and Integrated Model. In: Annual Review of Clinical Psychology, Vol 10, ed. T. D. Cannon and T. Widiger, pp. 393–423, Annual Review of Clinical Psychology, vol. 10. [13]

Pizzagalli, D. A., and A. C. Roberts. 2022. Prefrontal Cortex and Depression. *Neuropsychopharmacol.* **47**:225–246. [12, 13]

Pizzagalli, D. A., C. A. Webb, D. G. Dillon, et al. 2018. Pretreatment Rostral Anterior Cingulate Cortex Theta Activity in Relation to Symptom Improvement in Depression: A Randomized Clinical Trial. *JAMA Psych.* **75**:547–554. [13]

Platt, M., and C. Padoa-Schioppa. 2008. Neuronal Representations of Value. In: Neuroeconomics: Decision Making and the Brain, ed. P. W. Glimcher et al., p. 441. Amsterdam: Elsevier/Academic Press. [12]

Poldrack, R. A., T. O. Laumann, O. Koyejo, et al. 2015. Long-Term Neural and Physiological Phenotyping of a Single Human. *Nat. Commun.* **6**:8885. [12]

Poldrack, R. A., J. A. Mumford, T. Schonberg, et al. 2012. Discovering Relations between Mind, Brain, and Mental Disorders Using Topic Mapping. *PLoS Comput. Biol.* **8**:e1002707. [7]

Pollak, S. D., C. A. Nelson, M. F. Schlaak, et al. 2010. Neurodevelopmental Effects of Early Deprivation in Postinstitutionalized Children. *Child Dev.* **81**:224–236. [16]

Polyn, S. M., K. A. Norman, and M. J. Kahana. 2009. A Context Maintenance and Retrieval Model of Organizational Processes in Free Recall. *Psychol. Rev.* **116**:129–156. [12]

Pomberger, T., J. Loschner, and S. R. Hage. 2020. Compensatory Mechanisms Affect Sensorimotor Integration during Ongoing Vocal Motor Acts in Marmoset Monkeys. *Eur. J. Neurosci.* **52**:3531–3544. [16]

Pontzer, H., M. H. Brown, D. A. Raichlen, et al. 2016. Metabolic Acceleration and the Evolution of Human Brain Size and Life History. *Nature* **533**:390–392. [2]

Pope, M., M. Fukushima, R. F. Betzel, and O. Sporns. 2021. Modular Origins of High-Amplitude Cofluctuations in Fine-Scale Functional Connectivity Dynamics. *PNAS* **118**:e2109380118. [12]

Porrino, L. J., A. M. Crane, and P. S. Goldman-Rakic. 1981. Direct and Indirect Pathways from the Amygdala to the Frontal Lobe in Rhesus Monkeys. *J. Comp. Neurol.* **198**:121–136. [8]

Povysheva, N. V., A. V. Zaitsev, D. C. Rotaru, et al. 2008. Parvalbumin-Positive Basket Interneurons in Monkey and Rat Prefrontal Cortex. *J. Neurophysiol.* **100**:2348–2360. [2]

Power, J. D., A. L. Cohen, S. M. Nelson, et al. 2011. Functional Network Organization of the Human Brain. *Neuron* **72**:665–678. [7, 11]

Power, J. D., B. L. Schlaggar, and S. E. Petersen. 2015. Recent Progress and Outstanding Issues in Motion Correction in Resting State fMRI. *Neuroimage* **105**:536–551. [11]

Preuss, T. M. 1995. Do Rats Have Prefrontal Cortex? The Rose-Woolsey-Akert Program Reconsidered. *J. Cog. Neuro.* **7**:1–24. [2, 4]

Preuss, T. M., and S. P. Wise. 2022. Evolution of Prefrontal Cortex. *Neuropsychopharmacol.* **47**:3–19. [3, 4, 8]

Prévost, C., M. Liljeholm, J. M. Tyszka, and J. P. O'Doherty. 2012. Neural Correlates of Specific and General Pavlovian-to-Instrumental Transfer within Human Amygdalar Subregions: A High-Resolution fMRI Study. *J. Neurosci.* **32**:8383–8390. [4]

Price, J. L. 2007. Definition of the Orbital Cortex in Relation to Specific Connections with Limbic and Visceral Structures and Other Cortical Regions. *Ann. N. Y. Acad. Sci.* **1121**:54–71. [4]

Price, R. B., K. Gates, T. E. Kraynak, M. E. Thase, and G. J. Siegle. 2017. Data-Driven Subgroups in Depression Derived from Directed Functional Connectivity Paths at Rest. *Neuropsychopharmacol.* **42**:2623–2632. [13]

Price, R. B., C. M. Gillan, C. Hanlon, et al. 2021. Effect of Experimental Manipulation of the Orbitofrontal Cortex on Short-Term Markers of Compulsive Behavior: A Theta Burst Stimulation Study. *Am. J. Psych.* **178**:459–468. [16]

Priebe, N. J., and D. Ferster. 2012. Mechanisms of Neuronal Computation in Mammalian Visual Cortex. *Neuron* **75**:194–208. [4]

Prosperini, L., and M. Di Filippo. 2019. Beyond Clinical Changes: Rehabilitation-Induced Neuroplasticity in MS. *Mult. Scler.* **25**:1348–1362. [14]

Pucak, M. L., J. B. Levitt, J. S. Lund, and D. A. Lewis. 1996. Patterns of Intrinsic and Associational Circuitry in Monkey Prefrontal Cortex. *Journal of Comparative Neurology* **376**:614–630. [6]

Puelles, L., J. E. Sandoval, A. Ayad, et al. 2017. The Pallium in Reptiles and Birds in the Light of the Updated Tetrapartite Pallium Model. In: Evolution of Nervous Systems, ed. J. H. Kaas. Oxford: Elsevier. [5]

Puig, M. V., and E. K. Miller. 2012. The Role of Prefrontal Dopamine D1 Receptors in the Neural Mechanisms of Associative Learning. *Neuron* **74**:874–886. [6]

Pujara, M. S., N. K. Ciesinski, J. F. Reyelts, S. E. V. Rhodes, and E. A. Murray. 2022. Selective Prefrontal-Amygdala Circuit Interactions Underlie Social and Nonsocial Valuation in Rhesus Macaques. *J. Neurosci.* **42**:5593–5604. [8]

Putnam, P. T., C. J. Chu, N. A. Fagan, O. Dal Monte, and S. W. C. Chang. 2023. Dissociation of Vicarious and Experienced Rewards by Coupling Frequency within the Same Neural Pathway. *Neuron* **111**:2513–2522. [8]

Pyke, G. H. 1984. Optimal Foraging Theory: A Critical Review. *Ann. Rev. Ecol. Syst.* **15**:523–575. [2]

Pyke, G. H., H. R. Pulliam, and E. L. Charnov. 1977. Optimal Foraging: A Selective Review of Theory and Tests. *Q. Rev. Biol.* **52**:137–154. [2]

Rabut, C., S. Yoo, R. C. Hurt, et al. 2020. Ultrasound Technologies for Imaging and Modulating Neural Activity. *Neuron* **108**:93–110. [3]

Rac-Lubashevsky, R., and M. J. Frank. 2021. Analogous Computations in Working Memory Input, Output and Motor Gating: Electrophysiological and Computational Modeling Evidence. *PLoS Comput. Biol.* **17**:e1008971. [7]

Radhiyanti, P. T., A. Konno, Y. Matsuzaki, and H. Hirai. 2021. Comparative Study of Neuron-Specific Promoters in Mouse Brain Transduced by Intravenously Administered AAV-PHP.eB. *Neurosci. Lett.* **756**:135956. [2]

Radley, J. J., C. M. Arias, and P. E. Sawchenko. 2006. Regional Differentiation of the Medial Prefrontal Cortex in Regulating Adaptive Responses to Acute Emotional Stress. *J. Neurosci.* **26**:12967–12976. [16]

Rae, C. L., C. Nombela, P. V. Rodriguez, et al. 2016. Atomoxetine Restores the Response Inhibition Network in Parkinson's Disease. *Brain* **139**:2235–2248. [16]

Rahman, S., B. J. Sahakian, J. R. Hodges, R. D. Rogers, and T. W. Robbins. 1999. Specific Cognitive Deficits in Mild Frontal Variant Frontotemporal Dementia. *Brain* **122 (Pt 8)**:1469–1493. [4]

Rahmati, M., K. DeSimone, C. E. Curtis, and K. K. Sreenivasan. 2020. Spatially Specific Working Memory Activity in the Human Superior Colliculus. *J. Neurosci.* **40**:9487–9495. [4]

Raichle, M. E. 2015. The Brain's Default Mode Network. *Annu. Rev. Neurosci.* **38**:433–447. [4]

Rainer, G., W. F. Asaad, and E. K. Miller. 1998. Memory Fields of Neurons in the Primate Prefrontal Cortex. *PNAS* **95**:15008–15013. [8]

Ramnani, N., and A. M. Owen. 2004. Anterior Prefrontal Cortex: Insights into Function from Anatomy and Neuroimaging. *Nat. Rev. Neurosci.* **5**:184–194. [6, 15]

Ranti, C., C. H. Chatham, and D. Badre. 2015. Parallel Temporal Dynamics in Hierarchical Cognitive Control. *Cognition* **142**:205–229. [7]

Rapan, L., S. Froudist-Walsh, M. Niu, et al. 2021. Multimodal 3D Atlas of the Macaque Monkey Motor and Premotor Cortex. *Neuroimage* **226**:117574. [16]

Rapan et al. 2021 [16] See Chapter 16, Figure 16.3. Not in reference list in final version of Manuscript

Rapan, L., S. Froudist-Walsh, M. Niu, et al. 2023. Cytoarchitectonic, Receptor Distribution and Functional Connectivity Analyses of the Macaque Frontal Lobe. *eLife* **12**:e82850. [3, 4, 8, 16]

Rasmussen, S. A., and J. L. Eisen. 1992. The Epidemiology and Clinical Features of Obsessive Compulsive Disorder. *Psychiatr. Clin. N. Am.* **15**:743–758. [15]

———. 1997. Treatment Strategies for Chronic and Refractory Obsessive-Compulsive Disorder. *J. Clin. Psych.* **58 Suppl 13**:9–13. [16]

Rasmussen, S. A., and W. K. Goodman. 2022. The Prefrontal Cortex and Neurosurgical Treatment for Intractable OCD. *Neuropsychopharmacol.* **47**:349–360. [16]

Rasmussen, S. A., G. Noren, B. D. Greenberg, et al. 2018. Gamma Ventral Capsulotomy in Intractable Obsessive-Compulsive Disorder. *Biol. Psych.* **84**:355–364. [15, 16]

Rasmussen, S. A., and M. T. Tsuang. 1986. Clinical Characteristics and Family History in DSM-III Obsessive-Compulsive Disorder. *Am. J. Psych.* **143**:317–322. [15]

Ray, J. P., and J. L. Price. 1992. The Organization of the Thalamocortical Connections of the Mediodorsal Thalamic Nucleus in the Rat, Related to the Ventral Forebrain-Prefrontal Cortex Topography. *J. Comp. Neurol.* **323**:167–197. [4]

———. 1993. The Organization of Projections from the Mediodorsal Nucleus of the Thalamus to Orbital and Medial Prefrontal Cortex in Macaque Monkeys. *J. Comp. Neurol.* **337**:1–31. [4, 8]

Reber, J., J. S. Feinstein, J. P. O'Doherty, et al. 2017. Selective Impairment of Goal-Directed Decision-Making Following Lesions to the Human Ventromedial Prefrontal Cortex. *Brain* **140**:1743–1756. [8]

Redish, D. A., and J. A. Gordon. 2017. Computational Psychiatry: New Perspectives on Mental Illness. Strüngmann Forum Reports, J. R. Lupp, series ed. Cambridge, MA: MIT Press. [12]

Reekie, Y. L., K. Braesicke, M. S. Man, and A. C. Roberts. 2008. Uncoupling of Behavioral and Autonomic Responses after Lesions of the Primate Orbitofrontal Cortex. *PNAS* **105**:9787–9792. [4]

Reineberg, A. E., M. T. Banich, T. D. Wager, and N. P. Friedman. 2022. Context-Specific Activations Are a Hallmark of the Neural Basis of Individual Differences in General Executive Function. *Neuroimage* **249**:118845. [9]

Reineberg, A. E., D. E. Gustavson, C. Benca, M. T. Banich, and N. P. Friedman. 2018. The Relationship between Resting State Network Connectivity and Individual Differences in Executive Functions. *Front. Psychol.* **9**:1600. [11]

Reinhart, R. M. G., and G. F. Woodman. 2014. Causal Control of Medial-Frontal Cortex Governs Electrophysiological and Behavioral Indices of Performance Monitoring and Learning. *J. Neurosci.* **34**:4214–4227. [12]

Remington, E. D., D. Narain, E. A. Hosseini, and M. Jazayeri. 2018. Flexible Sensorimotor Computations through Rapid Reconfiguration of Cortical Dynamics. *Neuron* **98**:1005–1019. [4]

Rempel-Clower, N. L., and H. Barbas. 2000. The Laminar Pattern of Connections between Prefrontal and Anterior Temporal Cortices in the Rhesus Monkey Is Related to Cortical Structure and Function. *Cereb. Cortex* **10**:851–865. [2]

Resendez, S. L., J. H. Jennings, R. L. Ung, et al. 2016. Visualization of Cortical, Subcortical and Deep Brain Neural Circuit Dynamics during Naturalistic Mammalian Behavior with Head-Mounted Microscopes and Chronically Implanted Lenses. *Nat. Protoc.* **11**:566–597. [4]

Reser, M. P., R. Slikboer, and S. L. Rossell. 2019. A Systematic Review of Factors That Influence the Efficacy of Cognitive Remediation Therapy in Schizophrenia. *Aust. N. Z. J. Psych.* **53**:624–641. [14]

Restrepo-Martinez, M., J. Ramirez-Bermudez, J. Chacon-Gonzalez, et al. 2023. Defining Repetitive Behaviours in Frontotemporal Dementia. *Brain* [16]

Reuter-Lorenz, P. A., and K. A. Cappell. 2008. Neurocognitive Aging and the Compensation Hypothesis. *Curr. Dir. Psychol. Sci.* **17**:177–182. [14]

Reuter-Lorenz, P. A., and A. D. Iordan. 2018. From Cognitive Tasks to Cognitive Theories and Back Again: Fitting Data to the Real World. *J. Appl. Res. Mem. Cogn.* **7**:510–513. [14]

Reuter-Lorenz, P. A., and D. C. Park. 2014. How Does It Stac Up? Revisiting the Scaffolding Theory of Aging and Cognition. *Neuropsychol. Rev.* **24**:355–370. [14]

Reynolds, J. H., L. Chelazzi, and R. Desimone. 1999. Competitive Mechanisms Subserve Attention in Macaque Areas V2 and V4. *J. Neurosci.* **19**:1736–1753. [12]

Reynolds, J. H., and D. J. Heeger. 2009. The Normalization Model of Attention. *Neuron* **61**:168–185. [12]

Reynolds, J. H., T. Pasternak, and R. Desimone. 2000. Attention Increases Sensitivity of V4 Neurons. *Neuron* **26**:703–714. [12]

Reynolds, J. R., R. C. O'Reilly, J. D. Cohen, and T. S. Braver. 2012. The Function and Organization of Lateral Prefrontal Cortex: A Test of Competing Hypotheses. *PloS One* **7**:e30284. [7]

Rich, E. L., F. M. Stoll, and P. H. Rudebeck. 2018. Linking Dynamic Patterns of Neural Activity in Orbitofrontal Cortex with Decision Making. *Curr. Opin. Neurobiol.* **49**:24–32. [5]

Rich, E. L., and J. D. Wallis. 2016. Decoding Subjective Decisions from Orbitofrontal Cortex. *Nat. Neurosci.* **19**:973–980. [5]

———. 2017. Spatiotemporal Dynamics of Information Encoding Revealed in Orbitofrontal High-Gamma. *Nat. Commun.* **8**:1139. [5]

Rigotti, M., O. Barak, M. R. Warden, et al. 2013. The Importance of Mixed Selectivity in Complex Cognitive Tasks. *Nature* **497**:585–590. [2, 4, 6, 8, 12]

Rigotti, M., D. Ben Dayan Rubin, X.-J. Wang, and S. Fusi. 2010. Internal Representation of Task Rules by Recurrent Dynamics: The Importance of the Diversity of Neural Responses. *Front. Comput. Neurosci.* **4**:24. [9, 10, 12]

Rikhye, R. V., A. Gilra, and M. M. Halassa. 2018. Thalamic Regulation of Switching between Cortical Representations Enables Cognitive Flexibility. *Nat. Neurosci.* **21**:1753–1763. [3, 4]

Riley, M. R., and C. Constantinidis. 2016. Role of Prefrontal Persistent Activity in Working Memory. *Front. Syst. Neurosci.* **9**:181. [6]

Riley, M. R., X. L. Qi, and C. Constantinidis. 2017. Functional Specialization of Areas Along the Anterior-Posterior Axis of the Primate Prefrontal Cortex. *Cereb. Cortex* **27**:3683–3697. [5, 6, 8]

Riley, M. R., X. L. Qi, X. Zhou, and C. Constantinidis. 2018. Anterior-Posterior Gradient of Plasticity in Primate Prefrontal Cortex. *Nat. Commun.* **9**:3790. [8]

Rios-Florez, J. A., R. R. M. Lima, P. Morais, et al. 2021. Medial Prefrontal Cortex (A32 and A25) Projections in the Common Marmoset: A Subcortical Anterograde Study. *Sci. Rep.* **11**:14565. [3]

Ripley, D. L., C. E. Morey, D. Gerber, et al. 2014. Atomoxetine for Attention Deficits Following Traumatic Brain Injury: Results from a Randomized Controlled Trial. *Brain Inj.* **28**:1514–1522. [16]

Ritz, H., and A. Shenhav. 2024. Orthogonal Neural Encoding of Targets and Distractors Supports Multivariate Cognitive Control. *Nat. Hum. Behav.*, in press. [11, 12]

Robbins, T. W. 1996. Dissociating Executive Functions of the Prefrontal Cortex. *PNAS* **351**:1463–1470; discussion 1470–1461. [2]

Robbins, T. W., and A. F. Arnsten. 2009. The Neuropsychopharmacology of Fronto-Executive Function: Monoaminergic Modulation. *Annu. Rev. Neurosci.* **32**:267–287. [16]

Robbins, T. W., M. James, A. M. Owen, et al. 1994. Cognitive Deficits in Progressive Supranuclear Palsy, Parkinson's Disease, and Multiple System Atrophy in Tests Sensitive to Frontal Lobe Dysfunction. *J. Neurol. Neurosurg. Psych.* **57**:79–88. [16]

Robbins, T. W., G. H. Jones, and L. S. Wilkinson. 1996. Behavioural and Neurochemical Effects of Early Social Deprivation in the Rat. *J. Psychopharmacol.* **10**:39–47. [16]

Robbins, T. W., M. M. Vaghi, and P. Banca. 2019. Obsessive-Compulsive Disorder: Puzzles and Prospects. *Neuron* **102**:27–47. [15]

Roberts, A. C. 2011. The Importance of Serotonin for Orbitofrontal Function. *Biol. Psych.* **69**:1185–1191. [16]

Roberts, A. C., T. W. Robbins, and L. Weiskrantz, eds. 1998. The Prefrontal Cortex: Executive and Cognitive Functions. Oxford Oxford Univ. Press. [1]

Roberts, A. C., D. L. Tomic, C. H. Parkinson, et al. 2007. Forebrain Connectivity of the Prefrontal Cortex in the Marmoset Monkey (Callithrix jacchus): an Anterograde and Retrograde Tract-Tracing Study. *J. Comp. Neurol.* **502**:86–112. [3]

Robertson, B., L. Wang, M. T. Diaz, et al. 2007. Effect of Bupropion Extended Release on Negative Emotion Processing in Major Depressive Disorder: A Pilot Functional Magnetic Resonance Imaging Study. *J. Clin. Psychiatry* **68**:261–267. [13]

Robinson, D. A., and A. F. Fuchs. 1969. Eye Movements Evoked by Stimulation of Frontal Eye Fields. *J. Neurophysiol.* **32**:637–648. [8]

Robinson, E. S., D. M. Eagle, A. C. Mar, et al. 2008. Similar Effects of the Selective Noradrenaline Reuptake Inhibitor Atomoxetine on Three Distinct Forms of Impulsivity in the Rat. *Neuropsychopharmacol.* **33**:1028–1037. [16]

Robinson, M. F. 1946. What Price Lobotomy? *J. Abnorm. Psychol.* **41**:421–436. [15]

Rockland, K. S. 2022. Notes on Visual Cortical Feedback and Feedforward Connections. *Front. Syst. Neurosci.* **16**:784310. [8]

Roesch, M. R., and C. R. Olson. 2003. Impact of Expected Reward on Neuronal Activity in Prefrontal Cortex, Frontal and Supplementary Eye Fields and Premotor Cortex. *J. Neurophysiol.* **90**:1766–1789. [5, 8]

Roesch, M. R., A. R. Taylor, and G. Schoenbaum. 2006. Encoding of Time-Discounted Rewards in Orbitofrontal Cortex Is Independent of Value Representation. *Neuron* **51**:509–520. [4]

Rogers, J. C., and S. A. De Brito. 2016. Cortical and Subcortical Gray Matter Volume in Youths with Conduct Problems: A Meta-Analysis. *JAMA Psych.* **73**:64–72. [4]

Rogers, R. D., T. C. Andrews, P. M. Grasby, D. J. Brooks, and T. W. Robbins. 2000. Contrasting Cortical and Subcortical Activations Produced by Attentional-Set Shifting and Reversal Learning in Humans. *J. Cog. Neuro.* **12**:142–162. [4]

Roitman, J. D., and M. N. Shadlen. 2002. Response of Neurons in the Lateral Intraparietal Area during a Combined Visual Discrimination Reaction Time Task. *J. Neurosci.* **22**:9475–9489. [10]

Rollins, C. P. E., J. R. Garrison, M. Arribas, et al. 2020. Evidence in Cortical Folding Patterns for Prenatal Predispositions to Hallucinations in Schizophrenia. *Transl. Psych.* **10**:387. [4]

Rolls, E. T., F. Grabenhorst, and L. Franco. 2009. Prediction of Subjective Affective State from Brain Activations. *J. Neurophysiol.* **101**:1294–1308. [12]

Romo, R., C. D. Brody, A. Hernández, and L. Lemus. 1999. Neuronal Correlates of Parametric Working Memory in the Prefrontal Cortex. *Nature* **399**:470–473. [12]

Room, P., F. T. Russchen, H. J. Groenewegen, and A. H. M. Lohman. 1985. Efferent Connections of the Prelimbic (Area 32) and the Infralimbic (Area 25) Cortices: An Anterograde Tracing Study in the Cat. *J. Comp. Neurol.* **242**:40–55. [4]

Rose, D., G. Thornicroft, V. Pinfold, and A. Kassam. 2007. 250 Labels Used to Stigmatise People with Mental Illness. *BMC Health Serv. Res.* **7**:97. [16]

Rose, J. E., and C. N. Woolsey. 1948. The Orbitofrontal Cortex and Its Connections with the Mediodorsal Nucleus in Rabbit, Sheep and Cat. *Res. Publ. Assoc. Res. Nerv. Ment. Dis.* **27 (1 vol.)**:210–232. [2]

Roseman, L., L. Demetriou, M. B. Wall, D. J. Nutt, and R. L. Carhart-Harris. 2018. Increased Amygdala Responses to Emotional Faces after Psilocybin for Treatment-Resistant Depression. *Neuropharmacol.* **142**:263–269. [13]

Roseboom, P. H., S. A. L. Mueller, J. A. Oler, et al. 2021. Evidence in Primates Supporting the Use of Chemogenetics for the Treatment of Human Refractory Neuropsychiatric Disorders. *Mol. Ther.* **29**:3484–3497. [2]

Rosenbaum, J. F., J. Biederman, M. Gersten, et al. 1988. Behavioral Inhibition in Children of Parents with Panic Disorder and Agoraphobia. A Controlled Study. *Arch. Gen. Psych.* **45**:463–470. [15]

Rosenberg, A., T. Ngandu, M. Rusanen, et al. 2018. Multidomain Lifestyle Intervention Benefits a Large Elderly Population at Risk for Cognitive Decline and Dementia Regardless of Baseline Characteristics: The Finger Trial. *Alzheimers Dement.* **14**:263–270. [14]

Rosenberg, D. R., F. P. MacMaster, Y. Mirza, et al. 2005. Reduced Anterior Cingulate Glutamate in Pediatric Major Depression: A Magnetic Resonance Spectroscopy Study. *Biol. Psychiatry* **58**:700–704. [13]

Ross, S., A. Bossis, J. Guss, et al. 2016. Rapid and Sustained Symptom Reduction Following Psilocybin Treatment for Anxiety and Depression in Patients with Life-Threatening Cancer: A Randomized Controlled Trial. *J. Psychopharmacol.* **30**:1165–1180. [13]

Roth, B. L. 2016. DREADDs for Neuroscientists. *Neuron* **89**:683–694. [3, 4]

Rouault, M., J. Drugowitsch, and E. Koechlin. 2019. Prefrontal Mechanisms Combining Rewards and Beliefs in Human Decision-Making. *Nat. Commun.* **10**:301. [10, 15]

Rouault, M., and E. Koechlin. 2018. Prefrontal Function and Cognitive Control: From Action to Language. *Curr. Opin. Behav. Sci.* **21**:106–111. [10]

Rowe, J. B., I. Toni, O. Josephs, R. S. Frackowiak, and R. E. Passingham. 2000. The Prefrontal Cortex: Response Selection or Maintenance within Working Memory? *Science* **288**:1656–1660. [4]

Rowland, L. M., J. R. Bustillo, P. G. Mullins, et al. 2005. Effects of Ketamine on Anterior Cingulate Glutamate Metabolism in Healthy Humans: A 4-T Proton MRS Study. *Am. J. Psych.* **162**:394–396. [13].

Rück, C., S. Andréewitch, K. Flyckt, et al. 2003. Capsulotomy for Refractory Anxiety Disorders: Long-Term Follow-up of 26 Patients. *Am. J. Psych.* **160**:513–521. [15]

Rück, C., A. Karlsson, J. D. Steele, et al. 2008. Capsulotomy for Obsessive-Compulsive Disorder: Long-Term Follow-up of 25 Patients. *Arch. Gen. Psych.* **65**:914–921. [15]

Rudebeck, P. H., D. M. Bannerman, and M. F. Rushworth. 2008a. The Contribution of Distinct Subregions of the Ventromedial Frontal Cortex to Emotion, Social Behavior, and Decision Making. *Cogn. Affect. Behav. Neurosci.* **8**:485–497. [16]

Rudebeck, P. H., T. E. Behrens, S. W. Kennerley, et al. 2008b. Frontal Cortex Subregions Play Distinct Roles in Choices between Actions and Stimuli. *J. Neurosci.* **28**:13775–13785. [2, 8, 12, 16]

Rudebeck, P. H., M. J. Buckley, M. E. Walton, and M. F. S. Rushworth. 2006. A Role for the Macaque Anterior Cingulate Gyrus in Social Valuation. *Science* **313**:1310–1312. [4]

Rudebeck, P. H., and A. Izquierdo. 2022. Foraging with the Frontal Cortex: A Cross-Species Evaluation of Reward-Guided Behavior. *Neuropsychopharmacol.* **47**:134–146. [1, 2, 4]

Rudebeck, P. H., A. R. Mitz, R. V. Chacko, and E. A. Murray. 2013a. Effects of Amygdala Lesions on Reward-Value Coding in Orbital and Medial Prefrontal Cortex. *Neuron* **80**:1519–1531. [5, 8]

Rudebeck, P. H., and E. L. Rich. 2018. Orbitofrontal Cortex. *Curr. Biol.* **28**:R1083–R1088. [2, 4]

Rudebeck, P. H., J. A. Ripple, A. R. Mitz, B. B. Averbeck, and E. A. Murray. 2017a. Amygdala Contributions to Stimulus-Reward Encoding in the Macaque Medial and Orbital Frontal Cortex during Learning. *J. Neurosci.* **37**:2186–2202. [8]

Rudebeck, P. H., R. C. Saunders, D. A. Lundgren, and E. A. Murray. 2017b. Specialized Representations of Value in the Orbital and Ventrolateral Prefrontal Cortex: Desirability versus Availability of Outcomes. *Neuron* **95**:1208–1220. [2, 4, 8]

Rudebeck, P. H., R. C. Saunders, A. T. Prescott, L. S. Chau, and E. A. Murray. 2013b. Prefrontal Mechanisms of Behavioral Flexibility, Emotion Regulation and Value Updating. *Nat. Neurosci.* **16**:1140–1145. [4]

Rudebeck, P. H., M. E. Walton, B. H. P. Millette, et al. 2007. Distinct Contributions of Frontal Areas to Emotion and Social Behaviour in the Rat. *Eur. J. Neurosci.* **26**:2315–2326. [4]

Rush, A. J., M. H. Trivedi, S. R. Wisniewski, et al. 2006. Acute and Longer-Term Outcomes in Depressed Outpatients Requiring One or Several Treatment Steps: A STAR*D Report. *Am. J. Psych.* **163**:1905–1917. [13]

Rushworth, M. F. 2008. Intention, Choice, and the Medial Frontal Cortex. *Ann. N. Y. Acad. Sci.* **1124**:181–207. [1]

Rushworth, M. F. S., M. E. Walton, S. W. Kennerley, and D. M. Bannerman. 2004. Action Sets and Decisions in the Medial Frontal Cortex. *Trends Cogn. Sci.* **8**:410–417. [12]

Rusu, S. I., and C. M. A. Pennartz. 2020. Learning, Memory and Consolidation Mechanisms for Behavioral Control in Hierarchically Organized Cortico-Basal Ganglia Systems. *Hippocampus* **30**:73–98. [17]

Rylander, G. 1948. Personality Analysis before and after Frontal Lobotomy. *Res. Publ. Assoc. Res. Nerv. Ment. Dis.* **27**:691–705. [15]

Saber, G. T., F. Pestilli, and C. E. Curtis. 2015. Saccade Planning Evokes Topographically Specific Activity in the Dorsal and Ventral Streams. *J. Neurosci.* **35**:245–252. [4]

Sadaghiani, S., and M. D'Esposito. 2015. Functional Characterization of the Cingulo-Opercular Network in the Maintenance of Tonic Alertness. *Cereb. Cortex* **25**:2763–2773. [11]

Sadaghiani, S., R. Scheeringa, K. Lehongre, et al. 2012. Alpha-Band Phase Synchrony Is Related to Activity in the Fronto-Parietal Adaptive Control Network. *J. Neurosci.* **32**:14305–14310. [11]

Sadaghiani, S., and J. Wirsich. 2020. Intrinsic Connectome Organization across Temporal Scales: New Insights from Cross-Modal Approaches. *Netw. Neurosci.* **4**:1–29. [11]

Sakai, K. 2008. Task Set and Prefrontal Cortex. *Annu. Rev. Neurosci.* **31**:219–245. [12]

Sala, G., N. D. Aksayli, K. S. Tatlidil, et al. 2019. Near and Far Transfer in Cognitive Training: A Second-Order Meta-Analysis. *Collabra Psychology* **5**:18. [14]

Sallet, J., R. B. Mars, M. P. Noonan, et al. 2013. The Organization of Dorsal Frontal Cortex in Humans and Macaques. *J. Neurosci.* **33**:12255–12274. [8]

Salmi, J., L. Nyberg, and M. Laine. 2018. Working Memory Training Mostly Engages General-Purpose Large-Scale Networks for Learning. *Neurosci. Biobehav. Rev.* **93**:108–122. [14]

Salmi, J., A. Soveri, V. Salmela, et al. 2020. Working Memory Training Restores Aberrant Brain Activity in Adult Attention-Deficit Hyperactivity Disorder. *Hum. Brain Mapp.* **41**:4876–4891. [14]

Salvadore, G., B. R. Cornwell, V. Colon-Rosario, et al. 2009. Increased Anterior Cingulate Cortical Activity in Response to Fearful Faces: A Neurophysiological Biomarker That Predicts Rapid Antidepressant Response to Ketamine. *Biol. Psychiatry* **65**:289–295. [13]

Salzman, C. D., and S. Fusi. 2010. Emotion, Cognition, and Mental State Representation in Amygdala and Prefrontal Cortex. *Annu. Rev. Neurosci.* **33**:173–202. [8, 12]

Sämann, P. G., R. Wehrle, D. Hoehn, et al. 2011. Development of the Brain's Default Mode Network from Wakefulness to Slow Wave Sleep. *Cereb. Cortex* **21**:2082–2093. [11]

Samuels, B. A., K. M. Nautiyal, A. C. Kruegel, et al. 2017. The Behavioral Effects of the Antidepressant Tianeptine Require the Mu-Opioid Receptor. *Neuropsychopharmacol.* **42**:2052–2063. [13]

Sanacora, G., R. Gueorguieva, C. N. Epperson, et al. 2004. Subtype-Specific Alterations of Gamma-Aminobutyric Acid and Glutamate in Patients with Major Depression. *Arch. Gen. Psychiatry* **61**:705–713. [13]

Sandson, T. A., K. R. Daffner, P. A. Carvalho, and M. M. Mesulam. 1991. Frontal Lobe Dysfunction Following Infarction of the Left-Sided Medial Thalamus. *Arch. Neurol.* **48**:1300–1303. [8]

Sanides, F. 1962. Die Stirnhirnrinde. In: Die Architektonik des Menschlichen Stirnhirns, ed. F. Sanides, pp. 1–7, Monographien aus dem Gesamtgebiet der Neurologie und Psychiatrie, vol. 98. Berlin, Heidelberg: Springer. [3]

———. 1970. Functional Architecture of Motor and Sensory Cortices in Primates in the Light of a new Concept of Neocortex Evolution. In: The Primate Brain, ed. C. R. Noback and W. Montagna, pp. 137–208, Advances in Primatology, vol. 1. New York: Appleton-Century-Crofts. [5]

Sanides, F., and D. Sanides. 1972. The "Extraverted Neurons" of the Mammalian Cerebral Cortex. *Z Anat Entwicklungsgesch* **136**:272–293. [7]

Santarnecchi, E., D. Momi, L. Mencarelli, et al. 2021. Overlapping and Dissociable Brain Activations for Fluid Intelligence and Executive Functions. *Cogn. Affect. Behav. Neurosci.* **21**:327–346. [9]

Saper, C. B., and R. L. Stornetta. 2015. Central Autonomic System. In: The Rat Nervous System: Fourth Edition, ed. G. Paxinos, pp. 629–673. London: Elsevier. [4]

Sarafyazd, M., and M. Jazayeri. 2019. Hierarchical Reasoning by Neural Circuits in the Frontal Cortex. *Science* **364**:eaav8911. [4]

Satterthwaite, T. D., D. H. Wolf, G. Erus, et al. 2013. Functional Maturation of the Executive System during Adolescence. *J. Neurosci.* **33**:16249–16261. [16]

Saunders, R. C., M. Mishkin, and J. P. Aggleton. 2005. Projections from the Entorhinal Cortex, Perirhinal Cortex, Presubiculum, and Parasubiculum to the Medial Thalamus in Macaque Monkeys: Identifying Different Pathways Using Disconnection Techniques. *Exp. Brain Res.* **167**:1–16. [8]

Savaki, H. E., G. G. Gregoriou, S. Bakola, and A. K. Moschovakis. 2015. Topography of Visuomotor Parameters in the Frontal and Premotor Eye Fields. *Cereb. Cortex* **25**:3095–3106. [8]

Sawaguchi, T., and P. S. Goldman-Rakic. 1994. The Role of D1-Dopamine Receptor in Working Memory: Local Injections of Dopamine Antagonists into the Prefrontal Cortex of Rhesus Monkeys Performing an Oculomotor Delayed-Response Task. *J. Neurophysiol.* **71**:515–528. [6]

Sawiak, S. J., Y. Shiba, L. Oikonomidis, et al. 2018. Trajectories and Milestones of Cortical and Subcortical Development of the Marmoset Brain from Infancy to Adulthood. *Cereb. Cortex* **28**:4440–4453. [16]

Saxena, S., and S. L. Rauch. 2000. Functional Neuroimaging and the Neuroanatomy of Obsessive-Compulsive Disorder. *Psychiatr. Clin. N. Am.* **23**:563–586. [15]

Sayali, C., J. Rubin-McGregor, and D. Badre. 2023. Policy Abstraction as a Predictor of Cognitive Effort Avoidance. *J. Exp. Psychol. Gen.* **152**:3440–3458. [7]

Schacter, D. L., D. R. Addis, and R. L. Buckner. 2008. Episodic Simulation of Future Events: Concepts, Data, and Applications. *Ann. N. Y. Acad. Sci.* **1124**:39–60. [12]

Schaeffer, D. J., Y. Hori, K. M. Gilbert, et al. 2020. Divergence of Rodent and Primate Medial Frontal Cortex Functional Connectivity. *PNAS* **117**:21681–21689. [2]

Schall, J. D., T. J. Palmeri, and G. D. Logan. 2017. Models of Inhibitory Control. *Philos. Trans. R. Soc. Lond. B Biol. Sci.* **372**:20160193. [4]

Schall, J. D., W. Zinke, J. D. Cosman, et al. 2020. On the Evolution of the Frontal Eye Field: Comparisons of Monkeys, Apes, and Humans. In: Evolutionary Neuroscience (Second Edition), ed. J. H. Kaas, pp. 861–890. London: Academic Press. [8]

Scheidegger, M., M. Walter, M. Lehmann, et al. 2012. Ketamine Decreases Resting State Functional Network Connectivity in Healthy Subjects: Implications for Antidepressant Drug Action. *PloS One* **7**:e44799. [13]

Schluppeck, D., C. E. Curtis, P. W. Glimcher, and D. J. Heeger. 2006. Sustained Activity in Topographic Areas of Human Posterior Parietal Cortex during Memory-Guided Saccades. *J. Neurosci.* **26**:5098–5108. [4]

Schmaal, L., D. P. Hibar, P. G. Sämann, et al. 2017. Cortical Abnormalities in Adults and Adolescents with Major Depression Based on Brain Scans from 20 Cohorts Worldwide in the Enigma Major Depressive Disorder Working Group. *Mol. Psychiatry* **22**:900–909. [13]

Schmaal, L., E. Pozzi, T. C. Ho, et al. 2020. Enigma MDD: Seven Years of Global Neuroimaging Studies of Major Depression through Worldwide Data Sharing. *Transl. Psychiatry* **10**:1–19. [13]

Schmiedek, F., A. Hildebrandt, M. Lövdén, et al. 2009. Complex Span versus Updating Tasks of Working Memory: The Gap Is Not That Deep. *J. Exp. Psychol. Learn. Mem. Cogn.* **35**:1089–1096. [9]

Schmitt, L. I., R. D. Wimmer, M. Nakajima, et al. 2017. Thalamic Amplification of Cortical Connectivity Sustains Attentional Control. *Nature* **545**:219–223. [3, 4, 7]

Schoenbaum, G., A. A. Chiba, and M. Gallagher. 1999. Neural Encoding in Orbitofrontal Cortex and Basolateral Amygdala during Olfactory Discrimination Learning. *J. Neurosci.* **19**:1876–1884. [4]

Schoenbaum, G., B. Setlow, S. L. Nugent, M. P. Saddoris, and M. Gallagher. 2003. Lesions of Orbitofrontal Cortex and Basolateral Amygdala Complex Disrupt Acquisition of Odor-Guided Discriminations and Reversals. *Learn Mem* **10**:129–140. [4]

Schoenbaum, G., A. Takahashi, T. L. Liu, and M. A. Mcdannald. 2011. Does the Orbitofrontal Cortex Signal Value? *Ann. N. Y. Acad. Sci.* **1239**:87–99. [4]

Schonberg, T., J. P. O'Doherty, D. Joel, et al. 2010. Selective Impairment of Prediction Error Signaling in Human Dorsolateral but Not Ventral Striatum in Parkinson's Disease Patients: Evidence from a Model-Based fMRI Study. *Neuroimage* **49**:772–781. [7]

Schuck, N. W., M. B. Cai, R. C. Wilson, and Y. Niv. 2016. Human Orbitofrontal Cortex Represents a Cognitive Map of State Space. *Neuron* **91**:1402–1412. [7, 12]

Schuck, N. W., R. Wilson, and Y. Niv. 2018. A State Representation for Reinforcement Learning and Decision-Making in the Orbitofrontal Cortex. In: Goal-Directed Decision Making: Computations and Neural Circuits, ed. R. R. Morris et al., pp. 259–278. Philadelphia, PA: Elsevier. [12]

Schultz, W. 2015. Neuronal Rewards and Decision Signals: From Theories to Data. *Psychol. Rev.* **95**:853–951. [10]

Schultz, W., P. Dayan, and P. R. Montague. 1997. A Neural Substrate of Prediction and Reward. *Science* **275**:1593–1599. [12]

Schumacher, F. K., L. V. Schumacher, F. Amtage, et al. 2021. The Rostro-Caudal Gradient in the Prefrontal Cortex and Its Modulation by Subthalamic Deep Brain Stimulation in Parkinson's Disease. *Sci. Rep.* **11**:2138. [7]

Schumacher, F. K., L. V. Schumacher, B. O. Schelter, and C. P. Kaller. 2019. Functionally Dissociating Ventro-Dorsal Components within the Rostro-Caudal Hierarchical Organization of the Human Prefrontal Cortex. *Neuroimage* **185**:398–407. [7]

Schumann, G., E. Loth, T. Banaschewski, et al. 2010. The Imagen Study: Reinforcement-Related Behaviour in Normal Brain Function and Psychopathology. *Mol. Psych.* **15**:1128–1139. [17]

Schwartz, J. M., P. W. Stoessel, L. R. Baxter Jr., K. M. Martin, and M. E. Phelps. 1996. Systematic Changes in Cerebral Glucose Metabolic Rate after Successful Behavior Modification Treatment of Obsessive-Compulsive Disorder. *Arch. Gen. Psych.* **53**:109–113. [15]

Schwartz, M. L., J. J. Dekker, and P. S. Goldman-Rakic. 1991. Dual Mode of Corticothalamic Synaptic Termination in the Mediodorsal Nucleus of the Rhesus Monkey. *J. Comp. Neurol.* **309**:289–304. [8]

Scott, J. A., D. Grayson, E. Fletcher, et al. 2016. Longitudinal Analysis of the Developing Rhesus Monkey Brain Using Magnetic Resonance Imaging: Birth to Adulthood. *Brain Struct. Funct.* **221**:2847–2871. [16]

Seamans, J. K., D. Durstewitz, B. R. Christie, C. F. Stevens, and T. J. Sejnowski. 2001. Dopamine D1/D5 Receptor Modulation of Excitatory Synaptic Inputs to Layer V Prefrontal Cortex Neurons. *PNAS* **98**:301–306. [6]

Seamans, J. K., C. C. Lapish, and D. Durstewitz. 2008. Comparing the Prefrontal Cortex of Rats and Primates: Insights from Electrophysiology. *Neurotox. Res.* **14**:249–262. [2, 4]

Seeley, W. W. 2019. The Salience Network: A Neural System for Perceiving and Responding to Homeostatic Demands. *J. Neurosci.* **39**:9878–9882. [4, 15]

Seeley, W. W., R. K. Crawford, J. Zhou, B. L. Miller, and M. D. Greicius. 2009. Neurodegenerative Diseases Target Large-Scale Human Brain Networks. *Neuron* **62**:42–52. [11]

Seeley, W. W., V. Menon, A. F. Schatzberg, et al. 2007. Dissociable Intrinsic Connectivity Networks for Salience Processing and Executive Control. *J. Neurosci.* **27**:2349–2356. [11, 15]

Segalin, C., J. Williams, T. Karigo, et al. 2021. The Mouse Action Recognition System (Mars) Software Pipeline for Automated Analysis of Social Behaviors in Mice. *eLife* **10**:e63720. [2]

Seger, C. A., and E. K. Miller. 2010. Category Learning in the Brain. *Annu. Rev. Neurosci.* **33**:203–219. [12]

Seidemann, E., Y. Chen, Y. Bai, et al. 2016. Calcium Imaging with Genetically Encoded Indicators in Behaving Primates. *eLife* **5**:e16178. [4]

Seidlitz, J., A. Nadig, S. Liu, et al. 2020. Transcriptomic and Cellular Decoding of Regional Brain Vulnerability to Neurogenetic Disorders. *Nat. Commun.* **11**:3358. [16]

Seitzman, B. A., C. Gratton, T. O. Laumann, et al. 2019. Trait-Like Variants in Human Functional Brain Networks. *PNAS* **116**:22851–22861. [11, 12]

Seitzman, B. A., C. Gratton, S. Marek, et al. 2020. A Set of Functionally-Defined Brain Regions with Improved Representation of the Subcortex and Cerebellum. *Neuroimage* **206**:116290. [11]

Selemon, L. D., and P. S. Goldman-Rakic. 1988. Common Cortical and Subcortical Targets of the Dorsolateral Prefrontal and Posterior Parietal Cortices in the Rhesus Monkey: Evidence for a Distributed Neural Network Subserving Spatially Guided Behavior. *J. Neurosci.* **8**:4049–4068. [5, 11]

Sellitto, M., E. Ciaramelli, and G. di Pellegrino. 2010. Myopic Discounting of Future Rewards after Medial Orbitofrontal Damage in Humans. *J. Neurosci.* **30**:16429–16436. [4]

Seo, D., C. M. Lacadie, K. Tuit, et al. 2013. Disrupted Ventromedial Prefrontal Function, Alcohol Craving, and Subsequent Relapse Risk. *JAMA Psych.* **70**:727–739. [16]

Seo, D., K. A. Tsou, E. B. Ansell, M. N. Potenza, and R. Sinha. 2014. Cumulative Adversity Sensitizes Neural Response to Acute Stress: Association with Health Symptoms. *Neuropsychopharmacol.* **39**:670–680. [16]

Seo, M., E. Lee, and B. B. Averbeck. 2012. Action Selection and Action Value in Frontal-Striatal Circuits. *Neuron* **74**:947–960. [8]

Sescousse, G., X. Caldú, B. Segura, and J. C. Dreher. 2013. Processing of Primary and Secondary Rewards: A Quantitative Meta-Analysis and Review of Human Functional Neuroimaging Studies. *Neurosci. Biobehav. Rev.* **37**:681–696. [4]

Sescousse, G., J. Redouté, and J. C. Dreher. 2010. The Architecture of Reward Value Coding in the Human Orbitofrontal Cortex. *J. Neurosci.* **30**:13095–13104. [4]

Setogawa, T., T. Mizuhiki, N. Matsumoto, et al. 2019. Neurons in the Monkey Orbitofrontal Cortex Mediate Reward Value Computation and Decision-Making. *Commun. Biol.* **2**:126. [5]

Shackman, A. J., T. V. Salomons, H. A. Slagter, et al. 2011. The Integration of Negative Affect, Pain and Cognitive Control in the Cingulate Cortex. *Nat. Rev. Neurosci.* **12**:154–167. [15]

Shafto, M. A., L. K. Tyler, M. Dixon, et al. 2014. The Cambridge Centre for Ageing and Neuroscience (Cam-Can) Study Protocol: A Cross-Sectional, Lifespan, Multidisciplinary Examination of Healthy Cognitive Ageing. *BMC Neurol.* **14**:204. [16]

Shallice, T., and P. W. Burgess. 1991. Deficits in Strategy Application Following Frontal Lobe Damage in Man. *Brain* **114**:727–741. [12]

Shao, L.-X., C. Liao, I. Gregg, et al. 2021. Psilocybin Induces Rapid and Persistent Growth of Dendritic Spines in Frontal Cortex *in Vivo*. *Neuron* **109**:2535–2544. e2534. [13]

Sharpe, M. J., H. M. Batchelor, L. E. Mueller, et al. 2020. Dopamine Transients Do Not Act as Model-Free Prediction Errors during Associative Learning. *Nat. Commun.* **11**:106. [12]

Shashidhara, S., D. J. Mitchell, Y. Erez, and J. Duncan. 2019. Progressive Recruitment of the Frontoparietal Multiple-Demand System with Increased Task Complexity, Time Pressure, and Reward. *J. Cogn. Neurosci.* **31**:1617–1630. [7]

Sheline, Y. I., J. L. Price, Z. Yan, and M. A. Mintun. 2010. Resting-State Functional MRI in Depression Unmasks Increased Connectivity between Networks via the Dorsal Nexus. *PNAS* **107**:11020–11025. [13]

Shen, C., S. Ardid, D. Kaping, et al. 2014. Anterior Cingulate Cortex Cells Identify Process-Specific Errors of Attentional Control Prior to Transient Prefrontal-Cingulate Inhibition. *Cereb. Cortex* **25**:2213–2228. [12]

Shen, C., Q. Luo, T. Jia, et al. 2020. Neural Correlates of the Dual-Pathway Model for ADHD in Adolescents. *Am. J. Psych.* **177**:844–854. [14]

Shenhav, A., M. M. Botvinick, and J. D. Cohen. 2013. The Expected Value of Control: an Integrative Theory of Anterior Cingulate Cortex Function. *Neuron* **79**:217–240. [12, 15]

Shenhav, A., J. D. Cohen, and M. M. Botvinick. 2016. Dorsal Anterior Cingulate Cortex and the Value of Control. *Nat. Neurosci.* **19**:1286–1291. [12, 15, 16]

Shenhav, A., S. Musslick, F. Lieder, et al. 2017. Toward a Rational and Mechanistic Account of Mental Effort. *Annu. Rev. Neurosci.* **40**:99–124. [12]

Shenhav, A., M. A. Straccia, S. Musslick, J. D. Cohen, and M. M. Botvinick. 2018. Dissociable Neural Mechanisms Track Evidence Accumulation for Selection of Attention versus Action. *Nat. Commun.* **9**:2485. [7, 11, 12]

Shenoy, K. V., M. Sahani, and M. M. Churchland. 2013. Cortical Control of Arm Movements: A Dynamical Systems Perspective. *Annu. Rev. Neurosci.* **36**:337–359. [12]

Shepard, R. N., and J. Metzler. 1971. Mental Rotation of Three-Dimensional Objects. *Science* **171**:701–703. [12]

Sheth, S. A., and H. S. Mayberg. 2023. Deep Brain Stimulation for Obsessive-Compulsive Disorder and Depression. *Annu. Rev. Neurosci.* **46**:341–358. [15]

Sheth, S. A., M. K. Mian, S. R. Patel, et al. 2012. Human Dorsal Anterior Cingulate Cortex Neurons Mediate Ongoing Behavioural Adaptation. *Nature* **488**:218–221. [15, 16]

Sheth, S. A., J. Neal, F. Tangherlini, et al. 2013. Limbic System Surgery for Treatment-Refractory Obsessive-Compulsive Disorder: A Prospective Long-Term Follow-up of 64 Patients. *J Neurosurg* **118**:491–497. [15]

Shiffrin, R. M., and W. Schneider. 1977. Controlled and Automatic Human Information Processing: II. Perceptual Learning, Automatic Attending, and a General Theory. *Psychol. Rev.* **84**:127–190. [14]

Shima, K., and J. Tanji. 1998. Role for Cingulate Motor Area Cells in Voluntary Movement Selection Based on Reward. *Science* **282**:1335–1338. [4]

Shin, E. J., Y. Jang, S. Kim, et al. 2021. Robust and Distributed Neural Representation of Action Values. *eLife* **10**:e53045. [12]

Shin, H., R. Law, S. Tsutsui, C. I. Moore, and S. R. Jones. 2017. The Rate of Transient Beta Frequency Events Predicts Behavior across Tasks and Species. *eLife* **6**:e29086. [8]

Shine, J. M., L. D. Lewis, D. D. Garrett, and K. Hwang. 2023. The Impact of the Human Thalamus on Brain-Wide Information Processing. *Nat. Rev. Neurosci.* **24**:416–430. [8]

Shirer, W. R., S. Ryali, E. Rykhlevskaia, V. Menon, and M. D. Greicius. 2012. Decoding Subject-Driven Cognitive States with Whole-Brain Connectivity Patterns. *Cereb. Cortex* **22**:158–165. [11]

Sibley, M. H. 2020. Motivational and Executive Functioning Considerations When Treating Adolescents with ADHD. In: ADHD in Adolescence: Development, Assessment, and Treatment, ed. S. P. Becker, pp. 306–329. New York, NY: Guilford Press. [14]

Siddiqi, S. H., and M. D. Fox. 2023. Targeting Symptom-Specific Networks with TMS. *Biol. Psych.* **95**:P502–509. [16]

Siddiqi, S. H., S. F. Taylor, D. Cooke, et al. 2020. Distinct Symptom-Specific Treatment Targets for Circuit-Based Neuromodulation. *Am. J. Psych.* **177**:435–446. [16]

Sierra-Mercado, D., N. Padilla-Coreano, and G. J. Quirk. 2011. Dissociable Roles of Prelimbic and Infralimbic Cortices, Ventral Hippocampus, and Basolateral Amygdala in the Expression and Extinction of Conditioned Fear. *Neuropsychopharmacol.* **36**:529–538. [4]

Sikora, M., J. Heffernan, E. T. Avery, et al. 2016. Salience Network Functional Connectivity Predicts Placebo Effects in Major Depression. *Biol. Psych. Cogn. Neurosci. Neuroimag.* **1**:68–76. [13]

Simmonds, D. J., M. N. Hallquist, M. Asato, and B. Luna. 2014. Developmental Stages and Sex Differences of White Matter and Behavioral Development through Adolescence: A Longitudinal Diffusion Tensor Imaging (DTI) Study. *Neuroimage* **92**:356–368. [16]

Simmonds, D. J., M. N. Hallquist, and B. Luna. 2017. Protracted Development of Executive and Mnemonic Brain Systems Underlying Working Memory in Adolescence: A Longitudinal fMRI Study. *Neuroimage* **157**:695–704. [16]

Sinha, R., C. M. Lacadie, R. T. Constable, and D. Seo. 2016. Dynamic Neural Activity during Stress Signals Resilient Coping. *PNAS* **113**:8837–8842. [16]

Skeberdis, V. A., V. Chevaleyre, C. G. Lau, et al. 2006. Protein kinase A Regulates Calcium Permeability of NMDA Receptors. *Nat. Neurosci.* **9**:501–510. [16]

Sklar, S., M. Walmer, P. Sacre, et al. 2017. Neuronal Activity in Human Anterior Cingulate Cortex Modulates with Internal Cognitive State during Multi-Source Interference Task. *Annu Int Conf IEEE Eng Med Biol Soc* **2017**:962–965. [15]

Sleezer, B. J., M. D. Castagno, and B. Y. Hayden. 2016. Rule Encoding in Orbitofrontal Cortex and Striatum Guides Selection. *J. Neurosci.* **36**:11223–11237. [5]

Smaers, J. B., A. Gómez-Robles, A. N. Parks, and C. C. Sherwood. 2017. Exceptional Evolutionary Expansion of Prefrontal Cortex in Great Apes and Humans. *Curr. Biol.* **27**:714–720. [4]

Smaers, J. B., C. S. Mongle, K. Safi, and D. K. N. Dechmann. 2019. Allometry, Evolution and Development of Neocortex Size in Mammals. *Prog. Brain Res.* **250**:83–107. [4]

Smaers, J. B., and D. R. Vanier. 2019. Brain Size Expansion in Primates and Humans Is Explained by a Selective Modular Expansion of the Cortico-Cerebellar System. *Cortex* **118**:292–305. [4]

Smith, D. M., D. C. Perez, A. Porter, A. Dworetsky, and C. Gratton. 2021. Light through the Fog: Using Precision fMRI Data to Disentangle the Neural Substrates of Cognitive Control. *Curr. Opin. Behav. Sci.* **40**:19–26. [11]

Smith, E. E., J. Jonides, and R. A. Koeppe. 1996. Dissociating Verbal and Spatial Working Memory Using PET. *Cerebral Cortex* **6**:11–20. [4]

Smith, S. M., P. T. Fox, K. L. Miller, et al. 2009. Correspondence of the Brain's Functional Architecture during Activation and Rest. *PNAS* **106**:13040–13045. [11]

Smucny, J., S. J. Dienel, D. A. Lewis, and C. S. Carter. 2022. Mechanisms Underlying Dorsolateral Prefrontal Cortex Contributions to Cognitive Dysfunction in Schizophrenia. *Neuropsychopharmacol.* **47**:292–308. [3]

Snyder, H. R., M. T. Banich, and Y. Munakata. 2014. All Competition Is Not Alike: Neural Mechanisms for Resolving Underdetermined and Prepotent Competition. *J. Cogn. Neurosci.* **26**:2608–2623. [12]

Sohal, V. S., F. Zhang, O. Yizhar, and K. Deisseroth. 2009. Parvalbumin Neurons and Gamma Rhythms Enhance Cortical Circuit Performance. *Nature* **459**:698–702. [4]

Sohn, H., N. Meirhaeghe, R. Rajalingham, and M. Jazayeri. 2021. A Network Perspective on Sensorimotor Learning. *Trends Neurosci.* **44**:170–181. [4]

Solmi, M., J. Radua, M. Olivola, et al. 2022. Age at Onset of Mental Disorders Worldwide: Large-Scale Meta-Analysis of 192 Epidemiological Studies. *Mol. Psych.* **27**:281–295. [16]

Soltani, A., and E. Koechlin. 2022. Computational Models of Adaptive Behavior and Prefrontal Cortex. *Neuropsychopharmacol.* **47**:58–71. [3, 7, 10]

Somel, M., H. Franz, Z. Yan, et al. 2009. Transcriptional Neoteny in the Human Brain. *PNAS* **106**:5743–5748. [4]

Somel, M., X. Liu, L. Tang, et al. 2011. MicroRNA-Driven Developmental Remodeling in the Brain Distinguishes Humans from Other Primates. *PLoS Biol* **9**:e1001214. [4]

Sommer, M. A., and E. J. Tehovnik. 1997. Reversible Inactivation of Macaque Frontal Eye Field. *Experimental Brain Research* **116**:229–249. [4]

Sommer, M. A., and R. H. Wurtz. 2001. Frontal Eye Field Sends Delay Activity Related to Movement, Memory, and Vision to the Superior Colliculus. *J. Neurophysiol.* **85**:1673–1685. [4]

Song, T., W. Chen, X. Chen, et al. 2021. Repeated Fluoxetine Treatment Induces Transient and Long-Term Astrocytic Plasticity in the Medial Prefrontal Cortex of Normal Adult Rats. *Prog. Neuropsychopharmacol. Biol. Psychiatry* **107**:110252. [13]

Sosa, J. L. R., D. Buonomano, and A. Izquierdo. 2021. The Orbitofrontal Cortex in Temporal Cognition. *Behav. Neurosci.* **135**:154–164. [4]

Soveri, A., J. Antfolk, L. Karlsson, B. Salo, and M. Laine. 2017. Working Memory Training Revisited: A Multi-Level Meta-Analysis of n-back Training Studies. *Psychon. Bull. Rev.* **24**:1077–1096. [14]

Spielberg, J. M., W. Heller, and G. A. Miller. 2013. Hierarchical Brain Networks Active in Approach and Avoidance Goal Pursuit. *Front. Hum. Neurosci.* **7**:284. [13]

Spielberg, J. M., G. A. Miller, S. L. Warren, et al. 2014. Transdiagnostic Dimensions of Anxiety and Depression Moderate Motivation-Related Brain Networks during Goal Maintenance. *Depress. Anxiety* **31**:805–813. [13]

Spitzer, B., and S. Haegens. 2017. Beyond the Status Quo: A Role for Beta Oscillations in Endogenous Content (Re)Activation. *eNeuro* **4**:ENEURO.0170. [12]

Sporns, O. 2016. Networks of the Brain. Cambridge, MA: The MIT Press. [11, 12]

Sporns, O., C. J. Honey, and R. Kötter. 2007. Identification and Classification of Hubs in Brain Networks. *PloS One* **2**:e1049. [11]

Sprague, T. C., E. F. Ester, and J. T. Serences. 2014. Reconstructions of Information in Visual Spatial Working Memory Degrade with Memory Load. *Curr. Biol.* **24**:2174–2180. [4]

Sprooten, E., B. Franke, and C. U. Greven. 2022. The P-Factor and Its Genomic and Neural Equivalents: an Integrated Perspective. *Mol. Psych.* **27**:38–48. [16]

Sprung-Much, T., N. Eichert, E. Nolan, and M. Petrides. 2022 Broca's Area and the Search for Anatomical Asymmetry: Commentary and Perspectives. *Brain Struct. Funct.* **227**:441–449. [4]

Srimal, R., and C. E. Curtis. 2008. Persistent Neural Activity during the Maintenance of Spatial Position in Working Memory. *Neuroimage* **39**:455–468. [4]

Srinivasan, L., W. F. Asaad, D. T. Ginat, et al. 2013. Action Initiation in the Human Dorsal Anterior Cingulate Cortex. *PloS One* **8**:e55247. [15]

Srirangarajan, T., L. Mortazavi, T. Bortolini, J. Moll, and B. Knutson. 2021. Multi-Band fMRI Compromises Detection of Mesolimbic Reward Responses. *Neuroimage* **244**:118617. [4]

Stamenova, V., and B. Levine. 2019. Effectiveness of Goal Management Training® in Improving Executive Functions: A Meta-Analysis. *Neuropsychol. Rehabil.* **29**:1569–1599. [14]

Stamm, J. S., and M. L. Weber-Levine. 1971. Delayed Alternation Impairments Following Selective Prefrontal Cortical Ablations in Monkeys. *Exp Neurol* **33**:263–278. [4]

Stawicka, Z. M., R. Massoudi, N. K. Horst, et al. 2020. Ventromedial Prefrontal Area 14 Provides Opposing Regulation of Threat and Reward-Elicited Responses in the Common Marmoset. *PNAS* **117**:25116–25127. [8]

Stawicka, Z. M., R. Massoudi, L. Oikonomidis, et al. 2022. Differential Effects of the Inactivation of Anterior and Posterior Orbitofrontal Cortex on Affective Responses to Proximal and Distal Threat, and Reward Anticipation in the Common Marmoset. *Cereb. Cortex* **32**:1319–1336. [8]

Stephan, K. E., L. Kamper, A. Bozkurt, et al. 2001. Advanced Database Methodology for the Collation of Connectivity Data on the Macaque Brain (CoCoMac). *Philos. Trans. R. Soc. Lond. B Biol. Sci.* **356**:1159–1186. [11]

Stephan, K. E., and C. Mathys. 2014. Computational Approaches to Psychiatry. *Curr. Opin. Neurobiol.* **25**:85–92. [12]

Stevens, M. C., A. Gaynor, K. L. Bessette, and G. D. Pearlson. 2016. A Preliminary Study of the Effects of Working Memory Training on Brain Function. *Brain Imag. Behav.* **10**:387–407. [14]

Stollstorff, M., J. Foss-Feig, E. H. Cook, Jr., et al. 2010. Neural Response to Working Memory Load Varies by Dopamine Transporter Genotype in Children. *Neuroimage* **53**:970–977. [7]

Stolyarova, A., M. Rakhshan, E. E. Hart, et al. 2019. Contributions of Anterior Cingulate Cortex and Basolateral Amygdala to Decision Confidence and Learning under Uncertainty. *Nat. Commun.* **10**:4704. [2, 4]

Strait, C. E., B. J. Sleezer, T. C. Blanchard, et al. 2016. Neuronal Selectivity for Spatial Positions of Offers and Choices in Five Reward Regions. *J. Neurophysiol.* **115**:1098–1111. [5]

Strange, B. A., R. N. A. Henson, K. J. Friston, and R. J. Dolan. 2001. Anterior Prefrontal Cortex Mediates Rule Learning in Humans. *Cereb. Cortex* **11**:1040–1046. [6]

Strawn, J. R., J. A. Mills, V. Suresh, et al. 2022. Combining Selective Serotonin Reuptake Inhibitors and Cognitive Behavioral Therapy in Youth with Depression and Anxiety. *J. Affect. Disord.* **298**:292–300. [16]

Striedter, G. F., and R. G. Northcutt. 2020. Brains through Time : A Natural History of Vertebrates. New York: Oxford Univ. Press. [5]

Stuss, D. T., and M. P. Alexander. 2000. Executive Functions and the Frontal Lobes: A Conceptual View. *Psychol. Res.* **63**:289–298. [11]

———. 2007. Is There a Dysexecutive Syndrome? *Philos. Trans. R. Soc. Lond. B Biol. Sci.* **362**:901–915. [8]

Stuss, D. T., and D. F. Benson. 1984. Neuropsychological Studies of the Frontal Lobes. *Psychol. Bull.* **95**:3–28. [4]

Stuss, D. T., and D. F. Benson. 1987. The Frontal Lobes and Control of Cognition and Memory. In: The Frontal Lobes Revisited, ed. E. Perecman, pp. 141–158. New York: The IRBN Press. [7]

Stuss, D. T., and R. T. Knight, eds. 2002. Principles of Frontal Lobe Function. Oxford: Oxford Univ. Press. [1]

———, eds. 2012. Principles of Frontal Lobe Function. Oxford: Oxford Univ. Press. [1]

Stuss, D. T., T. Shallice, M. P. Alexander, and T. W. Picton. 1995. A Multidisciplinary Approach to Anterior Attentional Functions. *Ann. N. Y. Acad. Sci.* **769**:191–212. [9]

Sutton, R. S., and A. G. Barto. 1998. Reinforcement Learning. Adaptive Computation and Machine Learning. Cambridge, MA: The MIT Press. [10, 12]

———. 2018. Reinforcement Learning: An Introduction, 2nd edition. Cambridge, MA: MIT Press. [4]

Suwaluk, A., and N. Chutabhakdikul. 2022a. Altered Development of Prefrontal GABAergic Functions and Anxiety-Like Behavior in Adolescent Offspring Induced by Prenatal Stress. *Brain Sci.* **12**:1015. [16]

———. 2022b. Long-Term Effects of Prenatal Stress on the Development of Prefrontal Cortex in the Adolescent Offspring. *J. Chem. Neuroanat.* **125**:102169. [16]

Suzuki, K., E. Nosyreva, K. W. Hunt, E. T. Kavalali, and L. M. Monteggia. 2017. Effects of a Ketamine Metabolite on Synaptic Nmdar Function. *Nature* **546**:E1–E3. [13]

Swanson, L. W. 2004. Brain Maps: Structure of the Rat Brain. Amsterdam: Elsevier. [3]

Swanson, L. W. 2018. Brain Maps 4.0-Structure of the Rat Brain: An Open Access Atlas with Global Nervous System Nomenclature Ontology and Flatmaps. *J. Comp. Neurol.* **526**:935–943. [4]

Sweis, B. M., S. V. Abram, B. J. Schmidt, et al. 2018. Sensitivity to "Sunk Costs" in Mice, Rats, and Humans. *Science* **361**:178–181. [2]

Szechtman, H., and E. Woody. 2004. Obsessive-Compulsive Disorder as a Disturbance of Security Motivation. *Psychol. Rev.* **111**:111–127. [15]

Tai, L. H., A. M. Lee, N. Benavidez, A. Bonci, and L. Wilbrecht. 2012. Transient Stimulation of Distinct Subpopulations of Striatal Neurons Mimics Changes in Action Value. *Nat. Neurosci.* **15**:1281–1289. [7]

Takahashi, D. Y., A. R. Fenley, Y. Teramoto, et al. 2015. Language Development: The Developmental Dynamics of Marmoset Monkey Vocal Production. *Science* **349**:734–738. [16]

Talishinsky, A., J. Downar, P. E. Vértes, et al. 2022. Regional Gene Expression Signatures Are Associated with Sex-Specific Functional Connectivity Changes in Depression. *Nat. Commun.* **13**:1–20. [13]

Tanaka, S. C., B. W. Balleine, and J. P. O'Doherty. 2008. Calculating Consequences: Brain Systems That Encode the Causal Effects of Actions. *J. Neurosci.* **28**:6750–6755. [4]

Tang, H., R. Bartolo, and B. B. Averbeck. 2021. Reward-Related Choices Determine Information Timing and Flow across Macaque Lateral Prefrontal Cortex. *Nat. Commun.* **12**:894. [5]

Tang, H., V. D. Costa, R. Bartolo, and B. B. Averbeck. 2022a. Differential Coding of Goals and Actions in Ventral and Dorsal Corticostriatal Circuits during Goal-Directed Behavior. *Cell Rep.* **38**:110198. [5, 8]

Tang, W., S. Jbabdi, Z. Zhu, et al. 2019. A Connectional Hub in the Rostral Anterior Cingulate Cortex Links Areas of Emotion and Cognitive Control. *eLife* **8**:e43761. [1, 16]

Tang, Y.-Y., R. Tang, M. I. Posner, and J. J. Gross. 2022b. Effortless Training of Attention and Self-Control: Mechanisms and Applications. *Trends Cogn. Sci.* **26**:567–577. [14]

Taren, A. A., V. Venkatraman, and S. A. Huettel. 2011. A Parallel Functional Topography between Medial and Lateral Prefrontal Cortex: Evidence and Implications for Cognitive Control. *J. Neurosci.* **31**:5026–5031. [12]

Tark, K.-J., and C. E. Curtis. 2009. Persistent Neural Activity in the Human Frontal Cortex When Maintaining Space That Is Off the Map. *Nat. Neurosci.* **12**:1463–1468. [4]

Tavor, I., O. P. Jones, R. B. Mars, et al. 2016. Task-Free MRI Predicts Individual Differences in Brain Activity during Task Performance. *Science* **352**:216–220. [11, 12]

Teh, Y. W., M. I. Jordan, M. J. Beal, and D. M. Blei. 2006. Hierarchical Dirichlet Processes. *J. Am. Stat. Assoc.* **101**:1566–1581. [10]

Teissier, A., C. Le Magueresse, J. Olusakin, et al. 2020. Early-Life Stress Impairs Postnatal Oligodendrogenesis and Adult Emotional Behaviour through Activity-Dependent Mechanisms. *Mol. Psych.* **25**:1159–1174. [16]

ten Brinke, L. F., J. C. Davis, C. K. Barha, and T. Liu-Ambrose. 2017. Effects of Computerized Cognitive Training on Neuroimaging Outcomes in Older Adults: A Systematic Review. *BMC Geriatrics* **17**:139. [14]

Tervo-Clemmens, B., F. J. Calabro, A. C. Parr, et al. 2023. A Canonical Trajectory of Executive Function Maturation from Adolescence to Adulthood. *Nat. Commun.* **14**:6922. [16]

Teuber, H.-L. 1972. Unity and Diversity of Frontal Lobe Functions. *Acta Neurobiol. Exp.* **32**:615–656. [9, 12]

Theves, S., D. A. Neville, G. Fernández, and C. F. Doeller. 2021. Learning and Representation of Hierarchical Concepts in Hippocampus and Prefrontal Cortex. *J. Neurosci.* **41**:7675–7686. [4]

Thiebaut de Schotten, M., M. Urbanski, B. Batrancourt, et al. 2016. Rostro-Caudal Architecture of the Frontal Lobes in Humans. *Cerebral Cortex* **27**:4033–4047. [7]

Thomas, C., F. Q. Ye, M. O. Irfanoglu, et al. 2014. Anatomical Accuracy of Brain Connections Derived from Diffusion MRI Tractography Is Inherently Limited. *PNAS* **111**:16574–16579. [8]

Thompson, T. W., M. L. Waskom, and J. D. E. Gabrieli. 2016. Intensive Working Memory Training Produces Functional Changes in Large-Scale Frontoparietal Networks. *J. Cog. Neuro.* **28**:575–588. [14]

Thoroughman, K. A., and R. Shadmehr. 2000. Learning of Action through Adaptive Combination of Motor Primitives. *Nature* **407**:742–747. [4]

Tian, L., and L. L. Looger. 2008. Genetically Encoded Fluorescent Sensors for Studying Healthy and Diseased Nervous Systems. *Drug Discov. Today Dis. Models* **5**:27–35. [3]

Timbie, C., and H. Barbas. 2015. Pathways for Emotions: Specializations in the Amygdalar, Mediodorsal Thalamic, and Posterior Orbitofrontal Network. *J. Neurosci.* **35**:11976–11987. [8]

Tokuda, T., O. Yamashita, Y. Sakai, and J. Yoshimoto. 2021. Clustering of Multiple Psychiatric Disorders Using Functional Connectivity in the Data-Driven Brain Subnetwork. *Front. Psych.* **12**:683280. [16]

Tomassini, A., F. H. Hezemans, R. Ye, et al. 2022. Prefrontal Cortical Connectivity Mediates Locus Coeruleus Noradrenergic Regulation of Inhibitory Control in Older Adults. *J. Neurosci.* **42**:3484–3493. [16]

Tomita, H., M. Ohbayashi, K. Nakahara, I. Hasegawa, and Y. Miyashita. 1999. Top-Down Signal from Prefrontal Cortex in Executive Control of Memory Retrieval. *Nature* **401**:699–703. [8]

Tomov, M. S., P. A. Tsividis, T. Pouncy, J. B. Tenenbaum, and S. J. Gershman. 2023. The Neural Architecture of Theory-Based Reinforcement Learning. *Neuron* **111**:1331–1344. [12]

Torres-Gomez, S., J. D. Blonde, D. Mendoza-Halliday, et al. 2020. Changes in the Proportion of Inhibitory Interneuron Types from Sensory to Executive Areas of the Primate Neocortex: Implications for the Origins of Working Memory Representations. *Cereb. Cortex* **30**:4544–4562. [6, 16]

Tosches, M. A., T. M. Yamawaki, R. K. Naumann, et al. 2018. Evolution of Pallium, Hippocampus, and Cortical Cell Types Revealed by Single-Cell Transcriptomics in Reptiles. *Science* **360**:881–888. [5]

Toyoizumi, T., H. Miyamoto, Y. Yazaki-Sugiyama, et al. 2013. A Theory of the Transition to Critical Period Plasticity: Inhibition Selectively Suppresses Spontaneous Activity. *Neuron* **80**:51–63. [16]

Trambaiolli, L. R., X. Xiaolong Peng, J. F. Lehman, et al. 2022. Anatomical and Functional Connectivity Support the Existence of a Salience Network Node within the Caudal Ventrolateral Prefrontal Cortex. *eLife* **11**: DOI: 10.7554/eLife.76334. [16]

Traut, H. J., R. M. Guild, and Y. Munakata. 2021. Why Does Cognitive Training Yield Inconsistent Benefits? A Meta-Analysis of Individual Differences in Baseline Cognitive Abilities and Training Outcomes. *Front. Psychol.* **12**:662139. [14]

Tremblay, L., and W. Schultz. 1999. Relative Reward Preference in Primate Orbitofrontal Cortex. *Nature* **398**:704–708. [5]

———. 2000. Reward-Related Neuronal Activity during Go-Nogo Task Performance in Primate Orbitofrontal Cortex. *J. Neurophysiol.* **83**:1864–1876. [5]

Tremblay, R., S. Lee, and B. Rudy. 2016. GABAergic Interneurons in the Neocortex: From Cellular Properties to Circuits. *Neuron* **91**:260–292. [10]

Tremblay, S., C. Testard, J. Inchauspé, and M. Petrides. 2023. Non-Necessary Activity in the Primate Cortex. https://www.biorxiv.org/content/10.1101/2022.09.12.506984v2. (accessed Jan. 30, 2024). [5]

Tsividis, P. A., J. Loula, J. Burga, et al. 2021. Human-Level Reinforcement Learning through Theory-Based Modeling, Exploration, and Planning. http://arxiv.org/abs/2107.12544. (accessed Jan. 30, 2024). [12]

Tsuchida, A., and L. K. Fellows. 2009. Lesion Evidence That Two Distinct Regions within Prefrontal Cortex Are Critical for n-back Performance in Humans. *J. Cogn. Neurosci.* **21**:2263–2275. [8]

———. 2013. Are Core Component Processes of Executive Function Dissociable within the Frontal Lobes? Evidence from Humans with Focal Prefrontal Damage. *Cortex* **49**:1790–1800. [8]

Tsuda, B., K. M. Tye, H. T. Siegelmann, and T. J. Sejnowski. 2020. A Modeling Framework for Adaptive Lifelong Learning with Transfer and Savings through Gating in the Prefrontal Cortex. *PNAS* **117**:29872–29882. [2]

Tsujimoto, S., A. Genovesio, and S. P. Wise. 2010. Evaluating Self-Generated Decisions in Frontal Pole Cortex of Monkeys. *Nat. Neurosci.* **13**:120–126. [7]

———. 2011. Comparison of Strategy Signals in the Dorsolateral and Orbital Prefrontal Cortex. *J. Neurosci.* **31**:4583–4592. [5]

Tsujimoto, S., A. Genovesio, and S. P. Wise. 2012. Neuronal Activity during a Cued Strategy Task: Comparison of Dorsolateral, Orbital, and Polar Prefrontal Cortex. *J. Neurosci.* **32**:11017–11031. [8]

Tsumura, K., R. Aoki, M. Takeda, K. Nakahara, and K. Jimura. 2021. Cross-Hemispheric Complementary Prefrontal Mechanisms during Task Switching under Perceptual Uncertainty. *J. Neurosci.* **41**:2197–2213. [9]

Tsutsui, K., F. Grabenhorst, S. Kobayashi, and W. Schultz. 2016a. A Dynamic Code for Economic Object Valuation in Prefrontal Cortex Neurons. *Nat. Commun.* **7**:12554. [5]

Tsutsui, K., T. Hosokawa, M. Yamada, and T. Iijima. 2016b. Representation of Functional Category in the Monkey Prefrontal Cortex and Its Rule-Dependent Use for Behavioral Selection. *J. Neurosci.* **36**:3038–3048. [5, 8]

Tufail, Y., A. Yoshihiro, S. Pati, M. M. Li, and W. J. Tyler. 2011. Ultrasonic Neuromodulation by Brain Stimulation with Transcranial Ultrasound. *Nat. Protoc.* **6**:1453–1470. [8]

Tullo, D., Y. Feng, A. Pahor, et al. 2023. Investigating the Role of Individual Differences in Adherence to Cognitive Training. *J. Cogn.* **6**:48. [14]

Tullo, D., and S. M. Jaeggi. 2022. Working Memory Training: Meta-Analyses and Clinical Implications. In: The Cambridge Handbook of Working Memory and Language, ed. J. W. Schwieter and Z. E. Wen, pp. 881–906, Cambridge Handbooks in Language and Linguistics. Cambridge, UK: Cambrudge University Press. [14]

Tzanoulinou, S., C. Garcia-Mompo, O. Riccio, et al. 2016. Neuroligin-2 Expression in the Prefrontal Cortex Is Involved in Attention Deficits Induced by Peripubertal Stress. *Neuropsychopharmacol.* **41**:751–761. [16]

Uddin, L. Q. 2016. Salience Network of the Human Brain. Cambridge MA: Academic Press. [15]

Uddin, L. Q., R. F. Betzel, J. R. Cohen, et al. 2022. Controversies and Progress on Standardization of Large-Scale Brain Network Nomenclature. *Netw. Neurosci.* **7**:864–905. [11]

Uddin, L. Q., B. T. T. Yeo, and R. N. Spreng. 2019. Towards a Universal Taxonomy of Macro-Scale Functional Human Brain Networks. *Brain Topogr.* **32**:926–942. [11]

Uhlhaas, P. J., C. G. Davey, U. M. Mehta, et al. 2023. Towards a Youth Mental Health Paradigm: A Perspective and Roadmap. *Mol. Psych.* **28**:3171–3181. [16]

Uliana, D. L., C. Diniz, L. A. da Silva, et al. 2023. Contextual Fear Expression Engages a Complex Set of Interactions between Ventromedial Prefrontal Cortex Cholinergic, Glutamatergic, Nitrergic and Cannabinergic Signaling. *Neuropharmacol.* **232**:109538. [16]

Umemoto, A., T. Drew, E. F. Ester, and E. Awh. 2010. A Bilateral Advantage for Storage in Visual Working Memory. *Cognition* **117**:69–79. [8]

Unger, K., L. Ackerman, C. H. Chatham, D. Amso, and D. Badre. 2016. Working Memory Gating Mechanisms Explain Developmental Change in Rule-Guided Behavior. *Cognition* **155**:8–22. [7]

Upright, N. A., and M. G. Baxter. 2020. Effect of Chemogenetic Actuator Drugs on Prefrontal Cortex-Dependent Working Memory in Nonhuman Primates. *Neuropsychopharmacol.* **45**:1793–1798. [2]

Upright, N. A., S. W. Brookshire, W. Schnebelen, et al. 2018. Behavioral Effect of Chemogenetic Inhibition Is Directly Related to Receptor Transduction Levels in Rhesus Monkeys. *J. Neurosci.* **38**:7969–7975. [2, 8]

Urban, K. R., and R. J. Valentino. 2017. Age- and Sex-Dependent Impact of Repeated Social Stress on Intrinsic and Synaptic Excitability of the Rat Prefrontal Cortex. *Cereb. Cortex* **27**:244–253. [16]

Uylings, H. B., H. J. Groenewegen, and B. Kolb. 2003. Do Rats Have a Prefrontal Cortex? *Behav. Brain Res.* **146**:3–17. [2, 4]

Vaghi, M. M., P. E. Vértes, M. G. Kitzbichler, et al. 2017. Specific Frontostriatal Circuits for Impaired Cognitive Flexibility and Goal-Directed Planning in Obsessive-Compulsive Disorder: Evidence from Resting-State Functional Connectivity. *Biol. Psych.* **81**:708–717. [4]

Vahid, A., M. Mückschel, S. Stober, A.-K. Stock, and C. Beste. 2020. Applying Deep Learning to Single-Trial EEG Data Provides Evidence for Complementary Theories on Action Control. *Commun. Biol.* **3**:112. [12]

———. 2022. Conditional Generative Adversarial Networks Applied to EEG Data Can Inform About the Inter-Relation of Antagonistic Behaviors on a Neural Level. *Commun. Biol.* **5**:148. [12]

Vai, B., C. Bulgarelli, B. R. Godlewska, et al. 2016. Fronto-Limbic Effective Connectivity as Possible Predictor of Antidepressant Response to SSRI Administration. *Eur. Neuropsychopharmacol.* **26**:2000–2010. [13]

Vaidya, A. R., and D. Badre. 2022. Abstract Task Representations for Inference and Control. *Trends Cogn. Sci.* **26**:484–498. [7]

Vaidya, A. R., H. M. Jones, J. Castillo, and D. Badre. 2021. Neural Representation of Abstract Task Structure during Generalization. *eLife* **10**:e63226. [7]

Vaidya, A. R., M. S. Pujara, M. Petrides, E. A. Murray, and L. K. Fellows. 2019. Lesion Studies in Contemporary Neuroscience. *Trends Cogn. Sci.* **23**:653–671. [8]

Vaillant-Tenzer, C., and E. Koechlin. 2023. A Statisitical Physics Theory of Adaptive Human Behavior. Technical Reports N°10_6_2023, Ecole Normale Supérieure, Paris. [10]

Valenstein, E. S., A. F. Mirsky, M. H. Orzack, et al. 1977. Appendix: Psychosurgery: The National Commission for the Protection of Human Subjects of Biomedical and Behavioral Research. Department of Health Education and Welfare. Washington D.C.: US Government Printing Office. [15]

Valentin, V. V., A. Dickinson, and J. P. O'Doherty. 2007. Determining the Neural Substrates of Goal-Directed Learning in the Human Brain. *J. Neurosci.* **27**:4019–4026. [15]

van Balkom, T. D., O. A. van den Heuvel, H. W. Berendse, Y. D. van der Werf, and C. Vriend. 2020. The Effects of Cognitive Training on Brain Network Activity and Connectivity in Aging and Neurodegenerative Diseases: A Systematic Review. *Neuropsychol. Rev.* **30**:267–286. [14]

Vancraeyenest, P., J. T. Arsenault, X. Li, et al. 2020. Selective Mesoaccumbal Pathway Inactivation Affects Motivation but Not Reinforcement-Based Learning in Macaques. *Neuron* **108**:568–581. [2, 4]

Van Dijk, K. R. A., T. Hedden, A. Venkataraman, et al. 2010. Intrinsic Functional Connectivity as a Tool for Human Connectomics: Theory, Properties, and Optimization. *J. Neurophysiol.* **103**:297–321. [11]

Van Essen, D. C., and M. F. Glasser. 2018. Parcellating Cerebral Cortex: How Invasive Animal Studies Inform Noninvasive Mapmaking in Humans. *Neuron* **99**:640–663. [5, 11]

Van Essen, D. C., S. M. Smith, D. M. Barch, et al. 2013. The WU-Minn Human Connectome Project: an Overview. *Neuroimage* **80**:62–79. [4]

van Heukelum, S., R. B. Mars, M. Guthrie, et al. 2020. Where Is Cingulate Cortex? A Cross-Species View. *Trends Neurosci.* **43**:285–299. [2]

van Lieshout, L. L. F., F. P. de Lange, and R. Cools. 2019. Motives Underlying Human Curiosity. *Nat. Hum. Behav.* **3**:550–551. [12]

van Lieshout, L. L. F., F. P. de Lange, and R. Cools. 2021a. Uncertainty Increases Curiosity, but Decreases Happiness. *Sci. Rep.* **11**:14014. [12]

van Lieshout, L. L. F., I. J. Traast, F. P. de Lange, and R. Cools. 2021b. Curiosity or Savouring? Information Seeking Is Modulated by Both Uncertainty and Valence. *PloS One* **16**:e0257011. [12]

Vartanian, O., V. Replete, S. A. Saint, et al. 2022. What Is Targeted When We Train Working Memory? Evidence from a Meta-Analysis of the Neural Correlates of Working Memory Training Using Activation Likelihood Estimation. *Front. Psychol.* **13**:868001. [14]

Vaswani, A., N. Shazeer, N. Parmar, et al. 2017. Attention Is All You Need. In: Advances in Neural Information Processing Systems 22 (Nips 2009), ed. Y. Bengio et al., p. 11. Red Hook, NY: Curran Associates Inc. [12]

Venkatraman, V., and S. A. Huettel. 2012. Strategic Control in Decision-Making under Uncertainty. *Eur. J. Neurosci.* **35**:1075–1082. [11, 12]

Venkatraman, V., J. W. Payne, J. R. Bettman, M. F. Luce, and S. A. Huettel. 2009a. Separate Neural Mechanisms Underlie Choices and Strategic Preferences in Risky Decision Making. *Neuron* **62**:593–602. [12]

Venkatraman, V., A. G. Rosati, A. A. Taren, and S. A. Huettel. 2009b. Resolving Response, Decision, and Strategic Control: Evidence for a Functional Topography in Dorsomedial Prefrontal Cortex. *J. Neurosci.* **29**:13158–13164. [7]

Veres, J. M., T. Andrasi, P. Nagy-Pal, and N. Hajos. 2023. CaMKIIα Promoter-Controlled Circuit Manipulations Target Both Pyramidal Cells and Inhibitory Interneurons in Cortical Networks. *eNeuro* **10**:ENEURO.0070. [2]

Vernet, M., R. Quentin, L. Chanes, A. Mitsumasu, and A. Valero-Cabre. 2014. Frontal Eye Field, Where Art Thou? Anatomy, Function, and Non-Invasive Manipulation of Frontal Regions Involved in Eye Movements and Associated Cognitive Operations. *Front. Integr. Neurosci.* **8**:66. [8]

Verstynen, T. D., D. Badre, K. Jarbo, and W. Schneider. 2012. Microstructural Organizational Patterns in the Human Corticostriatal System. *J. Neurophysiol.* **107**:2984–2995. [7]

Vertes, R. P. 2004. Differential Projections of the Infralimbic and Prelimbic Cortex in the Rat. *Synapse* **51**:32–58. [3]

———. 2006. Interactions among the Medial Prefrontal Cortex, Hippocampus and Midline Thalamus in Emotional and Cognitive Processing in the Rat. *Neurosci.* **142**:1–20. [3]

Vickery, C. E., and D. Dorjee. 2016. Mindfulness Training in Primary Schools Decreases Negative Affect and Increases Meta-Cognition in Children. *Front. Psychol.* **6**: [14]

Vijayraghavan, S., M. Wang, S. G. Birnbaum, G. V. Williams, and A. F. Arnsten. 2007. Inverted-U Dopamine D1 Receptor Actions on Prefrontal Neurons Engaged in Working Memory. *Nat. Neurosci.* **10**:376–384. [6]

Vita, A., S. Barlati, A. Ceraso, et al. 2021. Effectiveness, Core Elements, and Moderators of Response of Cognitive Remediation for Schizophrenia: A Systematic Review and Meta-Analysis of Randomized Clinical Trials. *JAMA Psych.* **78**:848–858. [14]

Vogt, B. A., E. A. Nimchinsky, L. J. Vogt, and P. R. Hof. 1995. Human Cingulate Cortex: Surface Features, Flat Maps, and Cytoarchitecture. *J. Comp. Neurol.* **359**:490–506. [4]

Vogt, B. A., and G. Paxinos. 2014. Cytoarchitecture of Mouse and Rat Cingulate Cortex with Human Homologies. *Brain structure & function* **219**:185–192. [2, 4]

Volger, S., M. L. Vetter, M. Dougherty, et al. 2012. Patients' Preferred Terms for Describing Their Excess Weight: Discussing Obesity in Clinical Practice. *Obesity* **20**:147–150. [16]

Volkow, N. D., J. A. Gordon, and G. F. Koob. 2021. Choosing Appropriate Language to Reduce the Stigma around Mental Illness and Substance Use Disorders. *Neuropsychopharmacol.* **46**:2230–2232. [16]

Volkow, N. D., G. F. Koob, R. T. Croyle, et al. 2018. The Conception of the ABCD Study: From Substance Use to a Broad NIH Collaboration. *Dev. Cogn. Neurosci.* **32**:4–7. [16]

Vollenweider, F. X., and M. Kometer. 2010. The Neurobiology of Psychedelic Drugs: Implications for the Treatment of Mood Disorders. *Nat. Rev. Neurosci.* **11**:642–651. [13]

von Bonin, G., and P. Bailey. 1947. The Neocortex of Macaca mulatta. Illinois Monographs in the Medical Sciences. Urbana: Univ. of Illinois Press. [5]

von Economo, C. F., and G. N. Koskinas. 1925. Die Cytoarchitektonik der Hirnrinde des Erwachsenen Menschen. Heidelberg: Springer. [3]

Voon, V., B. Reynolds, C. Brezing, et al. 2010. Impulsive Choice and Response in Dopamine Agonist-Related Impulse Control Behaviors. *Psychopharmacol.* **207**:645–659. [7]

Voorhies, W. I., J. A. Miller, J. K. Yao, S. A. Bunge, and K. S. Weiner. 2021. Cognitive Insights from Tertiary Sulci in Prefrontal Cortex. *Nat. Commun.* **12**:5122. [4, 12]

Vranic, A., A. M. Spanic, B. Carretti, and E. Borella. 2013. The Efficacy of a Multifactorial Memory Training in Older Adults Living in Residential Care Settings. *Int. Psychogeriatr.* **25**:1885–1897. [14]

Vyas, S., M. D. Golub, D. Sussillo, and K. V. Shenoy. 2020. Computation through Neural Population Dynamics. *Annu. Rev. Neurosci.* **43**:249–275. [4]

Wager, T. D., M. L. Davidson, B. L. Hughes, M. A. Lindquist, and K. N. Ochsner. 2008. Prefrontal-Subcortical Pathways Mediating Successful Emotion Regulation. *Neuron* **59**:1037–1050. [13]

Walker, A. E. 1940. A Cytoarchitectural Study of the Prefrontal Area of the Macaque Monkey. *J. Comp. Neurol* **73**:59–86. [4, 5]

Walker, S. C., T. W. Robbins, and A. C. Roberts. 2009. Differential Contributions of Dopamine and Serotonin to Orbitofrontal Cortex Function in the Marmoset. *Cereb. Cortex* **19**:889–898. [8]

Wallis, C. U., R. N. Cardinal, L. Alexander, A. C. Roberts, and H. F. Clarke. 2017. Opposing Roles of Primate Areas 25 and 32 and Their Putative Rodent Homologs in the Regulation of Negative Emotion. *PNAS* **114**:E4075–E4084. [3, 4]

Wallis, C. U., G. J. Cockcroft, R. N. Cardinal, A. C. Roberts, and H. F. Clarke. 2019. Hippocampal Interaction with Area 25, but Not Area 32, Regulates Marmoset Approach-Avoidance Behavior. *Cereb. Cortex* **29**:4818–4830. [8]

Wallis, J. D. 2007. Orbitofrontal Cortex and Its Contribution to Decision-Making. *Annu. Rev. Neurosci.* **30**:31–56. [4]

———. 2011. Cross-Species Studies of Orbitofrontal Cortex and Value-Based Decision-Making. *Nat. Neurosci.* **15**:13–19. [2, 4]

Wallis, J. D., K. C. Anderson, and E. K. Miller. 2001. Single Neurons in Prefrontal Cortex Encode Abstract Rules. *Nature* **411**:953–956. [3–5, 7, 8, 12]

Wallis, J. D., and E. K. Miller. 2003a. From Rule to Response: Neuronal Processes in the Premotor and Prefrontal Cortex. *J. Neurophysiol.* **90**:1790–1806. [7]

Wallis, J. D., and E. K. Miller. 2003b. Neuronal Activity in Primate Dorsolateral and Orbital Prefrontal Cortex during Performance of a Reward Preference Task. *Eur. J. Neurosci.* **18**:2069–2081. [4, 5, 8]

Walton, M. E., T. E. Behrens, M. J. Buckley, P. H. Rudebeck, and M. F. Rushworth. 2010. Separable Learning Systems in the Macaque Brain and the Role of Orbitofrontal Cortex in Contingent Learning. *Neuron* **65**:927–939. [4]

Wang, C., I. Ulbert, D. L. Schomer, K. Marinkovic, and E. Halgren. 2005. Responses of Human Anterior Cingulate Cortex Microdomains to Error Detection, Conflict Monitoring, Stimulus-Response Mapping, Familiarity, and Orienting. *J. Neurosci.* **25**:604–613. [12]

Wang, H., G. G. Stradtman III, X. J. Wang, and W. J. Gao. 2008. A Specialized NMDA Receptor Function in Layer 5 Recurrent Microcircuitry of the Adult Rat Prefrontal Cortex. *PNAS* **105**:16791–16796. [6]

Wang, M., D. Datta, J. Enwright, et al. 2019. A Novel Dopamine D1 Receptor Agonist Excites Delay-Dependent Working Memory-Related Neuronal Firing in Primate Dorsolateral Prefrontal Cortex. *Neuropharmacol.* **150**:46–58. [16]

Wang, M., S. Vijayraghavan, and P. S. Goldman-Rakic. 2004a. Selective D2 Receptor Actions on the Functional Circuitry of Working Memory. *Science* **303**:853–856. [6]

Wang, M., Y. Yang, C. J. Wang, et al. 2013. NMDA Receptors Subserve Persistent Neuronal Firing during Working Memory in Dorsolateral Prefrontal Cortex. *Neuron* **77**:736–749. [10]

Wang, S., R. Falcone, B. Richmond, and B. B. Averbeck. 2023. Attractor Dynamics Reflect Decision Confidence in Macaque Prefrontal Cortex. *Nat. Neurosci.* **26**:1–11. [4]

Wang, X., B. C. Bernhardt, T. Karapanagiotidis, et al. 2018. The Structural Basis of Semantic Control: Evidence from Individual Differences in Cortical Thickness. *Neuroimage* **181**:480–489. [12]

Wang, X.-J. 1999. Synaptic Basis of Cortical Persistent Activity: The Importance of NMDA Receptors to Working Memory. *J. Neurosci.* **19**:9587–9603. [6, 10, 12]

———. 2001. Synaptic Reverberation Underlying Mnemonic Persistent Activity. *Trends Neurosci.* **24**:455–463. [4, 6]

———. 2002. Probabilistic Decision Making by Slow Reverberation in Cortical Circuits. *Neuron* **36**:955–968. [10, 12]

———, ed. 2013. The Prefrontal Cortex as a Quintessential "Cognitive-Type" Neural Circuit: Working Memory and Decision Making. Principles of Frontal Lobe Function, vol. D. T. Stuss and R. T. Knight. New York: Cambridge University Press. [10, 12]

———. 2020. Macroscopic Gradients of Synaptic Excitation and Inhibition in the Neocortex. *Nat. Rev. Neurosci.* **21**:169–178. [10, 12, 16]

———. 2022. Theory of the Multiregional Neocortex: Large-Scale Neural Dynamics and Distributed Cognition. *Annu. Rev. Neurosci.* **45**:533–560. [11, 12]

Wang, X.-J., and J. H. Krystal. 2014. Computational Psychiatry. *Neuron* **84**:638–654. [10, 12, 16]

Wang, X.-J., U. Pereira, M. G. P. Rosa, and H. Kennedy. 2020. Brain Connectomes Come of Age. *Curr. Opin. Neurobiol.* **65**:152–161. [11]

Wang, X.-J., J. Tegner, C. Constantinidis, and P. S. Goldman-Rakic. 2004b. Division of Labor among Distinct Subtypes of Inhibitory Neurons in a Cortical Microcircuit of Working Memory. *PNAS* **101**:1368–1373. [6, 10, 12]

Wapstra, N. J., M. Ketola, S. Thompson, et al. 2022. Increased Basal Ganglia Modulatory Effective Connectivity Observed in Resting-State fMRI in Individuals with Parkinson's Disease. *Front. Aging Neurosci.* **14**:719089. [17]

Warren, D. E., J. D. Power, J. Bruss, et al. 2014. Network Measures Predict Neuropsychological Outcome after Brain Injury. *PNAS* **111**:14247–14252. [11]

Wasmuht, D. F., E. Spaak, T. J. Buschman, E. K. Miller, and M. G. Stokes. 2018. Intrinsic Neuronal Dynamics Predict Distinct Functional Roles during Working Memory. *Nat. Commun.* **9**:3499. [12]

Watakabe, A., M. Ohtsuka, M. Kinoshita, et al. 2015. Comparative Analyses of Adeno-Associated Viral Vector Serotypes 1, 2, 5, 8 and 9 in Marmoset, Mouse and Macaque Cerebral Cortex. *Neurosci. Res.* **93**:144–157. [2]

Watakabe, A., H. Skibbe, K. Nakae, et al. 2023. Local and Long-Distance Organization of Prefrontal Cortex Circuits in the Marmoset Brain. *Neuron* **111**:2258–2273. [8]

Watanabe, K., S. Igaki, and S. Funahashi. 2006. Contributions of Prefrontal Cue-, Delay-, and Response-Period Activity to the Decision Process of Saccade Direction in a Free-Choice Odr Task. *Neural Netw.* **19**:1203–1222. [5]

Watanabe, M. 1996. Reward Expectancy in Primate Prefrontal Neurons. *Nature* **382**:629–632. [5, 8]

Webb, T. L., E. Miles, and P. Sheeran. 2012. Dealing with Feeling: A Meta-Analysis of the Effectiveness of Strategies Derived from the Process Model of Emotion Regulation. *Psychol. Bull.* **138**:775–808. [14]

Weber, J., G. Iwama, A.-K. Solbakk, et al. 2023. Subspace Partitioning in the Human Prefrontal Cortex Resolves Cognitive Interference. *PNAS* **120**:e2220523120. [12]

Weiner, K. S. 2019. The Mid-Fusiform Sulcus (Sulcus Sagittalis Gyri Fusiformis). *Anat. Rec.* **302**:1491–1503. [4]

———. 2023. The Hypothesis of Fundal Cognition. *Nat. Rev. Neurosci.* **24**:521. [4, 8]

Weiner, K. S., and E. H. Willbrand. 2023. Is There an Association between Tuber Involvement of the Fusiform Face Area in Autism Diagnosis? *Annals of Neurology* **93**:1218–1220. [4]

Weiner, K. S., and K. Zilles. 2016. The Anatomical and Functional Specialization of the Fusiform Gyrus. *Neuropsychologia* **83**:48–62. [4]

Welker, W. 1990. Why Does Cerebral Cortex Fissure and Fold? In: Cereb Cortex, ed. E. G. Jones and A. Peters, pp. 3–136, vol. 8B. Boston: Springer. [4]

Wendiggensen, P., N. Adelhöfer, R. Jamous, et al. 2022. Processing of Embedded Response Plans Is Modulated by an Interplay of Frontoparietal Theta and Beta Activity. *J. Neurophysiol.* **128**:543–555. [12]

Wendiggensen, P., A. Prochnow, C. Pscherer, et al. 2023. Interplay between Alpha and Theta Band Activity Enables Management of Perception-Action Representations for Goal-Directed Behavior. *Commun. Biol.* **6**:494. [12]

Westbrook, A., and T. S. Braver. 2015. Cognitive Effort: A Neuroeconomic Approach. *Cogn. Affect. Behav. Neurosci.* **15**:395–415. [12]

Westbrook, A., D. Kester, and T. S. Braver. 2013. What Is the Subjective Cost of Cognitive Effort? Load, Trait, and Aging Effects Revealed by Economic Preference. *PloS One* **8**:e68210. [12]

Westbrook, A., B. Lamichhane, and T. Braver. 2019. The Subjective Value of Cognitive Effort Is Encoded by a Domain-General Valuation Network. *J. Neurosci.* **39**:3934–3947. [12]

Whelan, R., P. J. Conrod, J. B. Poline, et al. 2012. Adolescent Impulsivity Phenotypes Characterized by Distinct Brain Networks. *Nat. Neurosci.* **15**:920–925. [16]

Whitaker, K. J., P. E. Vertes, R. Romero-Garcia, et al. 2016. Adolescence Is Associated with Genomically Patterned Consolidation of the Hubs of the Human Brain Connectome. *PNAS* **113**:9105–9110. [16]

White, I. M., and S. P. Wise. 1999. Rule-Dependent Neuronal Activity in the Prefrontal Cortex. *Exp. Brain Res.* **126**:315–335. [5, 12]

White, J. K., E. S. Bromberg-Martin, S. R. Heilbronner, et al. 2019. A Neural Network for Information Seeking. *Nat. Commun.* **10**:5168. [2]

Whitfield-Gabrieli, S., S. S. Ghosh, A. Nieto-Castanon, et al. 2016. Brain Connectomics Predict Response to Treatment in Social Anxiety Disorder. *Mol. Psych.* **21**:680–685. [16]

Whittington, J. C. R., T. H. Muller, S. Mark, et al. 2020. The Tolman-Eichenbaum Machine: Unifying Space and Relational Memory through Generalization in the Hippocampal Formation. *Cell* **183**:1249–1263. [10]

Whitton, A. E., J. M. Reinen, M. Slifstein, et al. 2020. Baseline Reward Processing and Ventrostriatal Dopamine Function Are Associated with Pramipexole Response in Depression. *Brain* **143**:701–710. [13]

Wiecki, T. V., and M. J. Frank. 2013. A Computational Model of Inhibitory Control in Frontal Cortex and Basal Ganglia. *Psychol. Rev.* **120**:329–355. [12]

Wikenheiser, A. M., and G. Schoenbaum. 2016. Over the River, through the Woods: Cognitive Maps in the Hippocampus and Orbitofrontal Cortex. *Nat. Rev. Neurosci.* **17**:513–523. [12]

Wilder, B. G. 1881. A Partial Revision of Anatomical Nomenclature, with Especial Reference to That of the Brain. *Science* **2**:133–138. [4]

———. 1896. Neural Terms, International and National. *Journal of Comparative Neurology* **6**:216–352. [4]

Willbrand, E. H., S. A. Bunge, and K. S. Weiner. 2023a. Neuroanatomical and Functional Dissociations between Variably Present Anterior Lateral Prefrontal Sulci. *J. Cog. Neuro.* **35**:1846–1867. [4, 8]

Willbrand, E. H., S. Jackson, S. Chen, et al. 2023b. Sulcal Variability in Anterior Lateral Prefrontal Cortex Contributes to Variability in Reasoning Performance among Young Adults. https://www.biorxiv.org/content/10.1101/2023.02.10.528061v2. [4]

Willbrand, E. H., S. A. Maboudian, J. P. Kelly, et al. 2023c. Sulcal Morphology of Posteromedial Cortex Substantially Differs between Humans and Chimpanzees. *Commun. Biol.* **6**:586. [4]

Willbrand, E. H., B. J. Parker, W. I. Voorhies, et al. 2022. A New Tripartite Landmark in Posterior Cingulate Cortex. *Sci. Adv.* **8**:eabn9516. [4]

Willbrand, E. H., Y.-H. Tsai, T. Gagnant, and K. S. Weiner. 2023d. Updating the Sulcal Landscape of the Human Lateral Parieto-Occipital Junction Provides Anatomical, Functional, and Cognitive Insights. *eLife* **12**:RP90451. [4]

Williams, G. V., and P. S. Goldman-Rakic. 1995. Modulation of Memory Fields by Dopamine D1 Receptors in Prefrontal Cortex. *Nature* **376**:572–575. [6]

Williams, L. M. 2016. Precision Psychiatry: A Neural Circuit Taxonomy for Depression and Anxiety. *Lancet Psych.* **3**:472–480. [13]

Williams, N. R., B. D. Heifets, B. S. Bentzley, et al. 2019. Attenuation of Antidepressant and Antisuicidal Effects of Ketamine by Opioid Receptor Antagonism. *Mol. Psychiatry* **24**:1779–1786. [13]

Williams, N. R., B. D. Heifets, C. Blasey, et al. 2018. Attenuation of Antidepressant Effects of Ketamine by Opioid Receptor Antagonism. *Am. J. Psych.* **175**:1205–1215. [13]

Williams, R. S., N. E. Adams, L. E. Hughes, et al. 2023. Syndromes Associated with Frontotemporal Lobar Degeneration Change Response Patterns on Visual Analogue Scales. *Sci. Rep.* **13**:8939. [16]

Wilson, C. R., D. Gaffan, P. G. Browning, and M. G. Baxter. 2010. Functional Localization within the Prefrontal Cortex: Missing the Forest for the Trees? *Trends Neurosci.* **33**:533–540. [5]

Wilson, F. A. W., S. P. Scalaïdhe, and P. S. Goldman-Rakic. 1993. Dissociation of Object and Spatial Processing Domains in Primate Prefrontal Cortex. *Science* **260**:1955–1958. [11]

Wilson, N. R., C. A. Runyan, F. L. Wang, and M. Sur. 2012. Division and Subtraction by Distinct Cortical Inhibitory Networks *in Vivo*. *Nature* **488**:343–348. [12]

Wilson, R. C., Y. K. Takahashi, G. Schoenbaum, and Y. Niv. 2014. Orbitofrontal Cortex as a Cognitive Map of Task Space. *Neuron* **81**:267–279. [7, 12]

Wiltschko, A. B., M. J. Johnson, G. Iurilli, et al. 2015. Mapping Sub-Second Structure in Mouse Behavior. *Neuron* **88**:1121–1135. [12]

Wimmer, K., D. Q. Nykamp, C. Constantinidis, and A. Compte. 2014. Bump Attractor Dynamics in Prefrontal Cortex Explains Behavioral Precision in Spatial Working Memory. *Nat. Neurosci.* **17**:431–439. [8]

Wimmer, R. D., L. I. Schmitt, T. J. Davidson, et al. 2015. Thalamic Control of Sensory Selection in Divided Attention. *Nature* **526**:705–709. [3, 4]

Winstanley, C. A., D. E. Theobald, R. N. Cardinal, and T. W. Robbins. 2004. Contrasting Roles of Basolateral Amygdala and Orbitofrontal Cortex in Impulsive Choice. *J. Neurosci.* **24**:4718–4722. [4]

Wise, R. A. 2004. Dopamine, Learning and Motivation. *Nat. Rev. Neurosci.* **5**:483–494. [14]

Wise, S. P. 2008. Forward Frontal Fields: Phylogeny and Fundamental Function. *Trends Neurosci.* **31**:599–608. [2, 4]

Wisniewski, D., C. Reverberi, A. Tusche, and J.-D. Haynes. 2015. The Neural Representation of Voluntary Task-Set Selection in Dynamic Environments. *Cereb. Cortex* **25**:4715–4726. [12]

Wojcik, M. J., J. P. Stroud, D. Wasmuht, et al. 2023. Learning Shapes Neural Geometry in the Prefrontal Cortex. https://www.biorxiv.org/content/10.1101/2023.04.24.538054v1. (accessed Feb. 2, 2024). [8]

Wolff, M., S. J. Gibb, J. C. Cassel, and J. C. Dalrymple-Alford. 2008. Anterior but Not Intralaminar Thalamic Nuclei Support Allocentric Spatial Memory. *Neurobiol. Learn. Mem.* **90**:71–80. [3]

Wong, F. K., K. Bercsenyi, V. Sreenivasan, et al. 2018. Pyramidal Cell Regulation of Interneuron Survival Sculpts Cortical Networks. *Nature* **557**:668–673. [2]

Wong, P., and J. H. Kaas. 2009a. An Architectonic Study of the Neocortex of the Short-Tailed Opossum (*Monodelphis Domestica*). *Brain Behav. Evol.* **73**:206–228. [3]

———. 2009b. Architectonic Subdivisions of Neocortex in the Tree Shrew (Tupaia Belangeri). *Anat. Rec.* **292**:994–1027. [4]

Woo, C. W., L. Schmidt, A. Krishnan, et al. 2017. Quantifying Cerebral Contributions to Pain Beyond Nociception. *Nat. Commun.* **8**:14211. [16]

Wood, C. M., L. Alexander, J. Alsio, et al. 2023. Chemogenetics Identifies Separate Area 25 Brain Circuits Involved in Anhedonia and Anxiety in Marmosets. *Sci. Transl. Med.* **15**:eade1779. [2, 4, 13, 16]

Wood, J. L., and D. E. Nee. 2023. Cingulo-Opercular Subnetworks Motivate Frontoparietal Subnetworks during Distinct Cognitive Control Demands. *J. Neurosci.* **43**:1225–1237. [7, 16]

Woody, E. Z., V. Lewis, L. Snider, et al. 2005. Induction of Compulsive-Like Washing by Blocking the Feeling of Knowing: an Experimental Test of the Security-Motivation Hypothesis of Obsessive-Compulsive Disorder. *Behav. Brain Funct.* **1**:11. [15]

Woolf, C., A. Lampit, Z. Shahnawaz, et al. 2022. A Systematic Review and Meta-Analysis of Cognitive Training in Adults with Major Depressive Disorder. *Neuropsychol. Rev.* **32**:419–437. [14]

Woolgar, A., J. Jackson, and J. Duncan. 2016. Coding of Visual, Auditory, Rule, and Response Information in the Brain: 10 Years of Multivoxel Pattern Analysis. *J. Cog. Neuro.* **28**:1433–1454. [9]

Woolgar, A., A. Parr, R. Cusack, et al. 2010. Fluid Intelligence Loss Linked to Restricted Regions of Damage within Frontal and Parietal Cortex. *PNAS* **107**:14899–14902. [9]

Woolsey, C. N. 1963. Comparative Studies on Localization in Precentral and Supplementary Motor Areas. *Int. J. Neurol.* **4**:13–20. [8]

Woolsey, C. N., P. H. Settlage, D. R. Meyer, et al. 1952. Patterns of Localization in Precentral and "Supplementary" Motor Areas and Their Relation to the Concept of a Premotor Area. *Res. Publ. Assoc. Res. Nerv. Ment. Dis.* **30**:238–264. [8]

Wu, C. W., and J. H. Kaas. 2003. Somatosensory Cortex of Prosimian Galagos: Physiological Recording, Cytoarchitecture, and Corticocortical Connections of Anterior Parietal Cortex and Cortex of the Lateral Sulcus. *J. Comp. Neurol.* **457**:263–292. [3]

Wu, Z., D. Lin, and Y. Li. 2022. Pushing the Frontiers: Tools for Monitoring Neurotransmitters and Neuromodulators. *Nat. Rev. Neurosci.* **23**:257–274. [3]

Xia, C. H., Z. Ma, R. Ciric, et al. 2018. Linked Dimensions of Psychopathology and Connectivity in Functional Brain Networks. *Nat. Commun.* **9**:3003. [13, 16]

Xia, C.-Y., N.-N. Zhang, H. Jiang, et al. 2023. Gap Junction Is Essential for the Antidepressant Effects of Fluoxetine. *J. Pharm. Pharmacol.* **75**:686–692. [13]

Xiao, D., B. Zikopoulos, and H. Barbas. 2009. Laminar and Modular Organization of Prefrontal Projections to Multiple Thalamic Nuclei. *Neurosci.* **161**:1067–1081. [8]

Xiao, Y., J. J. Mann, J. C. Chow, et al. 2023. Patterns of Social Determinants of Health and Child Mental Health, Cognition, and Physical Health. *JAMA Pediatr.* **177**:1294–1305. [16]

Xu, F., Y. Shen, L. Ding, et al. 2021. High-Throughput Mapping of a Whole Rhesus Monkey Brain at Micrometer Resolution. *Nat. Biotechnol.* **39**:1521–1528. [2]

Xu, T., K. H. Nenning, E. Schwartz, et al. 2020. Cross-Species Functional Alignment Reveals Evolutionary Hierarchy within the Connectome. *Neuroimage* **223**:117346. [4]

Xu, Y., P. Zou, and A. E. Cohen. 2017. Voltage Imaging with Genetically Encoded Indicators. *Curr. Opin. Chem. Biol.* **39**:1–10. [3]

Yamada, M., M. C. Pita, T. Iijima, and K. Tsutsui. 2010. Rule-Dependent Anticipatory Activity in Prefrontal Neurons. *Neurosci. Res.* **67**:162–171. [5, 8]

Yamins, D. L. K., H. Hong, C. F. Cadieu, et al. 2014. Performance-Optimized Hierarchical Models Predict Neural Responses in Higher Visual Cortex. *PNAS* **111**:8619–8624. [12]

Yan, C.-G., X. Chen, L. Li, et al. 2019. Reduced Default Mode Network Functional Connectivity in Patients with Recurrent Major Depressive Disorder. *PNAS* **116**:9078–9083. [13]

Yang, C. R., and J. K. Seamans. 1996. Dopamine D1 Receptor Actions in Layers V-VI Rat Prefrontal Cortex Neurons *in Vitro*: Modulation of Dendritic-Somatic Signal Integration. *J. Neurosci.* **16**:1922–1935. [6]

Yang, G. R., M. R. Joglekar, H. F. Song, W. T. Newsome, and X. J. Wang. 2019. Task Representations in Neural Networks Trained to Perform Many Cognitive Tasks. *Nat. Neurosci.* **22**:297–306. [10, 12]

Yang, G. R., J. D. Murray, and X.-J. Wang. 2016. A Dendritic Disinhibitory Circuit Mechanism for Pathway-Specific Gating. *Nat. Commun.* **7**:12815. [12]

Yang, G. R., and X.-J. Wang. 2021. Artificial Neural Networks for Neuroscientists: A Primer. *Neuron* **109**:739. [12]

Yang, S., D. Datta, W. Elizabeth, et al. 2022. Inhibition of Glutamate-Carboxypeptidase-II in Dorsolateral Prefrontal Cortex: Potential Therapeutic Target for Neuroinflammatory Cognitive Disorders. *Mol. Psych.* **27**:4252–4263. [16]

Yangüez, M., B. Bediou, J. Chanal, and D. Bavelier. 2023. In Search of Better Practice in Executive Functions Assessment: Methodological Issues and Potential Solutions. *Psychol. Rev.*, in press. [14]

Yao, J. K., W. I. Voorhies, J. A. Miller, S. A. Bunge, and K. S. Weiner. 2022. Sulcal Depth in Prefrontal Cortex: A Novel Predictor of Working Memory Performance. *Cereb. Cortex* **33**:1799–1813. [4]

Yao, Z. F., and S. Hsieh. 2022. Age Differences of the Hierarchical Cognitive Control and the Frontal Rostro-Caudal Functional Brain Activation. *Cereb. Cortex* **32**:2797–2815. [7]

Yaple, Z. A., S. Tolomeo, and R. Yu. 2021. Mapping Working Memory-Specific Dysfunction Using a Transdiagnostic Approach. *Neuroimage Clin.* **31**:102747. [14]

Ye, R., F. H. Hezemans, C. O'Callaghan, et al. 2023a. Locus Coeruleus Integrity Is Linked to Response Inhibition Deficits in Parkinson's Disease and Progressive Supranuclear Palsy. *J. Neurosci.* **43**:7028–7040. [16]

Ye, T., J. L. Romero-Sosa, A. Rickard, et al. 2023b. Theta Oscillations in Anterior Cingulate Cortex and Orbitofrontal Cortex Differentially Modulate Accuracy and Speed in Flexible Reward Learning. *Oxf. Open Neurosci.* **2**:kvad005. [2, 4]

Yeo, B. T., F. M. Krienen, J. Sepulcre, et al. 2011. The Organization of the Human Cerebral Cortex Estimated by Intrinsic Functional Connectivity. *J. Neurophysiol.* **106**:1125–1165. [7, 11]

Yeterian, E. H., D. N. Pandya, F. Tomaiuolo, and M. Petrides. 2012. The Cortical Connectivity of the Prefrontal Cortex in the Monkey Brain. *Cortex* **48**:58–81. [5, 7]

Yin, Y.-Y., Y.-H. Wang, W.-G. Liu, et al. 2021. The Role of the Excitation:Inhibition Functional Balance in the mPFC in the Onset of Antidepressants. *Neuropharmacol.* **191**:108573. [13]

Yip, S. W., D. Scheinost, M. N. Potenza, and K. M. Carroll. 2019. Connectome-Based Prediction of Cocaine Abstinence. *Am. J. Psych.* **176**:156–164. [16]

Yizhar, O., L. E. Fenno, M. Prigge, et al. 2011. Neocortical Excitation/Inhibition Balance in Information Processing and Social Dysfunction. *Nature* **477**:171–178. [4]

Yoo, S. B. M., B. J. Sleezer, and B. Y. Hayden. 2018. Robust Encoding of Spatial Information in Orbitofrontal Cortex and Striatum. *J. Cogn. Neurosci.* **30**:898–913. [5]

Yoo, S. S., A. Bystritsky, J. H. Lee, et al. 2011. Focused Ultrasound Modulates Region-Specific Brain Activity. *Neuroimage* **56**:1267–1275. [8]

Yoon, G., I. L. Petrakis, and J. H. Krystal. 2019. Association of Combined Naltrexone and Ketamine with Depressive Symptoms in a Case Series of Patients with Depression and Alcohol Use Disorder. *JAMA Psych.* **76**:337–338. [13]

Young, K. S., R. T. LeBeau, A. N. Niles, et al. 2019. Neural Connectivity during Affect Labeling Predicts Treatment Response to Psychological Therapies for Social Anxiety Disorder. *J. Affect. Disord.* **242**:105–110. [16]

Young, S., J. Bramham, K. Gray, and E. Rose. 2008. The Experience of Receiving a Diagnosis and Treatment of ADHD in Adulthood: A Qualitative Study of Clinically Referred Patients Using Interpretative Phenomenological Analysis. *J. Atten. Disord.* **11**:493–503. [16]

Yu, S., S. Rempel, N. Gholamipourbarogh, and C. Beste. 2022. A Ventral Stream-Prefrontal Cortex Processing Cascade Enables Working Memory Gating Dynamics. *Commun. Biol.* **5**:1086. [12]

Yu, Y. H., C. Y. Ou, B. C. Shyu, and A. C. W. Huang. 2020. Basolateral Amygdala but Not Medial Prefrontal Cortex Contributes to Chronic Fluoxetine Treatments for PTSD Symptoms in Mice. *Behav. Neurol.* **2020**:8875087. [13]

Yuen, E. Y., J. Wei, W. Liu, et al. 2012. Repeated Stress Causes Cognitive Impairment by Suppressing Glutamate Receptor Expression and Function in Prefrontal Cortex. *Neuron* **73**:962–977. [16]

Yuste, R. 2015. From the Neuron Doctrine to Neural Networks. *Nat. Rev. Neurosci.* **16**:487–497. [15]

Yuste, R., M. Hawrylycz, N. Aalling, et al. 2020. A Community-Based Transcriptomics Classification and Nomenclature of Neocortical Cell Types. *Nat. Neurosci.* **23**:1456–1468. [2]

Zald, D. H. 2007. Orbital versus Dorsolateral Prefrontal Cortex: Anatomical Insights into Content versus Process Differentiation Models of the Prefrontal Cortex. *Ann. N. Y. Acad. Sci.* **1121**:395–406. [5]

Zald, D. H., C. Curtis, L. A. Chernitsky, and J. V. Pardo. 2005. Frontal Lobe Activation during Object Alternation Acquisition. *Neuropsychol.* **19**:97–105. [4]

Zanos, P., R. Moaddel, P. J. Morris, et al. 2016. NMDAR Inhibition-Independent Antidepressant Actions of Ketamine Metabolites. *Nature* **533**:481–486. [13]

Zarahn, E., G. K. Aguirre, and M. D'Esposito. 1999. Temporal Isolation of the Neural Correlates of Spatial Mnemonic Processing with fMRI. *Cogn. Brain Res.* **7**:255–268. [4]

Zarate, C. A., Jr., N. E. Brutsche, L. Ibrahim, et al. 2012. Replication of Ketamine's Antidepressant Efficacy in Bipolar Depression: A Randomized Controlled Add-on Trial. *Biol. Psychiatry* **71**:939–946. [13]

Zarate, C. A., Jr., J. B. Singh, P. J. Carlson, et al. 2006. A Randomized Trial of an N-Methyl-D-Aspartate Antagonist in Treatment-Resistant Major Depression. *Arch. Gen. Psychiatry* **63**:856–864. [13]

Zeisler, Z. R., L. London, W. G. Janssen, et al. 2023. High-Throughput Sequencing of Macaque Basolateral Amygdala Projections Reveals Dissociable Connectional Motifs with Frontal Cortex. https://www.ncbi.nlm.nih.gov/pubmed/36711708. [2, 4]

Zelazo, P. D., and S. M. Carlson. 2020. The Neurodevelopment of Executive Function Skills: Implications for Academic Achievement Gaps. *Psychol. Neurosci.* **13**:273–298. [14]

Zeredo, J. L., S. K. L. Quah, C. U. Wallis, et al. 2019. Glutamate Within the Marmoset Anterior Hippocampus Interacts with Area 25 to Regulate the Behavioral and Cardiovascular Correlates of High-Trait Anxiety. *J. Neurosci.* **39**:3094–3107. [16]

Zhang, C., Y. Chen, S. Tian, et al. 2017. Effects of Anterior Capsulotomy on Decision Making in Patients with Refractory Obsessive-Compulsive Disorder. *Front. Psychol.* **8**:1814. [15]

Zhang, J., T. Rittman, C. Nombela, et al. 2016. Different Decision Deficits Impair Response Inhibition in Progressive Supranuclear Palsy and Parkinson's Disease. *Brain* **139**:161–173. [16]

Zhang, J., N. Zhang, J. Lei, et al. 2022. Fluoxetine Shows Neuroprotective Effects against LPS-Induced Neuroinflammation via the Notch Signaling Pathway. *Int. Immunopharmacol.* **113**:109417. [13]

Zhang, Q., M. A. Weber, and N. S. Narayanan. 2021. Medial Prefrontal Cortex and the Temporal Control of Action. *Int. Rev. Neurobiol.* **158**:421–441. [12]

Zhao, H., C. Jiang, M. Zhao, et al. 2023. Comparisons of Accelerated Continuous and Intermittent Theta-Burst Stimulation for Treatment-Resistant Depression and Suicidal Ideation. *Biol. Psych.* **S0006–3223**:01788–01792. [16]

Zhao, W., L. Huang, Y. Li, et al. 2020. Evidence for the Contribution of COMT Gene Val158/108met Polymorphism (Rs4680) to Working Memory Training-Related Prefrontal Plasticity. *Brain Behav.* **10**:e01523. [14]

Zhong, P., and Z. Yan. 2011. Differential Regulation of the Excitability of Prefrontal Cortical Fast-Spiking Interneurons and Pyramidal Neurons by Serotonin and Fluoxetine. *PloS One* **6**:e16970. [13]

Zhou, J., C. Jia, M. Montesinos-Cartagena, et al. 2021a. Evolving Schema Representations in Orbitofrontal Ensembles during Learning. *Nature* **590**:606–611. [7]

Zhou, J., W. Zong, C. Jia, M. P. H. Gardner, and G. Schoenbaum. 2021b. Prospective Representations in Rat Orbitofrontal Ensembles. *Behav. Neurosci.* **135**:518–527. [7]

Zhou, X., F. Katsuki, X. L. Qi, and C. Constantinidis. 2012. Neurons with Inverted Tuning during the Delay Periods of Working Memory Tasks in the Dorsal Prefrontal and Posterior Parietal Cortex. *J. Neurophysiol.* **108**:31–38. [6]

Zick, J. L., R. K. Blackman, D. A. Crowe, et al. 2018. Blocking Nmdar Disrupts Spike Timing and Decouples Monkey Prefrontal Circuits: Implications for Activity-Dependent Disconnection in Schizophrenia. *Neuron* **98**:1243–1255. [6]

Zika, O., K. Wiech, A. Reinecke, M. Browning, and N. W. Schuck. 2023. Trait Anxiety Is Associated with Hidden State Inference during Aversive Reversal Learning. *Nat. Commun.* **14**:4203. [12]

Zikopoulos, B., M. Höistad, Y. John, and H. Barbas. 2017. Posterior Orbitofrontal and Anterior Cingulate Pathways to the Amygdala Target Inhibitory and Excitatory Systems with Opposite Functions. *J. Neurosci.* **37**:5051–5064. [3]

Zilcha-Mano, S., Z. Wang, B. S. Peterson, et al. 2019. Neural Mechanisms of Expectancy-Based Placebo Effects in Antidepressant Clinical Trials. *J. Psychiatr. Res.* **116**:19–25. [13]

Zilles, K., and N. Palomero-Gallagher. 2017. Comparative Analysis of Receptor Types That Identify Primary Cortical Sensory Areas. In: Evolution of Nervous Systems, pp. 225–245. Elsevier. [13]

Zola-Morgan, S. 1995. Localization of Brain Function: The Legacy of Franz Joseph Gall (1758-1828). . *Annu Rev Neurosci.* **18**:359-383. [16]

Zuo, C., Y. Ma, B. Sun, et al. 2013. Metabolic Imaging of Bilateral Anterior Capsulotomy in Refractory Obsessive Compulsive Disorder: an FDG PET Study. *J. Cereb. Blood Flow Metab.* **33**:880–887. [15]

Subject Index

Strüngmann Forum Report Series*

Digital Ethology: Human Behavior in Geospatial Context
Edited by Tomáš Paus and Hye-Chung Kum
ISBN: 9780262548137

Migration Stigma: Understanding Prejudice, Discrimination, and Exclusion
Edited by Lawrence H. Yang, Maureen A. Eger and Bruce G. Link
ISBN: 9780262548120

Exploring and Exploiting Genetic Risk for Psychiatric Disorders
Edited by Joshua A. Gordon and Elisabeth Binder
DOI: https://doi.org/10.7551/mitpress/15380.001.0001
ISBN electronic: 9780262377423

Intrusive Thinking: From Molecules to Free Will
Edited by Peter W. Kalivas and Martin P. Paulus
ISBN: 9780262542371

Deliberate Ignorance: Choosing Not to Know
edited by Ralph Hertwig and Christoph Engel
ISBN 9780262045599

Youth Mental Health: A Paradigm for Prevention and Early Intervention
Edited by Peter J. Uhlhaas and Stephen J. Wood
ISBN: 9780262043977

The Neocortex
Edited by Wolf Singer, Terrence J. Sejnowski and Pasko Rakic
ISBN: 9780262043243

Interactive Task Learning: Humans, Robots, and Agents Acquiring New Tasks through Natural Interactions
Edited by Kevin A. Gluck and John E. Laird
ISBN: 9780262038829

Agrobiodiversity: Integrating Knowledge for a Sustainable Future
Edited by Karl S. Zimmerer and Stef de Haan
ISBN: 9780262038683

Rethinking Environmentalism: Linking Justice, Sustainability, and Diversity
Edited by Sharachchandra Lele, Eduardo S. Brondizio, John Byrne, Georgina M. Mace and Joan Martinez-Alier
ISBN: 9780262038966

Emergent Brain Dynamics: Prebirth to Adolescence
Edited by April A. Benasich and Urs Ribary
ISBN: 9780262038638

---

* Available at https://mitpress.mit.edu/books/series/strungmann-forum-reports

The Cultural Nature of Attachment: Contextualizing Relationships and Development
Edited by Heidi Keller and Kim A. Bard
ISBN (Hardcover): 9780262036900 ISBN (ebook): 9780262342865
Winner of the Ursula Gielen Global Psychology Book Award

Investors and Exploiters in Ecology and Economics: Principles and Applications
edited by Luc-Alain Giraldeau, Philipp Heeb and Michael Kosfeld
ISBN (Hardcover): 9780262036122 ISBN (eBook): 9780262339797

Computational Psychiatry: New Perspectives on Mental Illness
edited by A. David Redish and Joshua A. Gordon
ISBN: 9780262035422

Complexity and Evolution: Toward a New Synthesis for Economics
edited by David S. Wilson and Alan Kirman
ISBN: 9780262035385

The Pragmatic Turn: Toward Action-Oriented Views in Cognitive Science
edited by Andreas K. Engel, Karl J. Friston and Danica Kragic
ISBN: 9780262034326

Translational Neuroscience: Toward New Therapies
edited by Karoly Nikolich and Steven E. Hyman
ISBN: 9780262029865

Trace Metals and Infectious Diseases
edited by Jerome O. Nriagu and Eric P. Skaar
ISBN 9780262029193

Pathways to Peace: The Transformative Power of Children and Families
edited by James F. Leckman, Catherine Panter-Brick and Rima Salah,
ISBN 9780262027984

Rethinking Global Land Use in an Urban Era
edited by Karen C. Seto and Anette Reenberg
ISBN 9780262026901

Schizophrenia: Evolution and Synthesis
edited by Steven M. Silverstein, Bita Moghaddam and Til Wykes,
ISBN 9780262019620

Cultural Evolution: Society, Technology, Language, and Religion
edited by Peter J. Richerson and Morten H. Christiansen,
ISBN 9780262019750

Language, Music, and the Brain: A Mysterious Relationship
edited by Michael A. Arbib
ISBN 9780262019620

Evolution and the Mechanisms of Decision Making
edited by Peter Hammerstein and Jeffrey R. Stevens
ISBN 9780262018081

Cognitive Search: Evolution, Algorithms, and the Brain
edited by Peter M. Todd, Thomas T. Hills and Trevor W. Robbins,
ISBN 9780262018098

Animal Thinking: Contemporary Issues in Comparative Cognition
edited by Randolf Menzel and Julia Fischer
ISBN 9780262016636

Disease Eradication in the 21st Century: Implications for Global Health
edited by Stephen L. Cochi and Walter R. Dowdle
ISBN 9780262016735

Better Doctors, Better Patients, Better Decisions: Envisioning Health Care 2020
edited by Gerd Gigerenzer and J. A. Muir Gray
ISBN 9780262016032

Dynamic Coordination in the Brain: From Neurons to Mind
edited by Christoph von der Malsburg, William A. Phillips and Wolf Singer,
ISBN 9780262014717

Linkages of Sustainability
edited by Thomas E. Graedel and Ester van der Voet
ISBN 9780262013581

Biological Foundations and Origin of Syntax
edited by Derek Bickerton and Eörs Szathmáry
ISBN 9780262013567

Clouds in the Perturbed Climate System: Their Relationship to Energy Balance, Atmospheric Dynamics, and Precipitation
edited by Jost Heintzenberg and Robert J. Charlson
ISBN 9780262012874
Winner of the Atmospheric Science Librarians International Choice Award

Better Than Conscious? Decision Making, the Human Mind, and Implications For Institutions
edited by Christoph Engel and Wolf Singer
ISBN 978-0-262-19580-5